Advances in Soft Computing

Springer
London
Berlin
Heidelberg
New York
Barcelona
Hong Kong
Milan
Paris
Santa Clara
Singapore
Tokyo

R. Roy, T. Furuhashi and P.K. Chawdhry (Eds)

Advances in Soft Computing

Engineering Design and Manufacturing

Springer

Rajkumar Roy, PhD
The CIM Institute, Cranfield University, Cranfield, Bedford, MK43 0AL,UK

Takeshi Furuhashi, PhD
Department of Information Electronics, Graduate School of Engineering,
Nagoya University, Furo-cho, Chikusa-ku, Nagoya 464-8603, JAPAN

Pravir K. Chawdhry, PhD
School of Mechanical Engineering, University of Bath, Bath, BA2 7AY, UK

ISBN-13:978-1-85233-062-0 Springer-Verlag London Berlin Heidelberg

British Library Cataloguing in Publication Data
Advances in soft computing: : engineering design and manufacturing
1. Soft computing 2. Soft computing - Design
I. Roy, Rajkumar II. Furuhashi, T. (Takeshi) III. Chawdhry, Pravir K.
620'. 0042'0285

ISBN-13:978-1-85233-062-0 e-ISBN-13:978-1-4471-0819-1

DOI: 10.1007/978-1-4471-0819-1

Library of Congress Cataloging-in-Publication Data
Advances in soft computing : engineering design and manufacturing /
Rajkumar Roy, Takeshi Furuhashi, and Pravir K. Chawdhry, eds.
p. cm.
ISBN-13:978-1-85233-062-0
1. Engineering design--Data processing. 2. Soft computing.
3. Intelligent control systems. I. Roy, R. (Rajkumar), 1966- .
II. Furuhashi, T. (Takeshi) III. Chawdhry, Pravir K, 1958- .
TA174.A417 1998
620'.0042'028563--dc21 98-41923

69/3830-543210 Printed on acid-free paper

Preface

This book contains the most recent developments in the field of soft computing in engineering design and manufacturing. Soft computing is a collection of methodologies, which aim to exploit tolerance for imprecision, uncertainty, and partial truth to achieve tractability, robustness, and low solution cost. The principal constituents of soft computing are fuzzy computing, neuro-computing, genetic computing and probabilistic computing. The area of applications of soft computing is very wide, and the topics in this book are well focused on the area of engineering design and manufacturing. The papers on evolutionary design and optimisation highlight the distinguished feature of this book.

This book is a collection of 43 research papers that were presented in June 1998 at the 3rd On-line World Conference on Soft Computing in Engineering Design and Manufacturing (WSC3). These include four invited papers by World-renowned researchers in the field. Seven different universities from six different countries hosted the conference on the Internet. The sponsors of the conference include IEEE Industrial Electronics Society (IES) and British Telecommunications (BT). The conference was a great success and has set a standard for cyber-conference. There were more than 600 registered participants of the conference. The papers represent researches from 22 countries around the world. Therefore, the book presents a good overview of current advances in soft computing research around the world. Another feature of the conference was on-line tutorial. Brief descriptions of two tutorials are also included in the book. A summary of discussion is also presented in the book.

Publication of this book was possible due to the successful organisation of WSC3. Thus we would like to thank the organising committee, the authors, the reviewers, the tutorial organisers, the sponsors, and the advertisers of the conference. This book is the final result of hard work of many people across the World. We hope the book will promote the research and development in soft computing for engineering design and manufacturing and justify the hard work.

Cyberspace
2nd September 1998

Dr. Rajkumar Roy
Prof. Takeshi Furuhashi
Dr. Pravir K. Chawdhry

Table of Contents

Introduction

Soft Computing is a computing paradigm that is tolerant of imprecision, uncertainty and partial truth. The motivation behind the paradigm is human mind. Conventional computing is considered to be 'hard', because it is limited to represent human like 'vagueness'. Soft computing provides the opportunity to represent 'ambiguity' in human thinking with the real life 'uncertainty'. The concept is proved to be useful for many real life problems. The principal constituents of soft computing are Fuzzy, Neuro, Evolutionary and Probabilistic computing techniques. This is a partnership, where each technique contributes a unique methodology for addressing problems in its domain.

The area of application of soft computing is very wide. The book presents advances in fuzzy, neural, and evolutionary computing techniques and its application in engineering design and manufacturing. The book deals with a wide range of applications, such as evolutionary design, and intelligent control to data mining. The book also presents new frontiers for soft computing. Very innovative techniques on evolutionary design and opimisation are one of the main topics of the book. New and advanced research results in these fields are also presented.

The book is divided into eleven parts, the first nine parts present 43 papers from the 3rd On-line World Conference on Soft Computing in Engineering Design and Manufacturing (WSC3). The nine parts are Keynote Papers, Design Support Systems, Intelligent Control, Identification and Modelling, Data Mining, Optimisation, Optimisation for Industrial Applications, New Topics in EA Basics, and New Frontiers for Soft Computing. The tenth and eleventh parts are the summary of tutorials and discussion about the papers. There are four keynote papers covering different topics of soft computing. The design decision support part concentrates on application of evolutionary computation to enhance creativity in design. The third part on intelligent control reports state of the art applications of fuzzy logic technique for control problems. The identification and modelling part presents several applications of neural and fuzzy computing in the area. Application of fuzzy and evolutionary computing techniques in data mining (part 5) demonstrates the strengths of soft computing to extract hidden knowledge from data. The next two chapters are on optimisation and its industrial applications. Different new optimisation techniques are presented. The industrial application part discusses the challenges for industrial and real life problems. The eighth part presents new topics in evolutionary computing, such as 'a study of altruism by genetic algorithm'. New frontier in soft computing includes granular computing and DNA computing (part 9). Therefore, the book presents a nice balance between recent advances in soft computing techniques and its recent applications in engineering design and manufacturing.

Part 1: Keynote Papers

Papers:

The NIST Design Repository Project

Simon Szykman, Ram D. Sriram, Christophe Bochenek and Janusz Racz

National Institute of Standards and Technology
Manufacturing Systems Integration Division
Building 304, Room 6
Gaithersburg, MD 20899
szykman@cme.nist.gov, sriram@cme.nist.gov, bchrist@cme.nist.gov and jwracz@cme.nist.gov

Abstract

Modern engineering industry is relying more and more on the use of knowledge in product development. This paper advocates design repositories as a natural progression from traditional design databases to systems that are created to more actively support knowledge-based design. In contrast to traditional design databases, design repositories serve not only as archives, but as repositories of heterogeneous information that are designed to enable representation, capture, sharing, and reuse of corporate design knowledge. This paper describes the NIST Design Repository Project, an ongoing project within the Engineering Design Technologies Group at the National Institute of Standards and Technology (NIST). The project objectives are to develop a computational framework for the creation of design repositories, and a proof-of-concept prototype to demonstrate their benefits. A number of research issues associated with the envisioned role of design repositories in industry are addressed. The current state of the project and its implementation are presented.

1 Introduction

Design of complex engineering systems is increasingly becoming a collaborative task among designers or design teams that are physically, geographically, and temporally distributed. The complexity of modern products means that a single designer or design team can no longer manage the complete product development effort. Developing products without sufficient expertise in a broad set of disciplines can result in extended product development cycles, higher development costs, and quality problems. On the other hand, ensuring comprehensive technical proficiency in a world where trends are toward more multidisciplinary design can become a costly undertaking for a company.

Driven by such issues, companies are increasingly staffing only their core competencies in-house and depending on other firms to provide the complementary design knowledge and design effort needed for a complete product. Designers are no longer merely exchanging geometric data, but more general knowledge about design and design process, including specifications, design rules, constraints, rationale, etc. As design becomes increasingly knowledge-intensive and collaborative, the need for computational design frameworks to support the representation and use of knowledge among distributed designers becomes more critical.

In addition to sharing and exchanging information, pressure to reduce product development times has resulted in an increased focus on methods for representing and storing engineering artifact knowledge in a way that facilitates its retrieval and subsequent reuse. Merely providing access to schematics and computer-aided design (CAD) models of artifacts is inadequate for this purpose. In order to support reuse of engineering knowledge, a representation must convey additional information that answers not only "what?" questions about a design, but also "how?" and "why?" questions. Mappings from form to function have often been pointed to as just one example of the kind of information that is needed for effective reuse of design knowledge, but which is absent from traditional CAD models. The emerging research area of design repositories is aimed at addressing these industry needs. Design repositories make use of research in knowledge-based design to facilitate the representation, capture, sharing, and reuse (search and retrieval) of corporate design knowledge.

It should be noted that although the term *design repositories* has not yet found its way into daily usage in industry, many companies are migrating from traditional design databases to design repositories. While companies may still be referring to their corporate knowledge stores as design databases, in many cases these stores would fall under the definition of design repositories as characterized in this paper. Although design repositories can in general terms be thought of as design databases, and indeed will most often be implemented using database management systems, design repositories are distinguished from traditional design databases in several significant ways:

- Traditional design databases are typically more data-centric than knowledge-centric, and contain only a limited representation of an artifact such as drawings and/or CAD models, version information, and often related documentation. Design repositories attempt to capture a more complete design representation that may include characterization of function, behavior, design rules, simulation models, and so on. It should be noted, however, that a fully comprehensive representation of every aspect of a design may simply not be possible.

- Design databases are generally more homogeneous in the kinds of information they contain. In addition to containing images (drawings), file pointers (to CAD models), and unstructured text (documentation), design repositories may contain formal data/information models, structured text (specialized languages for representing function, design rules, logical expressions), mathematical simulation models, animations, video, and other types of information.

- Design databases tend to be static sources of information (though their contents may grow with time). While they are used for storage and retrieval of design data, capabilities for supporting the design process are not built into traditional database systems. Such capabilities might include search for components/assemblies that satisfy required functions, explicit representation of physical and functional decompositions and the mappings between them, simulation of behavior and performance, (partially) automated reasoning about a design, and more. Since design databases have not been designed specifically for these purposes, they are limited in their ability to meet needs for design of large-scale engineering systems.

This paper describes the NIST Design Repository Project, which has the objective of developing a framework to enable the creation of design repositories. This project is part of a larger effort called the Engineering Design and Process Planning Testbed, being conducted in the Manufacturing Systems Integration Division of the Manufacturing Engineering Laboratory at NIST. The following section describes the project objectives and identifies a set of related research issues. Within the context of these various research issues, Section 3 presents related work in the area along with the current state of the project. Section 4 discusses conclusions and future directions for this research effort.

2 The NIST Design Repository Project: Objectives and Research Issues

The NIST Design Repository Project is an ongoing multi-year project with the objective of developing a framework to enable the creation of design repositories. In contrast to traditional design databases, design repositories serve not only as archives, but as repositories of heterogeneous knowledge and data that are designed to support representation, capture, sharing, and reuse of corporate design knowledge.

This research is driven by industry needs that were identified at a workshop held at NIST in November, 1996 [1]. The intent of the project is to impact industry in two ways: (1) by developing basic technological infrastructure to support the creation and use of design repositories in industry, and (2) by providing industry with access to design information that companies can use in their own design processes in order to demonstrate the value of these repositories before significant effort is invested in their development. Thus, in addition to performing basic research in the area of design repositories, the scope of this project also includes the development of sets of domain-specific prototype repositories that will be made available online via the World Wide Web.

Through the course of this project, a variety of research issues have arisen that will in the long term affect the way in which design repositories are implemented and used. These issues include[1]:

- Development of an information modeling framework to support modeling of engineering artifacts, that provides a more comprehensive knowledge representation than traditional CAD systems.
- The use of standard representations, when possible, to leverage research efforts and maximize interoperability with existing software used in industry, and contribution to long-term standards development where standards currently do not exist. The latter consists of the development of information models for representation of knowledge that is generally not found in traditional CAD systems, such as function, behavior, etc.
- A need for representations that are both human interpretable and machine interpretable so that information stored in a repository is accessible and usable by both human designers and knowledge-based design systems that might be used for (partially) automating part of a design process.
- Implementation of interfaces for creating, editing, and browsing design repositories that are easy to use and effective in conveying information that is desired.
- Development of taxonomies of standardized terminology to help provide consistency in, and across, design repositories, as well as to facilitate indexing, search, and retrieval of information from them.

Because of limitations on project resources, the degree to which these issues have been considered in the NIST Design Repository Project varies. However, these issues are all important within our view of the role of design repositories in industry, and ultimately all will have to be addressed by the research community before design repositories can successfully transition into engineering industry.

The goal of this research is not to provide commercial-quality software for creating design repositories, but to provide prototypes that show the utility of the design repository concept. By highlighting and starting to address these issues, this work will speed up the transition and adoption of this kind of technology into industry. The following section outlines the current state of the project and elaborates on each of the above issues within that context. The existing implementation is also described.

---

[1] In addition to these research issues, technological issues such as database and Internet access issues also impact this project. Other pragmatic issues such as security of communications and protection of intellectual property when sharing or exchanging design knowledge will also have to be resolved before design repositories become widely used. Such issues are not the focus of this paper.

3 Current State of the Project

3.1 Information Modeling Framework

Traditional CAD systems are limited to representation of geometric data and other types of information relating to geometry that may include constraints, parametric information, features, and so on. Because of industry's increasing dependence on other types of knowledge in the design process, new classes of tools to support knowledge-based design, product data management (PDM), and concurrent engineering have begun emerging in the engineering software marketplace.

When contrasted with traditional CAD tools, these new systems are making progress toward the next generation of engineering design support tools. However, these systems have been focusing primarily on database-related issues and do not place a primary emphasis on information models for artifact representation (e.g., [2], [3], [4], [5], and [6]). Furthermore, although these systems can often represent non-geometric knowledge (e.g. information about the design process, manufacturing process, bills of materials, etc.), representation of the *artifact* itself is still generally limited to geometry.

Research in the area of intelligent design systems has taken an approach to artifact representation where modeling is typically divided into three main areas of study: the physical layout of the artifact (form), an indication of the overall effect that the artifact creates (function), and a causal account of the operation of the artifact (behavior). Examples of these areas include the qualitative simulation work in [7], behavioral and functional representation in [8], functional representation in [9] and its successor "SBF models" from projects such as KRITIK [10] and INTERACTIVE KRITIK [11], the YMIR project [12], CONGEN [13], and others.

While there are major differences in the implementations of such models, the top level division into representation of form, behavior and function is a common one. The NIST Design Repository Project utilizes a three-tiered approach to artifact representation which incorporates these three concepts. This project, building on previous design representation research, has resulted in an object-oriented representation format that provides a high level division into *form*, *function*, and *behavior* [13], [14].

This modeling language represents artifacts as sets of *objects*, and *relationships*. *Objects* represent physical entities such as assemblies, subassemblies, and components[2], as well as non-physical concepts such as function and behavior. *Relationships* are used to represent the relationship between sets of objects, including a physical decomposition (of an assembly into subassemblies), a functional decomposition (of a function into subfunctions), and other kinds of relationships between ob-

---

[2] In the remainder of this paper, the term *artifact* is used in a generic sense and may refer to assemblies, subassemblies or individual components.

jects. The overall artifact representation is comprised not only of the collection of the objects that represent physical entities, but also the other objects, relationships, the interconnections between them, as well as various attributes and their values.

As is true in general for object-oriented representations, objects and relationships are instantiated from classes that contain attributes and other information that are transferred via inheritance mechanisms. Thus, *a priori* development of useful generic classes of objects and relationships makes modeling of a specific artifact easier since much of the required attributes are already represented in the class schemata. The artifact modeling language and associated object-oriented representation are described in greater detail in [15].

3.2 Knowledge Representation

In order to leverage research efforts and maximize interoperability with existing software used in industry, this work makes use of standard representations when possible. For representation of geometry, this work uses ISO 10303 which is more commonly known as STEP (Standard for the Exchange of Product Model Data) [16] (specifically AP (application protocol) 203 [17]). As more and more CAD companies incorporate the ability to import and export STEP data, the use of STEP AP 203 as a neutral format will make the NIST design repositories more widely accessible than would a proprietary, CAD system-specific format.

In addition to the STEP AP 203 format, geometry is also maintained in VRML (Virtual Reality Modeling Language) format [18] to facilitate visualization of repositories that may be accessed remotely via the World Wide Web. Although STEP AP 203 viewers are available, they are neither as pervasive nor as well integrated with web browsers as are VRML viewers. VRML is not suitable as a complete replacement for STEP AP 203; however, because it was designed for graphical display of geometry and not as a general geometric representation, VRML is not able to represent all kinds of geometric information needed by designers.

Within the standards development community, the focus to date has been primarily on representation of geometric data. This is clearly essential, but it is no longer sufficient. As industry's reliance on non-geometric information and knowledge-based design increases, so does the need for standards for design knowledge such as function, behavior, and other kinds of non-geometric data. In addition to geometry, the information modeling framework that has been developed within this project presently includes representation of function and behavior [15]; if required, the framework will be extended to include other kinds of non-geometric information in the future. Although formal standards for this kind of information are not yet being explored in the standards community, it is hoped that work done within the National Institute of Standards and Technology will lay a foundation for long-term standards development.

3.3 Interpretability of Knowledge Representations

Design repositories are intended to serve as information-rich stores of corporate design knowledge. They are not, however, intended simply to be used as sophisticated part catalogs where a designer can search for final parts or subassemblies that can be dropped into a new design. In most cases, artifact knowledge retrieved from previous designs will not be completely applicable to a similar problem. A design may require further modification before being usable in a new design. In many instances, even a modified design may not be applicable. A designer can still benefit from retrieval of knowledge about previous designs by abstracting information and applying it to a new design, or by gaining insight into how an earlier related artifact was designed. Another way in which stored design knowledge can be used is to support design automation. Knowledge-based design systems are becoming more prevalent in industry because of their ability to automate (or partially automate) design reasoning and in some cases to take on the role of designer for selected portions of the design process.

From the knowledge representation standpoint, these two uses produce conflicting requirements. Although natural language is the most expressive and easily comprehended format for humans, for the purposes of storage and retrieval, as well as computer-based reasoning, natural language is a poor choice and formal representations are preferable. Conversely, a formal representation may be of little use to a human designer if extracting information from it is too difficult. Thus a key objective of this research has been to provide knowledge representations that are both human and machine interpretable, so that information stored in a design repository is accessible and usable by both human designers and knowledge-based design systems. The information modeling framework [15] makes use of a formal knowledge representation, but one that uses data structures that are comprehensible by a human designer, and which lends itself to browsing an artifact representation in a meaningful manner given suitable interfaces.

3.4 Interfaces

The development of extended knowledge representations is necessary but not sufficient in order to successfully use design repositories in industry. Initially, an artifact must be modeled and from this model a design repository must be created. For a design repository to be successfully used for knowledge reuse once it exists, information stored in it must be accessible and easily retrievable by a designer. In many instances, a repository may also be edited and updated through its lifetime. To address these needs, another focus of the current research effort is in the development of interfaces for creating, editing, and browsing design repositories.

There are two ways to create a design repository within the existing implementation: a batch process and an interactive editor. In the batch process, the designer models an artifact by creating an ASCII text file in a specified format that follows the schemata for the data structures in the information modeling framework described previously (objects, relationships, etc.). A compiler process can then be executed from the command line. The compiler parses all the information in this text file and transfers it into a design repository, creating object-oriented data structures and linking them together appropriately. Although the command-line compiler does not have (or need) an interface, in some cases it may be convenient for the designer to have the ability to do artifact modeling "off line" using nothing more than a text editor, and then create a design repository in a batch mode.

The other method for creating a design repository is interactively through a web-based application. This application provides the designer with an interface to the design repository database. Built into this interface are functions for artifact modeling (creating objects and relationships by instantiating them from predefined classes, specifying connections between objects and relationships, filling in attribute values, and so on), as well as some basic consistency maintenance functions.

The other available interface is a web-based artifact browser (see Figure 1). The artifact browser provides a point-and-click interface that allows the designer to browse an existing design repository. The object-oriented data structures (objects, relationships, and their classes) are presented in table form, where hyperlinks allow navigation to connected data structures. From the artifact object shown in the figure, one can click on one hyperlink to examine the artifact's function, or another to view the relationships for that artifact (one indicating the artifact of which this object is a subassembly, and the other specifying the further decomposition of this artifact into subassemblies and components). Following an artifact's form hyperlink brings up an object that has a link to a 3D model of that artifact, in either STEP AP 203 format or in VRML format for web-based viewing. The different types of data structures as well as the hyperlinks are color-coded to easily distinguish between them.

In addition to navigating from object to object as described, the bottom right portion of the interface provides the overall hierarchical physical decomposition of an artifact into assemblies, subassemblies, down to individual components. The user can quickly jump to any artifact object in the hierarchy by following one of those hyperlinks. As different parts of the design are visited, they are added to the history in the top right portion of the interface so that the user can quickly return to any object or relationship that was previously viewed during a browsing session.

3.5 Taxonomies and Terminology

Issues of taxonomy and terminology are another important area of focus within this project. One of the main barriers to retrieving and reusing knowledge effectively is problems with terminology: from domain to domain (and sometimes even within a domain) designers use different terms to mean the same thing, or use the same term to mean different things. In order to help provide consistency in and across design repositories, as well as to facilitate indexing, search, and retrieval of information from them, there is a need for standardized terminology and vocabulary. Ideally, this would be done at the research stage of the project and users would have to work

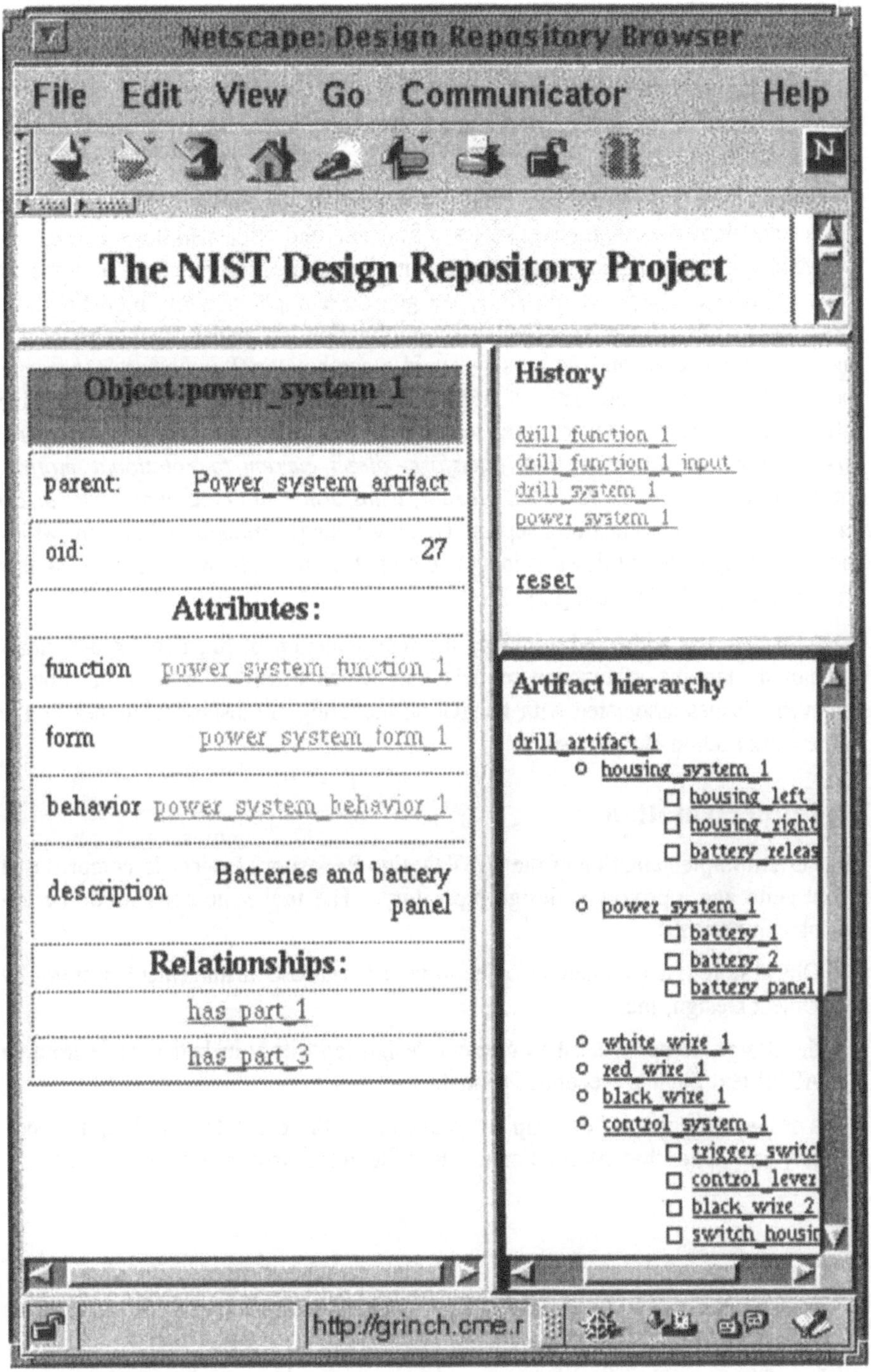

Figure 1. Web-based artifact browser interface.

within the prescribed terminology. Realistically, there are too many conflicts to be resolved independently of domain or application, but allowing a company to develop its own taxonomies will still be beneficial. Extensive work in supporting construction of taxonomies and even more formal ontologies has been done by [19], among others.

A portion of the current research effort is focusing on functional representation, and the development of a comprehensive set of functions and associated flows[3]. The objective is to generate a pair of taxonomies one for engineering functions and one for flows, which is as small as possible, yet generic enough to allow modeling of a broad variety of engineering artifacts. The motivation for creating two separate taxonomies for functions and flows is to avoid a combinatorial explosion in concepts required for modeling artifacts. If flows were considered as part of a function, four different functions would be required to represent the following functions: *transform direct current to linear motion, transform direct current to rotational motion, transform linear motion to direct current, transform rotational motion to direct current.* By considering flows separately, only a single *transform* function, which can have many different flows as input and output, is needed. Related work in this area includes [20], [21], and [22].

Work to date has addressed primarily the representation of function for an artifact and not a mapping of function to behavior, which describes how a function is achieved. Issues associated with behavioral modeling are discussed further in the future work section of the paper.

3.6 Implementation

The current implementation of the NIST Design Repository Project is comprised of a tool suite and a prototype design repository. The tool suite consists of the following components:

- ObjectStore[4], a commercial object-oriented database management system by Object Design, Inc.,
- the design compiler used to create a design repository in batch mode from an ASCII text file in a predefined format,
- a design extractor or "decompiler" that extracts the contents of a design repository and writes that information to a text file in that same predefined format,

---

[3] For example, if an artifact converts direct current to rotational motion, the function is to transform and the associated flows are direct current and rotational motion.

[4] Use of any commercial product or company names in this paper is intended to provide readers with information regarding the implementation of the research described, and does not imply recommendation or endorsement by the National Institute of Standards and Technology.

- STEP/Works, a STEP AP 203 viewer from TechneGroup, Inc., for visualization of three-dimensional geometry,
- the web-based artifact browser and the web-based design repository editor/design modeler, used in conjunction with any of several web browsers,
- any of several VRML viewers or plug-ins, also for visualization of three-dimensional geometry.

The web-based interfaces are client applications that communicate with the database server containing a design repository via common gateway interface (CGI) scripts. The client can be run through most web browsers, allowing remote access of a design repository from any machine on the Internet that can communicate with the machine running the server.

The prototype design repository contains the artifact representation of a Black & Decker cordless power drill. The drill artifact repository contains 28 artifact objects (assemblies, subassemblies, and components) and over 250 additional objects and relationships belonging to 64 (object) classes and 6 relationship classes that describe their form, functions, and behaviors. In addition to representations of form, function, and behavior, the artifact representation includes information about many other attributes (including color, material, and control settings) and parameters (such as battery voltage, output torque, and direction of rotation).

4 Conclusions and Future Directions

The NIST Design Repository Project is working toward the development of an information modeling framework to support the creation of design repositories. This project is driven by industry needs for technology to support the increasing role of knowledge-based design, including the representation, capture, sharing, and reuse of corporate design knowledge. This paper has presented the project objectives, associated research issues, and the current status of the project in the context of those issues. The prototype implementation was also described.

Focus of work to date has been in four areas: (1) the development of a design modeling language for representing design artifact knowledge, (2) the implementation of interfaces for creating and browsing artifact repositories, (3) the creation of a prototype design repository, and (4) work toward identifying a taxonomy of functions and associated flows that is small, yet generic enough to model a broad variety of engineering artifacts. In addition to continuing these aspects of the project and populating the design repositories with new design artifacts, there are several important areas of research that will be addressed by future project milestones.

The functionality of existing interfaces will be extended and new interfaces will be developed. The current artifact browser interface provides the designer with a hierarchical physical decomposition of the artifact. For browsing an artifact from the perspective of function, providing a functional decomposition would also be useful. Additional session management enhancements will also be made, to allow multiple browsing sessions to exist concurrently, to allow a designer to end a session and return to the same state at a later point in time rather than initiating a new session, and allowing definition of multiple users that can set preferences or personalize the interface in various ways. Regarding new interfaces, one important requirement that is not currently addressed is to provide a class editor for creation of new object and relationship classes. The current artifact modeling interface allows the user to create a new repository by instantiating new objects and relationships from predefined classes. However, no set of predefined classes will be universally applicable to all design problems and the ability to extend this set will be important to the designer.

The main aspects of the design modeling language are the representation of form, function, and behavior (as well as relationships between their associated objects). The representation of form (geometry) and function in this project. The representation of behavior within the existing framework is a relatively simple descriptive text-based approach. Another objective for future work is to incorporate more a formal representation of behavior to allow composable simulations. Work in this area is being done as part of two projects that are funded by the Defense Advanced Research Projects Agency (DARPA) Rapid Design Exploration and Optimization (RaDEO) program, for which NIST is a funding agent: the Model-Based Support of Distributed Collaborative Design (previously How Things Work) project at the Stanford University Knowledge Systems Laboratory [23], [24], [25] and the Active Catalog project at the University of Southern California Information Sciences Institute [26], [27]. Discussions are underway to explore the possibility of incorporating an existing behavior modeling language into the NIST Design Repository Project rather than developing an entirely new one.

These attempts at formalizing representations of function and behavior, as well as the other taxonomic issues discussed previously, will facilitate the tasks of indexing design knowledge for efficient search and retrieval. However, aside from formal representations and a standardized vocabulary, the indexing and search mechanisms must themselves be developed. This is an area of long term future research, as much work remains to be done before this problem can be adequately addressed.

Once design repositories are more broadly used by industry, the question of how to exchange information between repositories will become more important. If a standardized design modeling language is developed (possibly, but not necessarily, using the framework developed in this project as a basis), repositories themselves may not be interchangeable between systems. If two design repository development systems are built on top of two different object-oriented database software systems, one will not be able to use a repository created by the other without some kind of translation step. The ASCII text file format used as part of this work is inadequate for that purpose because there is no mechanism to include a computer interpretable description of the knowledge schemata or the file content within the file itself.

The Extensible Markup Language (XML) [28] and Resource Definition Framework (RDF) [29], [30] are being examined as a possible solution to this problem. Though other work has been done in developing formal approaches for the exchange of knowledge (e.g. Knowledge Interchange Format (KIF) [31] and Open Knowledge Base Connectivity (OKBC) [32], [33]), XML and RDF have the main advantage of widespread adoption in the information technology world; XML (and likely RDF) support is expected in upcoming versions of several commercial web browsers and word processing applications. This research direction is being explored in order to provide a more broad-based solution to an industry which is increasingly looking towards purchase of off-the-shelf software over in-house development when possible.

The NIST Design Repository Project was motivated by needs identified at an industry workshop held at NIST in November 1996 [1] and, as such, it is expected that industry will benefit from the technical foundation for the development of design repositories provided by this research. Another potential use for this work is to enhance the content of patent databases residing at the U.S. Patent and Trademark Office. Currently, patent information consists of text and two-dimensional images. Providing a formal artifact representation that extends to function and behavior not only leads to a more comprehensive description of a device, it also provides additional types of information for meaningful indexing, which can lead to more efficient search and retrieval of patent information.

References

[1] Szykman, S., R. D. Sriram, and S. J. Smith (eds.) (1998), *Proceedings of the NIST Design Repository Workshop*, NISTIR 6159, National Institute of Standards and Technology, Gaithersburg, MD, November, 1996.

[2] Hardwick, M. and D. Loffredo (1995), "Using EXPRESS to Implement Concurrent Engineering Databases," *Proceedings of the 1995 ASME Computers in Engineering Conference and Engineering Database Symposium*, Boston, MA, September.

[3] Bliznakov, P. I., J. J. Shah, and S. D. Urban (1996), "Integration Infrastructure to Support Concurrence and Collaboration in Engineering Design," *Proceedings of the 1996 ASME Design Engineering Technical Conferences and Computers in Engineering Conference*, Paper No. 96-DETC/EIM-1420, Irvine, CA, August.

[4] Kim, T. S., S.-H. Han, and Y. J. Shin (1996), "Product Data Management Using AP203 of STEP Standard," *Proceedings of the 1996 ASME Design Engineering Technical Conferences and Computers in Engineering Conference*, Paper No. 96-DETC/DAC-1069, Irvine, CA, August.

[5] Shah, J. J., D. K. Jeon, S. D. Urban, P. Bliznakov, and M. Rogers (1996), "Database Infrastructure for Supporting Engineering Design Histories," *Computer-Aided Design*, **28**(5).

[6] Wood III, W. H. and A. M. Agogino (1996), "Case-Based Conceptual Design Information Server for Concurrent Engineering," *Computer-Aided Design*, **28**(5).

[7] de Kleer, J. and J. S. Brown (1983), "Assumptions and Ambiguities in Mechanistic Mental Models," *Mental Models*, D. Gentner and A. L. Stevens (eds.), Lawrence Erlbaum Associates.

[8] Iwasaki, Y. and B. Chandrasekaran (1992), "Design Verification through Function and Behavior-Oriented Representations: Bridging the Gap between Function and

Behavior," *Artificial Intelligence in Design '92*, J. S. Gero (ed.), Kluwer Academic Publishers, Boston.

[9] Chandrasekaran, B., A. Goel, and Y. Iwasaki (1993), "Functional Representation as Design Rationale" *IEEE Computer*, January, pp. 48-56.

[10] Goel, A., S. Bhatta, and E. Stroulia (1996), "Kritik: An Early Case-Based Design System," To appear in *Issues in Case-Based Design*, M. Maher and P. Pu (eds.), Hillsdale, NJ, Erlbaum.

[11] Goel, A., A. Gomez, N. Grue, J. W. Murdock, M. Recker, and T. Govindaraj (1996), "Explanatory Interface in Interactive Design Environments," *Artificial Intelligence in Design '96*, J. S. Gero (ed.), Kluwer Academic Publishers, Boston.

[12] Alberts, L. K. and F. Dikker (1994), "Integrating Standards and Synthesis Knowledge Using the YMIR Ontology," *Artificial Intelligence in Design '94*, J. S. Gero (ed.), Kluwer Academic Publishers, Boston.

[13] Gorti, S. R., A. Gupta, G. J. Kim, R. D. Sriram, and A. Wong (1998), "An Object-Oriented Representation for Product and Design Process", To appear in *Computer-Aided Design*.

[14] Wong, A. and D. Sriram (1993), "SHARED: An Information Model for Cooperative Product Development," *Research in Engineering Design*, **5**(1):21-39. Extended version: *Shared Workspaces for Computer-Aided Collaborative Engineering*, MIT Intelligent Engineering Systems Laboratory, Document 93-06 (technical report), Massachusetts Institute of Technology, Cambridge, MA.

[15] Murdock, J. W., S. Szykman, and R. D. Sriram (1997), "An Information Modeling Framework to Support Design Databases and Repositories," *Proceedings of the 1997 ASME Design for Manufacturing Conference*, Paper No. DETC97/DFM-4373, Sacramento, California, September.

[16] ISO 10303, *Industrial Automation Systems and Integration – Product Data Representation and Exchange*, ISO TC184, SC4 (ISO Technical Committee 184, Subcommittee 4).

[17] ISO 10303-203:1994, *Industrial Automation Systems and Integration – Product Data Representation and Exchange -- Part 203: Application Protocol: Configuration Controlled Design*, ISO TC184, SC4 (ISO Technical Committee 184, Subcommittee 4).

[18] ISO/IEC 14772, *Information technology – Computer graphics and Image Processing - The Virtual Reality Modeling Language*, ISO/IEC JTC1, SC24 (ISO/IEC, Joint Technical Committee 1, Subcommittee 24).

[19] Farquhar, A., R. Fikes, and J. Rice (1996), "The Ontolingua Server: a Tool for Collaborative Ontology Construction," *Proceedings of the Knowledge Acquisition Workshop*, B. Gaines (ed.), Banff, Canada.

[20] Kirschman, C. F., G. M. Fadel, and C. C. Jara-Almonte (1996), "Classifying Functions for Mechanical Design," *Proceedings of the 8th ASME International Conference on Design Theory and Methodology*, Paper No. 96-DETC/DTM-1504, Irvine, California, August.

[21] Little, A., K. L. Wood, and D. McAdams (1997), "Functional Analysis: A Fundamental Empirical Study for Reverse Engineering, Benchmarking and Redesign,"

Proceedings of the 9th ASME International Conference on Design Theory and Methodology, Paper No. DETC97/DTM-3879, Sacramento, CA, September.

[22] Stone, R. B., K. L. Wood, and R. H. Crawford (1998), "A Heuristic Method to Identify Modules from a Functional Description of a Product," *Proceedings of the 10th ASME International Conference on Design Theory and Methodology*, Paper No. DETC98/DTM-5642, Atlanta, GA, September.

[23] Falkenhainer, B., A. Farquhar, D. Bobrow et al. (1994), *CML: A compositional modeling language*, KSL-94-16 (technical report). Stanford Knowledge Systems Laboratory, Stanford University, Stanford, CA.

[24] Iwasaki, Y., A. Farquhar, R. Fikes, and J. Rice (1997), "A Web-Based Compositional Modeling System for Sharing of Physical Knowledge," *Proceedings of the 15th International Joint Conference on Artificial Intelligence*, AAAI Press, pp. 494-500.

[25] Chouiery, B. Y., S. McIlraith, Y. Iwasaki et al. (1998), "Thoughts on a Practical Theory of Reformulation for Reasoning about Physical Systems," *Proceedings of the Symposium on Abstraction, Reformulation, and Approximation (SARA '98)*, pp. 25-36; revised version to appear in *Proceedings of the 12th International Workshop on Qualitative Reasoning (QR '98)*.

[26] Ling, S. R, J. Kim, P. Will, and P. Luo (1997), "Active Catalog: Searching and Using Catalog Information in Internet-Based Design," *Proceedings of the 1997 ASME Computers in Engineering Conference*, Paper No. DETC97/CIE-4292, Sacramento, California, September.

[27] Luo, P. and P. Will (1997), "Active Catalog: A Knowledge-Rich Design Library Facilitating Information Consumption," *Knowledge Intensive CAD, Volume 2 - Proceedings of the IFIP TC5 WG5.2 International Conference on Knowledge Intensive CAD*, Mäntylä, M., S. Finger, and T. Tomiyama, (eds.), Pittsburgh, PA, September, 1996.

[28] World Wide Web Consortium (1998), *Extensible Markup Language (XML) 1.0*, World Wide Web Consortium (W3C) Recommendation, February.

[29] Lassila, O. and R. R. Swick (1998), *Resource Description Framework (RDF) Model and Syntax Specification*, World Wide Web Consortium (W3C) Working Draft, February.

[30] Brickley, D., R. V. Guha, and A. Layman (1998), *Resource Description Framework (RDF) Schemas*, World Wide Web Consortium (W3C) Working Draft, April.

[31] Genesereth, M. R. and R. E. Fikes (1992), *Knowledge Interchange Format, Version 3.0 Reference Manual*, KSL-92-86 (technical report), Stanford Knowledge Systems Laboratory, Stanford University, Stanford, CA.

[32] V. K. Chaudhri, A. Farquhar, R. Fikes, P. D. Karp, and J. P. Rice (1998), "OKBC: A Programmatic Foundation for Knowledge Base Interoperability," To appear in: *Proceedings of the 15th National Conference on Artificial Intelligence – AAAI-98*, Madison, WI.

[33] V. K. Chaudhri, A. Farquhar, R. Fikes, P. D. Karp, and J. P. Rice (1998), *Open Knowledge Base Connectivity 2.0*, KSL-98-06 (technical report), Stanford Knowledge Systems Laboratory, Stanford University, Stanford, CA.

Evolving Connectionist and Fuzzy-Connectionist Systems for On-line Adaptive Decision Making and Control

Nikola Kasabov

Department of Information Science
University of Otago, P.O Box 56, Dunedin, New Zealand
Phone: +64 3 479 8319, fax: +64 3 479 8311
nkasabov@otago.ac.nz

Keywords: evolving neural networks; fuzzy neural networks; intelligent information systems; on-line adaptive control; on-line decision making.

Abstract

The paper contains a discussion material and preliminary experimental results on a new approach to building on-line, adaptive decision making and control systems. This approach is called evolving connectionist systems (ECOS). ECOS evolve through incremental, on-line learning. They can accommodate any new input data, including new features, new classes, etc. New connections and new neurons are created during operation. The ECOS framework is illustrated here on a particular type of evolving neural networks - evolving fuzzy neural networks. ECOS are three to six orders of magnitude faster than the multilayer perceptrons, or fuzzy neural networks, trained with the backpropagation algorithm or with a genetic algorithm. ECOS are appropriate techniques to use for creating on-line, real-time, adaptive intelligent systems. This is illustrated on a case study problem of on-line, wastewater time-series flow prediction and control. Possible real world applications of this approach are discussed.

1. Introduction: On-line, Adaptive Decision Making and Control

The complexity and the dynamics of many real-world problems, especially in engineering and manufacturing, require using sophisticated methods and tools for building on-line, adaptive decision making and control systems (OLADECS). Such systems should be able to 'grow' as they work. They should be able to build-up their knowledge and refine the model through interaction with the environment.

There are different methods and tools, based on soft computing that have been developed for decision making and control so far. They include: traditional statistical and probabilistic methods; neural networks [1,2,3,4,14]; fuzzy systems

[22]; genetic algorithms and evolutionary computing [20]; fuzzy neural networks and hybrid systems [5,6,9,12,13,16,21]. The traditional approach to building decision making and control systems assumes that a model is preliminary designed based on existing knowledge and/or data. The above methods and the resulting systems are usually concerned with the precision of the control or decision making model and only few of them have the ability to partially adapt during operation. But in many cases existing knowledge or data are either not sufficient to build an adequate and precise model of the process, or they change during the system's operation due to changes in the environment. In this case OLADECS are needed.

The following is a list of requirements to the contemporary OLADECS:

- to learn fast in an incremental, on-line mode;
- to accommodate both data and a priory knowledge (e.g., existing rules)
- to produce explanation about their decision on (e.g., extract rules);
- to adjust to changes in the operating environment, introducing new variables and features if needed
- if past data is available, to learn quickly from such data.

In this section a general framework of OLADECS, that addresses all the above requirements, is presented. Its realisation with the use of evolving connectionist systems (ECOS) and evolving fuzzy neural networks EFuNNs in particular is illustrated on a case study problem after the principles of ECOS and EFuNNs are introduced in the following sections.

A block diagram of OLADECS is given in fig.1. It consists of the following blocks:

- Pre-processing (filtering) block: data is processed (filtered) in this block (e.g. checking for consistency; feature extraction, calculating moving averages, selecting time-lags for a time-series).
- Neural network (NN), multi-modular learning block: it consists of many NNs that are continuously trained with data (both old, historical data, and new incoming data)
- Rule-based block for final decision - this block takes the produced by the NN outputs and applies expert rules. The rules take into account some other variables as inputs for which there might not be data available.
- Adaptation block - this block compares the output of the system with the desired-, or the real data, obtained over certain periods of time. The error is used to adjust/adapt the NN modules in a continuous mode.
- Rule extraction, explanation block - this block uses both extracted from the NN modules rules and rules from the final DM block to explain: (1) *what* the system currently 'knows' about the problem it is solving; (2) *why* a particular decision for a concrete input vector has been made.

The framework allows: feature selection, modelling, final multivariable decision-making, rule extraction and explanation and adaptation to changes in the operating environment and to new incoming data.

There are some neural network methods and techniques for adaptive learning developed so far. They include: incremental learning [3]; on-line learning [7]; real-time learning; life-long learning; growing neural networks [4]; neural networks trained with forgetting and pruning [8,15,19]. None of them though address all the listed above requirements to OLADECS.

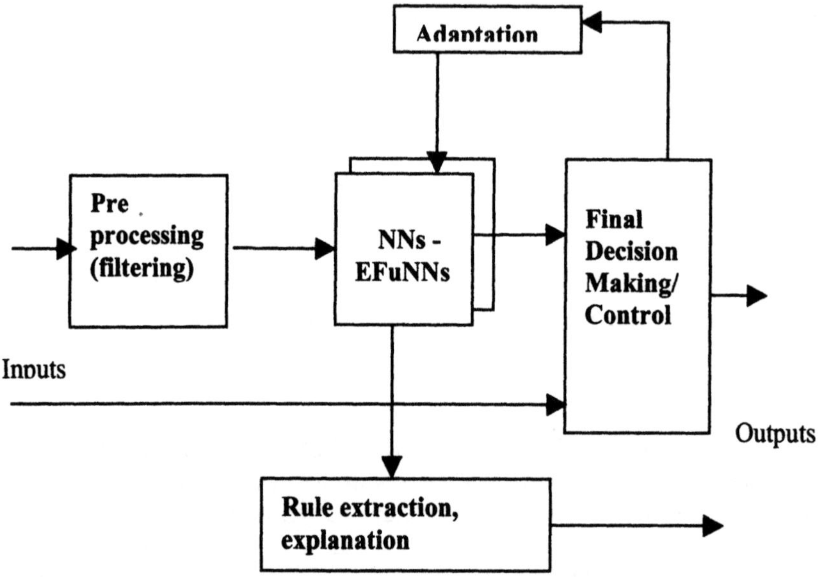

Figure 1: A block diagram of OLADECS

In the next sections the principles of ECOS and EFuNNs are presented and then applied to a case study OLADECS.

2. ECOS - Evolving Connectionist Systems

ECOS are systems that evolve in time trough interaction with the environment, i.e. an ECOS adjusts its structure with a reference to the environment (fig.2).

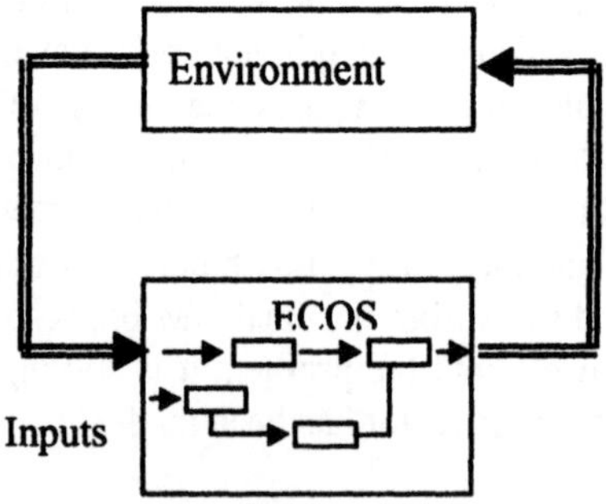

Fig.2. ECOS evolve through an interaction with the environment

A block diagram of the ECOS framework is given in [10]. ECOS are multi-level, multi-modular structures where many modules are connected with inter- and intra-connections. The evolving connectionist system does not have to have a 'clear' multi-layer structure. It has a modular 'open' structure.

Initially an ECOS contains nodes (neurons) with very little connections between them pre-defined through *prior* knowledge or 'genetic' information. These connections mainly connect modules of the initial connectionist structure. An initial set of rules can be inserted in this structure. Gradually, through self-organisation, the system becomes more and more 'wired'. The network stores different patterns (exemplars) from the training examples. A node is created and designated to represent an individual example if it is significantly different from the previous ones (with a level of differentiation set through dynamically changing parameters. The functioning of ECOS is based on the following general principles [10]:

(1) Input patterns are presented one by one, in a pattern mode, having not necessarily the same input feature sets. After each input example is presented, the ECOS either associates this example with an already existing rule (case) node, or creates a new one. A NN module, or a neuron is created when needed at any time of the functioning of the whole system. After the presentation of each new input example, the system is able to react properly on both new and old examples.

(2) The representation module evolves in two phases. In phase one input vector **x** is passed through the representation module and the case nodes become activated based on the similarity between the input vector and their input connection weights. If there is no node activated above a certain *sensitivity threshold (Sthr)* a new rule neuron (*rn*) is created and its input weights are set equal to the values of the input vector **x** and the output weights - to the desired output vector. In phase two, activation from either the winning case neuron ("one-out of-n" mode), or from all case neurons with activation above an *activation threshold (Athr)* ("many-of-on" mode) is passed to the next level of neurons. Evolving can be achieved in both supervised and unsupervised modes. In a supervised mode the final decision which class (e.g., phoneme) the current vector **x** belongs to, is made in the higher-level decision module that may activate an adaptation process. Then the connections of the representation nodes to the class output nodes, and to the input nodes are updated with the use of *learning rate coefficients lr1 and lr2,* correspondingly. If the class activated is not the desired one, then a new case node is created. The feedback from the higher level decision module goes also to the feature selection and filtering part. New features may be involved in the current adaptation and evolving phase. In an unsupervised mode a new case node is created if there is no existing case node or existing output node that are activated above *Sthr* and an *output threshold Othr* respectively. The parameters *Sthr, lr1, lr2, Errthr, Athr and Othr* can change dynamically during learning.

(3) Along with growing, an ECOS has a pruning rule defined. It allows for removing neurons and their corresponding connections that are not actively involved in the functioning of the ECOS thus making space for new input patterns. Pruning is based on local information kept in the neurons. Each neuron in ECOS keeps a 'track' of its 'age', its average activation over the whole life span, the error

it contributes to, and the density (in Euclidean sense) of the surrounding area of neurons. Pruning is performed through the fuzzy rule:

IF case node (j) is OLD, and average activation of (j) is LOW, and the density of the neighbouring area of neurons is HIGH or MODERATE, and the sum of the incoming or outgoing connection weights is LOW, THEN the probability of pruning node (j) is HIGH.

(4) The case neurons are spatially organised and each neuron has its relative spatial dimensions in regards to the rest of the neurons based on their reaction to the input patterns. If a new neuron is created when the input vector **x** was presented, the new neuron is allocated close to the neuron which had the highest activation to the input vector **x**.

(5) There are two global modes of learning in ECOS:

(a) *Active learning mode* - learning is performed when a stimulus (input pattern) is presented and kept active.

(b) *Eco learning mode*: learning is performed when there is no input pattern presented at the input of the ECOS. In this case the process of further elaboration of the connections in ECOS is done in a passive learning phase, when existing connections that store previously 'seen' input patterns are used as eco training examples. The connection weights that represent stored input patterns are now used as compressed input patterns for training other modules in ECOS. This type of learning with the use of 'echo' data is called here eco training. There are two types of eco training that are applicable to classification tasks: (1) cascade eco training; (2) sleep eco training. In *cascade eco training* a new NN module is created when a new class data is presented. The module is trained on the positive examples of this class, plus the negative examples of the following different class data, and on the negative examples of previously stored patterns in previously created modules. In the sleep eco training, modules are created with positive examples only when data is presented. Then the modules are trained on the stored in the other modules patterns as negative examples.

(6) ECOS provide explanation information extracted from the structure of the NNs. Each case (rule) node is interpreted as an IF-THEN rule as it is the case in the FuNN fuzzy neural networks [12,13].

(7) ECOS are biologically inspired. Some biological motivations for evolving systems are given in [17,18].

(8) The ECOS framework can be applied to different types of NNs, neurons, activation functions, etc. One realisation that uses the ECOS framework is the evolving fuzzy neural networks EFuNNs and the EFuNN algorithm as given in [11].

3. Fuzzy Neural Networks FuNNs and Evolving Fuzzy Neural Networks EFuNNs

3.1. The FuNN Architecture and its Functionality

Fuzzy neural networks are neural networks that realise a set of fuzzy rules and a fuzzy inference machine in a connectionist way [5,6,9,12,13,16,21]. FuNN is a fuzzy neural network introduced in [12] and then developed as FuNN/2 in [13]. It is a connectionist feed-forward architecture with five layers of neurons and four layers of connections. The first layer of neurons receives the input information. The second layer calculates the fuzzy membership degrees to which the input values belong to predefined fuzzy membership functions, e.g. small, medium, large. The third layer of neurons represents associations between the input and the output variables, fuzzy rules. The forth layer calculates the degrees to which output membership functions are matched by the input data and the fifth layer does defuzzification and calculates values for the output variables. A FuNN has both the features of a neural network and a fuzzy inference machine. A simple FuNN structure is shown in Figure 2. The number of neurons in each of the layers can potentially change during operation through growing or shrinking. The number of connections is also modifiable through learning with forgetting, zeroing, pruning and other operations [12].

The membership functions, used in FuNN to represent fuzzy values, are of triangular type, the centres of the triangles being attached as weights to the corresponding connections. The membership functions can be modified through learning.

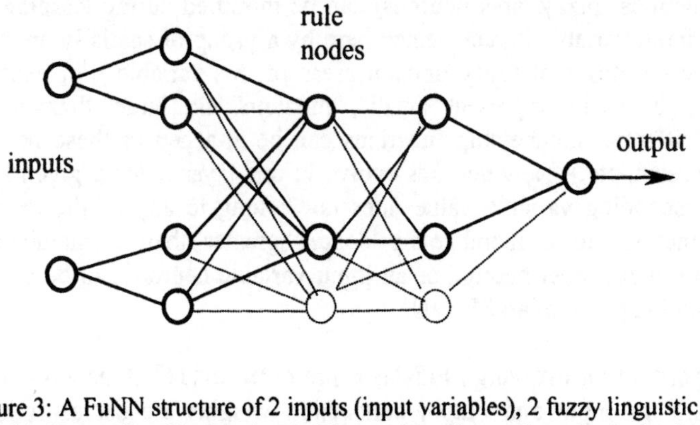

Figure 3: A FuNN structure of 2 inputs (input variables), 2 fuzzy linguistic terms for each variable (2 membership functions). The number of the rule nodes is fixed in a FuNN structure and is evolving in the EFuNN structure.

Several training algorithms have been developed for FuNN [12,13]:

(a) A modified back-propagation (BP) algorithm that does not change the input and the output connections representing the membership functions.

(b) A modified BP algorithm that utilises structural learning with forgetting, i.e. a small forgetting ingredient, e.g. 10^{-5}, is used when the connection weights are updated (see [8]).
(c) A modified BP algorithm that updates both the inner connection layers and the membership layers. This is possible when the derivatives are calculated separately for the two parts of the triangular membership functions. These are also the non-monotonic activation functions of the neurons in the condition element layer.
(d) A genetic algorithm for training [20].
(e) A combination of any of the methods above used in different time intervals as part of a single training procedure.

Several algorithms for rule extraction from FuNNs have been developed and applied [12,13]. One of them represents each rule node of a trained FuNN as an IF-THEN fuzzy rule.

FuNNs have several advantages when compared with the traditional connectionist systems or with the fuzzy systems:
(a) They are both statistical and knowledge engineering tools.
(b) They are robust to catastrophic forgetting, i.e. when they are further trained on new data, they keep a reasonable memory of the old data.
(c) They interpolate and extrapolate well in regions where data is sparse.
(d) They accept both real input data and fuzzy input data represented as singletons (centres of gravity of the input membership functions))

3.2. EFuNNs - Evolving FuNNs

EFuNNs are FuNN structures that evolve according to the ECOS principles. All nodes in an EFuNN are created during learning. The nodes representing membership functions (fuzzy label neurons) can be modified during learning. As in FuNN, each input variable is represented here by a group of spatially arranged neurons to represent different fuzzy domain areas of this variable. For example three neurons can be used to represent "small", "medium" and "large" fuzzy values of a variable. Different membership functions can be attached to these neurons (triangular, Gaussian, etc.). New neurons evolve in this layer if for a given input vector the corresponding variable value does not belong to any of the existing membership functions to a membership degree greater than a membership threshold. A new fuzzy label neuron, or an input variable neuron, can be created during the adaptation phase of an EFuNN.

The EFuNN algorithm for evolving EFuNNs is presented in [11]. A new rule node rn is created and its input and output connection weights are set as follows: W1(rn)=EX; W2(rn) = TE, where TE is the fuzzy output vector for the current example EX. In case of "one-of-n" EFuNNs, the maximum activation of a rule node is propagated to the next level. Saturated linear functions are used as activation functions of the fuzzy output neurons. In case of "many-of-n" mode, all the activation values of rule (case) nodes that are above an activation threshold *Athr* are propagated.

4. Applying ECOS and EFuNNs to Building OLADECS: A Case Study Problem

A case study problem is taken here to illustrate the potential of the ECOS and the EFuNNs. The problem is to predict a waste water flow coming from three pumps into a sewage plant (see [12] for a description of the problem and the WWW http://divcom.otago.ac.nz:800/com/infosci/KEL/home.htm for the data set). The flow is measured every hour. It is important to be able to predict the flow as the collecting tank has a limited capacity (in this case it is 650 cubic meters) and a sudden overflow will cause bacteria, that clean the water, to be thrown away. As there is very little data available before the control system is installed and put in operation, the control system has to be adaptive and learn the dynamics of the flow as it operates. This is a task for an OLADECS.

Here one EFuNN that has 4 inputs, F(t), F(t-1), MA12h(t), MA24h(t), and one output F(t+1) is evolved from the time series data. The resulting EFuNN after 500 data points, has 397 rule nodes (Sthr=0.9; Errthr=0.05; lr=0; no pruning). The MSSE over a test set of the last 100 data points is 0.0068 (normalised data is used) - see fig.4.

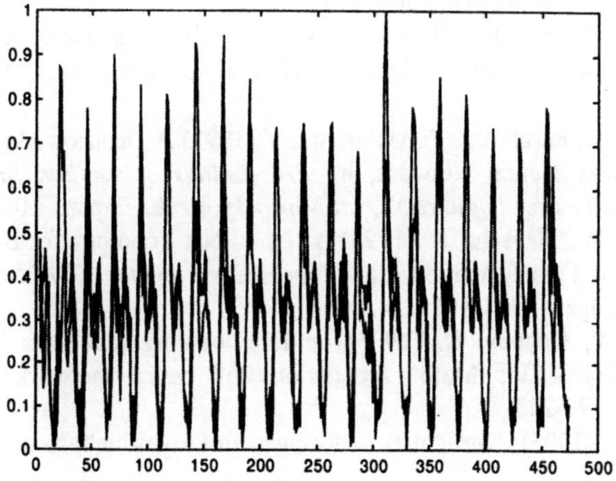

Figure 4: AnEFuNNs is evolved from the flow F(t), the moving average 12 hours MA12h(t) and the moving average 24 hours MA24h(t) data. Here the real versus the predicted one day ahead values of the flow are plotted.

It is seen from fig.4 that in the beginning the EFuNN could not generalise well on the next hour flow value, but after learning (accommodating) about 400 data points, it produces a generalisation that is much better than the generalisation on the same test data when a MLP is used that had 5 inputs, two hidden layers with 6 nodes in each of them and one output, trained with the BP algorithm for 20,000 epochs (MSSE=0.044, see [12]). The EFuNN used 4 orders of magnitude less time for training per example at average than the MLP.

5. Conclusions and Directions for Further Research

This paper applies a framework ECOS for evolving connectionist systems, and an evolving fuzzy neural network EFuNN to building on-line adaptive decision making and control systems (OLADECS). A case study, real problem has been used to illustrate the potential of this approach. Several applications of ECOS and EFuNNs, some of them already shown, will be explored in the future: adaptive speech recognition [11]; OLADECS for financial applications; OLADECS for mobile robots; OLADECS for multi-modal information retrieval.

References

1. Amari, S. and Kasabov, N. eds (1997) Brain-like computing and intelligent information systems, Springer Verlag
2. Arbib, M. (ed) (1995) *The Handbook of Brain Theory and Neural Networks.* The MIT Press
3. Carpenter, G.A., Grossberg, S., Markuzon, N., Reynolds, J.H., Rosen, D.B., FuzzyARTMAP: A neural network architecture for incremental supervised learning of analog multi-dimensional maps, IEEE Trans. on Neural Networks,vol.3, No.5 (1991), 698-713
4. Fritzke, B., Growing cell structures - a self- organising network for unsupervised and supervised learning, Neural Networks, vol.7, No.9 (1994) 1441-1460.
5. Hashiyama, T., Furuhashi, T., Uchikawa, Y.(1992) A Decision Making Model Using a Fuzzy Neural Network, in: *Proceedings of the 2nd International Conference on Fuzzy Logic & Neural Networks*, Iizuka, Japan, 1057-1060.
6. Hauptmann, W., Heesche, K. (1995) A Neural Net Topology for Bidirectional Fuzzy-Neuro Transformation, in: *Proceedings of the FUZZ-IEEE/IFES*, Yokohama, Japan, 1511-1518.
7. Heskes, T.M., Kappen, B. (1993) On-line learning processes in artificial neural networks, in: Math. foundations of neural networks, Elsevier, Amsterdam, 199-233
8. Ishikawa, M. (1996) "Structural Learning with Forgetting", Neural Networks 9, 501-521.
9. Jang, R. (1993) ANFIS: adaptive network-based fuzzy inference system, *IEEE Trans. on Syst.,Man, Cybernetics*, 23(3), May-June 1993, 665-685
10. Kasabov, N. ECOS: A framework for evolving connectionist systems and the eco learning paradigm, Proc. of ICONIP'98, Kitakyushu, Oct. 1998
11. Kasabov, N. Evolving Fuzzy Neural Networks - Algorithms, Applications and Biological Motivation, in Proc. of Iizuka'98, Iizuka, Japan, Oct.1998
12. Kasabov, N.(1996) *Foundations of Neural Networks, Fuzzy Systems and Knowledge Engineering*, The MIT Press, CA, MA.

13. Kasabov, N., Kim J S, Watts, M., Gray, A (1997) FuNN/2- A Fuzzy Neural Network Architecture for Adaptive Learning and Knowledge Acquisition, *Information Sciences - Applications,* 1997, 3(2)
14. Kohonen, T. (1990) The Self-Organizing Map. Proceedings of the IEEE, vol.78, N-9, pp.1464-1497.
15. Le Cun, Y., J.S. Denker and S.A. Solla (1990) "Optimal Brain Damage", in: Touretzky,ed.,*Advances in Neural Inform. Proc. Systems*, Morgan Kaufmann, 2, 598-605.
16. Lin, C.T. and C.S. G. Lee, "Neuro Fuzzy Systems", Prentice Hall (1996).
17. McClelland, J., B.L. McNaughton, and R.C. Reilly (1994) "Why there are Complementary Learning Systems in the Hippocampus and Neocortx: Insights from the Successes and Failures of Connectionist Models of Learning and Memeory", CMU Technical Report PDP.CNS.94.1, March
18. Quartz, S.R., and Senowski, T.J. The neural basis of cognitive development: a constructivist manifesto, Behavioral and Brain Science, in print
19. Reed, R. (1993) "Pruning algorithms - a survey", IEEE Trans. Neural Networks, 4 (5) 740-747.
20. Watts, M., and Kasabov, N. Genetic algorithms for the design of fuzzy neural networks, in Proc. of ICONIP'98, Kitakyushu, Oct. 1998
21. Yamakawa, T., H. Kusanagi, E. Uchino and T.Miki, (1993) "A new Effective Algorithm for Neo Fuzzy Neuron Model", in: Proceedings of Fifth IFSA World Congress, 1017-1020
22. Zadeh, L. 1965. Fuzzy Sets, *Information and Control*, vol.8, 338-353.

Recent New Development in Evolutionary Programming

Xin Yao
Computational Intelligence Group
School of Computer Science
University College, The University of New South Wales
Australian Defence Force Academy
Canberra, ACT 2600, Australia
Email: xin@cs.adfa.oz.au, URL: http://www.cs.adfa.oz.au/~xin

Abstract

Evolutionary programming (EP) is one of the major branches of evolutionary computation. It has been applied to many learning and optimisation problems with success in recent years. This paper gives an overview of the latest results on evolutionary programming. In particular, the paper will analyse why the recently proposed fast EP performs better than classical EP for most benchmark functions, discuss the scalability of EP, and demonstrate how EP has been used to solve the function optimisation problem and neural network design problem.

Keywords — Evolutionary programming, global optimisation, evolutionary learning, neural networks, mutation.

1 Introduction

Although evolutionary programming (EP) was first proposed as an approach to artificial intelligence [1], it has been recently applied with success to many numerical and combinatorial optimisation problems [2, 3, 4]. Optimisation by EP can be summarised into two major steps:

1. Mutate all the solutions in the current population, and

2. Select the next generation from the mutated and the current solutions.

These two steps can be regarded as a population-based version of the classical generate-and-test method [5], where mutation is used to **generate** new solutions (offspring) and selection is used to **test** which of the newly generated solutions should survive to the next generation. Formulating EP as a special case of the generate-and-test method establishes a bridge between EP and other search algorithms, such as evolution strategies, genetic algorithms, simulated annealing (SA), tabu search (TS), and others, and thus facilitates cross-fertilisation among different research areas.

One of the features of EP is its use of self-adaptive parameters. Fogel [3] and Bäck and Schwefel [6] have indicated that CEP with self-adaptive mutation usually performs better than CEP without self-adaptive mutation. This paper will only investigate EP with self-adaptive parameters. According to Bäck and Schwefel [6], the classical EP (CEP) can be implemented as follows in this study:[1]

1. Generate the initial population of μ individuals, and set $k = 1$. Each individual is taken as a pair of real-valued vectors, $(\mathbf{x}_i, \eta_i)$, $\forall i \in \{1, \cdots, \mu\}$, where $\mathbf{x}_i$'s are objective variables and η_i's are standard deviations for Gaussian mutations (also known as strategy parameters in self-adaptive evolutionary algorithms).

2. Evaluate the fitness score for each individual $(\mathbf{x}_i, \eta_i)$, $\forall i \in \{1, \cdots, \mu\}$, of the population based on the objective function, $f(\mathbf{x}_i)$.

3. Each parent $(\mathbf{x}_i, \eta_i)$, $i = 1, \cdots, \mu$, creates a single offspring $(\mathbf{x}_i', \eta_i')$ by: for $j = 1, \cdots, n$,

$$x_i'(j) = x_i(j) + \eta_i(j)N(0,1), \quad (1)$$
$$\eta_i'(j) = \eta_i(j)\exp(\tau' N(0,1) + \tau N_j(0,1)) \quad (2)$$

   where $x_i(j)$, $x_i'(j)$, $\eta_i(j)$ and $\eta_i'(j)$ denote the j-th component of the vectors $\mathbf{x}_i$, $\mathbf{x}_i'$, η_i and η_i', respectively. $N(0,1)$ denotes a normally distributed one-dimensional random number with mean 0 and standard deviation 1. $N_j(0,1)$ indicates that the random number is generated anew for each value of j. The factors τ and τ' have commonly set to $\left(\sqrt{2\sqrt{n}}\right)^{-1}$ and $\left(\sqrt{2n}\right)^{-1}$ [6, 8].

4. Calculate the fitness of each offspring $(\mathbf{x}_i', \eta_i')$, $\forall i \in \{1, \cdots, \mu\}$.

5. Conduct pairwise comparison over the union of parents $(\mathbf{x}_i, \eta_i)$ and offspring $(\mathbf{x}_i', \eta_i')$, $\forall i \in \{1, \cdots, \mu\}$. For each individual, q opponents are chosen uniformly at random from all the parents and offspring. For each comparison, if the individual's fitness is no smaller than the opponent's, it receives a "win."

6. Select the μ individuals out of $(\mathbf{x}_i, \eta_i)$ and $(\mathbf{x}_i', \eta_i')$, $\forall i \in \{1, \cdots, \mu\}$, that have the most wins to be parents of the next generation.

7. Stop if the halting criterion is satisfied; otherwise, $k = k + 1$ and go to Step 3.

---

[1] A recent study by Gehlhaar and Fogel [7] showed that swapping the order of Eq.1 and Eq.2 may improve CEP's performance.

The rest of this paper is organised as follows. Section 2 discusses global optimisation by EP. Two EP algorithms, i.e., fast EP (FEP) and improved fast EP (IFEP), will be introduced. The analysis of two different mutation operators, i.e., Gaussian and Cauchy mutation, will be carried out. The analysis shows that there is an optimal value for the step size. This section also investigates the scalability of different EP algorithms for global optimisation. The experimental results show that IFEP scales better than other EP algorithms. Section 3 describes the application of an EP algorithm in evolving neural network architectures. An evolutionary system, EPNet, is introduced. EPNet can evolve the neural network architecture and weights simultaneously. Extensive testing of EPNet on many data sets was carried out. Very good results have been obtained. Finally, Section 4 concludes this paper with a brief summary and some remarks.

2 Optimisation by Evolutionary Programming

2.1 Fast Evolutionary Programming

One disadvantage of CEP in solving some of the multimodal optimisation problems is its slow convergence to a good near-optimum. The generate-and-test formulation of EP indicates that mutation is a key search operator which generates new solutions from the current ones. We introduced the Cauchy mutation and FEP to increase the rate of convergence of EP [9, 10]. The motivation of introducing Cauchy mutation into EP comes from fast simulated annealing [11, 12, 13].

FEP [9, 10] is exactly the same as the CEP described in Section 1 except for Eq.(1) which is replaced by the following:

$$x_i'(j) = x_i(j) + \eta_i(j)\delta_j \tag{3}$$

where δ_j is an Cauchy random variable with the scale parameter $t = 1$, and is generated anew for each value of j. It is worth indicating that we leave Eq.(2) unchanged in FEP in order to keep our modification of CEP to a minimum. η in FEP plays the role of the scale parameter t, not the variance in the Cauchy distribution. It is not unreasonable to expect improvement of FEP's performance if a different mutation scheme for η is used in FEP since Eq.(2) is designed to work with the Gaussian not Cauchy mutation operator.

Extensive empirical studies of both FEP and CEP were carried out for a large number of benchmark functions in order to evaluate the relative strength and weakness of FEP and CEP for different problems. The 23 benchmark functions used in our studies are given in the Appendix. The experimental results are summarised in Tables 1, 2 and 3. It is apparent that FEP outperforms CEP for most benchmark functions, especially for functions $f_8 - -f_{13}$ which are high dimensional functions with many local optimum.

In all experiments, the same self-adaptive method (i.e., Eq.(2)), the same population size $\mu = 100$, the same tournament size $q = 10$ for selection, the same initial standard deviations 3.0, and the same initial population were used for both CEP and FEP. These parameters follow the suggestions from Bäck and Schwefel [6] and Fogel [14]. The initial population was generated uniformly at random in the range as specified in the Appendix.

Table 1: Comparison between CEP and FEP on f_1–f_7. All results have been averaged over 50 runs, where "Mean Best" indicates the mean best function values found in the last generation, and "Std Dev" stands for the standard deviation.

F's	# of Gen's	FEP Mean Best	FEP Std Dev	CEP Mean Best	CEP Std Dev	FEP−CEP t-test
f_1	1500	5.7×10^{-4}	1.3×10^{-4}	2.2×10^{-4}	5.9×10^{-4}	$4.06^{\dagger}$
f_2	2000	8.1×10^{-3}	7.7×10^{-4}	2.6×10^{-3}	1.7×10^{-4}	$49.83^{\dagger}$
f_3	5000	1.6×10^{-2}	1.4×10^{-2}	5.0×10^{-2}	6.6×10^{-2}	$-3.79^{\dagger}$
f_4	5000	0.3	0.5	2.0	1.2	$-8.25^{\dagger}$
f_5	20000	5.06	5.87	6.17	13.61	−0.52
f_6	1500	0	0	577.76	1125.76	$-3.67^{\dagger}$
f_7	3000	7.6×10^{-3}	2.6×10^{-3}	1.8×10^{-2}	6.4×10^{-3}	$-10.72^{\dagger}$

†The value of t with 49 degrees of freedom is significant at $\alpha = 0.05$ by a two-tailed test.

2.2 Analysis of Evolutionary Programming

It is known that large jumps in evolutionary search are beneficial [12. 9, 15]. However, little theoretical work had been done to analyse why this is the case until recently [16]. This section shows that large jumps are beneficial only when the global optimum is sufficiently far away from the current search point, i.e., when the distance between the current point and the global optimum is larger than the "step size" of the mutation. The optimal search step size is the distance between the current point and the global optimum.

For the sake of convenience, only one dimensional case is considered in the following analysis.

Take the Gaussian mutation in CEP as an example, which uses the following distribution with expectation 0 (which implies the current search point is located at 0) and variance σ^2,

$$f_{G(0,\sigma^2)}(x) = \frac{1}{\sigma\sqrt{2\pi}} e^{-\frac{x^2}{2\sigma^2}}, \quad -\infty < x < +\infty$$

The probability of generating a point in the neighbourhood of the global optimum x^* is given by

$$P_{G(0,\sigma^2)}(|x - x^*| \le \epsilon) = \int_{x^*-\epsilon}^{x^*+\epsilon} f_{G(0,\sigma^2)}(x)dx \quad (4)$$

Table 2: Comparison between CEP and FEP on f_8–f_{13}. The results are averaged over 50 runs, where "Mean Best" indicates the mean best function values found in the last generation, and "Std Dev" stands for the standard deviation.

F's	# of Gen's	FEP Mean Best	FEP Std Dev	CEP Mean Best	CEP Std Dev	FEP–CEP t-test
f_8	9000	-12554.5	52.6	-7917.1	634.5	$-51.39^{\dagger}$
f_9	5000	4.6×10^{-2}	1.2×10^{-2}	89.0	23.1	$-27.25^{\dagger}$
f_{10}	1500	1.8×10^{-2}	2.1×10^{-3}	9.2	2.8	$-23.33^{\dagger}$
f_{11}	2000	1.6×10^{-2}	2.2×10^{-2}	8.6×10^{-2}	0.12	$-4.28^{\dagger}$
f_{12}	1500	9.2×10^{-6}	3.6×10^{-6}	1.76	2.4	$-5.29^{\dagger}$
f_{13}	1500	1.6×10^{-4}	7.3×10^{-5}	1.4	3.7	$-2.76^{\dagger}$

†The value of t with 49 degrees of freedom is significant at $\alpha = 0.05$ by a two-tailed test.

Table 3: Comparison between CEP and FEP on f_{14}–f_{23}. The results are averaged over 50 runs, where "Mean Best" indicates the mean best function values found in the last generation, and "Std Dev" stands for the standard deviation.

F's	# of Gen's	FEP Mean Best	FEP Std Dev	CEP Mean Best	CEP Std Dev	FEP–CEP t-test
f_{14}	100	1.22	0.56	1.66	1.19	$-2.21^{\dagger}$
f_{15}	4000	5.0×10^{-4}	3.2×10^{-4}	4.7×10^{-4}	3.0×10^{-4}	0.49
f_{16}	100	-1.03	4.9×10^{-7}	-1.03	4.9×10^{-7}	0.0
f_{17}	100	0.398	1.5×10^{-7}	0.398	1.5×10^{-7}	0.0
f_{18}	100	3.02	0.11	3.0	0	1.0
f_{19}	100	-3.86	1.4×10^{-5}	-3.86	1.4×10^{-2}	-1.0
f_{20}	200	-3.27	5.9×10^{-2}	-3.28	5.8×10^{-2}	0.45
f_{21}	100	-5.52	1.59	-6.86	2.67	$3.56^{\dagger}$
f_{22}	100	-5.52	2.12	-8.27	2.95	$5.44^{\dagger}$
f_{23}	100	-6.57	3.14	-9.10	2.92	$4.24^{\dagger}$

†The value of t with 49 degrees of freedom is significant at $\alpha = 0.05$ by a two-tailed test.

where $\epsilon > 0$ is the neighbourhood size and σ is often regarded as the step size of the Gaussian mutation. Figure 1 illustrates the situation.

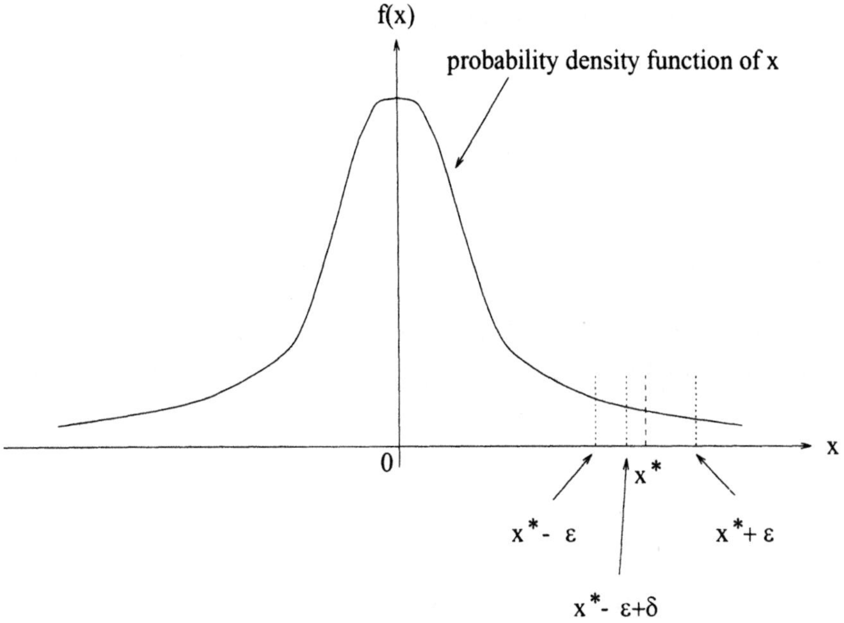

Figure 1: Evolutionary search as neighbourhood search, where x^* is the global optimum and $\epsilon > 0$ is the neighbourhood size. δ is a small positive number $(0 < \delta < 2\epsilon)$.

The derivative $\frac{\partial}{\partial\sigma} P_{G(0,\sigma^2)}(|x - x^*| \leq \epsilon)$ can be used to evaluate the impact of σ on $P_{G(0,\sigma^2)}(|x - x^*| \leq \epsilon)$. According to the mean value theorem for definite integrals [17](pp.322), there exists a number δ $(0 < \delta < 2\epsilon)$ such that

$$\int_{x^*-\epsilon}^{x^*+\epsilon} f_{G(0,\sigma^2)}(x)dx = 2\epsilon f_{G(0,\sigma^2)}(x^* - \epsilon + \delta)$$

Hence,

$$\begin{aligned}
& \frac{\partial}{\partial\sigma} P_{G(0,\sigma^2)}(|x - x^*| \leq \epsilon) \\
= & \frac{\partial}{\partial\sigma} \int_{x^*-\epsilon}^{x^*+\epsilon} f_{G(0,\sigma^2)}(x)dx \\
= & \frac{\partial}{\partial\sigma} \left(2\epsilon f_{G(0,\sigma^2)}(x^* - \epsilon + \delta)\right) \\
= & 2\epsilon \frac{\partial}{\partial\sigma} \left(\frac{1}{\sigma\sqrt{2\pi}} e^{-\frac{(x^*-\epsilon+\delta)^2}{2\sigma^2}}\right) \\
= & 2\epsilon \left(\frac{(x^* - \epsilon + \delta)^2}{\sigma^4\sqrt{2\pi}} e^{-\frac{(x^*-\epsilon+\delta)^2}{2\sigma^2}} - \frac{1}{\sigma^2\sqrt{2\pi}} e^{-\frac{(x^*-\epsilon+\delta)^2}{2\sigma^2}}\right)
\end{aligned}$$

$$= \frac{2\epsilon}{\sigma^2\sqrt{2\pi}} e^{-\frac{(x^*-\epsilon+\delta)^2}{2\sigma^2}} \left(\frac{(x^*-\epsilon+\delta)^2}{\sigma^2} - 1 \right)$$

It is apparent from the above equation that

$$\frac{\partial}{\partial\sigma} P_{G(0,\sigma^2)}(|x-x^*| \leq \epsilon) > 0 \quad \text{if} \quad \sigma < |x^* - \epsilon + \delta| \tag{5}$$

$$\frac{\partial}{\partial\sigma} P_{G(0,\sigma^2)}(|x-x^*| \leq \epsilon) < 0 \quad \text{if} \quad \sigma > |x^* - \epsilon + \delta| \tag{6}$$

That is, the larger σ is, the larger $P_{G(0,\sigma^2)}(|x-x^*| \leq \epsilon)$ will be, if $\sigma < |x^* - \epsilon + \delta|$. However, if $\sigma > |x^* - \epsilon + \delta|$, the larger σ is, the smaller $P_{G(0,\sigma^2)}(|x-x^*| \leq \epsilon)$ will be.

Similar analysis can be carried out for Cauchy mutation in FEP. Denote the Cauchy distribution as $f_{C(t)}(x) = \frac{1}{\pi}\frac{t}{t^2+x^2}, -\infty < x < \infty$,. Then we have

$$\begin{aligned}
& \frac{\partial}{\partial t} P_{C(t)}(|x-x^*| \leq \epsilon) \\
= & \frac{\partial}{\partial t} \int_{x^*-\epsilon}^{x^*+\epsilon} f_{C(t)}(x)dx \\
= & \frac{\partial}{\partial t} \left(2\epsilon f_{C(t)}(x^* - \epsilon + \delta)\right) \\
= & \frac{2\epsilon}{\pi}\frac{\partial}{\partial t} \left(\frac{t}{t^2 + (x^* - \epsilon + \delta)^2} \right) \\
= & \frac{2\epsilon}{\pi} \left(\frac{1}{t^2 + (x^* - \epsilon + \delta)^2} - \frac{2t^2}{(t^2 + (x^* - \epsilon + \delta)^2)^2} \right) \\
= & \frac{2\epsilon}{\pi} \frac{(x^* - \epsilon + \delta)^2 - t^2}{(t^2 + (x^* - \epsilon + \delta)^2)^2}
\end{aligned}$$

where δ $(0 < \delta < 2\epsilon)$ may not be the same as that in Eqs.(5) and (6). It is obvious that

$$\frac{\partial}{\partial t} P_{C(t)}(|x-x^*| \leq \epsilon) > 0 \quad \text{if} \quad t < |x^* - \epsilon + \delta| \tag{7}$$

$$\frac{\partial}{\partial t} P_{C(t)}(|x-x^*| \leq \epsilon) < 0 \quad \text{if} \quad t > |x^* - \epsilon + \delta| \tag{8}$$

That is, the larger t is, the larger $P_{C(t)}(|x-x^*| \leq \epsilon)$ will be, if $t < |x^* - \epsilon + \delta|$. However, if $t > |x^* - \epsilon + \delta|$, the larger t is, the smaller $P_{C(t)}(|x-x^*| \leq \epsilon)$ will be.

Since σ and t can be regarded as search step sizes for Gaussian and Cauchy mutations, the above analyses show that a large step size is beneficial (i.e., increases the probability of finding a near-optimal solution) only when the distance between the neighbourhood of x^* and the current search point (at 0) is larger than the step size. Or else a large step size may be detrimental to finding a near-optimal solution. The above analyses also show the rates of probability increase/decrease by deriving the explicit expressions for $\frac{\partial}{\partial \sigma} P_{G(0,\sigma^2)}(|x - x^*| \leq \epsilon)$ and $\frac{\partial}{\partial t} P_{C(t)}(|x - x^*| \leq \epsilon)$.

The analytical results explain why FEP achieved better results than CEP for most of the benchmark problems we tested, because the initial population was generated uniformly at random in a relatively large space and was far away from the global optimum on average. Cauchy mutation is more likely to generate larger jumps than Gaussian mutation and thus better in such cases. However, FEP would be less effective than CEP near the small neighbourhood of the global optimum because Gaussian mutation's step size is smaller (smaller is better in this case), which explains why FEP did not perform as well as CEP for some other functions. Further empirical studies [16] have confirmed our analytical results.

2.3 Improved Fast Evolutionary Programming

The previous analyses show the benefits of FEP and CEP in different situations. Generally speaking, Cauchy mutation performs better when the current search point is far away from the global minimum, while Gaussian mutation would be better at finding a local optimum in a good region. It would be ideal if Cauchy mutation is used when search points are far away from the global optimum and Gaussian mutation is adopted when search points are in the neighbourhood of the global optimum. Unfortunately, the global optimum is usually unknown in practice, making the ideal switch from Cauchy to Gaussian mutation very difficult. Self-adaptive Gaussian mutation [6, 14, 3, 18] is an excellent technique to partially address the problem. That is, the evolutionary algorithm itself will learn when to "switch" from one step size to another. However, there is room for further improvement to self-adaptive algorithms like CEP or even FEP.

An improved FEP (IFEP) [16] based on *mixing* (rather than switching) different mutation operators has been shown to be quite robust and efficient for a large number of benchmark functions. The idea is to mix different search biases of Cauchy and Gaussian mutations. The importance of search biases has been pointed out by some earlier studies [19](pp.375-376). The implementation of IFEP is very simple. It differs from FEP and CEP only in Step 3 of the algorithm described in Section 1. Instead of using Eq.(1) (for CEP) *or* Eq.(3) (for FEP) alone, IFEP generates two offspring from each parent, one by Cauchy mutation and the other by Gaussian. The better one is then chosen as the offspring. The rest of the algorithm is exactly the same as FEP and CEP.

2.3.1 Experimental Studies

In order to carry out fair comparison among IFEP, FEP, and CEP, the population size of IFEP was reduced to half of that of FEP or CEP in all the following experiments, since each individual in IFEP generates two offspring. However, reducing IFEP's population size by half actually disadvantages IFEP slightly because it does not double the time for any operators (such as selection) other than mutations. Nevertheless, such comparison offers a good and simple compromise.

IFEP was tested in the same experimental setup as before. For the sake of clarity and brevity, only some representative functions (out of 23) from each group were tested. Functions f_1 and f_2 are typical unimodal functions. Functions f_{10} and f_{11} are multimodal functions with many local minima. Functions f_{21}–f_{23} are multimodal functions with only a few local minima, and are particularly challenging to FEP. Table 4 summarises the final results of IFEP in comparison with FEP and CEP.

Table 4: Comparison among IFEP, FEP, and CEP on functions f_1, f_2, f_{10}, f_{11}, f_{21}, f_{22}, f_{23}. All results have been averaged over 50 runs, where "Mean Best" indicates the mean best function values found in the last generation.

F's	# of Gen's	IFEP Mean Best	FEP Mean Best	CEP Mean Best	IFEP−FEP t-test	IFEP−CEP t-test
f_1	1500	4.16×10^{-5}	5.72×10^{-4}	1.91×10^{-4}	$-28.06^{\dagger}$	$-2.39^{\dagger}$
f_2	2000	2.44×10^{-2}	7.60×10^{-2}	2.29×10^{-2}	$-51.61^{\dagger}$	$3.47^{\dagger}$
f_{10}	1500	4.83×10^{-3}	1.76×10^{-2}	8.79	$-48.54^{\dagger}$	$-21.26^{\dagger}$
f_{11}	2000	4.54×10^{-2}	2.49×10^{-2}	8.13×10^{-2}	$2.16^{\dagger}$	$-2.19^{\dagger}$
f_{21}	100	−6.46	−5.50	−6.43	$-2.19^{\dagger}$	$-5.46^{\dagger}$
f_{22}	100	−7.10	−5.73	−7.62	$-2.25^{\dagger}$	0.84
f_{23}	100	−7.80	−6.41	−8.86	$-2.14^{\dagger}$	$1.73^{\dagger}$

†The value of t with 49 degrees of freedom is significant at $\alpha = 0.05$ by a two-tailed test.

It is very clear from Table 4 that IFEP has improved FEP's performance significantly for all test functions except for f_{11}. In the case of f_{11}, IFEP was better than FEP for 25 out of 50 runs although the t-test indicates a statistically significant difference. In other words, IFEP's performance is still rather close to FEP's and certainly better than CEP's (35 out of 50 runs) on f_{11}. These results show that IFEP continues to perform at least as well as FEP on multimodal functions with many minima, and also performs very well on unimodal functions and multimodal functions with only a few local minima with which FEP has difficulty handling. IFEP achieved performance similar to CEP's on these functions.

2.4 Scalability of Evolutionary Programming

Although evolutionary algorithms have enjoyed great success in many applications, little is known about their computational complexity in solving a problem. The study of computational complexity is important because it tells us how well an algorithm scales as the problem becomes larger. We have studies the scalability of EP empirically [20]. Functions up to 300 dimensions have been tested using CEP, FEP and IFEP. This is one of the few attempts in the evolutionary computation field in attacking problems of such a size. Table 5 shows the preliminary results of our study using one unimodal (f_1)and one multimodal (f_{10}) functions.

Table 5: Comparison among CEP, FEP and IFEP on functions f_1 and f_{10} with $n = 100$, 200 and 300, respectively. All results have been averaged over 10 independent runs, where "Mean Best" indicates the mean best function values found in the last generation.

F's	n	# of Gen's	CEP Mean Best	FEP Mean Best	IFEP Mean Best
f_1	100	15000	4.2×10^{-4}	4.7×10^{-3}	7.0×10^{-4}
f_1	200	30000	2.7×10^{-3}	9.1×10^{-2}	0.24
f_1	300	60000	8.2×10^{-3}	0.46	2.3×10^{-2}
f_{10}	100	15000	15.4	3.7×10^{-2}	1.1×10^{-2}
f_{10}	200	50000	16.3	15.2	2.1×10^{-2}
f_{10}	300	75000	17.1	20.7	7.6×10^{-2}

It can be seen from the table that CEP performed well for the simple unimodal function, but failed badly for the multimodal function. On the other hand, IFEP appeared to handle both unimodal and multimodal functions quite well. For problems of 300 dimensions, IFEP was able to achieve the average result in the order of 10^{-2}, which was an excellent result when we consider that IFEP achieved the average result in the order of 10^{-3} for 30 dimensional f_{10}. The time for IFEP to find an approximate solution to the multimodal function f_{10} appears to be $O(n^2)$. However, more studies are necessary before any convincing conclusion can be drawn.

3 Evolving Artificial Neural Networks

Artificial neural networks (ANNs) have been used widely in many application areas in recent years. Most applications use feedforward ANNs and the back-propagation (BP) training algorithm. There are numerous variants of the classical BP algorithm and other training algorithms. All these training algorithms assume a fixed ANN architecture. They only train weights in the fixed architecture that includes both connectivity and node transfer functions. The problem of designing a near optimal ANN architecture for an application remains unsolved. However, this is an important issue because there are strong biological and engineering evidences to support that the function, i.e., the information processing capability of an ANN is determined by its architecture.

There have been many attempts in designing ANN architectures (especially connectivity[2]) automatically, such as various constructive and pruning algorithms [22]. Roughly speaking, a constructive algorithm starts with a minimal network (i.e., a network with a minimal number of hidden layers, nodes, and connections) and adds new layers, nodes, and connections if necessary during training, while a pruning algorithm does the opposite, i.e., deletes unnecessary layers, nodes, and connections during training. However, as indicated by Angeline *et al.* [23], "Such *structural hill climbing* methods are susceptible to becoming trapped at structural local optima." In addition, they "only investigate restricted topological subsets rather than the complete class of network architectures."

Design of a near optimal ANN architecture can be formulated as a search problem in the architecture space where each point represents an architecture. Given some performance (optimality) criteria, e.g., minimum error, fastest learning, lowest complexity, etc., about architectures, the performance level of all architectures forms a surface in the space [24, 25, 26, 27]. The optimal architecture design is equivalent to finding the highest point on this surface. There are several characteristics with such a surface, as indicated by Miller *et al.* [28], which make evolutionary algorithms better candidates for searching the surface than those constructive and pruning algorithms mentioned above.

There are two major approaches to evolving ANN architectures [29]. One is the evolution of "pure" architectures (i.e., architectures without weights). Connection weights will be trained after a near optimal architecture has been found. The other is the simultaneous evolution of both architectures and weights.

In the first approach, genotypes do not contain any weight information. In order to evaluate them, they have to be trained from a random set of initial weights using a training algorithm like BP. Unfortunately, such fitness evaluation of the genotypes is very noisy because a phenotype's fitness is used to represent the genotype's fitness. There are two major sources of noise:

1. The first source is the random initialisation of the weights. Different random initial weights may produce different training results. Hence, the same genotype may have quite different fitness due to different random initial weights used by the phenotypes.

2. The second source is the training algorithm. Different training algorithms may produce different training results even from the same set of initial weights. This is especially true for multimodal error functions. For example, a BP may reduce an ANN's error to 0.05 through training, but an EP could reduce the error to 0.001 due to its global search capability.

Such noise can mislead the evolution because of the fact that the fitness of a phenotype generated from genotype G_1 is higher than that generated from

[2]This paper is only concerned with connectivity and will use architecture and connectivity interchangeably. The work on evolving both connectivity and node transfer functions was reported elsewhere [21].

genotype G_2 does not mean that G_1 has higher fitness than G_2. In order to reduce such noise, an architecture usually has to be trained many times from different random initial weights. The average results will then be used to estimate the genotype's fitness. This method increases the computation time for fitness evaluation dramatically.

One way to alleviate the noisy fitness evaluation problem is to have a one to one mapping between genotypes and phenotypes. That is, both architecture and weight information are encoded in individuals and are evolved simultaneously [23, 30, 31, 3]. This approach matches nicely with the framework of EP.

3.1 EPNet

This section describes a new evolutionary system, i.e., EPNet [32, 29], for evolving feedforward ANNs. It combines the architectural evolution with the weight learning. The evolutionary algorithm used to evolve ANNs is based on Fogel's evolutionary programming (EP) [1, 14, 3]. EP's emphasis on the behavioral link between parents and offspring can increase the efficiency of ANN's evolution [33].

EPNet is different from previous work on evolving ANNs [34] on a number of aspects. First, EPNet emphasises the evolution of ANN behaviours by EP and uses a number of techniques, such as partial training after each architectural mutation and node splitting, to maintain the behavioural link between a parent and its offspring effectively. While some of previous EP systems [35, 31, 23, 36, 37, 3], acknowledged the importance of evolving behaviours, few techniques have been developed to maintain the behavioural link between parents and their offspring. The common practice in architectural mutations was to add or delete hidden nodes or connections uniformly at random. In particular, a hidden node was usually added to a hidden layer with full connections. Random initial weights were attached to these connections. Such an approach tends to destroy the behaviour already learned by the parent and create poor behavioural link between the parent and its offspring.

Second, EPNet encourages parsimony of evolved ANNs by attempting different mutations sequentially. That is, node or connection deletion is always attempted before addition. If a deletion is "successful", no other mutations will be made. Hence, a parsimonious ANN is always preferred. This approach is quite different from existing ones which add a network complexity (regularisation) term in the fitness function to penalise large ANNs (i.e., the fitness function would look like $f = f_{error} + \alpha f_{complexity}$). The difficulty in using such a function in practice lies in the selection of suitable coefficient α, which often involves tedious trial-and-error experiments. Evolving parsimonious ANNs by sequentially applying different mutations provides a novel and simple alternative which avoids the problem.

Third, EPNet has been tested on a number of benchmark problems [29, 38, 39, 40], including the parity problem of various sizes, the Australian credit card accessment problem, four medical diagnosis problems (breast cancer, diabetes, heart disease, and thyroid), and the Mackey-Glass time series prediction problem. It was also tested on the two-spiral problem [41]. Few evolutionary systems have been tested on a similar range of benchmark problems. The experimental results obtained by EPNet are better than those obtained by other systems in terms of generalisation and the size of ANNs.

The major steps of EPNet can be described by Figure 2, which are explained further as follows [29]:

1. Generate an initial population of M networks at random. The number of hidden nodes and the initial connection density for each network are uniformly generated at random within certain ranges. The random initial weights are uniformly distributed inside a small range.

2. Partially train each network in the population on the training set for a certain number of epochs using a modified BP (MBP) with adaptive learning rates. The number of epochs, K_0, is specified by the user. The error value E of each network on the validation set is checked after partial training. If E has not been significantly reduced, then the assumption is that the network is trapped in a local minimum and the network is marked with 'failure'. Otherwise the network is marked with 'success'.

3. Rank the networks in the population according to their error values, from the best to the worst.

4. If the best network found is acceptable or the maximum number of generations has been reached, stop the evolutionary process and go to Step 11. Otherwise continue.

5. Use the rank-based selection to choose one parent network from the population. If its mark is 'success', go to Step 6, or else go to Step 7.

6. Partially train the parent network for K_1 epochs using the MBP to obtain an offspring network and mark it in the same way as in Step 2, where K_1 is a user specified parameter. Replace the parent network with the offspring in the current population and go to Step 3.

7. Train the parent network with a simulated annealing (SA) algorithm to obtain an offspring network. If the SA algorithm reduces the error E of the parent network significantly, mark the offspring with 'success', replace its parent by it in the current population, and then go to Step 3. Otherwise discard this offspring and go to Step 8.

8. First decide the number of hidden nodes N_{hidden} to be deleted by generating a uniformly distributed random number between 1 and a user-specified maximum number. N_{hidden} is normally very small in the experiments, no more than 3 in most cases. Then delete N_{hidden} hidden nodes from the parent network uniformly at random. Partially train the pruned network by the MBP to obtain an offspring network. If the

offspring network is better than the worst network in the current population, replace the worst by the offspring and go to Step 3. Otherwise discard this offspring and go to Step 9.

9. Calculate the approximate importance of each connection in the parent network using the nonconvergent method. Decide the number of connections to be deleted in the same way as that described in Step 8. Randomly delete the connections from the parent network according to the calculated importance. Partially train the pruned network by the MBP to obtain an offspring network. If the offspring network is better than the worst network in the current population, replace the worst by the offspring and go to Step 3. Otherwise discard this offspring and go to Step 10.

10. Decide the number of connections and nodes to be added in the same way as that described in Step 8. Calculate the approximate importance of each virtual connection with zero weight. Randomly add the connections to the parent network to obtain *Offspring 1* according to their importance. Addition of each node is implemented by splitting a randomly selected hidden node in the parent network. The new grown network after adding all nodes is *Offspring 2*. Partially train *Offspring 1* and *Offspring 2* by the MBP to obtain a survival offspring. Replace the worst network in the current population by the offspring and go to Step 3.

11. After the evolutionary process, train the best network further on the combined training and validation set until it "converges".

The above evolutionary process appears to be rather complex, but its core is an EP algorithm with five mutations: hybrid training, node deletion, connection deletion, connection addition and node addition.

3.2 Use Population Information to Improve Performance

Although there have been many studies on how to evolve ANNs more effectively and efficiently [25, 26, 27], the issue of how to generate the final result from the evolutionary process has been overlooked [42]. The final result was always the best individual in the last generation or among all the generations. Few attempts have been made to exploit the population information. Such practice in essence treats evolutionary learning as the same as evolutionary optimisation, although the two are different in practice. We argue in this section that the difference between learning and optimisation can be exploited in evolutionary computation where both learning and optimisation are population-based. We show that a population contains more information than any single individual in the population. Such information can be used to improve generalisation of learned systems [43, 44].

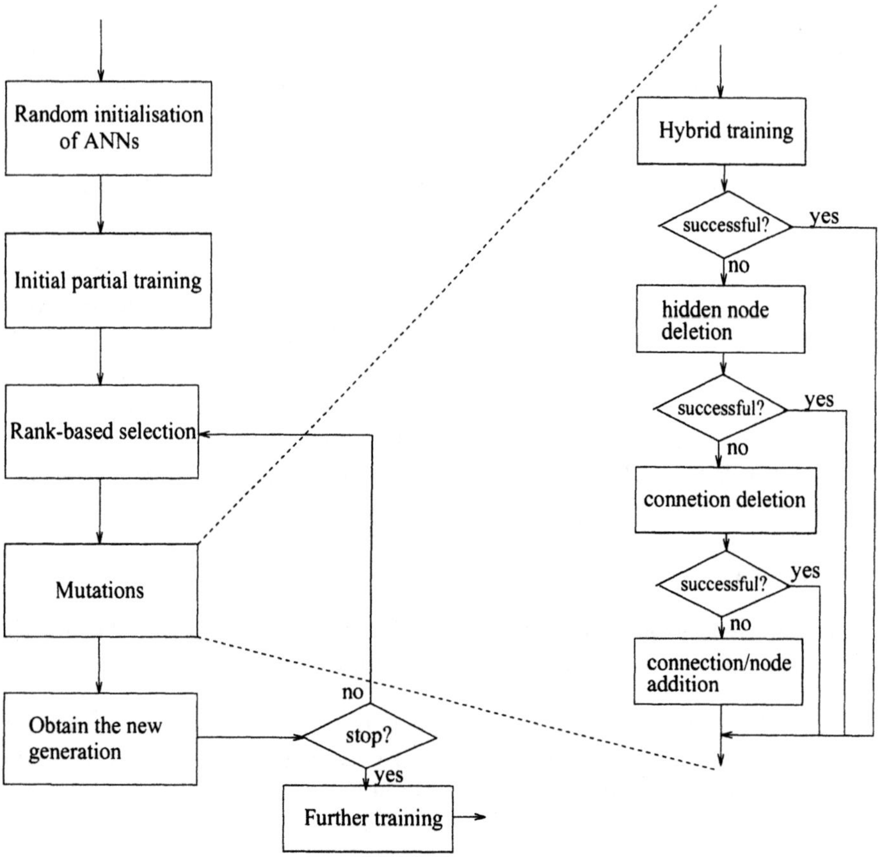

Figure 2: Major steps of EPNet [29].

Learning is different from optimisation because we want the learned system to have best generalisation, which is different from minimising an error function. The ANN with the minimum error does not necessarily mean that it has best generalisation unless there is an equivalence between generalisation and the error function. Unfortunately, measuring generalisation exactly and accurately is almost impossible in practice [45], although there are many theories and criteria on generalisation, such as the minimum description length (MDL) [46], Akaike information criteria (AIC) [47], and minimum message length (MML) [48]. In practice, these criteria are often used to define better error functions in the hope that minimising the functions will maximise generalisation. While better error functions often lead to better generalisation of learned systems, there is no guarantee. Regardless of the error functions used, BP or other more advanced learning algorithms are still used as *optimisation* algorithms. They just optimise different error functions. The nature of the problem is unchanged.

Evolutionary learning is a population-based learning method. Most people use an evolutionary algorithm to maximise a fitness function or minimise an error function, and thus face the same problem as that described above. Maximising a fitness function is different from maximising generalisation. The evolutionary algorithm is actually used as an optimisation, not learning, algorithm. While little can be done for traditional non-population-based learning, there are opportunities for improving population-based learning, e.g., evolutionary learning.

Since the maximum fitness is not equivalent to best generalisation in evolutionary learning, the best individual with the maximum fitness in a population may not be the one we want. Other individuals in the population may contain some useful information that will help to improve generalisation of learned systems. It is thus beneficial to make use of the whole population rather than any single individual. A population always contains at least as much information as any single individual. Hence, combining different individuals in the last generation to form an integrated system is expected to produce better results.

In order to integrate useful information in different individuals in the last generation, we can treat each individual as a module and linearly combine them together. We will call this combined system as an ensemble of ANNs. The reason we consider linear combination is its simplicity. The purpose here is not to find the best combination method, but to show the importance of using population information and the advantage of combining EANNs. Better results would be expected if we had used nonlinear combination methods.

We have studied four different combination methods [43]. The simplest one is majority voting. That is, the output of the most number of EANNs will be the output of the ensemble. The ensemble in this case is the whole population. All individuals in the last generation participate in voting. The second combination method is based on the ranking (fitness) of individuals in the last generation. The third combination method uses the well-known recursive least square (RLS) algorithm [49](pp.31–33) to compute the linear combination weights. The last combination method uses a genetic algorithm to reduce the size of the neural network ensemble.

We have carried out a number of experiments to show the effectiveness of our proposed ensemble approach to evolving neural networks [43]. We use the same evolutionary system, EPNet [29], as described before in our experiments. But the final output was combined according to the aforementioned methods. Three benchmark data sets were used in our studies, i.e., the Australian credit card data set, the diabetes data set and and the heart disease data set. All of them were obtained from the UCI machine learning repository. Table 6 shows the results generated by the combination and the single best individual. It is clear from the table that the ensemble results are better than those produced by the best individual.

Table 6: Accuracies of the ensemble formed by the RLS algorithm. "Best Individual" is the best one from the final generation of EPNet. *Mean, SD, Min* and *Max* indicate the mean value, standard deviation, minimum and maximum value respectively. All results were averaged over 30 runs.

Problems		Testing Error Rate		*t*-test
		RLS Ensemble	Best Individual	Ensemble−Best
Card	Mean	0.093	0.100	−2.7882
	SD	0.011	0.013	
	Min	0.076	0.081	
	Max	0.116	0.128	
Diabetes	Mean	0.226	0.232	−1.9046
	SD	0.021	0.018	
	Min	0.193	0.198	
	Max	0.260	0.271	
Heart	Mean	0.151	0.154	−0.7862
	SD	0.033	0.028	
	Min	0.088	0.103	
	Max	0.221	0.235	

4 Conclusion

This paper presents a number of recent results in EP studies. It shows that EP can be made quite efficient and effective in tackling either global optimisation or learning problems. It shows that Cauchy mutation can perform quite well in dealing with multimodal functions with many local optima. The newly proposed IFEP appears to have combined the advantages of both CEP and FEP, and scale very well for problems up to 300 dimensions (variables). The paper also describes an EP-based system for evolving NN architectures and connection weights simultaneously. Compact NNs with good generalisation can be evolved using the system.

Although learning has often been formulated as an optimisation problem, it is different from optimisation. This paper argues that there is useful information in a population that we should use in evolutionary learning. Some empirical results which demonstrate the benefit of exploiting population information are given in this paper.

Appendix: Benchmark Functions

4.1 Sphere Model

$$f_1(x) = \sum_{i=1}^{30} x_i^2$$

$$-100 \leq x_i \leq 100, \quad \min(f_1) = f_1(0,\ldots,0) = 0$$

4.2 Schwefel's Problem 2.22

$$f_2(x) = \sum_{i=1}^{30} |x_i| + \prod_{i=1}^{30} |x_i|$$

$$-10 \leq x_i \leq 10, \quad \min(f_2) = f_2(0,\ldots,0) = 0$$

4.3 Schwefel's Problem 1.2

$$f_3(x) = \sum_{i=1}^{30} \left(\sum_{j=1}^{i} x_j \right)^2$$

$$-100 \leq x_i \leq 100, \quad \min(f_3) = f_3(0,\ldots,0) = 0$$

4.4 Schwefel's Problem 2.21

$$f_4(x) = \max_i \{|x_i|, 1 \leq i \leq 30\}$$

$$-100 \leq x_i \leq 100, \quad \min(f_4) = f_4(0,\ldots,0) = 0$$

4.5 Generalised Rosenbrock's Function

$$f_5(x) = \sum_{i=1}^{29} [100(x_{i+1} - x_i^2)^2 + (x_i - 1)^2]$$

$$-30 \leq x_i \leq 30, \quad \min(f_5) = f_5(1,\ldots,1) = 0$$

4.6 Step Function

$$f_6(x) = \sum_{i=1}^{30} (\lfloor x_i + 0.5 \rfloor)^2$$

$$-100 \leq x_i \leq 100, \quad \min(f_6) = f_6(0, \ldots, 0) = 0$$

4.7 Quartic Function with Noise

$$f_7(x) = \sum_{i=1}^{30} i x_i^4 + random[0, 1)$$

$$-1.28 \leq x_i \leq 1.28, \quad \min(f_7) = f_7(0, \ldots, 0) = 0$$

4.8 Generalised Schwefel's Problem 2.26

$$f_8(x) = -\sum_{i=1}^{30} \left(x_i \sin \left(\sqrt{|x_i|} \right) \right)$$

$$-500 \leq x_i \leq 500, \quad \min(f_8) = f_8(420.9687, \ldots, 420.9687) = -12569.5$$

4.9 Generalised Rastrigin's Function

$$f_9(x) = \sum_{i=1}^{30} [x_i^2 - 10 \cos(2\pi x_i) + 10)]$$

$$-5.12 \leq x_i \leq 5.12, \quad \min(f_9) = f_9(0, \ldots, 0) = 0$$

4.10 Ackley's Function

$$f_{10}(x) = -20 \exp \left(-0.2 \sqrt{\frac{1}{30} \sum_{i=1}^{30} x_i^2} \right) - \exp \left(\frac{1}{30} \sum_{i=1}^{30} \cos 2\pi x_i \right) + 20 + e$$

$$-32 \leq x_i \leq 32, \quad \min(f_{10}) = f_{10}(0, \ldots, 0) = 0$$

4.11 Generalised Griewank Function

$$f_{11}(x) = \frac{1}{4000} \sum_{i=1}^{30} x_i^2 - \prod_{i=1}^{30} \cos \left(\frac{x_i}{\sqrt{i}} \right) + 1$$

$$-600 \leq x_i \leq 600, \quad \min(f_{11}) = f_{11}(0, \ldots, 0) = 0$$

4.12 Generalised Penalised Functions

$$f_{12}(x) = \frac{\pi}{30}\left\{10\sin^2(\pi y_1) + \sum_{i=1}^{29}(y_i - 1)^2[1 + 10\sin^2(\pi y_{i+1})] + (y_n - 1)^2\right\} + \sum_{i=1}^{30} u(x_i, 10, 100, 4)$$

$$-50 \le x_i \le 50, \quad \min(f_{12}) = f_{12}(1, \ldots, 1) = 0$$

$$f_{13}(x) = 0.1\{\sin^2(\pi 3x_1 + \sum_{i=1}^{29}(x_i - 1)^2[1 + \sin^2(3\pi x_{i+1})] + (x_n - 1)[1 + \sin^2(2\pi x_{30})]\} + \sum_{i=1}^{30} u(x_i, 5, 100, 4)$$

$$-50 \le x_i \le 50, \quad \min(f_{13}) = f_{13}(1, \ldots, 1) = 0$$

where

$$u(x_i, a, k, m) = \begin{cases} k(x_i - a)^m, & x_i > a, \\ 0, & -a \le x_i \le a, \\ k(-x_i - a)^m, & x_i < -a. \end{cases}$$

$$y_i = 1 + \frac{1}{4}(x_i + 1)$$

4.13 Shekel's Foxholes Function

$$f_{14}(x) = \left[\frac{1}{500} + \sum_{j=1}^{25} \frac{1}{j + \sum_{i=1}^{2}(x_i - a_{ij})^6}\right]^{-1}$$

$$-65.536 \le x_i \le 65.536, \quad \min(f_{14}) = f_{14}(-32, -32) \approx 1$$

where

$$(a_{ij}) = \begin{pmatrix} -32 & -16 & 0 & 16 & 32 & -32 & \cdots & 0 & 16 & 32 \\ -32 & -32 & -32 & -32 & -32 & -16 & \cdots & 32 & 32 & 32 \end{pmatrix}$$

4.14 Kowalik's Function

$$f_{15}(x) = \sum_{i=1}^{11}\left[a_i - \frac{x_1(b_i^2 + b_i x_2)}{b_i^2 + b_i x_3 + x_4}\right]^2$$

$$-5 \le x_i \le 5, \quad \min(f_{15}) \approx f_{15}(0.1928, 0.1908, 0.1231, 0.1358) \approx 0.0003075$$

Table 7: Kowalik's Function f_{15}

i	a_i	b_i^{-1}
1	0.1957	0.25
2	0.1947	0.5
3	0.1735	1
4	0.1600	2
5	0.0844	4
6	0.0627	6
7	0.0456	8
8	0.0342	10
9	0.0323	12
10	0.0235	14
11	0.0246	16

4.15 Six-hump Camel-Back Function

$$f_{16} = 4x_1^2 - 2.1x_1^4 + \frac{1}{3}x_1^6 + x_1x_2 - 4x_2^2 + 4x_2^4$$

$$-5 \le x_i \le 5$$

$$x_{min} = (0.08983, -0.7126),\ (-0.08983, 0.7126)$$

$$\min(f_{16}) = -1.0316285$$

4.16 Branin Function

$$f_{17}(x) = \left(x_2 - \frac{5.1}{4\pi^2}x_1^2 + \frac{5}{\pi}x_1 - 6\right)^2 + 10\left(1 - \frac{1}{8\pi}\right)\cos x_1 + 10$$

$$-5 \le x_1 \le 10, \quad 0 \le x_2 \le 15$$

$$x_{min} = (-3.142, 12.275),\ (3.142, 2.275),\ (9.425, 2.425)$$

$$\min(f_{17}) = 0.398$$

4.17 Goldstein-Price Function

$$\begin{aligned} f_{18}(x) &= [1 + (x_1 + x_2 + 1)^2(19 - 14x_1 + 3x_1^2 - 14x_2 + 6x_1x_2 + 3x_2^2)] \\ &\quad \times[30 + (2x_1 - 3x_2)^2(18 - 32x_1 + 12x_1^2 + 48x_2 - 36x_1x_2 + 27x_2^2)] \end{aligned}$$

$$-2 \le x_i \le 2, \quad \min(f_{18}) = f_{18}(0, -1) = 3$$

4.18 Hartman's Family

$$f(x) = -\sum_{i=1}^{4} c_i \exp\left[-\sum_{j=1}^{n} a_{ij}(x_j - p_{ij})^2\right]$$

with $n = 3, 6$ for $f_{19}(x)$ and $f_{20}(x)$, respectively, $0 \leq x_j \leq 1$. The coefficients are defined by Tables 8 and 9, respectively.

Table 8: Hartman Function f_{19}

i	$a_{ij}, j = 1,2,3$			c_i	$p_{ij}, j = 1,2,3$		
1	3	10	30	1	0.3689	0.1170	0.2673
2	0.1	10	35	1.2	0.4699	0.4387	0.7470
3	3	10	30	3	0.1091	0.8732	0.5547
4	0.1	10	35	3.2	0.038150	0.5743	0.8828

Table 9: Hartman Function f_{20}

i	$a_{ij}, j = 1, \cdots, 6$						c_i
1	10	3	17	3.5	1.7	8	1
2	0.05	10	17	0.1	8	14	1.2
3	3	3.5	1.7	10	17	8	3
4	17	8	0.05	10	0.1	14	3.2

i	$p_{ij}, j = 1, \cdots, 6$					
1	0.1312	0.1696	0.5569	0.0124	0.8283	0.5886
2	0.2329	0.4135	0.8307	0.3736	0.1004	0.9991
3	0.2348	0.1415	0.3522	0.2883	0.3047	0.6650
4	0.4047	0.8828	0.8732	0.5743	0.1091	0.0381

For $f_{19}(x)$ the global minimum is equal to -3.86 and it is reached at the point $(0.114, 0.556, 0.852)$. For $f_{20}(x)$ the global minimum is -3.32 at the point $(0.201, 0.150, 0.477, 0.275, 0.311, 0.657)$.

4.19 Shekel's Family

$$f(x) = -\sum_{i=1}^{m} [(x - a_i)(x - a_i)^T + c_i]^{-1}$$

with $m = 5, 7$ and 10 for $f_{21}(x)$, $f_{22}(x)$ and $f_{23}(x)$, respectively, $0 \leq x_j \leq 10$.

These functions have 5, 7 and 10 local minima for $f_{21}(x)$, $f_{22}(x)$, and $f_{23}(x)$, respectively. $x_{local\_opt} \approx a_i$, $f(x_{local\_opt}) \approx 1/c_i$ for $1 \leq i \leq m$. The coefficients are defined by Table 10.

Table 10: Shekel Functions f_{21}, f_{22}, f_{23}

i	$a_{ij}, j = 1, \cdots, 4$				c_i
1	4	4	4	4	0.1
2	1	1	1	1	0.2
3	8	8	8	8	0.2
4	6	6	6	6	0.4
5	3	7	3	7	0.4
6	2	9	2	9	0.6
7	5	5	3	3	0.3
8	8	1	8	1	0.7
9	6	2	6	2	0.5
10	7	3.6	7	3.6	0.5

Acknowledgement

The work described in this paper is partially supported by the Australian Research Council through its small grant scheme and by a University College Special Research Grant.

References

[1] L. J. Fogel, A. J. Owens, and M. J. Walsh, *Artificial Intelligence Through Simulated Evolution.*
New York, NY: John Wiley & Sons, 1966.

[2] D. B. Fogel, "Applying evolutionary programming to selected traveling salesman problems," *Cybernetics and Systems*, vol. 24, pp. 27–36, 1993.

[3] D. B. Fogel, *Evolutionary Computation: Towards a New Philosophy of Machine Intelligence.*
New York, NY: IEEE Press, 1995.

[4] K.-H. Liang, X. Yao, C. Newton, and D. Hoffman, "Solving cutting stock problems by evolutionary programming," in *Evolutionary Programming VII: Proc. of the 7th Annual Conference on Evolutionary Programming*, Lecture Notes in Computer Science, (Berlin), Springer-Verlag, 1998.

[5] X. Yao, "An overview of evolutionary computation," *Chinese Journal of Advanced Software Research (Allerton Press, Inc., New York, NY 10011)*, vol. 3, no. 1, pp. 12–29, 1996.

[6] T. Bäck and H.-P. Schwefel, "An overview of evolutionary algorithms for parameter optimization," *Evolutionary Computation*, vol. 1, no. 1, pp. 1–23, 1993.

[7] D. K. Gehlhaar and D. B. Fogel, "Tuning evolutionary programming for conformationally flexible molecular docking," in *Evolutionary Programming V: Proc. of the Fifth Annual Conference on Evolutionary Programming* (L. J. Fogel, P. J. Angeline, and T. Bäck, eds.), pp. 419–429, MIT Press, Cambridge, MA, 1996.

[8] D. B. Fogel, "An introduction to simulated evolutionary optimisation," *IEEE Trans. on Neural Networks*, vol. 5, no. 1, pp. 3–14, 1994.

[9] X. Yao and Y. Liu, "Fast evolutionary programming," in *Evolutionary Programming V: Proc. of the Fifth Annual Conference on Evolutionary Programming* (L. J. Fogel, P. J. Angeline, and T. Bäck, eds.), (Cambridge, MA), pp. 451–460, The MIT Press, 1996.

[10] X. Yao and Y. Liu, "Fast evolution strategies," *Control and Cybernetics*, vol. 26, no. 3, pp. 467–496, 1997.

[11] H. H. Szu and R. L. Hartley, "Nonconvex optimization by fast simulated annealing," *Proceedings of IEEE*, vol. 75, pp. 1538–1540, 1987.

[12] X. Yao, "Simulated annealing with extended neighbourhood," *Int. J. of Computer Math.*, vol. 40, pp. 169–189, 1991.

[13] X. Yao, "A new simulated annealing algorithm," *Int. J. of Computer Math.*, vol. 56, pp. 161–168, 1995.

[14] D. B. Fogel, *System Identification Through Simulated Evolution: A Machine Learning Approach to Modeling.* Needham Heights, MA 02194: Ginn Press, 1991.

[15] C. Kappler, "Are evolutionary algorithms improved by large mutations?," in *Parallel Problem Solving from Nature (PPSN) IV* (H.-M. Voigt, W. Ebeling, I. Rechenberg, and H.-P. Schwefel, eds.), vol. 1141 of *Lecture Notes in Computer Science*, (Berlin), pp. 346–355, Springer-Verlag, 1996.

[16] X. Yao, G. Lin, and Y. Liu, "An analysis of evolutionary algorithms based on neighbourhood and step sizes," in *Evolutionary Programming VI: Proc. of the Sixth Annual Conference on Evolutionary Programming* (P. J. Angeline, R. G. Reynolds, J. R. McDonnell, and R. Eberhart, eds.), vol. 1213 of *Lecture Notes in Computer Science*, (Berlin), pp. 297–307, Springer-Verlag, 1997.

[17] R. A. Hunt, *Calculus with Analytic Geometry.* New York, NY 10022-5299: Harper & Row Publ., Inc., 1986.

[18] K.-H. Liang, X. Yao, Y. Liu, C. Newton, and D. Hoffman, "An experimental investigation of self-adaptation in evolutionary programming," in *Evolutionary Programming VII: Proc. of the 7th Annual Conference on Evolutionary Programming*, Lecture Notes in Computer Science, (Berlin), Springer-Verlag, 1998.

[19] X. Yao (ed.), "Special issue on evolutionary computation," *Informatica*, vol. 18, pp. 375–450, 1994.

[20] X. Yao and Y. Liu, "Scaling up evolutionary programming algorithms," in *Evolutionary Programming VII: Proc. of the 7th Annual Conference on Evolutionary Programming*, Lecture Notes in Computer Science, (Berlin), Springer-Verlag, 1998.

[21] Y. Liu and X. Yao, "Evolutionary design of artificial neural networks with different nodes," in *Proc. of the 1996 IEEE Int'l Conf. on Evolutionary Computation (ICEC'96), Nagoya, Japan*, pp. 670–675, IEEE Press, New York, NY 10017-2394, 1996.

[22] R. Reed, "Pruning algorithms — a survey," *IEEE Trans. on Neural Networks*, vol. 4, no. 5, pp. 740–747, 1993.

[23] P. J. Angeline, G. M. Sauders, and J. B. Pollack, "An evolutionary algorithm that constructs recurrent neural networks," *IEEE Trans. on Neural Networks*, vol. 5, no. 1, pp. 54–65, 1994.

[24] X. Yao, "Evolution of connectionist networks," in *Preprints of the Int'l Symp. on AI, Reasoning & Creativity* (T. Dartnall, ed.), (Queensland, Australia), pp. 49–52, Griffith University, 1991.

[25] X. Yao, "A review of evolutionary artificial neural networks," *International Journal of Intelligent Systems*, vol. 8, no. 4, pp. 539–567, 1993.

[26] X. Yao, "Evolutionary artificial neural networks," *International Journal of Neural Systems*, vol. 4, no. 3, pp. 203–222, 1993.

[27] X. Yao, "Evolutionary artificial neural networks," in *Encyclopedia of Computer Science and Technology* (A. Kent and J. G. Williams, eds.), vol. 33, pp. 137–170, New York, NY 10016: Marcel Dekker Inc., 1995.

[28] G. F. Miller, P. M. Todd, and S. U. Hegde, "Designing neural networks using genetic algorithms," in *Proc. of the Third Int'l Conf. on Genetic Algorithms and Their Applications* (J. D. Schaffer, ed.), pp. 379–384, Morgan Kaufmann, San Mateo, CA, 1989.

[29] X. Yao and Y. Liu, "A new evolutionary system for evolving artificial neural networks," *IEEE Transactions on Neural Networks*, vol. 8, no. 3, pp. 694–713, 1997.

[30] V. Maniezzo, "Genetic evolution of the topology and weight distribution of neural networks," *IEEE Trans. on Neural Networks*, vol. 5, no. 1, pp. 39–53, 1994.

[31] J. R. McDonnell and D. Waagen, "Evolving recurrent perceptrons for time-series modeling," *IEEE Trans. on Neural Networks*, vol. 5, no. 1, pp. 24–38, 1994.

[32] X. Yao and Y. Liu, "Evolving artificial neural networks through evolutionary programming," in *Evolutionary Programming V: Proc. of the Fifth Annual Conference on Evolutionary Programming* (L. J. Fogel, P. J. Angeline, and T. Bäck, eds.), (Cambridge, MA), pp. 257–266, The MIT Press, 1996.

[33] X. Yao, "The importance of maintaining behavioural link between parents and offspring," in *Proc. of the 1997 IEEE Int'l Conf. on Evolutionary Computation (ICEC'97), Indianapolis, USA*, pp. 629–633, IEEE Press, New York, NY, April 1997.

[34] X. Yao and Y. Shi, "A preliminary study on designing artificial neural networks using co-evolution," in *Proc. of the IEEE Singapore Intl Conf on Intelligent Control and Instrumentation*, (Singapore), pp. 149–154, IEEE Singapore Section, June 1995.

[35] D. B. Fogel, L. J. Fogel, and V. W. Porto, "Evolving neural networks," *Biological Cybernetics*, vol. 63, pp. 487–493, 1990.

[36] J. R. McDonnell and D. Waagen, "Neural network structure design by evolutionary programming," in *Proc. of the Second Annual Conference on Evolutionary Programming* (D. B. Fogel and W. Atmar, eds.), pp. 79–89, Evolutionary Programming Society, La Jolla, CA, 1993.

[37] D. B. Fogel, "Using evolutionary programming to create neural networks that are capable of playing tic-tac-toe," in *Proc. of the 1993 Int'l Joint Conf. on Neural Networks (IJCNN'93)*, pp. 875–880, IEEE Press, New York, NY 10017-2394, 1993.

[38] X. Yao and Y. Liu, "EPNet for chaotic time-series prediction," in *Selected Papers from the Frist Asia-Pacific Conference on Simulated Evolution and Learning (SEAL'96)* (X. Yao, J.-H. Kim, and T. Furuhashi, eds.), vol. 1285 of *Lecture Notes in Artificial Intelligence*, (Berlin), pp. 146–156, Springer-Verlag, 1997.

[39] Y. Liu and X. Yao, "A population-based learning algorithm which learns both architectures and weights of neural networks," *Chinese Journal of Advanced Software Research (Allerton Press, Inc., New York, NY 10011)*, vol. 3, no. 1, pp. 54–65, 1996.

[40] X. Yao and Y. Liu, "Evolutionary artificial neural networks that learn and generalise well," in *1996 IEEE International Conference on Neural Networks, Washington, DC, USA, Volume on Plenary, Panel and Special Sessions*, pp. 159–164, IEEE Press, New York, NY, 3-6 June 1996.

[41] X. Yao and Y. Liu, "Towards designing artificial neural networks by evolution," *Applied Mathematics and Computation*, vol. 91, no. 1, pp. 83–90, 1998.

[42] X. Yao and Y. Liu, "Ensemble structure of evolutionary artificial neural networks," in *Proc. of the 1996 IEEE Int'l Conf. on Evolutionary Computation (ICEC'96), Nagoya, Japan*, pp. 659–664, IEEE Press, New York, NY 10017-2394, 1996.

[43] X. Yao and Y. Liu, "Making use of population information in evolutionary artificial neural networks," *IEEE Trans. on Systems, Man, and Cybernetics, Part B: Cybernetics*, vol. 28, no. 3, pp. 417–425, June 1998.

[44] X. Yao, Y. Liu, and P. Darwen, "How to make best use of evolutionary learning," in *Complex Systems: From Local Interactions to Global Phenomena* (R. Stocker, H. Jelinek, and B. Durnota, eds.), (Amsterdam), pp. 229–242, IOS Press, 1996.

[45] D. H. Wolpert, "A mathematical theory of generalization," *Complex Systems*, vol. 4, pp. 151–249, 1990.

[46] J. Rissanen, "Modeling by shortest data description," *Automatica*, vol. 14, pp. 465–471, 1978.

[47] H. Akaike, "A new look at the statistical model identification," *IEEE Trans. Appl. Comp.*, vol. AC-19, pp. 716–723, 1974.

[48] C. S. Wallace and J. D. Patrick, "Coding decision trees," Tech. Rep. 91/153, Dept. of Computer Science, Monash University, Clayton, Victoria 3168, Australia, August 1991.

[49] B. Mulgrew and C. F. N. Cowan, *Adaptive Filters and Equalisers.* Boston: Kluwer Academic Publ., 1988.

Emotional Image Retrieval with Interactive Evolutionary Computation[1]

Sung-Bae Cho and Joo-Young Lee

Department of Computer Science, Yonsei University
134 Shinchon-dong, Sudaemoon-ku, Seoul 120-749, Korea
Tel: +82-2-361-2720 Fax: +82-2-365-2579
E-mail: sbcho@csai.yonsei.ac.kr, jylee@candy.yonsei.ac.kr

Keywords: Interactive Evolutionary Computation, Subjective Evaluation, Content-based Image Retrieval

Abstract

Recently, it has been actively exploited to combine several softcomputing techniques such as fuzzy logic, neural networks and evolutionary computation to develop better systems that can match human ability, but more work is still required to be able to match human performance. In order to remedy this shortcoming, several approaches to incorporate human himself into the softcomputing framework have been devised, and among them is interactive evolutionary computation in which the genetic optimization is conducted interactively with human evaluation. This paper presents the potential of interactive evolutionary computation in an engineering problem, called content-based image retrieval. In this system, each individual in population is evaluated by user and the next generation is produced based on the user's evaluation, by which it might incorporate human preference into the process of image retrieval. Experiments with a small database indicate that the interactive evolutionary computation is useful in content-based image retrieval.

1 Introduction

As the power of computer improves, computer users request the facility for information retrieval from multimedia database. In particular, they have interests in the methods that allow to retrieve information based

[1]This work was supported in part by a grant no. 981-0919-099-2 from the Korea Science and Engineering Foundation (KOSEF).

on the stored contents. Several working systems have already been developed: QBIC system of IBM [1], QVE (Query by Visual Example) of Hirata and Kato [2], chabot of UC Berkeley [3], photobook of MIT, and Image Surfer of Interpix Software. Since these systems obtain several features from user by query, they perform better than the conventional methods based on keyword matching. However, most of the previous systems are based on the engineering approach that has little relevancy to human intuition and emotion.

Due to the lack of expressiveness, it is difficult to retrieve the images that user wants. Image retrieval based on human intuition could overcome the difficulty, and play an important role of reducing psychological burden to retrieve the images for which the user cannot think of the concrete query.

This paper presents interactive genetic algorithm that can perform content-based image retrieval using human intuition. In interactive genetic algorithm, human evaluates each individual in population and the next generation is produced based on the evaluation, by which the system might incorporate human preference into the process of image retrieval. In this method, wavelet transform is used to extract feature from images. This paper attempts to analyze the effectiveness of the wavelet features in the content-based image retrieval with interactive genetic algorithm.

2 Related Works

Interactive genetic algorithm measures the fitness based on user's judgment. There have been not so many works on this approach, and utilizing the property that user determines fitness has been largely attempted in the field of graphics or art. For an instance, Caldwell and Johnston develop a system to track criminal suspect with interactive genetic algorithm [4]. This system produces montages based on the fitness on the faces of criminal suspect. Baker presents a line drawing system based on user's aesthetic criteria, which are among attraction, amusement, and control [5].

3 Discrete Wavelet Transform

Wavelet transform is a mathematical function that decomposes data into various kinds of frequency components. Since it is better than

Fourier transform in the case of analyzing physical situation that contains sharp area, it is applied to many image processing areas such as edge preserving smoothing, de-noising, and compression.

Wavelet transform allows very good image approximation with just a few coefficients. This property has been exploited for lossy image compression. Two-dimensional wavelet transform is utilized to extract features from images and the result of the transform is used for searching the images. In particular, Haar wavelet transform is adopted because it is easily implemented and fast.

A standard two-dimensional Haar wavelet decomposition of an image can be implemented by pyramid algorithm [6] as follows.

```
// one-dimensional decomposition
Proc DecomposeArray (A: array [0..h-1] of color)
  A = A / sqrt(h);
  while h > 1 do
    h = h / 2;
    for i = 0 to h-1 do
      A'[i] = (A[2i] + A[2i+1]) / sqrt(2);
      A'[h+i] = (A[2i] - A[2i+1]) / sqrt(2);
    end for
    A = A';
  end while
end proc

// entire r*r image decomposition
Proc DecomposeImage (T: array [0..r-1, 0..r-1] of color)
  for row = 1 to r do
    DecomposeArray (T[row, 0..r])
  end for
  for col = 1 to r do
    DecomposeArray (T[0..r, col])
  end for
end proc
```

This algorithm operates on a finite set of N input data, and produces the wavelet transform coefficients efficiently. The input data are passed through two convolution function, each of which creates output stream that is half the length of the original input. One half of the output is produced by low pass filter, and the other half is produced by high pass filter. Low pass output contains information content and is used as an input of filtering in the next step. On the other hand, high pass output contains difference information between original input and reconstructed input. In general, high-order wavelets tend to put more information into the low-pass output. Wavelet based compression is caused by making the high-pass output to be nearly zero. The pyramid algorithm involves one-dimensional decomposition on each row of the image followed by one-dimensional decomposition on the column of the result.

4 Interactive Genetic Algorithm

Genetic algorithm is a model of machine learning derived by the procedure of evolution in nature [7]. This is performed by creating the population of individuals that are represented by chromosome. The chromosome is the string that can be thought of as human gene. The individuals in the population go through the evolution. This procedure takes the evolutionary procedure in which different individuals compete for resources in the environment. Some of better individuals are more likely to survive, and propagate their genetic material to offsprings. The procedure of a simple genetic algorithm is as follows.

```
t = 0;
InitializePopulation P(t);
Evaluate P(t);
while not done do
  t = t + 1;
  P'= SelectParents P(t);
  Recombine P'(t);
  Mutate P'(t);
  Evaluate P'(t);
  P = Survive P, P'(t);
end while
```

The algorithm starts with initial population. In the beginning, fitness value of each individual is evaluated to determine how appropriate the individual is for the given problem. In general, individuals in an initial population are randomly generated. Next, two individuals that have relatively high fitness value are selected from the population. Selected individuals can be regarded as parents. New individuals called children are created by recombinating the chromosomes of parents. Here, selection and crossover operators are used. Mutation replaces existing gene code with randomly generated code. It performs a function that puts potentially good genes or properties that may be lost during applying genetic operation into the population.

Interactive genetic algorithm adopts user's choice as fitness, when fitness function cannot be exactly determined. This property allows developing system based on human intuition or emotion. Therefore, in this paper, system obtains fitness values to images from user, which are used to select good individuals for the next generation.

4.1 Chromosome Representation

Wavelet coefficients are obtained by decomposing an image using wavelet transform. Through the above procedure r×r matrix, T, contains the average color of the image in entry $T[0,0]$ and wavelet coefficients in the remaining entries of T. We can reconstruct original image without loss using this information, but because there is no need to maintain all the information for searching, we extract the largest 50 coefficients in RGB channels and construct a chromosome in 3×50 array.

Jacob's work shows that storing the 40~50 largest-magnitude coefficients in each color works best and truncating the coefficients appears to improve the discrimination power of the metric [8]. Therefore, we only store the sign information of coefficient values in the chromosome.

4.2 Genetic Operators

The size of population is twelve, and the fitness values for shape and color are obtained from user. The selection strategy is governed by expected frequency of each individual, and one point horizontal and vertical crossover operators are used. Mutation is not adopted. Figure 1 shows the schematic diagrams of horizontal and vertical crossover operations.

4.3 Implementation

The entire system is constructed as shown in Figure 2. In preprocessing step, we perform wavelet transform to every image in the database and store the overall average color and the indices and signs of the m magnitude wavelet coefficients in a table.

To search images, we use interactive genetic algorithm. A system displays twelve images, obtains the fitness value from user, and selects

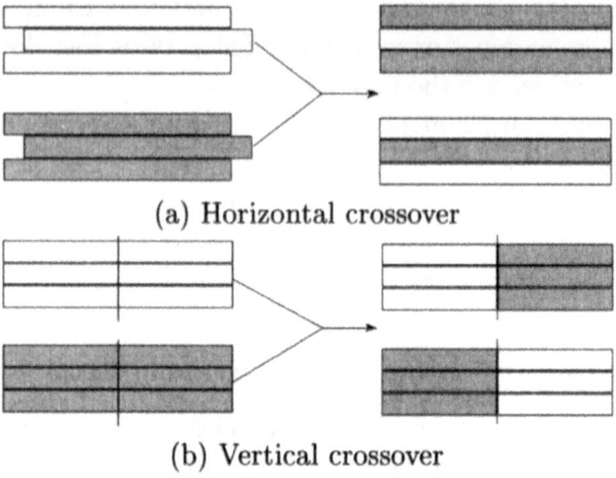

(a) Horizontal crossover

(b) Vertical crossover

Figure 1: Horizontal and vertical crossover operations.

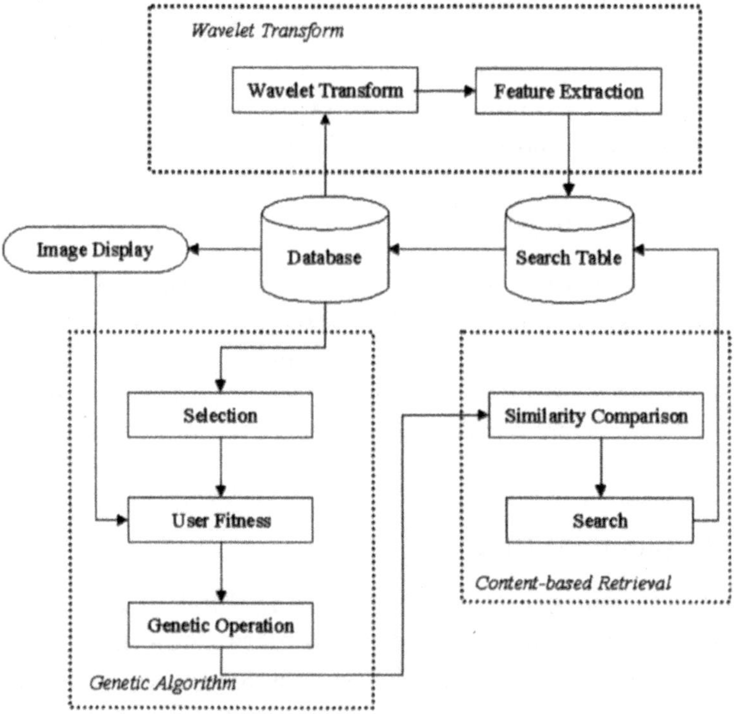

Figure 2: System structure.

candidates based on the fitness. A genetic operation, vertical or horizontal crossover, is applied to the selected candidates. To find best twelve images at the current generation, the stored image information is evaluated by each criteria. Twelve images of the higher magnitude value are provided as a result of the search. The similarity between potential target image and candidate image is calculated by the following equation.

$$||Q,T|| = w_{0,0}|Q[0,0] - T[0,0]| + \sum_{i,j} w_{i,j}|Q[i,j] - T[i,j]| \quad (1)$$

Here, $Q[i,j]$ and $T[i,j]$ represent single color channels of wavelet decomposition of the candidate and target images, and $Q[0,0]$ and $T[0,0]$ mean overall average intensities of those color channels. The system repeats this process to search new candidates until user finds the image that he wants.

Jacob's work compares feature of extracted data with those in the database and stores the degree of similarity between two features [8]. The average color value and the wavelet coefficients are evaluated by each criteria. An image of the highest magnitude value is provided as a result of the search.

Initial population consists of randomly selected images. After user determines fitness of images that are most similar to what he/she wants, the system creates the next generation using genetic algorithm and displays new and more similar images. This procedure is repeated until the user finds a wanted image. If the result of the next generation is worse, the system allows going back to the previous generation. Moreover, the user can increase or decrease the effectiveness of the color to search images. Mutation is not considered in the present implementation.

5 Experimental Results

5.1 Environments

The system is developed with the language of Microsoft Visual C++, and runs on Pentium PC. A searching table is constructed by a batch job over 256×256 JPEG images. The size of image database is 170. To evaluate a performance of this system, we requested ten graduate students to search cheerful, gloomy, and cool images and answer the questions given below: 1) How similar the result image is to what you want and 2) How long it has taken to find it.

Table 1: Response statistics from ten users for the three queries.

	cheerful	gloomy	cool
very good	2	1	3
good	3	0	4
normal	2	3	0
bad	0	2	2
very bad	3	4	1

Table 2: Comparison of number of trials until the target samples have been found.

samples	without crossover	with crossover
1	2	1
2	over 10	5
3	over 10	over 10
4	1	1
5	over 10	3

5.2 Analysis of Results

At first the satisfiability has been tested with the images searched and time taken for search over cheerful, gloomy, and cool images. Table 1 shows the results obtained from ten users with respect to the three queries. In this test, 70 percents of the subjects have been satisfied with finding cheerful and cool images, but in case of gloomy image most of the subjects have been unsatisfied. A major reason might be in the small size of database, which does not contain as many images as user wants. The retrieval of the cool image shows that the content-based image retrieval method by human intuition or emotion can have good performance if large database is given.

Next test is to inspect the usefulness of crossover. Test condition is divided into two cases: One is with crossover and the other is without crossover. The test starts with the same initial population and does not use back-tracking. The result is shown in Table 2.

Finally, we have tested whether wavelet coefficients are appropriate for genetic representation. To do this, the images are grouped into two classes: One is gloomy and the other is not. For each image, the similarities are calculated between images contained in the

Table 3: Within-class and inter-class similarities for gloomy images.

	within-class	inter-class	difference
mean	1154.692	1114.939	39.752
variance	27368.739	24196.524	3229.317
SD	165.434	155.552	56.827
t-test			5.508

same group (within-class similarity) and between images contained in different groups (inter-class similarity), respectively. The similarity is evaluated through the same method that is used to compare the target image with the candidate image.

The larger the value is, the closer the images are. High similarity would mean that the two images are near on the feature space of wavelet transform. Most of the gloomy images produce larger similarity value in the same class. The averages of within-class similarity and inter-class similarity are 1154.692 and 1114.939, respectively. It is shown in Table 3. Though the mean of within-class similarity is larger than that of inter-class, we need to determine whether this difference is statistically significant or not. This table also shows the result of paired t-test, which is useful for testing whether the difference of mean values is significant or not. We hypothesize that the mean values of within-class similarity and inter-class similarity are not different. The result of t-test is 5.508 and our hypothesis is rejected at a significance level of 99.995%. This means that images in the same class are closer on the feature space and we can assert that wavelet coefficients represent mood of images well.

6 Concluding Remarks

This paper has proposed an approach that searches an image using human intuition and emotion. This method allows to search not only explicitly expressed image, but also abstract image with feelings such as cheerful, gloomy and cool. Interactive genetic algorithm takes advantage of producing a solution that user's preference is reflected. However, since the result is very subjective, the evaluation of efficiency as well as quality is difficult. Takagi attempts to apply the subjective test based on a paired comparison to evaluate performance [9], whereas this paper has measured the efficiency using a statistical t-test.

As you can see from the experimental results, the performance of interactive genetic algorithm is subject to the size of database. Therefore, we need to prepare enough database and design better genetic operators including mutation. Even though there remain many things to do for improving the system, it is clear that the approach using interactive genetic algorithm based on human intuition and emotion allows developing a system to retrieve images effectively and intuitively.

References

[1] Niblack, W., Barber, R., Equitz, W., et al., 1993, "The QBIC project: Querying images by content using color, texture, and shape," *SPIE Proc. Storage and Retrieval for Image and Video Databases*, pp. 173–187.

[2] Hirata, K., and Kato, T., 1992, "Query by visual example: Content based image retrieval," *Proc. Int. Conf. Advances in Database Technology*, pp. 56–61.

[3] Ogel, V.E., and Stonebraker, M., September 1995, "Chabot: Retrieval from a relational database of images," *IEEE Computer*, **28**, 109–112.

[4] Caldwell, C., and Johnston, V. S., 1991, "Tracking a criminal suspect through face-space with a genetic algorithm," *Proc. Int. Conf. Genetic Algorithms*, pp. 416–421.

[5] Baker, E., and Seltzer, M., 1994, "Evolving line drawings," *Proc. Int. Conf. Graphics Interface*, pp. 533–536.

[6] Edwards, T., September 1991, "Discrete wavelet transform: Theory and implementation," *Technical Report*, Stanford University.

[7] Goldberg, D. E., 1989, *Genetic Algorithms in Search, Optimization, and Machine Learning*, Addison-Wesley Publishing Co.

[8] Jacobs, C. E., Findkelstein, A., and Salesin, D.H., 1995, "Fast multiresolution image querying," *Proc. SIGGRAPH 95*.

[9] Aoki, K., Takagi, H., and Fujimura, N., 1996, "Interactive GA-based design support system for lighting design in computer graphics," *Proc. Int. Conf. Soft Computing*, Iizuka, Japan, pp. 533–536.

Part 2: Design Support Systems

Papers:

Using Genetic Algorithms to Encourage Engineering Design Creativity

William A. Crossley

School of Aeronautics and Astronautics, Purdue University
1282 Grissom Hall, West Lafayette, Indiana 47907-1282
Tel. (765) 496-2872, Fax (765) 494-0307
E-mail: crossley@ecn.purdue.edu
URL: http://roger.ecn.purdue.edu/~crossley/

Keywords: genetic algorithms, engineering design, creativity, composite panel design

Abstract

This paper discusses how a genetic algorithm searches through a design space to generate highly fit designs for an example engineering design problem. The problem used herein is the design of a stiffened composite panel, combining discrete, integer and continuous variables. The GA process is examined and compared to an accepted view of creative engineering design practices. Examples from the stiffened composite panel problem are used to illustrate how the genetic algorithm may encourage creativity and / or act as a creative agent in engineering design. To be successful, a genetic algorithm search still requires intelligent, creative design engineers; this need is also discussed.

1. Introduction

With the proliferation of computational tools throughout the engineering disciplines, engineering designers have been given much assistance with the analysis tasks required to evaluate designs. The development and use of numerical optimization techniques have also assisted designers in improving design concepts. These achievements have improved the performance of designs and reduced the time needed to complete the engineering design process.

1.1 The Engineering Design Process

This process is often thought of in terms of several sequential phases: first conceptual design, then preliminary design, and finally detailed design. Using analysis and optimization tools requires a starting point or an initial design in order to proceed. Starting points used in the detailed design phase draw on the results of preliminary design phase, so analysis and optimization tools are often employed in the detailed design phase. The starting points used in the preliminary design phase draw on the results of the conceptual design phase, so optimization and analysis tools are used here as well. In fact, the successes of these computational "assistants" have inspired attempts to automate the design process.

The conceptual design phase, however, creates the most significant difficulties for attempts at design automation and for application of the traditional engineering analysis and optimization programs. During this initial phase of design, no starting point exists, so a set of requirements, which have no physical representation, must be transformed into an artifact or device, which has a physical representation, that will satisfy these requirements. The generation of this starting point requires the designer to use his or her engineering experience and intuition to device a scheme with a high likelihood of satisfying the requirements. For this to be truly successful, the designer must be "creative" or "innovative", so the design not only meets the requirements but also is original and has performance that exceeds other competing designs. Creativity, however, is a human trait that is not easily converted into a computational tool.

1.2 Creativity

The concept of creativity itself is quite difficult to define or understand. In his textbook about engineering design, Ullman [1] provides a summary of creativity and how it relates to other engineering attributes. This comparison may be summarized in the following seven points:

- There is little correlation to creativity and intelligence.
- Creative engineering designers have the ability to generate and manipulate visual images mentally.
- A creative designer must have knowledge of existing products and the design domain.
- New ideas are formed from the combination of existing knowledge, and creative engineering designers are able to decompose problems to partial solutions and can manipulate existing knowledge to find solutions.
- Creative designers are willing to take intellectual chances.
- Non-conformity is a trait of many creative designers.
- If an initial process does not yield solutions, a creative designer will turn to alternative techniques.

While this list does not provide a complete definition of creativity, it does provide some guidelines with which an engineering designer may be considered creative.

These characteristics will be used to consider how a genetic algorithm may be used to encourage creativity.

1.3 Genetic Algorithms

Genetic algorithms have been gaining recognition as solution techniques for many complex optimization problems and have been applied to numerous conceptual and preliminary engineering design studies. A genetic algorithm (GA) is not truly an optimization technique; rather it is a computational representation of the processes involved in natural selection as observed in biological populations. This includes representing many potential designs as co-existing individuals in a population. In a GA, as in nature, the fittest individuals (designs who better meet the requirements) are more likely to survive. Surviving individuals will mate and produce offspring, and these offspring inherit design features from their parents. In this process, features of good designs (the surviving parents) are combined to generate new designs (the offspring). If these new designs are better designs, they will then be more likely to survive and pass on their design features to future generations. This repeated combination of good design features leads towards better and better designs. This combination of information is often considered innovative or creative when performed by a human designer.

2. Stiffened Composite Panel Design

A challenging engineering design problem has been selected to represent complex problems, which may be encountered across a wide spectrum of engineering disciplines. The problem presented for this work is the design of a minimum weight stiffened composite panel of the type in a typical helicopter sub-floor. The panel is constructed of a flat plate with stiffeners attached to one face. The panel must support normal loading conditions without buckling and must also absorb energy under a specified crash loading condition. Crossley and Laananen [2] provides a fairly detailed description of this problem and the analysis techniques required for generating stiffened composite panels for energy absorption. For this paper, a small subset of the designs investigated in Ref. [2] was re-examined with some improvements to the genetic algorithm. The manner in which these designs are found was examined in detail for this paper.

One of the challenging tasks facing a structural designer working on helicopters is the design of sub-floor panels, which will support the loads normally associated with the helicopter's payload and will also crush in a vertical crash condition to absorb much of the energy of the impact. These vertically oriented stiffened panels act as one part of a crashworthy system that will help protect the helicopter's occupants from serious harm or death in an accident of a given severity; these panels may absorb as much as 80% of the aircraft's kinetic energy.

The use of composite materials for airframe structural components provides the opportunity for lighter-weight and lower-cost structures compared to components constructed from metals. The nature of composite materials, however, provides several design challenges. Searching for an optimal design requires a mix of discrete, integer, and continuous design variables. Continuous stiffener dimensions are combined with the integer number of stiffeners and number of plies; these are also combined with discrete choices of material type and stiffener shape. Further, for the crashworthy design problem, the composite materials require special approaches to predict energy absorption, because these materials are generally brittle and do not exhibit plastic behavior or elongation prior to failure. Figure 1 presents an illustration of a typical panel.

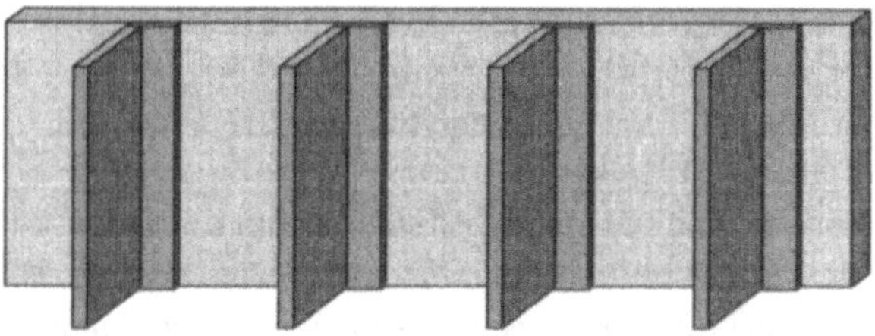

Figure 1. Typical stiffened composite panel with a plate and four stiffeners.

2.1 Design Variables

The design of a stiffened composite panel involves finding an optimal combination of several different design characteristics. For this paper, these include the number of stiffeners, stiffener shape, stiffener lay-up, panel material, stiffener dimensions, number of plies in the plate, and the plate lay-up. Traditionally, several characteristics are "selected" using qualitative decisions (based on an engineer's experience) during what would be considered the conceptual phase of design; these are generally discrete variables. Once these selections have been made, other variables (e.g. dimensions) are "sized" using quantitative decisions based on some form of analysis. Because the genetic algorithm uses a coding of the variables, rather than the actual value of the variables, in representing a design, a combination of discrete, integer and continuous variables can be simultaneously considered. By doing this, the genetic algorithm can "blur" some of the distinction between "selection" and "sizing".

In this example, the variables of interest are such a combination of variable types. The stiffener shape was treated as a discrete variable. Using a binary alphabet, six different possible shapes are represented in the chromosome using the mapping presented in Table 1. Because a binary alphabet is used, two of the six choices are given two binary representations. In the initial randomly generated population, more designs will likely have blade and channel section stiffeners. This will introduce some bias in the early generations. If these are undesirable traits, the fitness of designs with these sections will be poor, and the bias will be eventually eliminated from the population. The six possible shapes are presented in Figure 1.

Table 1. Binary mapping of stiffener cross-section shapes.

Binary string	Stiffener cross-sectional shape
000	Blade
001	Blade
010	Channel
011	Channel
100	I
101	Hat
110	J
111	Angle

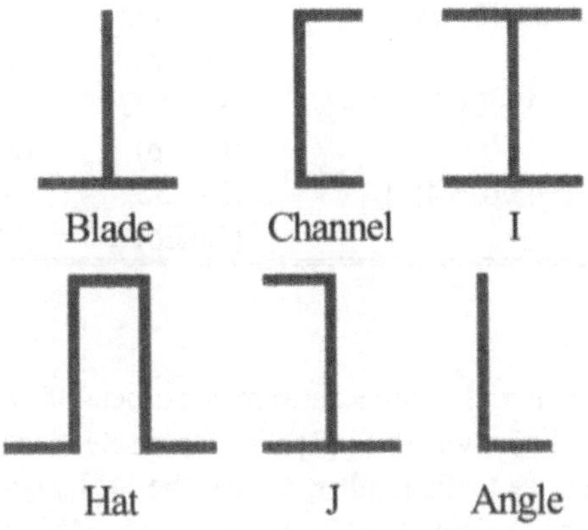

Figure 2. Stiffener cross-sectional shapes

As mentioned earlier, predicting the behavior of composite structures after failure is very difficult, which makes design for energy absorption challenging. A semi-empirical method developed by Bolukbasi and Laananen [3] provides an approach to predicting the sustained crushing stress of a composite stiffener using the results of simple, flat-plate specimen testing. This method requires the use of a material and lay up pattern for the stiffeners for which experimental crush test data exists. For this energy absorbing panel design problem, specimens constructed from two materials provide the flat-plate test data needed to predict stiffener sustained crushing stress. In this effort, the choice of material and stiffener lay up was treated as a discrete variable, with six potential material and lay up combinations. These material and lay up combinations, along with the associated flat-plate sustained crushing stress, were taken from Refs. [3] and [4]. Table 2 displays these.

Table 2. Binary mapping of stiffener lay up and material.

Binary string	Stiffener lay up	Material	Flat plate sustained crushing stress
000	$[45]_{10}$	T650-35/F854 (Gr/Ep)	18.56×10^3 psi (128.0×10^6 Pa)
001	$[45]_{10}$	T650-35/F854 (Gr/Ep)	18.56×10^3 psi (128.0×10^6 Pa)
010	$[45_2/0/45_2]_S$	T650-35/F854 (Gr/Ep)	21.47×10^3 psi (148.0×10^6 Pa)
011	$[45_2/0_2/45]_S$	T650-35/F854 (Gr/Ep)	22.05×10^3 psi (152.0×10^6 Pa)
100	$[\pm45/0_4/\pm45]_S$	AS4/3502 (Gr/Ep)	14.60×10^3 psi (100.7×10^6 Pa)
101	$[\pm45/0_4/\pm45]_S$	AS4/3502 (Gr/Ep)	14.60×10^3 psi (100.7×10^6 Pa)
110	$[45_2/-45_2/0_8/45_2/-45_2]_S$	AS4/3502 (Gr/Ep)	10.05×10^3 psi (69.29×10^6 Pa)
111	$[\pm45/0_4/\pm45_2]_{2S}$	AS4/3502 (Gr/Ep)	14.04×10^3 psi (96.80×10^6 Pa)

The T650-35/F854 material specimens were constructed from woven prepreg layers of graphite fibers and epoxy matrix. [3] For these selections, the material is woven with fibers at 90º angles to each other, so that the $[45]_{10}$ laminate has ten layers of the fabric with fibers oriented at both +45º and –45º degrees to the loading direction. The AS4/3502 specimens were constructed from unidirectional prepreg tape of graphite fibers in epoxy matrix. [4] It was assumed that the plate portion of the stiffened composite panel will be constructed of the same material as the stiffeners. In this manner, the material choice for the entire panel design is also implicitly included as a discrete variable.

The number of stiffeners in the composite panel design is an integer valued variable; for this problem, the stiffeners were assumed to be equally spaced along one face of the plate. Because the plate of the stiffened composite panel will be constructed from layers of prepreg material, the number of plies in the plate is also an integer variable. This variable determines the thickness of the plate as an integer multiple the ply thickness. To simplify this problem slightly, the plate was assumed to have a symmetric lay up, so the number of symmetric plies was used as the variable. With this representation, a value of two for the number of symmetric plies corresponds to four total plies in the plate.

The dimensions of the stiffeners were treated as a continuous variable for this problem. The stiffeners were assigned a "short" and a "long" dimension. The short dimension is the length of any stiffener element parallel to the plate; the long dimension, perpendicular to the plate. These "short" and "long" dimensions were also mapped to binary strings. Using binary strings requires that continuous variables are actually discretized with some level of resolution between each value. A smaller value of resolution requires a longer string length, so a reasonably large resolution was chosen for these dimensions to allow for a shorter binary chromosome. Table 3 presents the binary mapping of the integer and continuous variables used in the stiffened panel problem.

Table 3. Range, resolution and number of bits for integer and continuous variables.

Variable	Maximum	Minimum	Resolution	String length
Number of stiffeners	1	8	1	3 bits
Plate no. of symmetric plies	1	8	1	3 bits
Stiffener "long" dimension	0.500 in (12.700 mm)	1.375 in (34.925 mm)	0.125 in (3.175 mm)	3 bits
Stiffener "short" dimension	0.250 in (6.350 mm)	1.125 in (28.575 mm)	0.125 in (3.175 mm)	3 bits

Finally, the ply orientation angle of a composite laminate may be mathematically treated as a continuous variable, but in most manufacturing settings, the choice of angles is limited to a discrete set. For this work, the ply orientation angle was restricted to multiples of 45º, ranging from -45º to 90º. This set is readily adapted to binary coding, as each ply only needs two bits to represent the angle. The chromosome used to represent panel designs contains enough bits to represent the ply orientation angle of eight different pairs of symmetric plies. If there are fewer than eight ply pairs in the plate, then the extra information in that string will be ignored. With all design variables, the string length of the binary chromosome was 34 bits.

2.2 Problem Description

The stiffened composite panel design can be posed as an optimization problem. A more complete description of the basic problem description is found in Ref. [2]. Here, an abbreviated version is given. To describe the stiffened composite panel problem in a manner suitable for the genetic algorithm, a fitness function, $f(\mathbf{x})$ is defined as

$$f(\mathbf{x}) = \phi(\mathbf{x}) + P(\mathbf{x}) \tag{1}$$

where $\phi(\mathbf{x})$ is the objective function, here this is the panel weight, and $P(\mathbf{x})$ is a penalty function used to enforce constraints. The penalty function in this effort uses a linear form with a variable draw-down coefficient that increases with the number of generations evaluated. This has the form

$$P(\mathbf{x}) = (n_{gen} + 10)\sum_{i=1}^{n_{con}} \max[0, g_i(\mathbf{x})] \tag{2}$$

Here, n_{gen} represents the generation number, so that the "strength" of the penalty increases during the run of the genetic algorithm. This approach is based on the work described in Crossley and Williams [5] and has been shown to yield quality results for constrained optimization with the genetic algorithm.

The constraint functions, g_i, in Eq. 2 ensure that the design meets or exceeds requirements and limits imposed in the problem statement. In the genetic algorithm, an individual that violates one or more of the constraints will receive a fitness that is penalized and, therefore, less likely to survive. The constraint functions are posed so that the function is negative valued when the constraint is satisfied and positive valued when the constraint is violated. This problem has six constraints that are to be satisfied. Constraints dealing with the loading performance of the plate constitute the first four constraints, while constraints on the design space constitute the last two.

The first ensures that local buckling of plate between stiffeners will not occur under a given "normal loading condition" with some margin or safety factor. Buckling of the stiffeners is also undesirable, so the second constraint ensures that the buckling strength of the stiffeners is higher than that of the plate. For crashworthy considerations, the stiffener buckling stress should exceed the sustained crushing stress of the stiffeners; the third constraint imposes this condition so that crash energy absorption (crushing) occurs before buckling. The fourth constraint requires that the stiffener crushing stress equals that of a specified crash loading condition within a small level of tolerance.

The fifth and sixth constraints are imposed on the panel design space to address concerns of composite structure design. Constraint number five enforces a balanced plate laminate to prevent warping and load coupling; the constraint has a positive value if the number of -45º unidirectional plies does not equal the number of 45º unidirectional plies. Lastly, the sixth constraint is used to maintain a minimum contact dimension between the plate and each stiffener so that the stiffeners remain attached.

3. Generating Solutions with the GA

With the problem description and design variables described above, a genetic algorithm was employed to find a minimum weight panel design for energy absorption. While the problem is essentially that of Ref. [2], the genetic algorithm has been improved over the previous work. Additionally, information about how the GA searches through the design space was gathered for this paper.

3.1 Genetic Algorithm Operators and Parameters

A serial GA was used for this effort. Binary chromosomes represented the individual designs; the variables and coding were discussed in Section 2.1. Each chromosome had a string length of 34 bits. Tournament selection and uniform crossover operators were used. Guidelines from Ref. [6] were used to set the population size at 136 and the mutation rate at 0.018. A stopping criterion was imposed so that if the best individual had not improved for five consecutive generations, the GA halted its run. No elitism was used.

3.2 Resulting Designs

Because of the stochastic nature of the genetic algorithm, different runs of the GA can have different results. In the implementation used here, a pseudo-random number generator was used to perform all of the probabilistic functions; this requires an initial "seed" number to begin the process. Six runs, each with different initial seeds, were used to generate solutions for the panel design problem.

The GA was able to locate the same design in each of the six runs; this best panel design is summarized in Table 4. This panel is the "best ever" design. In some of the runs, this design was not present in the generation when the stopping criterion was enacted, but had been encountered earlier in the run of the GA.

4. The Genetic Algorithm and Creativity

The genetic algorithm generated designs that appear to have very high quality - they have very low weight and satisfy all of the design constraints. All of these designs were generated without the definition of an initial design point, as is required of many optimization approaches. It is this idea that has led many to compare the genetic algorithm's search to the creative process. Quoted in Levy's book, John Holland stated: "I could see evolution as a creative process, the essence of making something out of nothing." [7] Similarly, David Goldberg claimed in his text: "We have compared the genetic algorithm to human search processes commonly called creative or innovative." [8]

Table 4. Best panel design generated by the GA.

Parameter or objective	Value
Stiffener number	2
Stiffener shape	Blade
Stiffener lay-up	$[45]_{10}$
Material	T650-35/F854
"Long" dimension	0.500 in (12.70 mm)
"Short" dimension	0.250 in (6.35 mm)
Symmetric ply pairs	4
Plate lay-up	$[45]_8$
Weight (mass)	1.085 lb. (0.493 kg)

Using the stiffened composite panel problem as an example, the ability of the genetic algorithm to assist or encourage the creativity of a design engineer will be explored. As a framework for this discussion, the seven comparisons of creativity to engineering behavior presented by Ullman [1] are used. These comparisons will be mentioned in areas where the genetic algorithm can enhance creativity or act as a creative agent and where the engineering designer must still rely on his / her skills to be creative.

4.1 Encouraging Creativity

In his comparison of creativity and knowledge, Ullman [1] claims that a creative designer will "form their new ideas out of bits of old designs", and he further states that "part of being creative is being able to evaluate the viability of designs." In the comparison of creativity and partial solution manipulation, it is again emphasized that "new ideas are born from the combination of parts of existing knowledge." All these capacities are central to the GA's search, and the results of this search can encourage creativity in engineering design.

A simple illustration of this can be taken from the first run of the GA. The population of the initial generation is generated randomly, so designs are spread nearly evenly across the design space. Figure 3 displays the distribution of the population in two "planes" of the design space; the design space itself has fourteen variables. The plot on the left displays the plane with the integer number of stiffeners and the discrete stiffener shape, while the plot on the right displays the continuous "short" and "long" stiffener dimensions. In these planes, the individuals are fairly close to a uniform distribution. The population is distributed similarly in other planes.

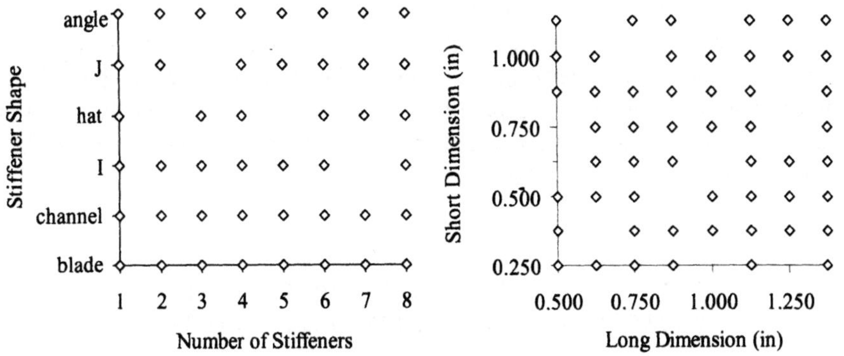

Figure 3. Initial population distribution for run 1.

To begin, the GA evaluates the fitness of each design in the population. These fitness values are used in the selection operator to determine if designs will survive to become parents for future generations. Two individuals are chosen at random from the initial population, and the one with the lower fitness value survives. During run 1, the tournament selection operator first randomly chose the 74th and the 23rd individual from the population as candidates to become parents for the first new generation. The fitness function value of the 74th individual was 20.536; the 23rd, 13.887. These fitness values include the weight of the panel as well as any penalties for constraint violations. Here, the 23rd individual survived to become a parent for the next generation. In this manner, the GA is able to evaluate the viability of one design when compared to another.

With one parent selected, two more individuals are chosen from the initial population as candidates. The next two selected from the initial population were the 124th and the 4th individuals. These had fitness values of 43.697 and 17.512, respectively. The selection operator chose the fourth individual to survive as the second parent. Once two parents have been selected as "good" designs (here, a good design is one that has won its tournament), these parents are crossed-over to form two children. This crossover occurs on the binary chromosome level; with the uniform crossover used in this GA implementation, the first child has a 50% chance of receiving its first bit from the first parent chromosome or from the second chromosome. If the first child "inherits" the first bit from the first parent, the second child will inherit the first bit from the second parent, and vice versa. This crossover mechanism acts very much like creative designers who form new ideas out of "bits of old designs".

Figure 4 displays the parent chromosomes selected above (23rd and 4th individuals from the initial population of run 1) and the child chromosomes formed by the crossover operator. The underlined locations show where the bits have been "exchanged" from parent 1 to child 2. If both parents have the same bit value at a given location, then both children will also have the same bit value; where this occurred, no bits were underlined. The two child designs are a combination of features, represented in the bit string, of their parents. Very occasionally, mutation occurs; the two bits that have mutated in this example are shown in italic font. Mutation will be discussed in a following paragraph.

Parents

0000000111111100101011000011110110
0101010100011000110011011100110010

Children

0001000110111100101011000110110110
01000101010110001100*0*1*1*11001110010

Figure 4. Parent and child chromosomes.

Table 5 presents the fitness and design variables of the two parent designs and the two resulting child designs. Because the crossover occurs on the chromosome level, design variables are not wholly "swapped" in the designs, but features of both parents can be seen in both children. In this example, the combination of bits of two good designs, along with the mutation operation, have formed two designs with even lower fitness. This is not a guarantee, but the selection operator does not allow poor child designs to survive as parents for the next generation. On average, this continually improves the population.

This process of selection and crossover, with infrequent mutation, continues for several generations. The GA ran for 20 generations before enacting the stopping criterion during run 1. During this time the design space is searched with some thoroughness. As presented in Figure 3, the initial population provides a nearly uniform distribution across the design space. During the run, the population gains "knowledge" about the domain, as designs with features that lead to good fitness values survive and designs with features that lead to poor fitness perish. Figure 5 presents the population distribution for generations 5, 10, and 20 for run 1 of the stiffened composite panel problem.

The population evolves towards promising areas of the domain. This is particularly evident in the stiffener number – stiffener shape plane of the design space, as the designs move towards the combination needed by the best individuals (two blade-section stiffeners). Each generation has 136 individuals; in these plots, several individuals may share the same combination of design variables. There is more

Table 5. Parent and children designs from selection and crossover.

Parameter	Parent 1	Parent 2	Child 1	Child 2
Fitness	13.887	17.512	13.018	10.066
Stiffener number	1	3	1	3
Stiffener shape	Blade	Hat	I	Blade
Stiffener lay-up and material	$[45_2/0_2/45]_S$ T650-35/F584	$[45_2/0/45_2]_S$ T650-35/F584	$[45_2/0_2/45]_S$ T650-35/F584	$[45_2/0/45_2]_S$ T650-35/F584
"Long" dimension	1.375 in (34.925 mm)	0.625 in (15.875 mm)	0.875 in (22.225 mm)	1.125 in (28.575 mm)
"Short" dimension	1.000 in (25.400 mm)	0.750 in (19.050 mm)	1.000 in (25.400 mm)	0.750 in (19.050 mm)
Symmetric ply pairs	3	4	3	4
Plate lay-up	$[45/0/45]_S$	$[45/0_3]_S$	$[45/0/45]_S$	$[45/0_2/45]_S$

variety in the long dimension – short dimension plane in late generations, but most of the population is centered near the best combination of 0.500 inches and 0.250 inches in the last generation.

In the context of knowledge and partial solution manipulation, it appears that the genetic algorithm can be a great aid to creativity. The GA's search embodies the facets of knowledge discussed by Ullman; it forms new designs from parts of old designs and evaluates the viability of designs during selection. While Ullman claims that "knowledge about a domain is only gained through hard work in that domain;" [1] the GA searches quickly without additional inputs from an engineer. The GA can do much of this hard work for the designer.

To a lesser extent, the genetic algorithm can act as a "risk taker". A non-creative designer would be unwilling to risk a mistake or to spend time looking at designs that might not solve the problem at hand. The genetic algorithm only uses the fitness value to determine the selection of parents; there is no guarantee that either of the child designs will be better than the parent designs. More traditional calculus-based optimization approaches (e.g. steepest descent, quasi-Newton, etc.) consider only new designs that are definite improvements over a current design; these approaches are also prone to finding local minima in multi-modal design spaces. Because the genetic algorithm is not "afraid" to generate and evaluate designs that may not be immediate improvements over the parents, it does not become "caught" in local minima. Further, through occasional mutation, the genetic algorithm may introduce design features not seen elsewhere in the population. The newly introduced features are not guaranteed to improve performance, but when

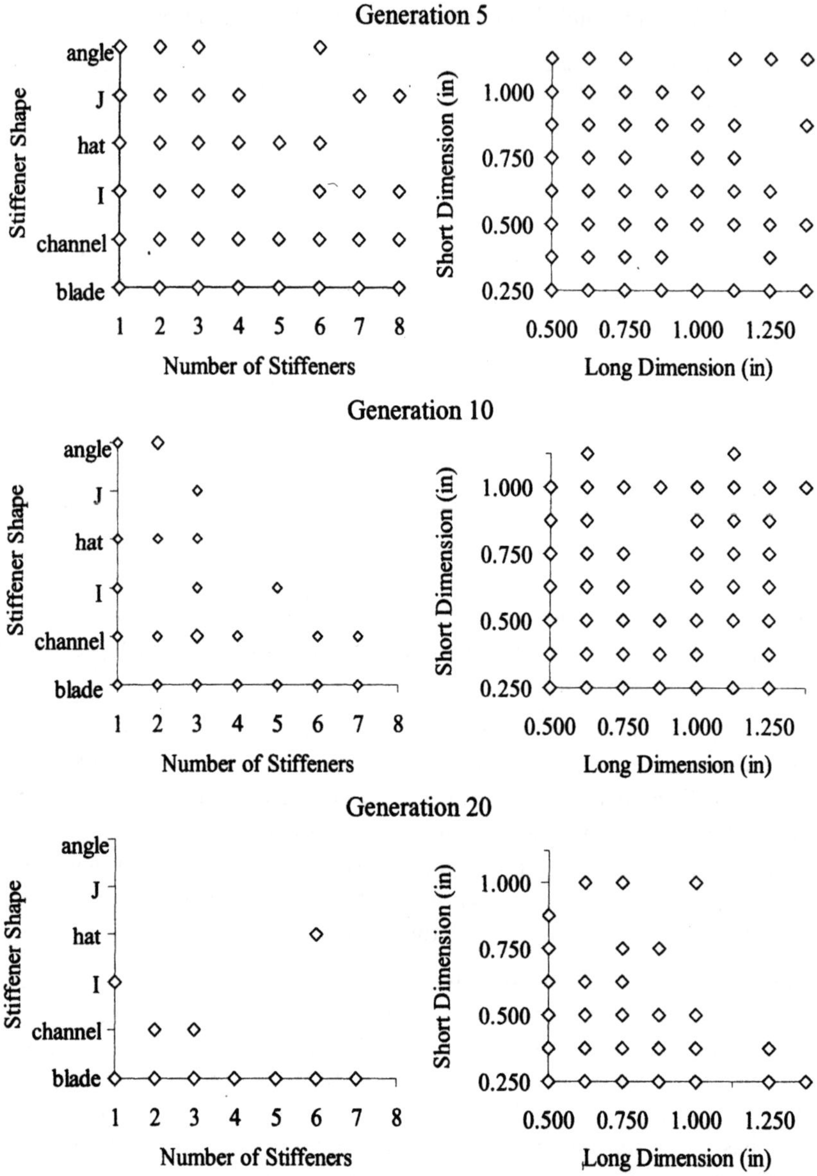

Figure 5 Population distribution for run 1; generations 5, 10 and 20.

they do, these features are passed on to other designs. In this regard, the genetic algorithm would also encourage creativity.

Similarly, the GA investigates an entire population of designs "in parallel" each generation. This not only allows a good coverage of the design space during early generations, but this also helps find nearly globally optimal designs. In Ullman's text, he claims that creative designers are willing to try more than one approach to solve a design problem. Again, the genetic algorithm's population of designs can be viewed as a number of different potential solutions to the design problem.

4.2 Creative Skills Needed

The genetic algorithm does not fulfill all of the characteristics of a creative designer. While successful engineers require a reasonably high level of intelligence, Ullman claims creative designers are not more intelligent than less creative engineers and that there is little correlation between creativity and intelligence. [1] A genetic algorithm itself has no inherent "intelligence". The GA requires an intelligent user to develop the fitness function and the computational routines needed to provide values for this function. Because the value assigned to the fitness function is the sole means by which the selection operator decides between "life and death" of a design, this function must reflect the objective and constraints adequately. Limitations of the accuracy and validity of the analyses used to compute a fitness value must be considered when examining results generated by the genetic algorithm.

The ability to visualize solutions to design problems is demonstrated by creative engineers; the GA has no such capability. Results of the genetic algorithm are generally a list of parameter values or choices. To interpret these as meaningful designs, the user must have the ability to visualize a physical representation of the GA's results. The genetic algorithm does not seem to aid creativity in this sense.

Ullman also observes that creative people are often non-conformists. [1] This is a behavior that cannot readily be captured by a computational tool. Because the genetic algorithm is still a relatively new tool for design and optimization, it may actually require a non-conforming person to introduce the use of GA in engineering settings where it is not currently in use.

5. Concluding Remarks

The genetic algorithm was able to successfully generate designs for the stiffened composite panel problem, which combined discrete, integer and continuous variables. The nature of the variables and the challenging nature of the problem suggested that some creativity would be needed to find solutions. Given that the GA found high quality solutions, the manner in which the GA generated these designs was compared to a description of the characteristics of a creative designer.

Based on this comparison, the GA appeared to encourage creativity when viewed in the comparison of creativity and knowledge and creativity and partial solution manipulation. The operation of tournament selection and crossover provide features that fit well with these characteristics of creative designers. The GA essentially acquires knowledge of the design domain without require significant interface with the user. To a lesser extent, the GA acts as a "risk taker" by sometimes creating children with fitness values poorer than their parents and occasionally introducing new design features through the mutation operator. Finally, the population-based search of the genetic algorithm is analogous to conduction design studies on several different configurations in parallel; this is also considered to be a characteristic of a creative designer.

The genetic algorithm does not remove the need for creative engineers in the design process. Intelligent, creative designers must construct the fitness evaluation routines and must define the design space in which the GA searches.

The fact that the genetic algorithm can act with several of the characteristics of a creative designer leads to the conclusion that the GA can be successfully used in engineering design problems to enhance and encourage the creativity of an engineer.

References

1. Ullman D G 1992 *The mechanical design process.* McGraw-Hill, New York

2. Crossley W A, Laananen D H 1996 Genetic algorithm-based optimal design of stiffened composite panels for energy absorption. In: *Proceedings of the 52nd annual forum of the American Helicopter Society.* American Helicopter Society, Washington, DC, pp 1367-1376

3. Bolukbasi A O, Laananen D H 1995 Energy absorption in composite stiffeners. *Composites* 26:291-301

4. Lavoie J A, Morton J 1993 Design and application of a quasistatic crush test fixture for investigating scale effects in energy absorbing composite plates, NASA Contractor Report 4526, NASA Langley Research Center, Hampton, VA

5. Crossley W A, Williams E A 1997 A study of adaptive penalty functions for constrained genetic algorithm based optimization. AIAA Paper 97-0083 AIAA 35th Aerospace Sciences Meeting and Exhibit

6. Williams E A, Crossley W A 1998 Empirically-derived population size and mutation rate guidelines for a genetic algorithm with uniform crossover. In: Chawdhry P K, Roy R, Pant R K (eds) 1998 *Soft Computing in Engineering Design and Manufacturing.* Springer-Verlag, London, pp 163-172

7. Levy S 1992 *Artificial Life: The Quest for a New Creation.* Pantheon Books, New York

8. Goldberg D E 1989 *Genetic Algorithms in Search, Optimization and Machine Learning.* Addison-Wesley, Reading, MA

Abduction Problem in Probabilistic Constraint Logic Programming

V.B.Valkovsky[1], K.O.Savvin[2], M.B.Gerasimov[3]

Software Engineering and Computer Application Department,
St.Petersburg Electrotechnical University,
Prof. Popov str., 5, St.Petersburg, 197376, Russia,
E-mail: [1]vlad@ailab.etu.spb.ru, [2]skonx@oisoft.ru,
[3]lloix@star.spb.ru

Keywords: reasoning with uncertainty, Probabilistic Logic Programming, abduction problem, Constraint Logic Programming.

Abstract

The paper deals with an abduction problem within the approach for handling probabilistic type of uncertainty in the framework of Constraint Logic Programming. The approach is intended for solving real-life logistic problems such as scheduling, manufacturing, planning, considering uncertainty of input information. The Monte-Carlo method used as background engine allows to combine logical inference with probability and straightforward statistical modeling features. We describe particular cases of abduction problem with different kinds of input and output data (probabilities of clauses and bounds of stochastic variables). We also discuss the problem of determining the dependence between input and output domain variables, and show that this dependence is multilinear in the particular case with independent uncertain clauses. Graph with uncertain edges exemplifies the discussion. At last, we demonstrate the use of some of described methods in the pilot application for solving the network based project planning problem in extended non-deterministic formulation.

1. Introduction

Solving different real-life logistic problems such as scheduling, manufacturing, planning is a topic of great interest for researchers. Constraint Logic Programming (CLP) is powerful technology to solve these problems in effective, rapid and flexible way [1][2][3]. Nowadays it becomes actual to solve such problems taking into account uncertainty and incompleteness of input information.

In the papers [4][5], we discussed the direct problem of probabilistic reasoning. However, the inverse task, the abduction problem, is no less important. For instance, it is necessary to solve the abduction problem to analyze the correctness of probabilistic knowledgebases during the design or modifications, to estimate how adequately the certain situation of problem domain is represented, to determine to what degree some queries/goals are sensitive to the probability of some events, etc.

Abduction is very important also in diagnostic systems, especially for analysis of those critical situations or events that must have occurred in a given object system but which are not directly observable.

In this paper we discuss the questions related to the abduction problem and its application in the framework of our approach for uncertainty handling. In the next section, we briefly describe the basics of handling uncertainty in probabilistic interpretation and introduce the example that will be used throughout the paper for illustration purposes. Then we consider particular cases of abduction problem with different kinds of input/output data. Namely, in the third section, we show how to estimate unknown probability of uncertain clause given probability of the goal. In the fourth section, we demonstrate the possibility of estimation of bounds for input stochastic variable given the value or range of output variable with the help of CLP specific features. Then we discuss the problem of determining the dependence between input and output domain variables. For the particular case, with independent uncertain clauses in probabilistic logic program (the case of maximal entropy) we show that the dependence between probability of the goal and unknown probabilities of uncertain clauses is multilinear. In the sixth section we demonstrate application of some of the described methods in the pilot system for solving network based project planning problem in extended non-deterministic formulation. At the end we summarize the results and mark the topics for future investigations.

2. Background

Probabilistic interpretation of uncertainty we dealing with is mathematically grounded on the theory of probabilistic logic suggested by N.Nilsson [6]. This model has clear semantics, described in the terms of *possible worlds*, operates with probabilities and uses strict methods of probability theory. The principles and semantics of probabilistic logic can be naturally used to generalize Logic Programming to Probabilistic Logic Programming (PLP)[7][8]. Here we summarize important points and refer the reader to relevant papers.

Probabilistic logic program consists of a set of certain clauses and set of uncertain clauses with their associated probabilities. The probability of the uncertain clause can be treated as the probability that the clause is present in the program. Suppose it is known whether each uncertain clause is present in the program or not. In this case, the uncertain program becomes an ordinary logic program and one can check whether some goal succeeds in this program or not. Such certain program is considered as *possible instance* of initial probabilistic program. This is a natural transfer of the *possible worlds* semantics from probabilistic logic to PLP.

Different possible combinations of presence of M uncertain clauses in the program imply 2^M different possible instances of the probabilistic program. The sum of the probabilities of possible probabilistic programs where the goal succeeds gives the desired probability. Obviously, the method is not suitable for large M. In [8] it is suggested to use Monte-Carlo method in such cases. On each trial of the method, a random instance of the probabilistic program with the corresponding probability is generated and an attempt to prove the goal is performed. The ratio between the number of successful attempts and total number of trials gives the estimation for the probability of the goal. The number of trials required to satisfy certain precision and validity of the estimation can be calculated by Hoeffding theorem [9]. So for $N \geq \frac{1}{2\varepsilon^2}\ln\frac{2}{\delta}$, the inequality $|\rho - \rho'| \leq \varepsilon$ is satisfied with probability not less than $1 - \delta$, where ρ - estimation for the probability of the goal success, ρ' - its exact value, N - number of iterations, ε - precision.

On the base of described scheme, we have developed a tool that extends CLP(R) language [10] in order to handle uncertainty in probabilistic interpretation. Its basic features aim to represent probabilistic logic programs and manage logical inference with probability in such programs. It is possible to deal with different kinds of logical dependencies between uncertain sentences and conditional probabilities. The tool also allows to describe stochastic parameters with predefined distribution laws, use them during inference to compute some other parameters and then perform statistical analysis of results. Some more details concerning tool implementation and its application can be found in [4][5].

Let's consider simple example that will be used throughout the paper to illustrate different cases of abduction problem. Suppose we have graph with uncertain edges (they are shown as dotted lines at Fig. 1) and we are interested in path from node **A** to node **C** of the graph. Correspondent Prolog program and goal are shown on Fig. 2. Each uncertain edge uedge(X,Y) has the probability of its presence in the graph: $\rho_{AB} = 0.5$, $\rho_{AD} = 0.7$, $\rho_{BC} = 0.7$, $\rho_{DE} = 0.7$, $\rho_{EB} = 0.9$, $\rho_{EC} = 0.5$. The problem of computing probability of the goal that path exists can be easily represented and solved by means of the mentioned tool. Thus, estimation for path probability ρ_{path} computed with 750 trials is 0.661 or with 5000 trials 0.67. Assuming validity 0.95 it means that $|\rho'_{path} - 0.661| < 0.05$ or $|\rho'_{path} - 0.67| < 0.02$ respectively, where ρ'_{path} – exact value of probability.

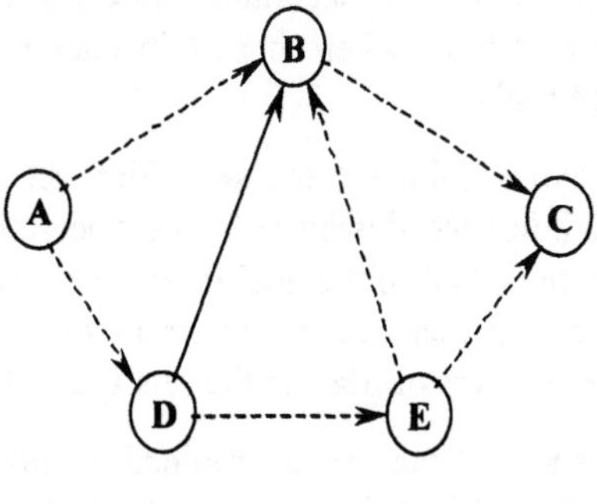

Figure 1. Graph with probabilistic edges.

```
% certain clauses (always present)
edge(d,b).
edge(X,Y)  :- uedge(X,Y).
path(X,Y)  :- edge(X,Y).
path(X,Y)  :- edge(X,Z),path(Z,Y).
% uncertain clauses
% (may present or not depending on a case)
uedge(a,b). uedge(b,c). uedge(a,d).
uedge(d,e). uedge(e,b). uedge(e,c).
% goal
?-path(a,c).
```

Figure 2. Probabilistic program.

The precision and validity of probabilistic estimations is determined by the number of trials performed. However each trial requires to start interpreter in order to check whether a goal succeeds or not. Obviously, the more trials are performed, the more computational efforts for calculating the result are made. The idea of using so-called *success formula* of the goal instead of running the interpreter was suggested in [7] with respect to pure logical programs. Success formula (SF) is the boolean formula, which defines invariably the truth of the goal, depending only on the uncertain clauses of probabilistic logic program. It assumes that all certain clauses are always inferable and thus they do not affect the success of the goal.

In respect of example described above SF(path(a,c)) = uedge(a, b) & uedge(b, c) ∨ uedge(a, d) & uedge(b, c) ∨ uedge(a, d) & uedge(d, e) & uedge(e, c) ∨ uedge(a, d) & uedge(d, e) & uedge(e, b) & uedge(b, c). Substituting uncertain sentences by correspondent arithmetic 0/1 variables **AB**, **AD**, **BC**, **DE**, **EB**, **EC** and boolean operations by arithmetic[1] ones, we produce CLP(R) constraints:

$$F = AB*BC + AD*BC + AD*DE*EC + AD*DE*EB*BC,\ F > 0 \qquad (1)$$

Variables take value 1 if the correspondent clause is present in program, and 0 otherwise. Satisfaction of these constraints is equivalent to the success of the goal. Therefore, on each trial we can only determine which clauses will be presented in the program and assign 0/1 values to variables. If the constraints (1) are satisfied, the goal succeeds and if not – the goal fails. In [5][11] we have demonstrated how this idea could be adapted to deal with real Prolog and constraint programs.

3. Estimation of unknown probability of the uncertain clause

Here we deal with the particular case of abduction problem, that is to estimate one unknown probability of uncertain clause (ρ_u), given the probability of the goal (ρ_{goal}). Additional condition is that all possible different cases of the goal success must be determined either only by positive appearances of uncertain clause or only by negative ones (but not by mixed). Let's consider in following only situation with positive appearances[2].

The approach used for solving the problem is based directly on principles of statistical trials method. Thus, estimation of probability of uncertain clause can be computed as $\rho_u = N_u/N$, where N_u is number of trials where uncertain clause is present in the program, and N - total number of trials.

It is clear that trials can be separated into two non-intersected sets. First set **D** contains trials where the result of the goal inference depends on the uncertain clause. The second set **R** contains trials where the result of the goal inference is the same regardless of whether uncertain clause is present in the program or not. Obviously, $N=N_d+N_r$, where N_d and N_r are the numbers of trials in the sets **D** and **R.**

Within the statistical trials approach, N_d and N_r can be determined without knowledge of the uncertain sentence probability. It is only necessary to perform inference of the goal on each trial two times: uncertain clause is present (or not present) in the program. If the results are different, the trial belongs to the set **D**,

---

[1] It is due to CLP(R) handles only arithmetic constraints.

[2] If all appearances are negative the resulting probability estimation will be $1-\rho_u$.

otherwise to **R**. In the last case we can count also the number of trials where the goal succeds N_{sr}.

Given the probability of the goal we calculate the total number of trials where the goal succeeds as $N \cdot \rho_{goal}$. Then we can calculate the number of trials from **D** where the goal has to succeed (N_{sd}) in order to provide given probability ρ_{goal}: $N_{sd} =$ $= N \cdot \rho_{goal} - N_{sr}$. Under the condition made at the beginning of the section, N_{sd} equals to the number of trials from **D** where the uncertain clause is present in the program N_{ud}. The total number of trials where uncertain clause is present in the program can be estimated assuming that in the limit $N_u/N = N_{ud}/N_d$. Thus, as soon as we get value of N_{sr} by counting successful trials in **R**, we have all necessary information for calculating the probability estimation of the uncertain clause.

The computations can be organized efficiently using SF of the goal represented as CLP(R) constraints like (1). On each trial, we initially keep variable corresponding to the uncertain clause with unknown probability non-grounded and assign 0/1 values to the variables corresponding to the remaining uncertain clauses depending on the case and their probabilities. Then we assign by turn values 0 and 1 to initially non-grounded variable and fix the outcomes (whether the constraints are satisfied or not, and whether the results are the same in both attempts). According to the outcomes of all trials, we compute N_{sr}, N_r, and N_d.

Table 1 presents the results of calculation for the sample from the 2-nd section. In the first of two situations considered, probability of edge AB is unknown, while in the other one probability of edge BC is unknown. Three different runs of the method are presented for each situation. The number of trials is 750. Probability of the goal '**path(a,c)**' is 0.7. Rows 1-3 contain data accumulated during the trials (N_d, N_r, N_{sr}), while rows 4-6 contain the result of computations as described above.

Table 1. Estimation of unknown probability of edge in the graph.

		Edge **AB**			edge **BC**		
		1	2	3	1	2	3
1	Path depends on edge existence, N_d	165	147	156	425	464	465
2	Path doesn't depend on edge existence, N_r	585	603	594	325	286	285
3	Path doesn't depend on edge existence and exists, N_{sr}	415	429	419	200	185	182
4	Edge exists and determines path existence, $N_{sd}(=N_{ud})$	110	96	106	325	340	343
5	Edge exists, N_u	500	490	510	574	550	553
6	Edge probability, ρ_u	0.667	0.653	0.679	0.765	0.733	0.738

4. Estimation of bounds of input stochastic variable

Our approach for handling uncertainty supports stochastic variables as input and can produce sampling distribution for stochastic variables as output by calculating their values during trials. Therefore, it is worth to solve abduction problem for stochastic parameters too. Here we will consider the particular case of the problem. Assume we have some domain parameter, which is represented by stochastic variable **A** in the program model. During Monte-Carlo trials, the variable **A** somehow participates in producing the values of some output stochastic variable **B** (by means of set of constraints). Suppose we know the range of values, which the output variable can belong to and have no information about input variable. The problem is to get some information about **A**.

The possibility to solve the problem is based on specific feature of CLP operational model - the set of active constraints associated with variables participating in the goal inference. Thus, on each trial, we can keep variable **A** and **B** non-grounded, constrain **B** to known range using inequality constraints and try to infer the goal. If the goal succeeds, we can access the set of active constraints and obtain information about **A**[1]. In the simplest case, this set contains a pair of inequality constraints which define the range for **A** and can be retrieved by build-in features of CLP(R). In general case retrieving bounds for **A** is a more complex procedure and one of the possible solutions is as follows. At first, we take some initial range which a priori contains bounds for **A**. Then we pick up values from initial range starting from its lower bound with some step and ground **A** to selected values. First value that satisfy the set of active constraints is a lower bound for **A**. Upper bound is retrieved in a similar way starting from the upper bound of initial range. In that way, on each trial we have range of admissible values for **A**. Having the results of all trials, we produce sampling distribution for lower and upper bounds[2].

Let's return to the example with graph at Fig. 1 and modify the formulation of the problem to illustrate the technology described above. Assume that all edges always exist and each edge except AD and DB has uncertain cost represented by stochastic variable with predefined distribution law: AB – N(6,0.5), BC – U(1,3), DE – U(0.5,2.5), EB – N(1.5,0.3), EC – N(2,0.4), where N(μ,σ) denotes normal distribution with average μ and standard deviation σ, U(a,b) denotes uniform distribution in the [a,b] range. Cost of edge DB is 2 and cost of edge AD is unknown. We are interested in minimal-cost path from node A to node C given that its cost is in interval [4;5]. Using the described method, we can compute the lower and upper bounds for admissible values of the edge cost assuming the initial range is [0.5;5]. Sampling distributions for the bounds are presented at Fig. 3. The number of trials is 750.

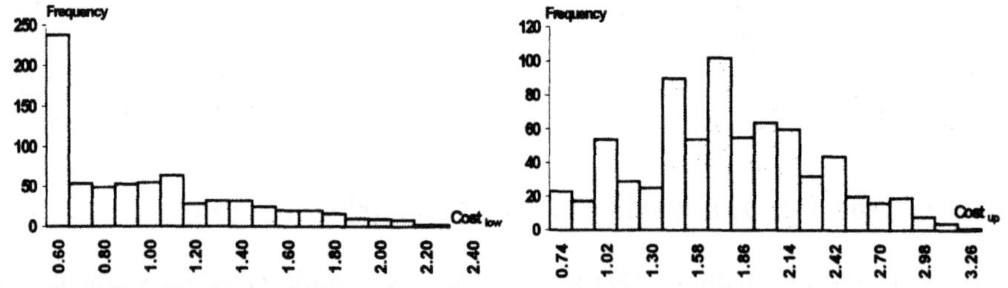

Sampling distribution of lower bound. Sampling distribution of upper bound.

Figure 3. Estimation of bounds of AD edge cost value.

---

[1] Here we assume that **B** depends on **A** so that all related constraints are in the active set.
[2] Sometimes it is more convenient to operate with middle and length of the range than with bounds.

5. Dependence between input and output domain variables

In previous sections, we have considered the cases of the abduction problem that operate with a single pair of input and output values. Certainly, the dependence between input and output variables can give us much more benefit. For instance, we could analyze how the changes of input affect the output or how sensitive is the output to changes of the input within the given range, that is we determine the importance of some of input parameters. When the dependence binds several input variables with the output one, we can also obtain information about relations between input variables by means of analysis of the certain projections of the dependence. Such information could be very useful during the development or modification of the knowledge base of the system or in the situations when some analyzed parameters of the system could not be directly observable.

An approximate empirical dependence for a pair of input and output variables can be obtained in a quite trivial way. Solving the direct problem for several different values of input variable, pairs of known input/output values are produced. These pairs can then be used to form an empirical dependence by curve fitting. The advantage of this method is that it is very generic, doesn't require any specific conditions for variables or the whole model to be satisfied and doesn't require any special support from the tool. However, this method is very resource consuming in the complex cases, while the dependence produced with a few number of pairs computed is only the rough approximation comparing to the actual one.

In some particular cases, the dependence can be obtained in the analytic form. Below we consider the case of pure probabilistic program with independent uncertain clauses and show that the dependence between the probabilities of the clauses and probability of the goal is a multilinear form [12].

We suppose that a probabilistic logic program ***P*** contains certain clauses and some uncertain clauses $U_1, \ldots, U_n$, which are assumed to be true with probabilities $\rho_1, \ldots, \rho_n$ respectively. Then a goal ***G*** succeeds with the probability q. The direct problem of probabilistic logic reasoning is: given ***P***, ***G*** and $\rho_1, \ldots, \rho_n$, find q. In the case of abductive reasoning all probabilities except some $\rho_1, \ldots, \rho_k$ (we enumerate them for convenience) are known. The goal ***G*** is required to succeeds with probability q. Find $\rho_1, \ldots, \rho_k$ such that ***G*** succeeds with probability q.

Let $\rho_1, \ldots, \rho_k$ be unknown probabilities associated with the uncertain clauses of the program *P*. Consider a probability distribution defined on the set of 2^k possible programs. This distribution agrees with probabilities $\rho_1, \ldots, \rho_k$ if for all i ρ_i is equal to the sum of the probabilities of those possible programs that contain U_i $(1 \leq i \leq k)$. Among all the distributions that agree with $\rho_1, \ldots, \rho_k$ there is only one distribution with a maximal entropy. This distribution, called the maximum entropy distribution, corresponds to the case when uncertain clauses are probabilistically independent. Therefore, each possible program has the probability $\rho_1' \cdot \ldots \cdot \rho_k'$, where $\rho_i' = \rho_i$ if the possible program contains U_i and $\rho_i' = 1 - \rho_i$ otherwise.

We denote the probability that G succeeds by $Prob(P,G)$; the probability that G succeeds on the possible program P_j by $Prob(P_j,G)$ and probability of P_j by $Prob(P_j)$ ($1 \le j \le 2^k$). If all possible programs are probabilistically independent, then: $Prob(P,G)=\sum_{j=1}^{2^k} Prob(P_j,G)\cdot Prob(P_j)$, that is $Prob(P,G)$ is a multilinear polynomial of $\rho_1, \ldots, \rho_k$.

In the particular case of two unknown probabilities ρ_1, ρ_2, we have:

$$Prob(P,G)=Prob(P_1,G)\cdot(1-\rho_1)\cdot(1-\rho_2)+Prob(P_2,G)\cdot(1-\rho_1)\cdot\rho_2+$$
$$+Prob(P_3,G)\cdot\rho_1\cdot(1-p_2)+Prob(P_4,G)\cdot\rho_1\cdot\rho_2$$

Let q, q_1, q_2, q_3, q_4 denote $Prob(P,G)$, $Prob(P_1,G)$, $Prob(P_2,G)$, $Prob(P_3,G)$, $Prob(P_4,G)$ respectively. Then $q = b_0+b_1\cdot\rho_1+b_2\cdot\rho_2+b_3\cdot\rho_1\cdot\rho_2$, where $b_0 = q_1$, $b_1 = q_3 - q_1$, $b_2 = q_2 - q_1$, $b_3 = q_1 - q_2 - q_3 + q_4$.

The following technology could be used for calculating the values of coefficients in a multilinear form. Solving the direct problem for arbitrary selected values for probabilities of uncertain clauses, we produce equation constraint involving unknown coefficients. Having several such equations produced with different sets of probabilities values, we make up the system of linear equations relative to the coefficients. The system is directly solved by CLP(R) solver.

Let's return to the example of graph at Fig. 1 and consider the formulation of the problem given in the second section. Now we are interested in dependencies between probabilities of uncertain edges and probability of the path. Note that in this case we deal with independent uncertain clauses, so the one-dimensional dependence $\rho_{path}(\rho_{edge})$ is linear: $\rho_{path}(\rho_{edge}) = \mathbf{a}\cdot\rho_{edge} + \mathbf{b}$. Computing two pairs of values $\rho_{path}(\rho_{edge})$ e.g. for ρ_{edge}=0.1 and ρ_{edge}=0.9 we get coefficients **a** and **b** as described above. Thus one-dimensional dependencies for different uncertain edges are: $\rho_{path}(\rho_{AB}) = 0.205\rho_{AB} + 0.566$, $\rho_{path}(\rho_{AD}) = 0.445\rho_{AD}+ + 0.359$, $\rho_{path}(\rho_{BC}) = 0.615\rho_{BC} + 0.241$, $\rho_{path}(\rho_{DE}) = 0.109\rho_{DE} + 0.594$, $\rho_{path}(\rho_{EB}) = 0.67$, $\rho_{path}(\rho_{EC}) = 0.149\rho_{EC} + 0.593$. These dependencies are shown at Fig. 4.

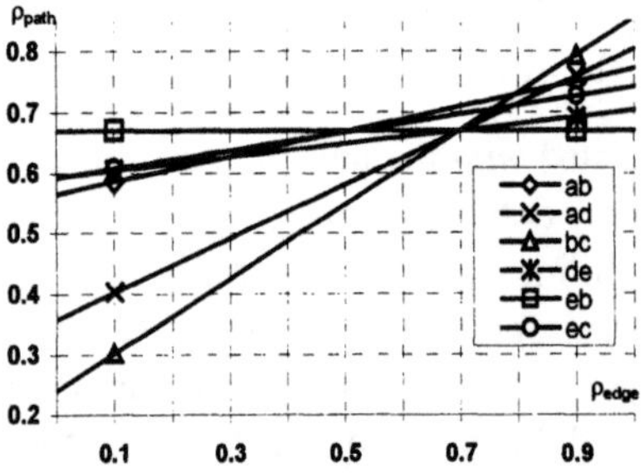

Figure 4. Dependencies between the probability of path and probabilities of uncertain edges.

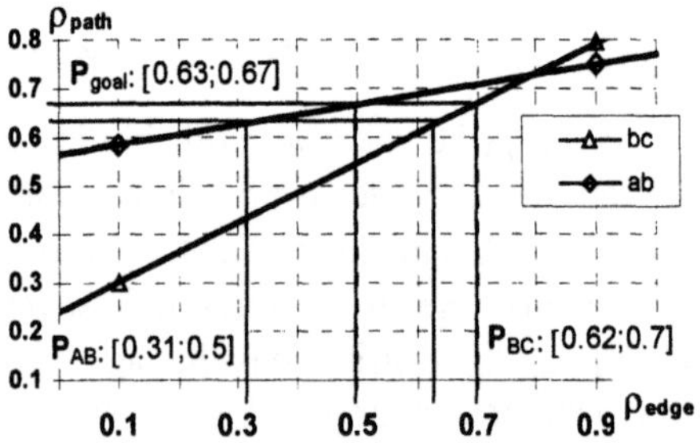

Figure 5. Abduction problem for range of values.

Inclination of the line correspondent to some edge at Fig. 4 indicates the importance of the edge for path to exist. The more inclination is, the more critical the existence of the edge for existence of the path is. It is occurred that probability of the path doesn't depend on probability of EB edge. Certainly, it is not quite trivial to detect this fact just by looking on the graph. Obviously, with this dependence it is possible to determine the value of edge probability for any given value of the path probability as well to calculate the range of admissible values of the edge probability for given range of path probability (like shown at Fig.5).

For two-dimensional case of dependency $\rho_{path}(\rho_{edge1}, \rho_{edge2})$, let's take the pair of edges that possibly conclude the path, namely the edges **BC** and **EC.** Multilinear form is $\rho_{path}=a\cdot\rho_{BC}+b\cdot\rho_{EC}+c\cdot\rho_{BC}\cdot\rho_{EC}+d$. To make up the system of equation with the unknown coefficients, we compute four sets of values $\rho_{path}(\rho_{BC}, \rho_{EC})$ by solving direct problem: $\rho_{path}(0.1,0.1)=0.1222$, $\rho_{path}(0.1,0.9)=0.4806$, $\rho_{path}(0.9,0.1)=0.7736$, $\rho_{path}(0.9,0.9)=0.8134$. Then we solve the system with the CLP(R) constraint solver:

```
f(P,P1,P2,A,B,C,D) :- P = P1*A + P2*B + P1*P2*C + D.

?:- f(0.1222,0.1,0.1,A,B,C,D), f(0.4806,0.1,0.9,A,B,C,D),
    f(0.7736,0.9,0.1,A,B,C,D), f(0.81,0.9,0.9,A,B,C,D).
```

The resulting coefficients are: D = -0.00905625, C = -0.503125, B = 0.498313, A = 0.864563. Figure 6 presents the chart of the dependence as well as the families of its projections $\rho_{path}(\rho_{EC})$ and $\rho_{BC}(\rho_{EC})$ for several fixed values of ρ_{BC} and ρ_{path} respectively. Analyzing such charts, we obtain more information in comparison with one-dimensional dependence. Namely, such information concerns the mutual relations between admissible values of the input probabilities for given value or range of the output one. For example, the presented charts show that high probability of the path is determined to the great extent by the high probability of edge BC, while the probability of edge EC is not very meaningful in this case.

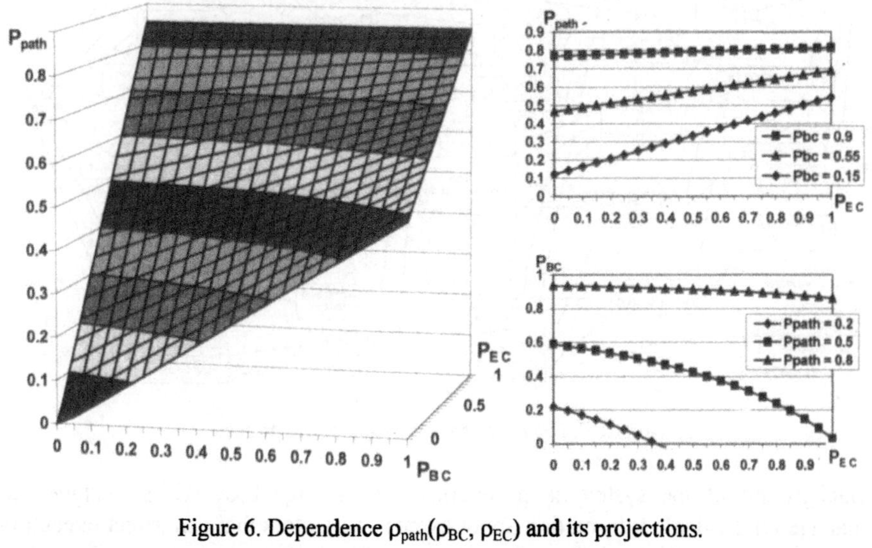

Figure 6. Dependence $\rho_{path}(\rho_{BC}, \rho_{EC})$ and its projections.

6. Abduction in the network-based project planning problem

The input of the classical problem is the set of tasks of some project to be accomplished with predefined durations. For each task it is known what tasks precede and what tasks follow it, so the project can be conveniently represented as a network (directed graph) of tasks according to their precedence. The main results of solving the problem are: time required to finish the project, set of tasks that belong to so-called *critical path*, and set of time parameters for each task that characterize the schedule of its execution. Basic time parameters are early start time, late finish time and free reserve. The last one is the time of possible delay in task execution that does not cause the delay of subsequent tasks. Tasks with no free reserve compose a critical path of project execution. The length of critical path is the minimal time required to complete the project. Delay of any of the tasks on a critical path leads to delay of the whole project. Classical method to solve the problem is the Critical Path Method (CPM). The detailed review of the classic network-based project planning problem can be found in [13].

In our pilot application we deal with extended formulation of the problem with regard to uncertainty of input information about project tasks. The skeleton of the system is still network representation of the project (see Fig. 7.) that allows to benefit from power and efficiency of CLP(R) solver for implementation of CPM-like algorithm. The nodes of the graph correspond to the tasks and links represent the order (left to right) of tasks execution. The project considered in the application is the software development project. The materials presented in [14] are used to prepare the input data for the system.

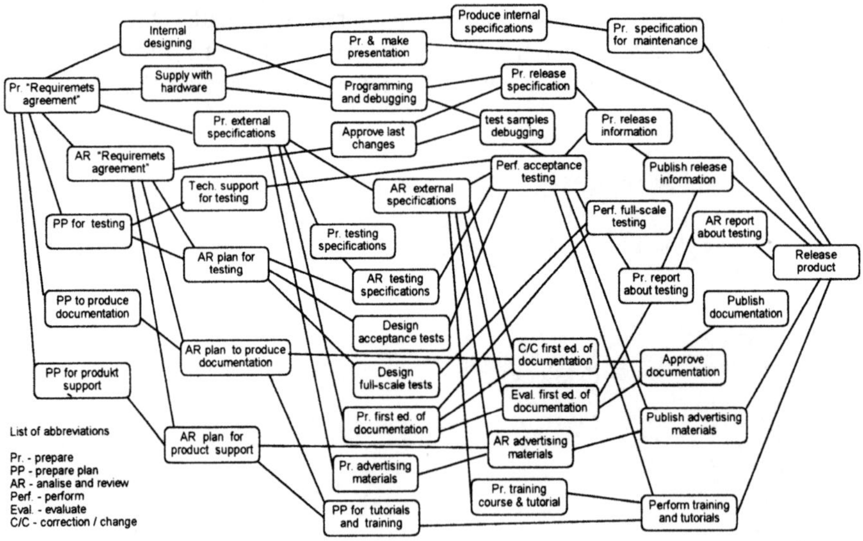

Figure 7. Network representation of the project.

Background of the system is probabilistic knowledge base (KB), designed to manage all available information about events and factors that can affect execution

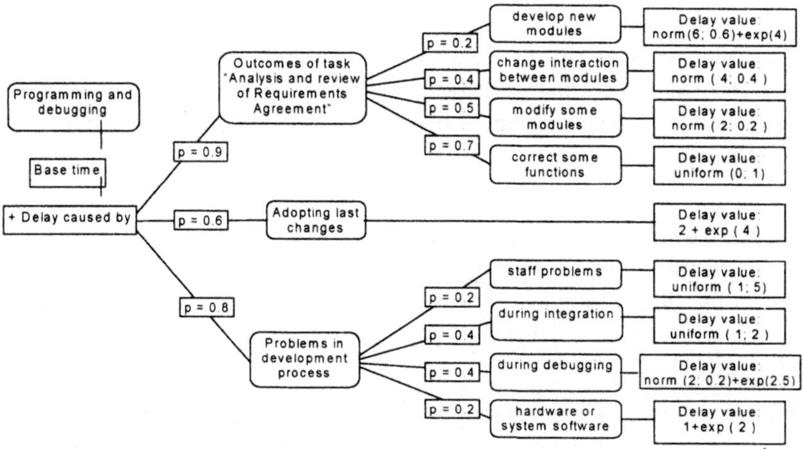

Figure 8. Model of the macro-task "Programming and debugging".

of tasks. Particularly, KB provides models of tasks execution and computes duration for them. Several tasks (about 10) are described as complex *macro-task*. These tasks may "unexpectedly" delay because of several reasons with certain probabilities. The appearance of these reasons, in turn, depends on several random factors with certain probabilities. These reasons and factors can be outcomes of other tasks or some random events that are not present in the project network directly. Delay values are considered as random variables with the predefined distribution laws. Thus, about 10-15 pseudo-random number generators are used to produce the instances of macro-tasks duration values. Note that the same random factors or outcomes of some tasks can affect execution of several tasks simultaneously and thus the correlation between the values of tasks duration takes place. Other tasks are represented by the more simple models using few random parameters. The example of the model for one of the macro-tasks ("Programming and debugging") is presented at Fig. 8.

The KB is integrated with the algorithm via variables that correspond to the durations of the tasks. The computations are organized in the following way. On each trial of the Monte-Carlo method the KB produces values of durations for all tasks or, in other words, the possible "instance" of the project execution. Then this

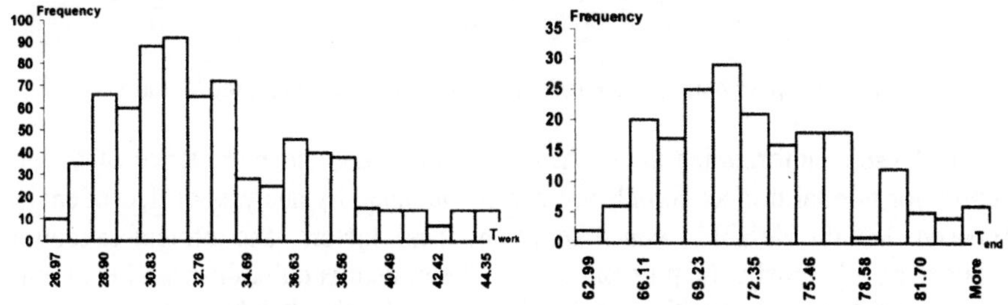

Duration of macro-task "Programming and debugging"

Project completion time

Figure 9. Some results of solving direct problem.

instance is evaluated by the algorithm to determine the project completion time and project critical path. During the computations, all information that is necessary for future statistical processing is logged. When all iterations are completed, the required statistical processing can be performed. Each trial takes about 0.07 sec (PC with Pentium-166 processor).

Figure 9 presents the results of solving direct problem with all parameters known in knowledge base. Namely they are sampling distributions for the duration of macro-task "Programming and debugging" which is defined in the base according to the scheme at Fig. 8 and sampling distribution for the project completion time. Many other details related to solving the direct problem for the pilot system could be found in [5][11].

As it follows from the system description, the top-level representation of the project is a directed graph dealing with stochastic variables. The situation is quite similar to the sample considered in section 4. The only difference is that in the example the variables have been associated with edges, while here variables are associated with nodes. Variables represent duration of tasks and participate in computing project completion time via the set of arithmetic and min/max constraints. Thereby, we can solve abduction problem for unknown duration of some of macro tasks if the value or the range of project completion time is given. The other useful treatment for this problem is: determine the admissible duration of the task in order to meet predefined project deadline. So, assume that project completion time $T_{end} \in [73, 79]$ and we need to compute admissible duration of the task "Programming and debugging". Assuming initial range for the duration $T_{work} \in [0, 50]$, we obtain upper bound of the task duration with sampling distribution shown at Fig. 10. The lower bound of the duration is always less than upper bound by the value of 6.

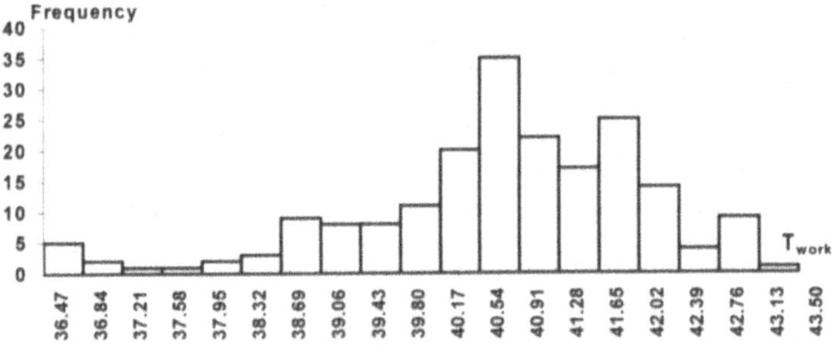

Figure 10. Upper bound of the task "Programming and debugging"duration.

A lot of useful information about the model adequacy and importance of certain factors for normal project completion can be obtained by analysis of dependencies between domain variables, which represent these factors, and project execution parameters. Obviously, the program used in the system is quite far from being pure logical and doesn't deal with probabilities on top level. Thereby, it is possible to work only with empirical dependencies, which produced as described in the beginning of section 5. The examples of such dependencies concerning the task "Programming and debugging" are presented at Fig.11.

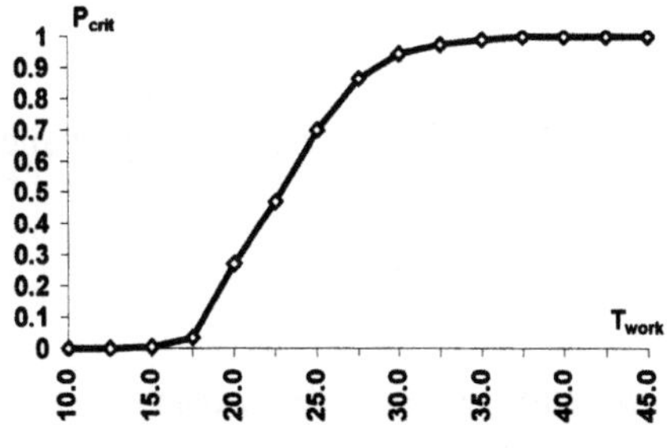

Dependence between the probability of event that task belongs to the project critical path and task duration.

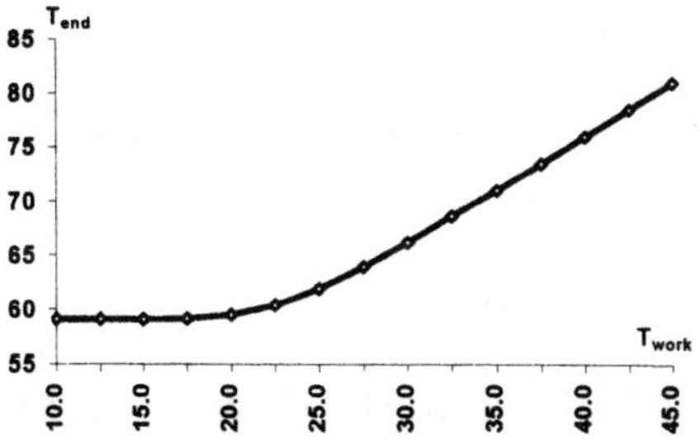

Dependence between the average project completion time and the task duration.

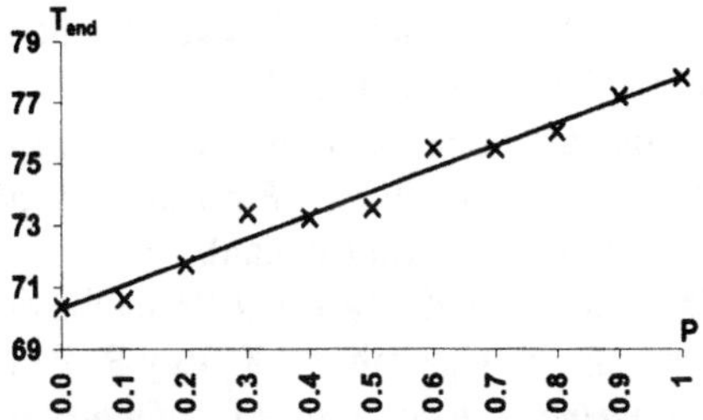

Dependence between the average project completion time and the probability of the outcome "develop new modules" of task "Analysis and review of Requirements Agreement" (see Fig.8) with linear trendline.

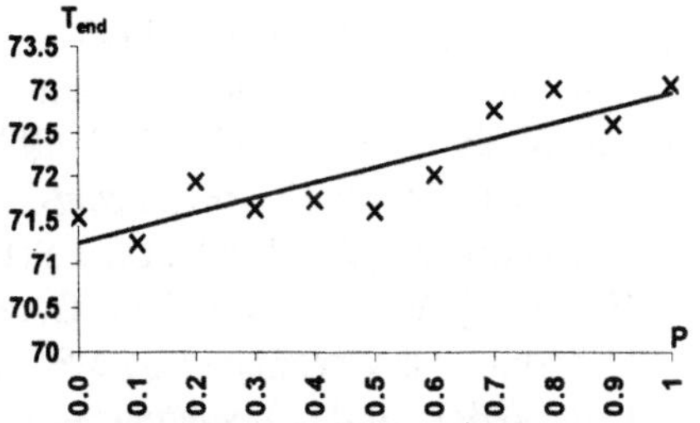

Dependence between the average project completion time and the probability of the delay in development process due to the problems at the debugging stage (see Fig.8) with linear trendline.

Figure 11. Empirical dependencies for macro-task "Programming and debugging".

7. Conclusions

In the paper, we have considered the methods for solving the abduction problem within the approach for handling uncertainty in probabilistic interpretation. First of them allows to estimate unknown value of uncertain clause probability given the goal probability value. The second one deals with stochastic variables and allows to estimate range of admissible values of unknown input variable given the range (or value) of output one. The methods of obtaining the dependence between input and output variables have been discussed as well. The use of some of the methods was demonstrated in pilot application for solving network based project planning problem in extended non-deterministic formulation.

Certainly, the presented methods require subsequent development. Currently, their applicability is limited by certain conditions. Additional efforts should be made for analysis of estimations accuracy. These topics are subject for future investigation. As for the pilot application, it would be quite interesting to consider a propagation of abduction results obtained on top-level to the underlying knowledge base level.

We think that methods described in the paper could be useful extension for the approach and the tool for handling uncertainty in probabilistic interpretation.

Acknowledgements

This work was done within the project INTAS-93-1702EXT "Efficient Symbolic Computing".

References

[1] Van Hentenryck P 1989 *Constraint Satisfaction in Logic Programming*, MIT Press, Cambridge, MA.

[2] Fruhwirth T, Herold A, Kuchenhoff V, et al. 1993 Constraint Logic Programming. An Informal Introduction. Tech. Report ECRC-93-5, Munich, Germany.

[3] Aggoun A, Beldiceanu N 1993 Extending CHIP in order to solve Complex Scheduling and Placement Problems. *Math Comput Modelling* 17(7):57-73.

[4] Valkovsky V B, Savvin K O 1995 Logical inference with probability in Constraint Prolog, *Proceedings of the St.Petersburg Electrotechnical University, Vol 489*. SPbETU, St.Petersburg. pp 37-49 (in Russian).

[5] Valkovsky V B, Savvin K O, Gerasimov M B 1998 Handling Probabilistic Uncertainty in Constraint Logic Programming. In: Chawdhry P K, Roy R, Pant R K (eds), *Soft Computing in Engineering Design and Manufacturing*, Springer-Verlag, UK.

[6] Nilsson N J 1986 Probabilistic Logic. *Artificial Intelligence* 18(1): 71-87

[7] Dantsin E 1992 Probabilistic Logic Programming and their Semantic. In: Voronkov A (eds) 1992 *Lecture Notes in Computer Science, Vol. 592*. Springer-Verlag, London, pp 152-164

[8] Dantsin E, Kreynovich V 1989 Probabilistic Inference in Forecasting Systems. In: 1989 *Reports of the Russian Academy of Science (mathematics)*, Vol. 307(1), pp 17-21 (in Russian).

[9] Hoeffding W 1963 Probability inequalities for sums of bounded random variables. *J Amer Stat Assoc*, 58(301):13-30.

[10] Jaffar J, Michaylov S, Stuckey P J, Yap R H C 1992 The CLP(R) Language and System. *ACM Transactions on Programming Languages and Systems* 14(3):339-395.

[11] Valkovsky V B, Gerasimov M B, Savvin K O 1996 CLP with Probability in Scheduling Problems. In: 1996 *Proceedings of the Second International Conference on the Practical Application of Constraint Technology*. London, UK, pp 299-315

[12] Dantsin E, Valkovsky V 1996 Abductive Reasoning in Probabilistic Prolog. Report. Workshop of INTAS-93-1702 project (Efficient Symbolic Computing), International Euler Mathematical Institute, St.Petersburg, Russia.

[13] Souder W E 1978 Project choice. Project Tasks Scheduling and Project Management. In: Moder J J, Elmaghraby S E (eds) 1978 *Handbook of Operational Research*. Van Nostrand Reinhold Company, pp 264-311

[14] Gunter R C 1979 *Management methodology for software product engineering*. A Wiley-Interscience publication, J.Wiley&Sons, NY-Chichester-Brisbane-Toronto.

Aspects of Evolutionary Design by Computers

Peter Bentley

Intelligent Systems Group, Department of Computer Science,
University College London, Gower St., London WC1E 6BT, UK.
P.Bentley@cs.ucl.ac.uk
Tel. +44 (0)171 391 1329 Fax. +44 (0)171 387 1397

ABSTRACT

This paper examines the four main types of Evolutionary Design by computers: Evolutionary Design Optimisation, Evolutionary Art, Evolutionary Artificial Life Forms and Creative Evolutionary Design. Definitions for all four areas are provided. A review of current work in each of these areas is given, with examples of the types of applications that have been tackled. The different properties and requirements of each are examined. Descriptions of typical representations and evolutionary algorithms are provided and examples of designs evolved using these techniques are shown. The paper then discusses how the boundaries of these areas are beginning to merge, resulting in four new 'overlapping' types of Evolutionary Design: Integral Evolutionary Design, Artificial Life Based Evolutionary Design, Aesthetic Evolutionary AL and Aesthetic Evolutionary Design. Finally, the last part of the paper discusses some common problems faced by creators of Evolutionary Design systems, including: interdependent elements in designs, epistasis, and constraint handling.

Keywords: evolutionary design, creative design, optimisation, computer art, artificial life

1. Introduction

The use of Evolutionary Computation to generate designs has taken place in many different guises over the last ten or fifteen years. Designers have optimised selected parts of their designs using evolution, artists have used evolution to generate aesthetically pleasing forms, architects have evolved new building plans from scratch, computer scientists have evolved the morphologies and control systems of artificial life.

However, work in these different application areas tends to be performed by researchers isolated within different fields, and whilst there is some cross-disciplinary communication, it seems that many work in ignorance of the highly related research performed by others. In order to remedy this situation, this paper attempts to bring together the varied aspects of this type of research under a single heading: *Evolutionary Design.*

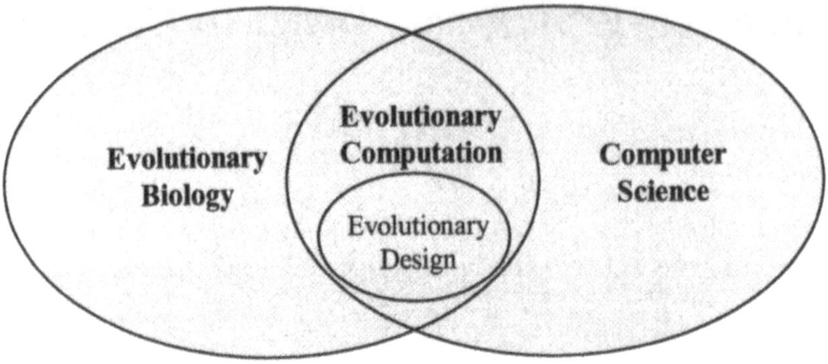

Figure 1 Evolutionary Computation and Evolutionary Design have their roots in Computer Science and Evolutionary Biology.

Cross-disciplinary research is hardly new to Evolutionary Computation. Indeed, the entire field only exists because of the efforts of those such as Holland [24], Rechenberg [32] and Fogel [16] to merge the boundaries of Evolutionary Biology and Computer Science, see figure 1.

Unfortunately, it does seem that the natural behaviour of many researchers is to entrench themselves within a single area, and then erect barriers around that area by trying to stress wider and wider distinctions - almost a form of intellectual territorialism. This paper tries to encourage some of these barriers to be removed, by identifying the common elements of research often considered as distinct and separate.

Consequently, this paper proposes that the branch of research entitled 'Evolutionary Design' should encompass four distinct aspects: Evolutionary Design Optimisation, Creative Evolutionary Design, Evolutionary Art, and Evolutionary Artificial Life Forms.

As is usually the case with any kind of classification system, the work of a few researchers does not fall neatly within one category, but may be included in two or more categories. This paper also examines such work. Particular attention is paid to the four 'overlapping' research areas, which this paper shall name: Integral Evolutionary Design, Aesthetic Evolutionary Design, Artificial Life-based Evolutionary Design, and Aesthetic Evolutionary AL. Figure 2 shows all of these areas of research and how they relate to each other.

The following sections of this paper summarise the scope of research in each area and describe the aims and objectives that researchers in each area typically have, examining some key contributions to each of these fields of research. For each aspect of Evolutionary Design, examples of the representations, evolutionary algorithms and evolved solutions are outlined. Finally, some typical problems encountered by creators of Evolutionary Design systems are discussed, together with ways to overcome them.

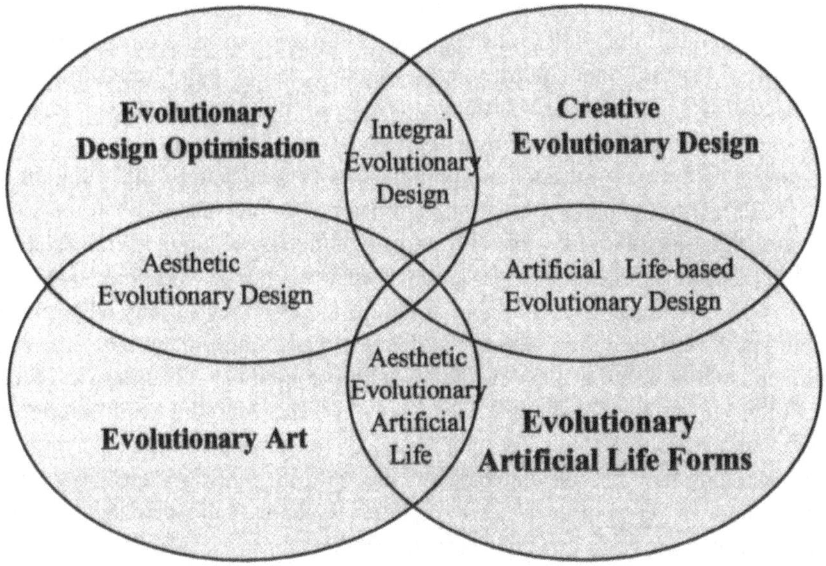

Figure 2 Aspects of Evolutionary Design by Computers

2. Four Aspects

2.1 Evolutionary Design Optimisation

The use of Evolutionary Computation to optimise existing designs was one of the first types of Evolutionary Design to be tackled. Over the last fifteen years, a huge variety of different engineering designs have been successfully optimised [3,19], from flywheels [15] to aircraft geometries [25].

Although the exact approach used by developers of such systems varies, typically this type of Evolutionary Design cannot be classed as *generative* or *creative* (see next section). Practitioners of Evolutionary Optimization usually begin the process with an existing design, and parameterise those parts of the design they feel need improvement. These parameters are then encoded as genes, the alleles (values) of which are then evolved by an evolutionary search algorithm. The designs are often judged by interfacing the system to analysis software, which is used to derive a fitness measure for each design. For example, Goodman [15] describes the optimisation of flywheels using genetic algorithms (GAs). Cross-sections of flywheels are parameterised (into a collection of height parameters) and evolved, see fig. 3 (top left). Evaluation of structure is performed using multiple Finite Element Models [15]. Alternatively, Keane has minimised the structural vibration of designs such as axially vibrating rods and satellite booms, using a GA to minimise fitness values returned by a Statistical Energy Analysis package [26]. For more examples of Evolutionary Optimisation of Designs, see Gen and Cheng's recent book: *Genetic Algorithms and Engineering Design* [19].

Phenotype representations for these design optimisation problems are application-specific, consisting of existing designs for that application, with the evolved parameter values simply inserted into the corresponding parameterised elements. Artificial embryologies (mapping or 'growth' stages from genotypes to phenotypes [14]) are often rudimentary or non-existent, simply because they are not necessary for such Evolutionary Design. Because of this, genotype representations may match phenotype representations closely, often with a one-to-one mapping between genes and parameters. Consequently, the addition or deletion of genes in genotypes, and parameters in phenotypes, is usually not performed by the evolutionary algorithm for Evolutionary Design Optimisation.

When optimising designs, great emphasis is placed upon finding a solution as close to the global optimal as possible - perhaps more so than for any other type of Evolutionary Design. Often designs that are already of good quality are to be improved, and it can be a challenge to improve them at all. Because of this motivation towards global optimality, researchers tend to concentrate on evolutionary search methods for reducing any tendencies towards convergence upon local optima. In addition, because the analysis software used to provide fitness functions for solutions can have heavy computational demands, there is often a strong emphasis towards reducing the number of evaluations required before a final solution is found.

Numerous techniques have been tried to achieve these goals. To improve performance, sometimes multiple genetic representations are used in parallel [15]. Many complex types of genetic algorithm are used [19]. For example, Husbands [25] describes the use of a distributed GA and a distributed GA hybridized with gradient descent techniques to evolve the cross-section of optimal aircraft wingboxes. The research found that hybrid GAs outperformed many other search algorithms for this problem [25].

2.2 Creative Evolutionary Design

Calling anything generated by computer 'creative' is fraught with ambiguity and controversy, so this section will begin by attempting to define, for the purposes of Evolutionary Design, what 'creative' actually means.

Writing about this very subject in his paper 'Computers and Creative Design' [20], Gero makes the distinction between cognitive and social views, i.e. an individual can display creativity when designing, and a design can have characteristics which may be regarded as being creative. Gero concentrates on the former definition, and concludes that a computer is designing creatively when it explores the set of possible design state spaces in addition to exploring individual design spaces. In other words, Gero indicates that by evolving the *number* of decision variables in addition to evolving the *values* of those variables, a computer is being creative [20]. In a similar vein, Boden suggests in her book *The Creative Mind* [9], that creativity is only possible by going beyond the bounds of a representation, and by finding a novel solution which simply could not have been defined by that representation. Boden, however, does not feel that computers are capable of such creativity [9].

Other definitions for creative design include: having ability to generate 'surprising and innovative solutions', or 'novel solutions that are qualitatively better than previous solutions' [21]. However, for the purposes of this paper, Rosenman's description seems most apt: 'the lesser the knowledge about existing relationships between the requirements and the form to satisfy those requirements, the more a design problem tends towards creative design' [33].

Consequently, the main feature that all Creative Evolutionary Design systems have in common, is the ability to generate new forms starting from little or nothing (i.e. random initial populations), and be guided purely by functional performance criteria. In achieving this, such systems often do vary the number of decision variables during evolution [4,21,33]. They can often generate surprising and innovative solutions, or novel solutions qualitatively better than others [6,23]. Whether this means that these systems are really 'designing creatively', or whether they simply generate 'creative designs', will be left for the reader to decide.

Research in the field of Creative Evolutionary Design is concerned with the preliminary stages of the design process. There are two main approaches. Both involve the use of evolutionary computation to generate entirely new designs from scratch, however the level at which these designs are represented is different:

Conceptual Evolutionary Design - *the production of high-level conceptual frameworks for designs.*

In this type of Evolutionary Design, the relationships and arrangements of high-level design concepts are evolved in an attempt to generate novel preliminary designs. A good example of this is the work of Pham, who describes his preliminary design system known as TRADES (TRAnsmission DESigner) [31]. TRADES uses a genetic algorithm to evolve the organisation of a set of conceptual building blocks (such as rack and pinion, worm gear, belt drive). When given the type of input (e.g. rotary motion) and the desired output (e.g. perpendicular linear motion), the system generates a suitable conceptual transmission system to convert the input into the output.

In these systems, evolution is used to search through the possible networks of interconnected conceptual building blocks. The genotype and phenotype representations of these systems are often simple, with rudimentary embryologies, if any. Basic evolutionary algorithms are usually sufficient for this type of problem. There are always exceptions to every rule, however, and the work of Parmee provides such an exception. Parmee [29] describes the use of structured GA to evolve a large-scale hydropower system (this is intended to take place at the feasibility/bid stages of the design process, after the conceptual design stage). His advanced GA manipulates a design hierarchy of 'sites', 'dam types', 'tunnel lengths', 'modes of operation', etc., and allows appropriate elements to be switched on or off by control genes during evolution [29].

Generative Evolutionary Design (or Genetic Design) - *the generation of the form of designs directly.*

Using computers to generate the form of designs rather than a collection of pre-defined high-level concepts has the advantage of giving greater freedom to the computer. Typically such systems are free to evolve any form capable of being represented, and the evolution of such forms may well result in the emergence of implicit design concepts [6,23]. However, the difficulty of this type of Creative Evolutionary Design is severe, since it often involves the creation of dynamically specified representations and complex evaluation routines (see below).

Often involving just the preliminary stages of design, the emphasis for this type of Evolutionary Design is on the generation of novelty and originality, and not the production of globally optimal solutions. Representations of form vary tremendously, but they do all share certain features. For example, because the emphasis is on the generation of new forms, phenotype representations are typically quite general, capable of representing vast numbers of alternative morphologies (this is in contrast to representations for optimisation, which can only define variations of a single form). Representations range from direct spatial partitioning (e.g. voxels), which have one-to-one mappings between genes in genotypes and elements of phenotypes [2], to highly indirect representations which use shape grammars or cellular automata (CA) with some advanced embryologies to map genotypes to phenotypes [11,33]. For example, Rosenman describes the use of a GA to evolve house plans, using a design grammar of rules to define how polygons should be constructed out of edge vectors [33]. Coates uses genetic programming (GP) to evolve abstract architectural forms, with L-Systems forming the basis of the representation [11]. Schnier and Gero even attempt to evolve new higher-level representations of form, suitable for subsequent evolution of house plans in a specific architectural style [35].

Simple genetic algorithms are often sufficient for the design systems that employ simpler representations [2]. However, for Creative Evolutionary Design systems with advanced representations and corresponding embryologies (e.g. to 'grow' phenotypes from a set of shape-grammar rules defined in the genotypes), more advanced GAs are essential. Typically, these GAs are used to evolve the larger structure of the representation (e.g. number and organisation of shape-rules) in addition to the detail (e.g. type and content of the individual rules). In other words, these GAs are capable of evolving designs which have representations of variable length - they explore new and different search spaces in addition to the parameter values within each space [7]. GAs and GP are used to evolve these highly variable forms [4,11,33], usually beginning from random, simple forms, and gradually improving the structure and detail of these designs until some functional criteria is met. The objective of this type of research is not normally to use computers to generate a single global optimal solution, but rather to generate a number of 'creative' alternatives. The evaluation of designs can be more difficult, since most off-the-shelf analysis packages are limited to judging specific types of designs - when presented with some of the initially random forms generated by these systems, they simply generate errors. Consequently, many of these systems rely on simplified custom-written evaluation routines, which can analyse everything presented to them, but perhaps not always with the desired accuracy [4].

However, this is one of the most recent and exciting types of Evolutionary Design, and is already showing great potential in a number of application areas.

Bentley [4,6,8] describes the evolution, from scratch, of a number of unusual and inventive designs for numerous applications, such as tables, heatsinks, optical prisms, aerodynamic and hydrodynamic forms, etc., see fig. 3 (top right). Pollack [17] demonstrates the use of a GA to evolve novel bridge and crane structures, which are subsequently built as LEGO™ models [17]. An application area currently receiving much media attention is 'Evolveable Hardware', where new logic circuits are evolved and evaluated in real silicon using FPGAs [23]. Already some surprising and novel electronic solutions have been found using these techniques [23].

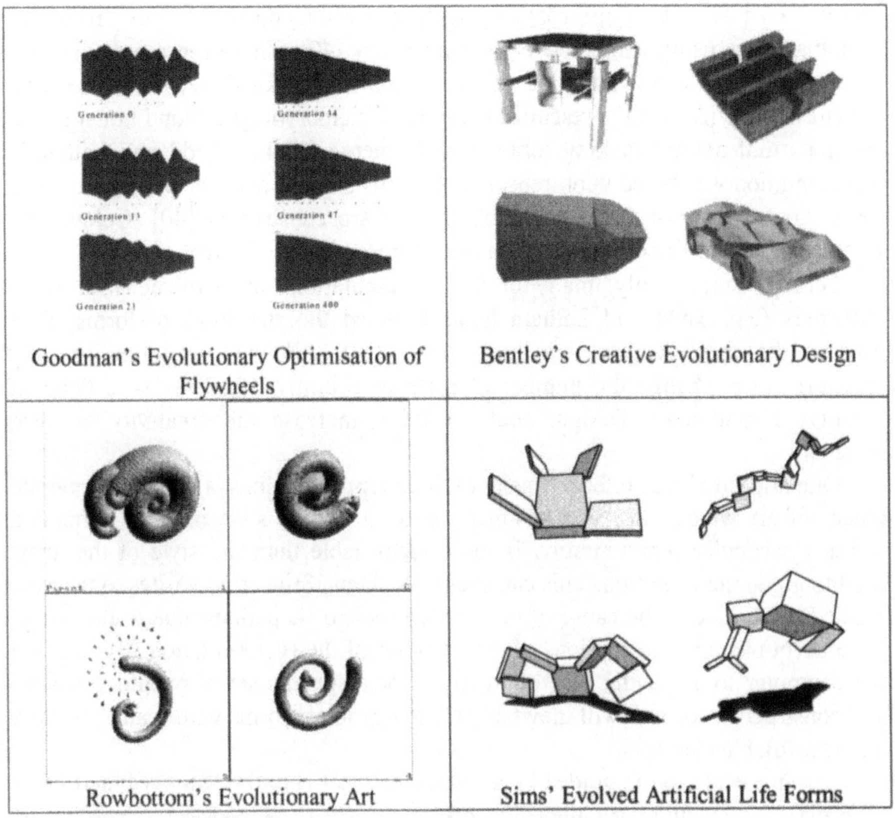

Figure 3 Examples of the four main types of Evolutionary Design.

2.3 Evolutionary Art

Evolutionary Art is perhaps the most commercially successful type of Evolutionary Design. Although academic research in this area is less common than in the other fields, there are perhaps more Evolutionary Art products available today than any other type of Evolutionary Design system.

All Evolutionary Art systems tend to resemble each other closely. They all generate new forms or images from scratch (random initial populations). They all rely completely upon a human evaluator to set fitnesses for each member of the population - normally based on aesthetic appeal. Population sizes are usually very small (often less than ten individuals), to allow them all to be quickly judged every generation. User-interfaces are often similar, with members of the current population shown on the screen in the form of a grid, allowing the user to rank them, or assign fitness scores by clicking on them with a mouse, see fig. 3, bottom left.

The main differences between these systems lie in their phenotype representations. A large variety of alternative representations have been employed, from fractal equations (such as John Mount's 'Interactive Genetic Art', which was shown on-line at http://robocop.modmath.cs.cmu.edu.8001), to recursive grammar-rules using constructive solid geometry [40]. These representations are created with different intentions. For example, Dawkins' recursive tree-like structures were intended to resemble the recursive embryologies found in nature, in the hope that natural looking forms would emerge [13,14]. Todd and Latham's representation was based upon repeated elements such as spheres and tori, used to form 'horns' and 'ribs' out of which images are constructed [40]. Colour and texture can also be incorporated into these representations [14,40].

Perhaps surprisingly, many of the representations are evolved with fixed structures (e.g. Todd and Latham hand-designed the structures of forms, then evolved the detail within these structures [40]). Allowing evolution to vary structures (e.g. change the number of rules or primitive shapes), as is done in Creative Evolutionary Design, could possibly increase the creativity of such systems.

One undesired side effect of many of these representations is that they generate pieces of art which have very distinct styles. Often the style of form generated using a particular representation is more identifiable than the style of the artist used to guide the evolution. This can cause problems if the artist wishes to take the credit for the piece. The cause of this 'style problem' is perhaps due to the initial preconceptions and assumptions of the designer of the representation. By limiting the computer to a specific type of structure, or a specific set of primitive shapes and constructive rules, it will inevitably always generate forms with many common and identifiable elements.

Because evolution is guided by a human selector (i.e. the 'fitness function' is an artist), the evolutionary algorithm does not have to be complex. Evolution is used more as a continuous novelty generator, not as an optimiser. The artist is likely to score designs highly inconsistently as he/she changes his/her mind about desirable features during evolution, so the continuous generation of new forms based on the fittest from the previous generation is essential. Consequently, an important element of the evolutionary algorithms used is *non-convergence.* If the populations of forms were ever to lose diversity and converge onto a single shape, the artist would be unable to explore any further forms. Because of this, most Evolutionary Art systems do not employ crossover within their evolutionary algorithms. Typically only mutation is used, with all offspring being mutated copies of their parents (and often only a single parent is used per generation). This mutation-driven evolution is similar to the approach used in evolution strategies, which are known to be excellent for finding solutions to problems with continuously changing fitness functions [1].

Examples of such systems include Dawkins' biomorphs [13,14], Todd & Latham's Evolutionary Art [40], and Sims' evolved computer graphics [37]. Today, numerous Evolutionary Art systems are available on-line (see http://www.netlink.co.uk/~snaffle/form/evolutio.html for an up-to-date list).

2.4 Evolutionary Artificial Life-Forms

Evolutionary Computation plays a significant role in many aspects of the new field of Computer Science known as Artificial Life (AL). Artificial 'brains', behaviour strategies, methods of communication, distributed problem solving and many other topics are commonly explored using genetic algorithms and other evolutionary search techniques [12].

Although all types of evolved AL could be described as aspects of Evolutionary Design, it is clear that certain topics within AL fall into the 'Evolutionary Design' category more comfortably than others. For the purposes of this paper, AL research that can be readily categorised as an aspect of Evolutionary Design will be defined as research which aims to evolve 'Artificial Life-Forms'. Examples of Evolutionary AL-Forms include: Lohn's Cellular Automata (CA) evolved to be capable of self-replication [27], Harvey's evolved layout and structure of neurons [22], and the evolved plant-like and animal-like morphologies of Dawkins and Sims [13,14,38,39].

Motivations for the creation of Evolutionary AL-Forms are usually theoretical. The goals of such research are often to discover more about the mechanisms of natural evolution, to find explanations of forms observed in nature, or to exploit the solutions proven in nature by attempting to duplicate them. Often the evolved AL-forms show the enormous potentials of this type of Evolutionary Design, but as yet, practical applications are still scarce. Figure 3 (bottom right) shows some of Sims' evolved 'virtual creatures'.

Representations are typically specific to each system. For example, Lohn and Reggia [27] use CA 'rule tables' within the chromosomes of their GA. Sims [38,39] uses a hierarchical chromosome structure to define both 'brain' and 'body'. Ventrella [41] combines ordered morphology and control parameters of his animats in a flat chromosome structure. Many of these representations are inspired by the genotype structure of natural organisms, and some researchers have attempted to evolve AL-forms with complex embryologies. Other researchers invent their own intricate coding schemes [39]. Most such representations are highly flexible and of variable length, requiring complex genetic operators with the evolutionary algorithms.

Because research in this field is still very much at the 'blue-sky' stage, evolutionary techniques are often used as exploration tools, in a similar way to Evolutionary Art. These algorithms can be used to generate multiple solutions, incorporating niching, speciation, parasitism, competition, co-operation and other advanced methods [12]. Evaluation usually consists of analysing behaviour in simulated virtual worlds (although some researchers do test solutions using real robots [22]), and can be very time-consuming. To try to shorten evolution run-times, advanced methods are commonly used, e.g. steady-state GAs, parallel GAs, hybrid GAs [12]. Many systems that evolve AL-forms also use changing fitness functions, which necessitate the use of other specialised genetic search techniques (Harvey) [22]. Most systems evolve the forms from scratch (the initial population is random), however some occasionally seed initial populations with the fittest individuals from previous runs [39].

3. Combining Good Ideas by Merging the Boundaries

Many researchers confine themselves to one of the four aspects of Evolutionary Design mentioned above, and seem loath to consider alternative approaches. However, more recently, some have begun to combine ideas from one or more of these areas in their work. This is leading to four more (still very new and relatively unexplored) 'overlapping' areas of research in Evolutionary Design (see fig. 2).

3.1 Integral Evolutionary Design

The evolution of engineering designs is becoming widespread today, with numerous academic engineering design centres exploring these ideas. Although most research seems to fall into either the Evolutionary Optimisation category, or the Creative Evolutionary Design category, some work does attempt to combine the two into unified, or *Integral* Evolutionary Design systems.

Parmee suggests that computers can be used within the entire design process, both the early conceptual design stages and the later, detailed design stages, by using a number of adaptive techniques. He discusses how the use of several different systems, each dedicated to a specific stage of design could be used in combination, thus 'integrating adaptive search at every stage of the design process' [29].

Alternatively, work performed by Bentley has investigated the use of a single Generic Evolutionary Design system, to perform the complete design process without making any distinction between the stages of design. It is 'generic' because it is capable of evolving designs for multiple different design tasks. This work has shown that it is possible to use a computer to evolve entirely new designs from scratch, and optimise them, such that they fulfil specific functional criteria [4,6,8].

With many optimisation systems beginning to be applied to more and more detailed parameterisations of designs, and many creative design systems beginning to be used to optimise the designs they generate, the field of Integral Evolutionary Design looks set to grow rapidly.

3.2 Artificial Life Based Evolutionary Design

Work in Artificial Life has generated forms of astonishing diversity and creativity, so some researchers are now using some of the techniques from AL in their Creative Evolutionary Design systems, in an attempt to improve the quality and originality of evolved engineering designs. For example, Parmee borrows distributed agent methods from AL to increase performance of search, in his 'Ant Colony' method, which he combines with evolutionary search [30]. Coates employs L-systems with GP to evolve new architectural forms [11].

Many other unexplored possibilities still exist in this area. For example, Bonabeau [10] describes an AL computer simulation of a swarm of artificial wasps, which build intricate three-dimensional nest architectures. One future avenue of research may be to evolve artificial wasps capable of building new engineering designs between them.

3.3 Aesthetic Evolutionary Artificial Life

The evolution of aesthetically pleasing AL was perhaps first performed by Dawkins, who hand-selected his artificial 'biomorphs' for reproduction in exactly the way artists select their forms using Evolutionary Art systems [13]. Ventrella has taken this one step further, and has evolved aesthetically pleasing animats which resemble animated stick-men. These are evolved for their ability to walk naturally in a virtual world, and also are guided by the aesthetic judgement of the user [41]. Alternatively, Lund [28] describes the use of neural networks which learn how to judge evolved pictures based on their aesthetics.

Although to date there have been few applications to benefit from this area of research, the use of computers to evolve amusing or attractive animated characters may well be lucrative in the computer games industry, or for television advertisements.

3.4 Aesthetic Evolutionary Design

The evolution of aesthetic designs is an obvious area of research, which should perhaps receive more attention than it does. Few designs are purely functional, most are chosen partly because of their aesthetics, and some functionally outstanding designs are discarded purely because of their ugly appearance. Furuta [18] describes an approach in which bridge designs can be optimised using GAs, based on their appearance, using 'psychovectors' to quantify the aesthetic factors of the structures. Husbands [25] describes the evolution of 3D solid objects resembling propellers, using a superquadric shape-description language and guided by 'the eye of the beholder'.

4. Epistasis, Constraints, and Other Problems

There are a number of common problems encountered when attempting to perform Evolutionary Design using computers. As is usual when applying evolutionary computation to anything, there are issues related to the fitness functions, such as noisy functions, discontinuous functions, multimodal functions, and multiobjective functions [3,5,24]. There are, however, some more specific problems that typically arise more often with Evolutionary Design [34].

4.1 Epistatic Representations

Epistasis means the 'degree of dependency' between multiple genes in a chromosome. A genetic representation with high epistasis may have many genes whose phenotypic effect relies to a large degree on the alleles of other genes. For example, a single shape-rule in a rule-based representation may have very different phenotypic effects, depending on which other shape-rules precede and succeed it. Conversely, a representation with low epistasis has few or no genes whose phenotypic effect relies on the alleles of other genes. For example, a simple voxel representation where every gene switches on or off a single voxel in a grid has zero epistasis.

Experiments investigating whether epistasis should be high or low have so far been inconclusive [36]. However, a simple thought experiment may shed light on this dilemma. Consider a (fictional) representation which uses, say, ten genes to represent the entire form of designs, and the phenotypic effect of every gene is *completely dependent* on all of the other genes through some embryology process. With this (maximum) amount of epistasis, these ten genes effectively become elements of a single overall gene (just as the separate binary digits in an ordinary gene are completely dependent on their neighbours). So our ten-gene representation with complete epistasis could be considered as a single-gene representation. Consider what effect varying any part of that gene would have on the phenotype. With every part of the design epistatically linked to every other part, any attempt to improve just one small area would result in changes to all of the rest of the design - making evolution to acceptable designs very difficult, if not impossible.

Alternatively, consider a representation with zero epistasis (e.g. a voxel representation using a 3D array). This requires no embryology, since every gene maps directly onto a specific area of the phenotype, and only that area of the phenotype. It should be clear that such a representation is well suited for evolution of small-scale detail, but evolution of large-scale characteristics becomes immensely difficult e.g., the scaling of the entire form in one dimension, or the duplication or mirroring of an existing feature in the design. (To duplicate or mirror an existing feature in the way segmentation or symmetry does, each new duplicate part would have to be re-evolved in entirety - highly unlikely to occur [8]).

Having examined the two extreme cases, it should be clear that both too much epistasis and too little epistasis in a representation is undesirable. Consequently, it seems that an ideal representation for Evolutionary Design should have a 'middling' amount of epistasis - just as nature seems to use.

4.2 Constraints

Many design problems are heavily constrained, and dealing with these constraints within an evolutionary algorithm can be difficult. If the creator of the Evolutionary Design system does not take care, constraints can limit the effectiveness of evolutionary search by blocking off useful areas of the search space. Alternatively, if soft constraints are specified, there can be no guarantee that all the constraints will be satisfied. There are eleven main ways in which a constraint can be implemented within an evolutionary system [42]:

C1: LEGAL SEARCHSPACE ***Design genotype representation.***
During the design of the evolutionary system, create a genotype representation that is only capable of representing legal solutions. Evolutionary search is then forced to consider only the space of legal solutions, where all constraints are satisfied. This method is frequently used, although designers who use it are often unaware that they are performing constraint handling of any kind. For example, in GAs, if the range of a problem parameter must be between 0 and 255, most designers would automatically use a binary gene consisting of eight bits - and this genotype representation would then ensure that the 0-255 range constraint was always satisfied.

C2: LEGAL SOLUTIONSPACE ***Design phenotype representation.***
During the design of the evolutionary system, create a new phenotype representation, so that only legal phenotypes can be defined. All genotypes are then mapped onto these phenotypes, which by definition, must always satisfy the constraints. Often great care can go into the design of suitable phenotypes. For example, practitioners of floor-planning problems have two important constraints: room-spaces should not overlap, and no space should be left unaccounted for. To ensure that the computer always evolves solutions that satisfy these constraints, designers of these systems tend to use phenotype representations which define the location of rooms indirectly, by defining the location and number of dividing walls [21].

C3: LEGAL SEED ***Seed with non-conflicting solutions.***
The initial population is seeded with solutions that do not conflict with the constraints and the crossover and mutation operators are designed so that they cannot generate illegal solutions. This method has been used in GP to ensure that only type-correct programs are considered during evolution [42]. The method ensures that all constraints are always satisfied, but may limit exploration by evolutionary search because of reduced initial diversity and limited genetic operations. (It may also be impractical or impossible to generate the initial population of individuals - particularly if the whole point of using evolutionary search was to find solutions that satisfy the constraints.)

C4: GENETIC REPAIR ***Correct illegal genotypes.***
If a new individual conflicts with a constraint, correct the genes that are responsible for the conflict to make it satisfy that constraint. For algorithms such as GP which make no explicit distinction between genotypes and phenotypes, this method modifies the solution, *and the modification is inherited by its offspring*.

This 'genetic engineering' approach ensures that all solutions will satisfy all constraints, but may damage epistatic genotypes, discarding the result of careful evolution over many generations. The design of the repair procedure may be a non-trivial task for some problems.

C5: LEGAL MAP *Correct illegal phenotypes.*

Map every genotype of an individual to a phenotype that satisfies the constraints using some form of simple embryology. This forces all solutions to satisfy all constraints, and also does not disrupt or constrain the genotypes in any way, allowing evolutionary search to continue unrestricted. For algorithms such as GP which make no distinction between genotypes and phenotypes, this method modifies the solution before fitness evaluation, *but the modification is not inherited by its offspring.*

Using a mapping stage to generate legal phenotypes is a very common approach to perform simple constraint handling. In his book, Goldberg describes perhaps the simplest: mapping the range of a gene to a specified interval. This permits constraints on parameter range and precision to be satisfied without the need to redesign the genotype representation and coding. More recently mapping stages have become more intricate and deserving of the term 'artificial embryology'. Researchers in GP have also reported that the use of an explicit genotype and mapping stage for constraint handling can increase diversity in populations [42].

C6: GENOTYPE PENALTY *Penalize illegal genotypes.*

Identify alleles or gene fragments within genotypes that seem to increase the chances of a solution conflicting the constraints, and reduce the fitness of any individual containing these fragments of genetic code. Although the identification of 'bad genes' may discourage solutions from conflicting constraints, it will not guarantee that all solutions satisfy all constraints. In addition, with epistatic genotypes, this approach may result in the discouragement of other, epistatically linked, useful features within solutions. To date, research has investigated the automatic identification of 'good genes' during evolution to encourage the evolution of solutions with higher fitnesses [21]. However, the author of this paper is unaware of any work which identifies 'bad genes' for constraint handling.

C7: PHENOTYPE PENALTY *Penalize illegal phenotypes.*

When a phenotype conflicts a constraint, reduce its fitness (ideally in direct proportion to the number of constraints conflicted, and the degree to which they are conflicted). This 'soft constraint' discourages all phenotypes that conflict the constraints, but does not force evolutionary search to generate legal solutions. In effect, the use of a penalty value becomes an additional criteria to be considered by the evolutionary algorithm, and multiobjective techniques should be used to ensure that all criteria are considered separately (otherwise one or more criteria may dominate the others) [5]. This is one of the most commonly used methods for constraint handling in evolutionary algorithms. (Indeed, it is the only one explicitly mentioned in Goldberg's book.)

C8: LEGAL SELECTION *Select legal parents only for reproduction.*

During reproduction, only select parent solutions which satisfy the constraints. This method should be used with a fitness-based replacement method to ensure that evolution is guided to evolve fit solutions in addition to legal solutions. However, the exclusion of potential parents discards potentially beneficial genetic material and so may be harmful to evolution. Only two recent investigations have been made on this method [42].

C9: LEGAL FERTILITY ***Increase the no. of offspring for legal parents.*** Having selected the parent genotypes (based on their fitnesses) this method allocates a larger *fertility* to parents which better satisfy the constraints. This method can be thought of as an implicit multiobjective method, allowing independent selection pressure to be exerted for high fitness and legal solutions. Being a 'soft constraint', there are no guarantees that all solutions will always satisfy the constraints. In addition, if legal parents are favoured excessively, it is possible that the diversity of the population could be reduced. To the author's knowledge, this idea has not been previously used for constraint handling.

C10: INFANTICIDE ***Stop illegal offspring from being born.***
If a new solution conflicts a constraint, discard it, and try generating another solution using the same parents. This brute-force method, which is sometimes used in GAs [42], forces all solutions to satisfy the constraints, but may discard useful genetic material. In addition, for heavily constrained problems, it may get stuck in an endless loop, generating and discarding illegal solutions forever, given two parents that can only have illegal offspring.

C11: ILLEGAL REPLACEMENT ***Replace illegal solutions with legal offspring.***
When replacing individuals with new offspring in the population, always replace the solutions that conflict constraints. (If all solutions satisfy the constraints, either replace randomly or replace the least fit.) This method should be used with a fitness-based selection method to ensure that evolution is guided to evolve fit solutions in addition to legal solutions. However, the replacement of potential parents discards potentially beneficial genetic material and so may be harmful to evolution.

Work is in progress to investigate and compare these different techniques, both separately and in combination [42].

4.3 Evolving Designs with Interdependent Elements

Opponents of Darwin's theory of natural section often give the example of the eye as a structure 'impossible to occur by evolution'. They state that the eye consists of many interdependent parts: the iris, the retina, the cornea, the lens, with each element relying on the correct functioning of all the other elements for the eye to work as a whole. Dawkins convincingly argues that there does exist a series of gradual evolutionary steps from *no eye* to *eye*, and that not only has the eye evolved, but it has evolved many times independently in different species [13]. However, although it is clear that such intricate designs have been evolved in nature, it is also clear that using evolutionary computation to generate designs with interdependent elements is a very difficult task.

For example, Bentley [6] describes (amongst other things) the evolution of a Penta-Prism. This design must bounce light twice using total internal reflection within its solid glass structure in order to reflect an image through 90 degrees whilst keeping the output image the right way up. Although it is a single component, it does have two interdependent elements: the two reflective parts. If either of these internally reflective sides are imprecisely oriented, or omitted, the design will not function correctly. Not only that, but the first reflection must direct the light in a direction almost opposite to the final, desired direction: so a design without the second reflective part is actually worse than a design with no reflective parts at all [6].

Evolution 'prefers' to use as few elements as possible, slowly building up the complexity of designs [14]. However, if care is not taken in the design of the representation and operators, evolution may commit itself too early to simple approximations of the desired design. In the example of the Penta-Prism, evolution normally began by evolving a simple right-angle prism (using a single reflection), which directed the light in the correct direction, but oriented the wrong way up. Having committed itself to this simple, but unsatisfactory type of solution, evolution was then unable to proceed to the more complex Penta-prism design. Perhaps because of limitations of the representation, the only way that evolution could be forced to abandon this unsatisfactory local optimum was by penalising all such designs with a fitness constraint [6].

Such examples of Evolutionary Design [6] illustrate that design problems requiring solutions with many interdependent elements can have large numbers of local optima - some of which do not resemble the functionally correct designs at all. Penalising all such unsatisfactory designs is not always a practical solution, so for design problems of this type, very careful consideration should be given to the creation of a representation and genetic operators that will always permit evolutionary paths from local optima to global optima.

4.4 Evolving Structure and Detail

As described in previous sections, some types of Evolutionary Design allow evolution to explore the *structure* (e.g. number of parameters/rules/primitive shapes) of designs in addition to the *detail* (e.g. parameter values) of designs. In most representations, varying the structure of designs has considerably more impact on the fitness of these designs, compared to varying the detail. Because of this, evolution typically converges on suitable design structures long before converging on design details. Although this is not always a disadvantage (e.g. the skeletal structure of all mammals seems to permit sufficient diversity, as does the cell structure of plants), in some instances evolution may prematurely converge onto an inappropriate structure. If this happens, the evolution of detail around this structure may be insufficient to allow the production of an acceptable design.

There are two potential solutions to this problem. First, an improved representation capable of allowing changes in structure without disastrous fitness changes would allow structure and detail to converge simultaneously. This could be accomplished if the evolution of detail could directly affect the degree to which changes in structure affected fitnesses. For example, in nature, a superfluous bone will be gradually reduced in size by evolution (a change in design detail), whilst at the same time improving other aspects of the organism because of that change. This reduction continues until at some point the bone becomes sufficiently small and redundant that evolution can remove the entire bone (structure mutation), without decreasing the fitness of the organism.

The second way to prevent premature convergence of structure is to reduce selection pressure once in a while, and permit less-fit designs to propagate. Dawkins [14] suggests that certain landmark mutations in structure such as the development of segmentation may have initially resulted in solutions with worse fitnesses. It seems likely that some of the more gross mutations are more likely to survive during 'good times' where food is plentiful, predators are scarce, and hence selection pressure low [14]. By reducing selection pressure in our Evolutionary Design systems in a similar way, we may well permit changes in structure to occur at later stages of evolution.

5. Conclusions

This paper has reviewed four highly related fields of research: *Evolutionary Design Optimisation*, *Evolutionary Art*, *Evolutionary Artificial Life-Forms* and *Creative Evolutionary Design*. In order to promote closer collaboration, it was proposed that the branch of research entitled *Evolutionary Design* should encompass these four distinct aspects. The different aims and techniques of each type of Evolutionary Design were then explored. For example, practitioners of Evolutionary Design Optimisation typically desire a single, global optimal design, in the minimum of time. In contrast, users of Evolutionary Art systems require a never-ending, non-converging, supply of novel forms to explore, and judge for themselves. Yet despite the differences between the approaches, researchers have now begun to combine these areas, forming four very new 'overlapping' types of Evolutionary Design: Integral Evolutionary Design, Artificial Life Based Evolutionary Design, Aesthetic Evolutionary AL and Aesthetic Evolutionary Design.

The paper then briefly described some common problems faced by creators of Evolutionary Design Systems: evolving designs with interdependent elements, evolving structure and detail, epistasis with representations, and constraint handling. Some solutions were discussed to these problems, but research is ongoing to discover conclusive answers.

Interest in Evolutionary Design is expanding from academia into industry, with the media now beginning to report its successes. Although Evolutionary Design is still in its infancy, it shows great promise in becoming an invaluable type of computer software tool for designers, artists and biologists.

ACKNOWLEDGEMENTS

My thanks to Erik Goodman, Karl Sims, and Andrew Rowbottom for their permission to reproduce images of their evolved designs. Thanks also to Tina Yu and Sanjeev Kumar for providing stimulating discussions and helpful criticism.

References

1. Bäck, T. 1996, *Evolutionary Algorithms in Theory and Practice.* Oxford University Press, New York.
2. Baron, P., Fisher, R. Mill, F. Sherlock, A. and Tuson, A. 1997. A Voxel-based Representation for the Evolutionary Shape Optimisation of a Simplified Beam: A Case-Study of a Problem-Centred Approach to Genetic Operator Design. *2nd On-line World Conference on Soft Computing in Engineering Design and Manufacturing (WSC2).*
3. Bentley, P. J. 1997. The Revolution of Evolution for Real-World Applications. *Emerging Technologies '97: Theory and Application of Evolutionary Computation,* 15th December, University College London.
4. Bentley, P. J. & Wakefield, J. P. 1997. Generic Evolutionary Design. Chawdhry, P.K., Roy, R., & Pant, R.K. (eds) *Soft Computing in Engineering Design and Manufacturing.* Springer Verlag, Part 6, 289-298.
5. Bentley, P. J. & Wakefield, J. P. 1997. Finding Acceptable Solutions in the Pareto-Optimal Range using Multiobjective Genetic Algorithms. Chawdhry, P.K., Roy, R., & Pant, R.K. (eds) *Soft Computing in Engineering Design and Manufacturing.* Springer Verlag London Limited, Part 5, 231-240.
6. Bentley, P. J. & Wakefield, J. P. 1997. Conceptual Evolutionary Design by Genetic Algorithms. *Engineering Design and Automation Journal* v3:2, John Wiley & Sons, Inc, 119-131.
7. Bentley, P J & Wakefield, J P 1996. Generic Representation of Solid Geometry for Genetic Search. *Microcomputers in Civil Engineering* 11:3, Blackwell Publishers, 153-161.
8. Bentley, P J & Wakefield, J P 1996. The Evolution of Solid Object Designs using Genetic Algorithms. In Rayward-Smith, V. (ed) *Modern Heuristic Search Methods.* Ch. 12, John Wiley & Sons Inc., 199-215.
9. Boden, M. A., 1992. *The Creative Mind : Myths & Mechanisms.* Basic Books Pub. ISBN: 0465014518.
10. Bonabeau, E., Theraulaz, G., Arpin, E., Sardet, E. 1994. The Building Behaviour of Lattice Swarms. Brooks, R., and Maes, P. (eds) *Artificial Live IV*, Proc. of the 4th Int. Workshop on the Synthesis and Simulation of Living Systems, MIT Press, 307-312.
11. Coates, P., 1997, Using Genetic Programming and L-Systems to explore 3D design worlds. *CAADFutures '97,* R. Junge (ed), Kluwer Academic Publishers, Munich.
12. Cliff, C.; Husbands, P.; Meyer, J.; Wilson, S.W. (eds) 1994. *From Animals to Animats 3.* Proceedings of the third international conference on simulation of adaptive behaviour, MIT Press.
13. Dawkins, R. 1986. The Blind Watchmaker. Longman Scientific & Technical Pub.
14. Dawkins, R. 1989. The Evolution of Evolvability. Artificial Life. The Proceedings of an Interdisciplinary Workshop on the Synthesis and Simulation of Living Systems, Vol. VI, Langton, C. G. (ed.), September, 1987, Los Alamos, New Mexico, Addison-Wesley Pub. Corp, pp.201-220.
15. Eby, D., Averill, R., Gelfand, B., Punch, W., Mathews, O., & Goodman, E. 1997. An Injection Island GA for Flywheel Design Optimization. In *5th European Congress on Intelligent Techniques and Soft Computing EUFIT '97*, vol 1, pp. 687-691.
16. Fogel, L. J., 1963. *Biotechnology: Concepts and Applications.* Englewood Cliffs, NJ: Prentice Hall.
17. Funes, P. & Pollack, J. 1997. Computer Evolution of Buildable Objects. *Brandeis University Computer Science Technical Report CS-97-191.*

18. Furuta, H., Maeda, K. & Watanabe, W. 1995. Apllication of Genetic Algorithm to Aesthetic Design of Bridge Structures. In *Microcomputers in Civil Engineering v10:6*. Blackwell Publishers, MA, USA, 415-421.
19. Gen, M. and Cheng, R. 1997. *Genetic Algorithms and Engineering Design*, John Wiley & Sons, USA.
20. Gero, J. S. 1996. Computers and Creative Design, In M.Tan and R.Teh (eds), *The Global Design Studio*, National University of Singapore, pp. 11-19.
21. Gero, J. S. & Kazakov, V. 1996. An exploration-based evolutionary model of generative design process. *Microcomputers In Civil Engineering* 11, 209-216.
22. Harvey, I. 1997. Cognition is not Computation: Evolution is not Optimisation. In *Artificial Neural Networks - ICANN97*, Proc. of 7th International Conference on Artificial Neural Networks, 7-10 October 1997, Lausanne, Switzerland, Gerstner, W., Germond, A., Hasler, M, and Nicoud, J-D (eds). Springer-Verlag LNCS 1327 pp. 685-690.
23. Harvey, I. and Thompson, A. 1997. Through the Labyrinth Evolution Finds a Way: A Slicon Ridge. *Proceedings of the 1st Int. Conf. on Evolveable Systems: from Biology to Hardware (ICES96)*, Higuchi, T. and Iwata, M. (eds), Springer Ver;ag LNCS 1259, pp. 406-422.
24. Holland, J. H., 1975, *Adaptation in Natural and Artificial Systems*. The University of Michigan Press, Ann Arbor.
25. Husbands, P., Jermy, G., McIlhagga, M., & Ives, R. 1996. Two Applications of Genetic Algorithms to Component Design. In *Selected Papers from AISB Workshop on Evolutionary Computing*. Fogarty, T. (ed.), Springer-Verlag, Lecture Notes in Computer Science, (pp. 50-61).
26. Keane, A. 1994. Experiences with optimizers in structural design. In Proceedings of the Conference on *Adaptive Computing in Engineering Design and Control '94*, ed. I. C. Parmee, Plymouth, pp. 14-27.
27. Lohn, J. and Reggia, J. 1995. Discovery of Self-Replicating Structures Using a Genetic Algorithm. *1995 IEEE Int. Conf. on Evolutionary Computation (ICEC '95)*, vol 1, Perth, Western Australia, pp. 678-683.
28. Lund, H., Pagliarini, L., Miglino, O., 1995, Artistic Design with GA and NN, *Proc. of the 1st Nordic Workshop on Genetic Algorithms and Their Applications (1NWGA)*, Uni. Vaasa, Finland, xiii+417 pp. 97-105.
29. Parmee, I. 1996. Towards an optimal Engineering Design Process using Appropriate Adaptive Search Strategies. *Journal of Engineering Design*, Vol 7:4, Carfax Pub.
30. Parmee, I. 1996. The Development of a Dual-Agent Strategy For Efficient Search Across Whole System Engineering Design Hierarchies. *4th Int. Conf. on Parallel Problem Solving From Nature*, Berlin, Germany, September 22-27.
31. Pham, D. T. & Yang, Y., 1993. A genetic algorithm based preliminary design system. *Journal of Automobile Engineers* v207:D2, 127-133.
32. Rechenberg, I., 1973. *Evolutionsstrategie: Optimierung Technisher Systeme nach Prinzipien der Biologischen Evolution.* Stuttgart: Fromman-Holzboog Verlag.
33. Rosenman, M. 1997. The Generation of form Using an Evolutionary Approach. Dasgupta, D. & Michalewicz (eds), *Evolutionary Algorithms in Engineering Applications*, Springer-Verlag Pub., 69-86.
34. Roston, G. 1997. Hazards in Genetic Design Methodologies. Dasgupta, D. & Michalewicz (eds), *Evolutionary Algorithms in Engineering Applications*, Springer-Verlag Pub., 135-154.

35. Schnier, T. & Gero, J. S. 1996. Learning genetic representatrions as an alternative to hand-coded shape grammars. In J. Gero & F. Sudweeks (eds), *Artificial Intelligence in Design '96*, Kluwer, Dordrecht, pp. 39-57.
36. Schoenauer, M. 1996. Shape Representations and Evolution Schemes. *Proc. of the 5th Annual Conf. on Evolutionary Programming*, MIT Press, Cambridge, MA, USA, viii+488pp., p.121-9.
37. Sims, K. 1991. Artificial Evolution for Computer Graphics. *Computer Graphics*, **25**, No.4, 319-328.
38. Sims, K. 1994. Evolving Virtual Creatures. In *Computer Graphics*, Annual Conference Series, (SIGGRAPH '94 Proceedings), July 1994 (pp. 15-22).
39. Sims, K. 1994. Evolving 3D Morphology and Behaviour by Competition. In *Artificial Life IV Proceedings*, Brooks, R. & Maes, P. (ed.s), MIT Press (pp. 28-39).
40. Todd, S. & Latham, W., 1992. *Evolutionary Art and Computers*. Academic Press.
41. Ventrella, J. 1994. Explorations in the Emergence of Morphology and Locomotion Behaviour in Animated Characters. Brooks, R., and Maes, P. (eds) *Artificial Live IV*, Proc. of the 4th Int. Workshop on the Synthesis and Simulation of Living Systems, MIT Press, 436-441.
42. Yu, T. and Bentley, P. 1998. Evolving Legal Phenotypes. Submitted to *Fifth Int. Conf. on Parallel Problem Solving From Nature*. Amsterdam, Sept 27-30, 1998.

Surface Optimisation within the CAD/CAM Environment using Genetic Algorithms

Riaz Mussa, Rajkumar Roy, and Graham Jared

The CIM Institute, School of Industrial and Manufacturing Science,
Cranfield University, Cranfield,
Bedfordshire, MK43 OAL, UK.
Tel: +44 (0)1234 754073, Fax: +44 (0)1234 750852
Email: R.Mussa@cranfield.ac.uk , R.Roy@cranfield.ac.uk,
G.Jared@cranfield.ac.uk

Keywords: CAD/CAM, Surface Optimisation, Genetic Algorithms, Shape Optimisation

Abstract

The design of surfaces often involves multiple criteria, which are governed by a number of geometry constraints. These constraints satisfy certain demands relating to curvature, and make the optimisation of surfaces quiet complex. A flexible approach to design optimisation is essential to handle the diverse criteria and constraints. The paper presents a framework for flexible surface optimisation within a CAD/CAM environment using the robust evolutionary computing technique of Genetic Algorithms (GAs). Particular needs for optimisation in a CAD/CAM environment are discussed with relation to current state of the art research and practice in surface optimisation. A simple GA is integrated in an existing CATIA CAD/CAM system, to provide a foundation for the flexible optimisation of freeform curves and surfaces. The paper presents some initial results on surface optimisation. Directions for future research are also identified.

1. Introduction

The competition in today's global markets is commanding rapid improvements in product performance and quality, with the reductions in costs and development time-scales. Forcing the product-oriented manufacturing enterprises to seek more advance technologies to gain profits by reducing the design development time, via the adoption of 'agile manufacturing strategies'. As a result, automobile and aerospace industries are introducing design automation in their organisations to improve the functionality of their designs and reduce the development time and thus reduce the cost. Both the industries are involved in designing complex surfaces. It is observed that 80% of the surface development time is spent on fine-tuning or optimising a surface. The CAD/CAM systems used for the designs provide limited facilities for automatic optimisation of surfaces. They support an iterative surface optimisation, which is very time consuming. The paper presents a framework to automate surface optimisation capability within an existing CAD/CAM system. A robust adaptive search technique, Genetic Algorithm (GA), is chosen for the optimisation. The optimisation algorithm is integrated with an existing CAD/CAM system, CATIA. The implementation is carried out using the APIs and other facilities available for the development of application programs integrated into CATIA environment. Case studies on ruled surface optimisation and free-form surface optimisation are also presented.

2. Definition of Problem

In practical industrial CAD/CAM applications, the surface definition has to be restrained by a number of geometry constraints. Often making the resulting surfaces extremely complex entities. The constraints help the surface to blend accurately with neighbouring surface definitions, to maintain product integrity and good aesthetics properties.

Automotive and aircraft design engineers conduct surface modelling in two distinct stages, definition of the primary core surface and final adjustment of the definition data. The time spent on this procedure can often be very long. In addition, the generated surface must have a certain shape (curvature) which the human eye considers to be smooth and fair, i.e. a true representation. This true representation of a surface is vital to provide acceptable quality of the final product, and is an important criterion used in automotive and aerospace industries.

The two main aspects behind this aesthetic criterion are the smoothness and fairness of the geometry as mentioned above. Smoothness implies that the curve/surface has no discontinuities in tangent or any huge variation in curvature, and fairness means that the undulation along the shape of the curve/surface is kept to a minimal. It may be relatively easy to grasp the notion of smoothness, but fairness requires a slightly more comprehension, as these two aspects do not occur in tandem. Fairness embodies more alterations of the surface geometry to comply with the criterion. This is illustrated in Figure 1, defining fairness with respect to two distinct curves.

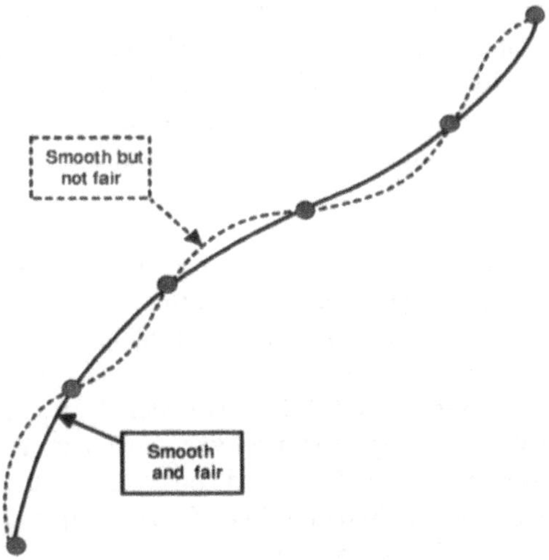

Figure 1: Smooth and fair curves.

This paper presents a flexible means of optimisation, to reduce the manual adjustment time for the latter stage of surface development. Flexible optimisation within a CAD/CAM environment enables the user to select appropriate techniques, and parameters for an optimisation [1]. This approach will help to handle different types of optimisation, such as multiobjective, multimodal and constrained within a CAD system, thereby allowing designers to define their criteria for surface evaluation within the CAD/CAM system. The research presented in the paper optimises the control curves of a surface for the surface optimisation. The system will then generate an optimised surface/curve solution for the designer, thus reducing the adjustment time. It is expected that the designer will perform limited final adjustments on the produced surface or curve.

3. Existing Surface and Shape Optimisation

CAD/CAM/CAE systems often obtain suitable designs through an iterative decision making process. The process is time consuming and involves lot of human intervention. This method of design improvement is dependent on the skill and experience of the designer, resulting in a need for a flexible optimisation approach [1] to be implemented, to limit the designer's intervention and produce improved designs.

Liden and Westberg [2] presented the FANGA (Formela ANGle Analysis) technique and analysis method. Developed by Saab-Scania Aircraft Division to optimise and asses the quality of surfaces used in their aircraft and car body design.

The fundamentals of which is a prolongation of Saab-Scania earlier development of FORMULA, given in Einar and Skappel [3]. This enables a surface to be refined by the response of a set of angles and directions which are easily computed from the surface model. These are then used in a standard optimisation technique to compute the necessary changes to the parameters of the surface to produce the required qualities.

Earlier research is reported by Kaufmann et al. [4], with regards to car body design. They use an algorithm to define reflection lines on splines and family of planes on the surface to represent surface irregularities. From which new reflection lines are produced and smoother spline and surface planes obtained. This method has been used in the CAD/CAM system Syrko at Daimler-Benz [4], with successful results in the improved quality of outer skin surfaces of car bodies. With the current increase in complexity and constraints on the shape of car body designs, coupled with the emphasis on time efficiency, this method is limited in simplicity and robustness that is required.

Watebe et al.[5] have introduced a methodology to generate a suitable shape automatically using Genetic Algorithms (GA), and by the application of Free-Form Deformation (FFD) technique [6]. Current related research is reported by Weinert et al. [7] at the University of Dortmund, Germany. Their research is aimed at generating optimal smooth surfaces from digitised point data using evolutionary algorithms. This includes in addition to GA, Evolutionary Programming (EP), Evolutionary Strategies and Simulated Annealing (SA). The work presents three solutions to the problem of reconstructing smooth surfaces using triangular tiles. It is observed that non-deterministic methods represent new ways of optimising triangulations for those surfaces.

Examples of other methods that improve quality of surfaces (smoothness and fairness) are those proposed by Higashi et al. [8] for the Toyota Technological Institute, and Szilvasi-Nagy [9]. A detailed list of procedures for the assessment and analysis of the surface quality and the detection of undesirable curvature are explained in Hoscheck [10] and Pottmann [11].

Ferguson et al. [12] and Andersson et al. [13] illustrate two mathematical means of smooth surface-control and constrained optimisation. These provide an complex mathematical approach to controlling the surface definition profile, which conforms to the design engineer's qualitative idea and true representation of the object under consideration. Thus, leading to an automatic mechanism for shape control, and creating convex surfaces with prescribed smoothness.

4. Genetic Algorithms

The basic principals of GAs were first laid down by Holland [14], where he presented them as an abstract of biological evolution. A simpler definition of the GAs and its parameters is found in Beasley, et al. [15] where GAs are described as adaptive methods which may be used to solve search and optimisation. Further from this definition, the basic principles and workings of a GAs are widely available and are extremely well explicated in texts by Goldberg [16], Davis [17] and Winston [18]. GAs are known as a robust optimisation or search technique. The research presented in this paper utilises GAs for the 'flexible optimisation' of surfaces. A simple GA [16] is used to prove the concept that the surface optimisation can be automated within an existing CAD/CAM system. In future the GA can easily be modified to handle multiobjective [19], multimodal [20] and constrained [21] optimisation.

A control curve of a surface is represented by several points on the curve. A binary representation (chromosome) is used to define the location of each point. The GA tries to define a combination of points that produce a smoother and fair curve. The size of the chromosome poses a limitation on the number of points that can be used to define a curve, and thus in some cases limits the accuracy of the curve. Each curve is evaluated using a curvature related objective function.

5. The Curve/Surface Evaluation

A simple method for gauging the undulations on a curve with regards to fairness has been derived, by employing the geometric characteristic of curvature, denoted as **R**. From this one can measure the undulation simply by calculating the curvature variation along the curve as shown in Figure 2. This illustrates the curvature variation plotting of the two curves shown formerly in Figure 1, where the undulation is represented by the dotted curve.

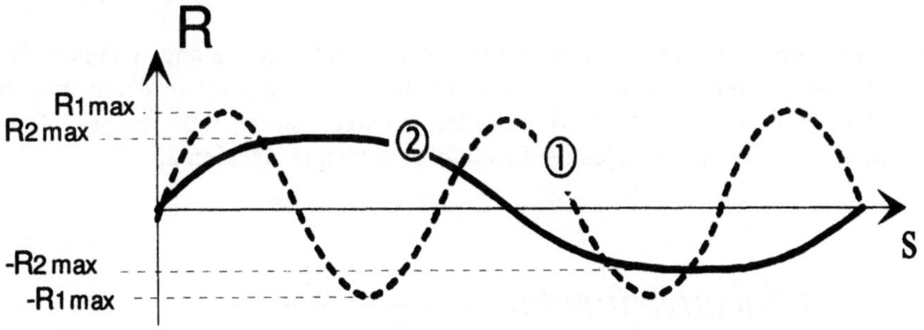

Figure 2: Curvature variation along two curves.

To simplify matters, an approximate measure of the curvature variation is determined by calculating the difference in curvature between consecutive evenly spaced points on the main curve, shown on the Figure 3. Where the calculation is implemented all along the curve's geometry, the quantity of variation is obtained by the sum of these differences.

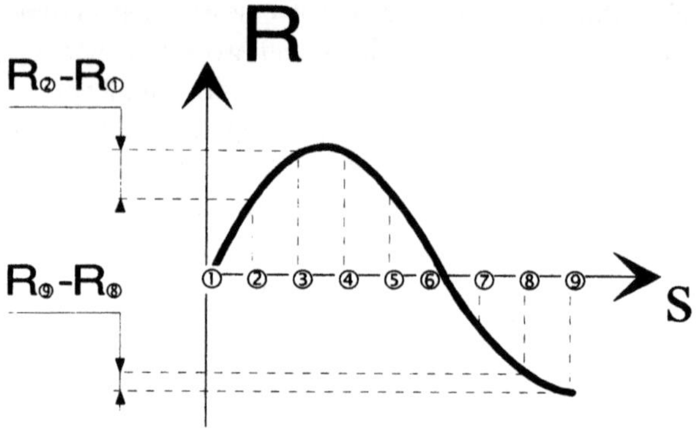

Figure 3: Curvature variation differences between points.

A flaw occurs when examining symmetric curves, where the absolute value of thc curvature difference must be obtained, otherwise the quantity of variation would be equal to zero. The resulting curvature variation measurement formula is given in Equation 1.

$$Curvature\ Variation = \sum_{i=1}^{i=npt-1} \left|\Re_{i+1} - \Re_{i}\right| \qquad \ldots(1)$$

npt : *number of points on the curve for which the curvature is calculated.*
R$_i$: *curvature on point i.*

The curvature variation equation measures the undulation along a curve. The purpose of the objective function is to minimise this amount of fluctuation to improve its fairness. The SGA performs a maximisation, thus inverse of the equation 1 is taken as the objective function, as shown in equation 2.

$$Objective\ function = \frac{1}{\sum_{i=1}^{i=npt-1} \left|\Re_{i+1} - \Re_{i}\right|} \qquad \ldots(2)$$

The GA procedure evaluates the quality of a curve and computes the curvature variation on different evenly spaced points, from which it applies Equation 2 in order to obtain the fitness score of the solution.

For the preliminary test, a spline curve is used for the optimisation as it embodies many advantages, like the ability to represent infinite slopes and ease of computation. Plus this type of curve is a commonly used free-form curve in the CAD environment.

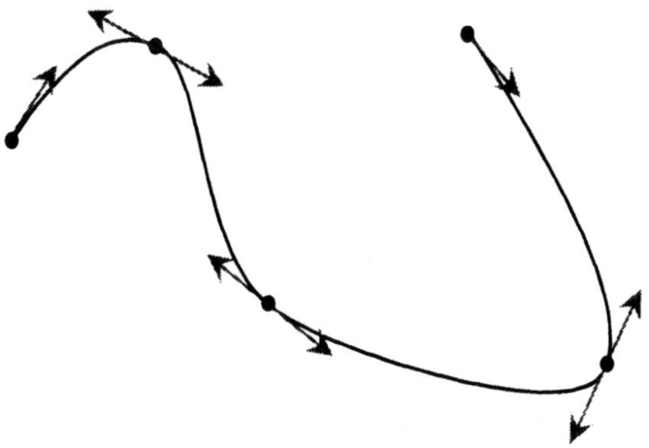

Figure 4: Spline curve with points and tangent controls.

The spline curve is controlled by two geometric elements, that are defined as variables for the GA to modify for the generation of optimum solutions via the fitness score obtained from the objective function. The first geometric element, are the control points in which the spline passes through, and the second element is the tangent directions on each of these points. Therefore each solution of the problem would be defined by a combination of a set of point positions, illustrated in Figure 4.

6. GA Integration into CATIA

There are several techniques to integrate applications within a CAD/CAM system, such as, using an *archive file* or via a *neutral file* format like STEP. The archive file method requires knowledge about the file format. The neutral file method does not allow enough interactions between the application and the CAD model, data flows in one direction only from CAD to application. Another limitation with this method is that if the application is to do any processing of provided geometric information, a considerable amount of code will have to be written within the application which duplicates coding contained within the CAD system. The limitations have contributed in adopting the approach reported here of interfacing directly to the CAD system through its Application Programming Interface (API). CATIA provides several API's to the applications programmer to use which are fairly modern in their approach. These API's hide much of the representation of the CATIA's internal data structures from the programmer with relation to its capacity and mechanism. Furthermore, the actual algorithms employed are also hidden to an extent, in that the API description is couched in terms of what a function does rather than how it is achieved. The concept is to provide access for programmers to

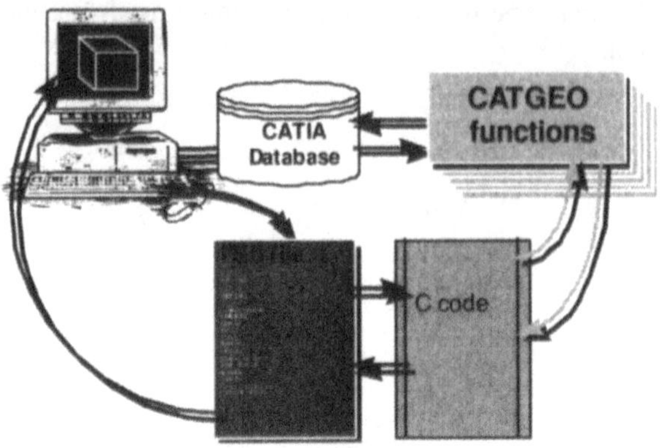

Figure 5: CATIA API structure.

integrate external applications into the system whilst maintaining the integrity of the CATIA interface. Thus, the programmer does not need to acquire too much knowledge of the internal workings of CATIA in order to access the functionality of the system exposed through various API's.

The API interface used here allows the development of code, which interacts directly with the CATIA user, employing the capabilities of the *Graphics Interactive Interface* (GII). That is controlled by a *Function Structure Description* (FSD) language, which supplies tools to customise the interaction between external programs and the end-user. Allowing the creation of codes that call external programs written in the C or FORTRAN language, depicted in Figure 5. Supplementary services provided by the CATIA's API is the CATGEO (*CATIA Geometry Interface*) and CATMSP (*CATIA Mathematical Subroutine*), that gives direct entry to the CAD model through basic mathematical and geometric functions. The CATGEO is a library of subroutines used to add, modify or read basic data in the CATIA data-structure, with access to all of the CATIA capabilities. The CATMSP library provides the capability to process mathematical and geometrical operations such as analyses and Boolean operations on geometric elements. The integration of GA within CATIA uses several CATGEO functions to access points on the curve, generate curve from points, etc.

7. Surface/Curve Optimisation within CATIA

The flexible GA optimisation approach is used to reduce the development time for surfaces. This technique is illustrated with two case studies, the optimisation of a ruled surface spline curve, and the optimisation of sectional control splines of a freeform surface. The control spline curves of both the surfaces are optimised individually and then the surfaces are regenerated with the optimised curves (Figure 6). The assumption made is by optimising the spline curve, the surface is also optimised.

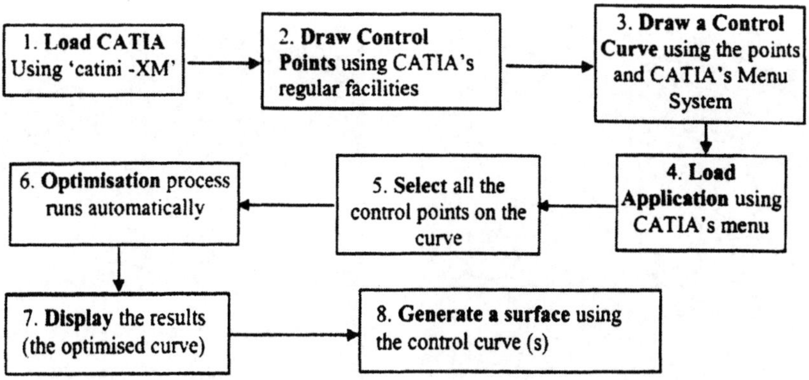

Figure 6: A step-by-step description of how the application is used.

To initiate the optimisation, a scenario is set where the designer using CATIA's regular facilities creates the control points, a curve is drawn connecting the points. Then the optimisation function is selected among the current CATIA utilities, prompting the designer to select the control points and accept the selection. The original curve is displayed with circles centred around the control points to visualise the displacement area. During the optimisation, the control points are allowed to vary within this displacement area. Each new point is defined by its polar co-ordinates (R, θ), where the origin is the centre of the displacement area. Thus a curve with 8 data points will be characterised by a model with 16 parameters: 8 values of R and 8 values of θ. The optimisation changes the co-ordinates of the control points, the new control point co-ordinates and the fitness value of the new curve are stored into a output text file. The optimised curve is generated from the information obtained from the GA and is displayed on CATIA, where it is used to re-create the surfaces. The optimised surface illustrates significant difference in appearance than the original surface, automating the necessary improvement in surface fairness. Due to time limitations, each spline control curve is optimised five times using the GA The best result from the five random runs are selected for the surface definition.

7. 1 Case study 1 - Ruled Surface Optimisation

To optimise a ruled surface, the spline control curve is optimised. The control curve is defined with eight control points. Thus, there are 8 Rs and 8 θs to be optimised. Each variable is represented with 8 bit binary strings. R is varied between 0 and 2 mm whereas, θ is varied between 0° and 360°. Five random GA runs are performed on the curve. The best set of points achieved is used to create the final optimised curve. An optimised ruled is generated using this curve. Figure 7, shows the control curves and surfaces, both original and optimised.

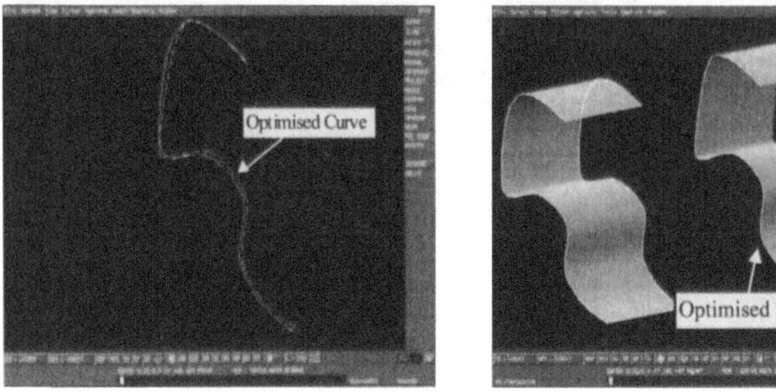

Figure 7: The original and optimised control curves along with the ruled surfaces.

7.2 Case study 2 - Free-Form Surface Optimisation

A set of three spline curves is defined using eight control points each. A free-form surface is generated using the three control curves. The surface is optimised by individually optimising the control curves. As before, each variable is represented by eight bit binary string. R is varied between 0 and 2 mm whereas, θ is varied between 0° and 360°. For each curve five random GA runs are carried out. The best curve out of the five is selected for the optimised surface generation. Figure 8 presents the spline curves and the free-form surfaces.

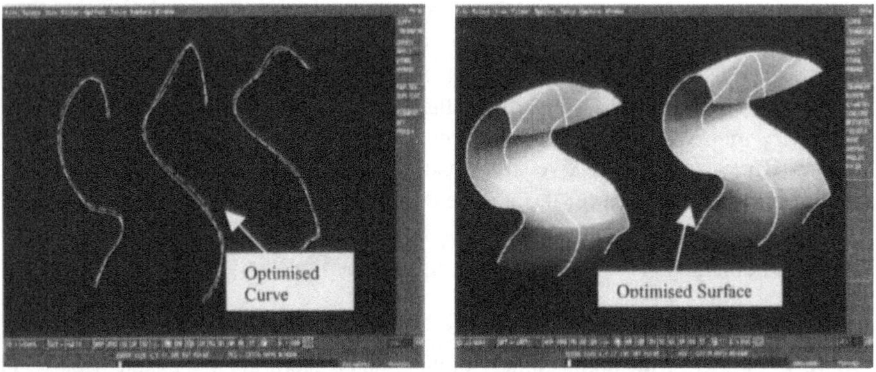

Figure 8: The original and the optimised control curves and the free-form surfaces.

8. DISCUSSION AND CONCLUSIONS

The framework presented in the paper uses only simple genetic algorithms as robust optimisation algorithms. It is observed in the case studies that the SGA has significantly improved the surfaces. It is important that a number of algorithms are integrated with the CAD/CAM environment, which will provide a choice of algorithms to the designer. The framework can be easily changed to handle multimodal and multiobjective optimisations. The current applications lack a graphical interface to update the optimisation parameters, such an interface would be very useful for the designers. The evaluation of control curves is simplistic, but representative. In future more comprehensive evaluation of the entire surface needs to be developed. The genetic algorithm is integrated within CATIA using its application programming interfaces (APIs). As a result, the designers do not require any prior knowledge about the internal representation of CATIA. The difficulty with this approach is that the application is API specific and thus not portable across different CAD/CAM systems. The use of APIs has helped to integrate the application within CATIA very nicely and the designers can now use the optimisation method within CATIA.

The surface optimisation within a CAD/CAM system is found to be very convenient for the designers. Designers can reduce their fine-tuning time by applying the 'flexible optimisation' approach. The framework is proved with two case studies, one on ruled surface optimisation and the other on a free-form surface optimisation. The optimisations take reasonably small time and are thus very useful for industrial applications. In future there will be many more automatic functionality's for the generation of high-quality surfaces in CAD/CAM systems. Integration of optimisation with Artificial Intelligence (AI) techniques such as Expert Systems (ES) will help in generating the required surface shape, by manipulating shape classifications and various surface-definitions. Making it feasible for even non-specialist CAD designers to develop good quality surface definitions. Further research is also necessary to implement the 'flexible optimisation' concept using an open interface, so that the application is not dependent on the API of the system. The surface optimisation research very much depends on the evaluation criterion used. A better surface evaluation technique is necessary that handles functional and geometric constraints. In case of complex surfaces with many more control curves, control points, constraints, and multiple criteria, an advanced GA would be essential.

9. References

1. Roy, R., Jared, G., Grau, J., and Buchannan, T., 1998, Adaptive Design Decision Support within CAD Environment, *Accepted for EDC'98 Conference.*

2. Liden, G., Westberg, S.K.E., 1993, Fairing of Surfaces with Optimization Techniques using FANGA curves as Quality Criterion. *Computer Aided Design,* Vol. 25, No. 7, pp.411-420.

3. Einar, H. and Skappel, E., 1973, FORMELA: A general design and production data system for sculptured products. *Computer Aided Design,* Vol. 5, No. 2, pp.68-76.

4. Kaufmann, E. and Klass, R., 1988, Smoothing surfaces using reflection lines for families of splines. *Computer Aided Design,* Vol. 20, No. 2, pp.73-78.

5. Watabe, H. and Okino, N., 1993, A Study on Genetic Shape Design'. In S.Forrest, ed., *Proceedings of the Fifth International Conference on Genetic Algorithms.* pp.445-450. Morgan Kaufmann.

6. Sederberg, T.W. and Parry, S.R., 1986, Free-Form Deformation of Solid Geometric Models. *SIGGRAPH '86 Vo.20 No.10.* pp.151-160.

7. Weinert, K., Mehnen, J. and Prestifilippo, 1997, Optimal Surface Reconstruction from Digitized Point Data using CI Methods. Technical Report No. CI-5/97. University of Dortmund, Germany.

8. Higashi, M., Kohji, K. and Hosaka, M., 1988, Generation of High-Quality Curve and Surface with Smoothly Varying Curvature. *EUROGRAPHICS '88 Conference,* pp.79-92.

9. Szilvasi-Nagy, M., 1997, Detecting regions with undesirable curvature. *Computer Aided Geometric Design,* Vol. 14, No. 8, pp. 699-706.

10. Hoscheck, J., 1984, Detecting regions with undesirable curvature. *Computer Aided Geometric Design,* Vol. 1, No. 2, pp. 183-193.

11. Pottmann, H., 1989, Visualizing Curvature Discontinuities of Free-Form Surfaces. *EUROGRAPHICS '89 Conference,* pp.529-536.

12. Freguson, D.R., Frank, P.D. and Jones, K.A., 1988, Surface shape control using constrained optimization on the B-spline representation. *Computer Aided Geometric Design,* Vol. 5, No. 2, pp. 87-103.

13. Andersson, E., Andersson, R., Boman, M., Elmroth, T., Dahlberg, B. and Johansson, B., 1988, Automatic construction of surface with prescribed shape. *Computer Aided Design,* Vol. 20, No. 6, pp. 317-324.

14. Holland, J.H., 1975, *Adaptation in Natural and Artificial Systems,* MIT Press.

15. Beasley, D., Bull, D.R. and Martin, R.R., 1993, An overview of genetic algorithms: Part 1 fundamentals. *University of Computing,* 15(2) pp. 58-69.

16. Goldberg, D.E., 1989, *Genetic Algorithms in search, Optimization and Machine Learning,* Addison-Wesley.

17. Davis, L., 1991, *Handbook of Genetic Algorithms*, Van Nostrand Reinhold.

18. Winston, H.P., 1992, *Artificial Intelligence.* 3rd Ed. pp. 505-531. Addison-Wesley.

19. Fonseca, C. and Fleming, P.J., 1995, An overview of evolutionary algorithms in multiobjective optimization, *Evolutionary Computation,* vol. 3, no. 1, The MIT Press, pp. 1-16.

20. Roy, R. and Parmee, I.C., 1997, An overview of evolutionary computing for multimodal function optimisation, *Soft Computing in Engineering Design and Manufacturing*, P.K. Chawdhry, R. Roy, and R.K. Pant (eds.), Springer, pp. 30-39

21. Michalewicz, Z., 1995, A survey of constraint handling techniques in evolutionary computation methods, *Proceedings of the Evolutionary Programming IV Conf.*, 1-3 March, CA, The MIT Press, pp. 135-155.

Part 3: Intelligent Control

Papers:

Adaptive Sugeno Fuzzy Control: A Case Study

J. Abonyi, L. Nagy, F. Szeifert

Department of Chemical Engineering Cybernetics,
University of Veszprém
Veszprém, POB. 158, H-8201 Hungary
Email: (abonyij,nagyl,szeifert)@kib.vein.hu

Keywords: Sugeno fuzzy control, direct learning, input-matching

Abstract

This paper proposes an inverse fuzzy-model-based controller. The gradient-descent algorithm can be used on-line to form adaptive fuzzy controllers. This ability allows the controller to be used in applications where the knowledge to control the process does not exist or the process is subject to changes in its dynamic characteristics. To demonstrate the feasibility of the method simulation and experimental control tests were evaluated in a non-linear level process, composed of mono-tank, which was submitted to reference change. The proposed adaptive Sugeno fuzzy logic controller is shown to be capable of compensating non-linear terms that affect the systems dynamics and providing better overall system performance than the optimal PI controller.

1. Introduction

The vast majority of conventional control techniques have been devised for linear-time invariant systems which are assumed to be completely known and well understood. In most practical instances, however, the systems to be controlled are nonlinear and the basic physical processes in it are not completely known *a priori*. These types of model uncertainties are extremely difficult to manage even with the conventional adaptive techniques.

The past few years have witnessed a rapid growth in the use of fuzzy logic controllers for the control of processes that are complex and badly defined. Most fuzzy controllers developed till now have been of the rule-based type [1], where the rules in the controller attempt to model the operator's response to particular process situations. An alternative approach uses fuzzy or inverse fuzzy model in process control [2-9] because it is often much easier to obtain information on how a process responds to particular inputs than to record how, and why, an operator responds to particular situations.

Learning ability is one of the essential attributes of an intelligent controller, which means that the controller can improve its future performance, based on experiential information it has gained in the past, through closed-loop interaction with the plant and its environment. Learning can be achieved by performance feedback and memory (or rule base) adjustment and can be used to compensate for a lack of a priori design information about plant and environment, improve the performance of the plant over a wide range of operating conditions, increase flexibility and reduce design costs.

Some learning control systems have been addressed using neural networks and fuzzy models in an inverse model compensation perspective [10-11].

There are two distinct architectures that have been formulated in the adaptive control field: direct and indirect schemes. A direct learning control scheme learns the control law based on the performance measure, whereas an indirect scheme produces a model of the plant and synthesises the control law using a predefined optimisation/inversion calculation [12].

The input matching technique adjusts the controller parameters directly according to the on-line identification of the inverse process model, which is subsequently used as a one-step-ahead controller [13].

This paper investigates the perspective of real-time application of input-matching technique based direct adaptive Sugeno fuzzy controller in a liquid level control problem.

The paper is organised as follows: The description and the architecture of the control system is in section 2. In section 3 the control problem is considered. The simulation and real-time experiments are presented in section 4. Section 5 offers some conclusions.

2. Description of the Control System

2.1. The Controller Architecture

In this paper, we present a simple iterative learning control scheme, which can be applied in a broad class of nonlinear systems. The tracking capability of the iterative learning process hinges upon the stability of the closed-loop system at each iteration. In order to ensure this, a proportional (P) feedback controller is used which keeps the control structure simple and limits the tracking error within a uniform bound.

The high-gain feedback technique offers a simple and useful approach to the nonlinear stability problem. However, the tracking error can not be made arbitrarily small unless the feedback gain is infinitely large. In practical applications the feedback gain can not be arbitrarily large due to actuator saturation, unmodelled dynamics, noise, etc. Our aim is to introduce a feed-forward fuzzy controller which reduces the control error at every iteration.

The fuzzy controller to be tuned is a feed-forward controller (Figure 2.1). Given the required set-point, it calculates the bias of the feedback P controller. The fuzzy and the P controller defined feedforward-feedback loop can be seen as a one-step-ahead controller which is identical with the inverse process model.

In the proposed control architecture the gradient-descent method is used to learn the inverse model of the plant by changing the parameters of the fuzzy part of the inverse model.

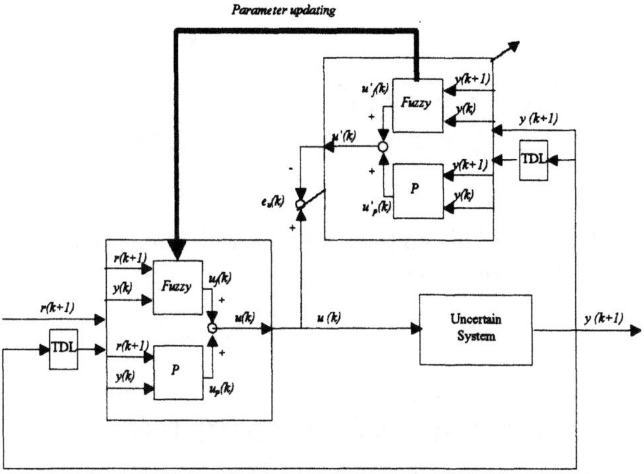

Figure 2.1 The controller architecture

2.2. The One-Step-Ahead Controller

This paper considers a class of nonlinear dynamic systems that can be described as follows:

$$y(k+1) = F\{y(k),\ldots, y(k-n+1), u(k),\ldots, u(k-m+1)\} \tag{1}$$

where $y(k)$ is the plant output at instance k, $F\{.\}$ is a nonlinear function, and $u(k)$ is a control signal. The constant n and m define the order of the plant. Without loss of generality in this study we deal with first-order systems (n, m= 1)

The inverse model of the first-order process is assumed in the following form:

$$L^{i,j}: \; IF \; y(k+1) \; is \; A_1^i \; and \; y(k) \; is \; A_2^j \; THEN \; u(k) = K\cdot\big(y(k+1)-y(k)\big)+b^{i,j} \tag{2}$$

where $L^{i,j}$ denotes the i,j-th implication, $i=1,2,\ldots,N_i$, $j=1,2,\ldots,N_j$, N_i and N_j are the number of the fuzzy sets on the i-th and j-th input domain, respectively; the symbols A_q^i and A_q^j are membership functions, $q=1,2$; $b^{i,j}$ are the rule consequent parameters (fuzzy singletons), and K is the gain of the proportional (P) controller.

This type of model can be used for one-step-ahead controller:

$$L^{i,j}: \text{ IF } r(k+1) \text{ is } A_1^i \text{ and } y(k) \text{ is } A_2^j \text{ THEN } u(k) = K\cdot\left(r(k+1)-y(k)\right)+b^{i,j} \quad (3)$$

where $r(k+1)$ is the set-point at the k-th discrete time step.

Using fuzzy inference, based on product-sum-gravity (T_p-norm) at given input $\left(r(k+1), y(k)\right)$, the final output of the fuzzy controller $u(k)$ is inferred by taking the weighted average of the $b^{i,j}$'s:

$$u(k) = u_p(k) + u_f(k) = K\left(r(k+1)-y(k)\right) + \frac{\sum_{i=1}^{N_i}\sum_{j=1}^{N_j} w(k)^{i,j} b^{i,j}}{\sum_{i=1}^{N_i}\sum_{j=1}^{N_j} w(k)^{i,j}} \quad (4)$$

where the weights, $w(k)^{i,j}>0$, implies the overall truth value of the i,j –th implication for input calculated as:

$$w(k)^{i,j} = A_1^i\left(r(k+1)\right)\times A_2^j\left(y(k)\right) \quad (5)$$

The proposed controller (inverse model) can be seen as a sum of a P and a Sugeno fuzzy controller or a specialised Takagi-Sugeno fuzzy model.

Output of the inverse fuzzy model can be calculated in a similar way:

$$u'(k) = u'_p(k) + u'_f(k) = K\left(y(k+1)-y(k)\right) + \frac{\sum_{i=1}^{N_i}\sum_{j=1}^{N_j} w'(k)^{i,j} b^{i,j}}{\sum_{i=1}^{N_i}\sum_{j=1}^{N_j} w'(k)^{i,j}} \quad (6)$$

where:

$$w'(k)^{i,j} = A_1^i\left(y(k+1)\right)\times A_2^j\left(y(k)\right) \quad (7)$$

In this paper, we employ triangular membership functions for each linguistic fuzzy value as illustrated in figure 2.2.

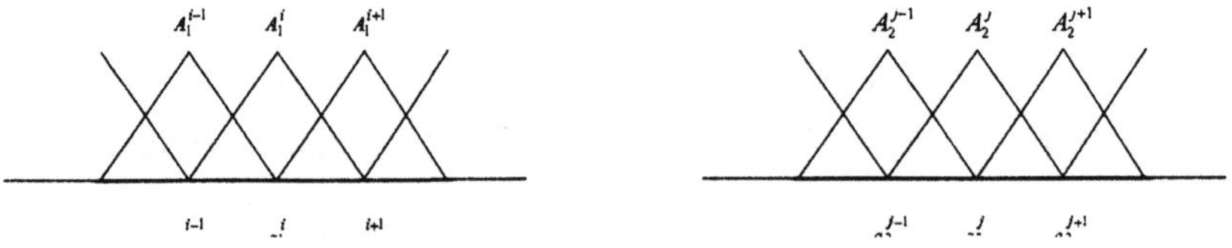

Figure 2.2 Membership functions

The cores of the adjacent fuzzy sets determine the support of a set thus ensuring that the fuzzy sets on a universe of discourse always form a fuzzy partition, as well as keeping the sum of the membership functions equal to 1. Constraints such as this are important in adaptive fuzzy control since they help to obtain an interpretable rule base [16], though they may cause loss of performance in control.

2.3. The Gradient-Descent Adaptation Method

The fuzzy part of the proposed inverse process model can be seen as an adaptive network [14].

For the identification of the proposed inverse fuzzy process model based on adaptive network we used gradient-descent method which several authors have applied to fuzzy systems [14-18]. What these methods have in common is that they minimise a similar objective function E:

$$\begin{aligned} E(k) &= \frac{1}{2}(e_u)^2 = \frac{1}{2}\left(\left(u_f(k)+u_p(k)\right)-\left(u'_f(k)+u'_p(k)\right)\right)^2 = \\ &= \frac{1}{2}\left(\left(u_f(k)-u'_f(k)\right)+K\left(r(k+1)-y(k+1)\right)\right)^2 \end{aligned} \tag{8}$$

The proposed architecture can be viewed as the combination of the input-matching technique and the specialised learning architecture, where the control error is propagated backwards through the plant.

The iterative algorithm seeks to decrease the value of the objective function. If $z_i(k)$ is the value of the i-th parameter at iteration, k, the steepest descent algorithm seeks to decrease the value of the objective function by modifying the parameters via

$$z_i(k+1) = z_i(k) - \alpha\frac{\partial E(k)}{\partial z_i(k)}, \qquad i = 1,\ldots,p \tag{9}$$

where p is the number of the parameters, α is a learning factor which controls how much the parameters are altered at each iteration.

In the proposed algorithm only the consequent parameters were adapted by:

$$\Delta b^{i,j} = \alpha \cdot w^{i,j} \cdot (u(k) - u'(k)) \tag{10}$$

3. The Control Problem

In this case study, the controller was configured to carry out liquid level control in a laboratory liquid level rig which is similar to that used by Graham and Newel [8-9], Posthlethwaite [4-6] and Linkens et al. [7]. The control problem is the level control of liquid (water) in a tank with manipulated inflow, when the outflow is dependent on the square root of the level in the tank. The process is highly non-linear with asymmetric and varying gain and time constant.

The schematic diagram of the level process is shown in figure 3.1. The experimental prototype of the level system consists of a 90cm x 15cm cylindrical tank. The water comes from the water pipeline into the tank through a pneumatic valve. An ultrasonic level transmitter attached to the top of the tank measures the level of the liquid. The plant is connected to a PC computer through a ADVANTECH LabCard PCL-812 data acquisition board. The GENIE 3.02 data acquisition and control software was used to filter and convert the input signals (0-5V). The control algorithm runs in MATLAB 4.0. The program gets the filtered and converted measured data through DDE in every 10 seconds according to [4].

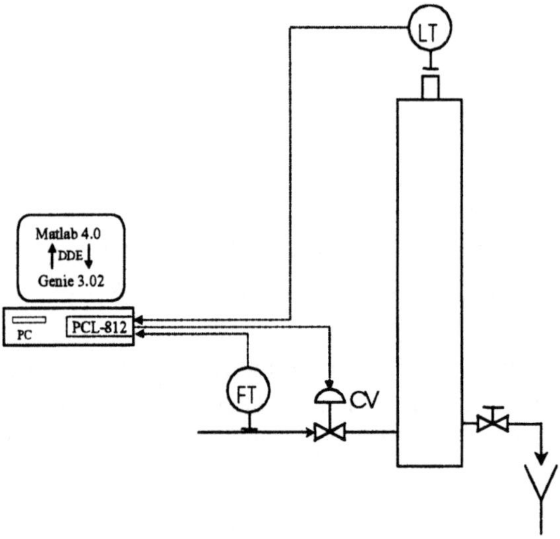

Figure 3.1 Illustration of the experimental control system

The model of the process can be characterised by the following differential equation:

$$A\frac{dy}{dt} = F - c\cdot\sqrt{y} \tag{11}$$

where A is the cross-sectional area in the tank, y (0-80 cm) is the liquid level in the tank, F is the inlet flow rate, c is a flow coefficient.

The parameters of the proposed process are in direct proportion with the benchmark problem presented in [4-9] ensuring similar dynamic properties. The representative values of the parameters are listed in table 3.1.

Table 3.1 The liquid level process parameter values

Parameter	*Physical system*	*Benchmark problem*
Inlet flow rate	0-140 cm^3/s	0-15 cm^3/s
Area	174 cm^2	10 cm^2
Flow coefficient	13.5	1

4. Simulation and Real-time experiments

4.1. Simulation Experiments

The simulator was built using MATLAB based on an *a priori* mathematical model (11) and the measured valve characteristic function, $K_v(u)$, representing the stationer valve characteristic, where u is the control signal restricted in the range [0-100].

In order to compare the control algorithms developed the following performance index was defined:

$$IAE = \sum_{k=1}^{N} |e(k)| \tag{12}$$

Where:

$e(k) = r(k) - y(k)$ the plan output error in the k-th discrete time step,

$N = \frac{t_{max}}{T_0}$ the maximal number of time steps, where T_0 is the sampling time (10s)

The performance of the learning (the input matching) was measured by:

$$IAEU = \sum_{k=1}^{N} |e_u(k)| \tag{13}$$

4.1.1. Conventional Solutions

Among the conventional solutions, the discrete PI controller has been examined formulated as:

$$u(k) = u(k-1) + q_0 \cdot e(k) + q_1 \cdot e(k-1) \tag{14}$$

where: $$q_0 = K \cdot \left(1 + \frac{T_0}{T_i} \right), \quad q_1 = -K \tag{15}$$

The parameters of the optimal PI controller (*K=6.33,T_i=152.91*) were determined by optimization with Sequential Quadratic Programming method with MATLAB *Optimization Toolbox's* **constr** function [19] (*IAE=1350.8*). The control task (set-point changes) is shown in figure 4.2.

4.1.2. Applying the Adaptive Sugeno Fuzzy Logic Controller

Since the system can be modelled approximately as a first-order system, the following Sugeno fuzzy controller structure was assumed:

$$L^{i,j}: \quad IF\ r(k+1)\ is\ A_1^i\ and\ y(k)\ is\ A_2^j\ THEN \quad u_f(k) = b^{i,j} \tag{16}$$

For a good control performance 10 antecedent fuzzy sets were used on each input universe. The initial consequent parameters of the fuzzy controller were set to zero and 100 according to the following algorithm:

$$\begin{aligned} &\text{if } a_1^i \geq a_2^j \quad \text{then } b^{i,j} = 0 \\ &\text{if } a_1^i < a_2^j \quad \text{then } b^{i,j} = 100 \end{aligned} \tag{17}$$

The adaptation factor, α, was set to 0.025. The feedback P controller parameter was equal to the gain of the optimal PI controller, $K = 6.33$. This parameters were experimentally found to yield good convergence and control performance. The consequent, $b^{i,j}$, parameters were restricted to the range 0-100. In order to compare the control algorithms developed the same set-point trajectory was introduced as in the case of the PI controller.

Figure 4.1 shows the effect of the gradient-descent adaptation to the controller output error and the performance of the control system.

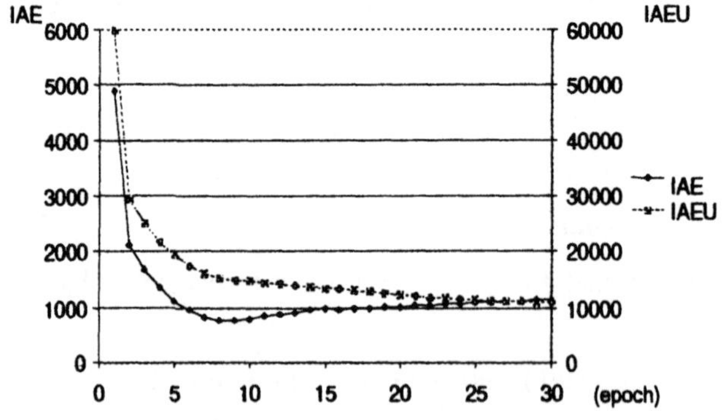

Figure 4.1 Change of IAE and IAEU in 30 simulation experiments

Note that the optimal inverse model does not ensure optimal control performance. The reason of this phenomenon is that the structure of the discrete time model is not exactly describes the structure of the real object.

4.2. Real-time Experiments

4.2.1. PI Control

The experiments on the laboratory liquid level rig have been constructed according to the simulation results. Therefore, the gain and the time constant of the PI controller were set based on the data presented in chapter 4.1.1. By means of these parameters we got a higher performance index (*IAE=1917.8*) than with the simulation results *(IAE=1350.8)*.

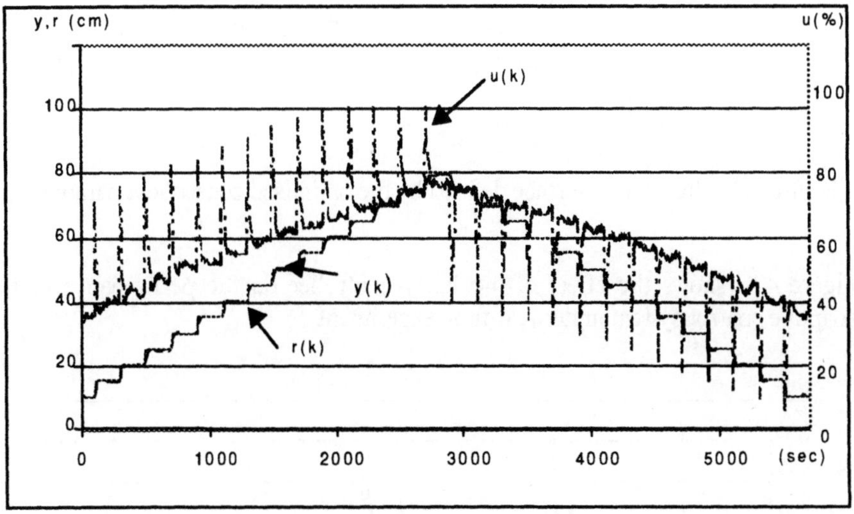

Figure 4.2 The control task (set-point changes) and the performance of the PI controller

4.2.2. Fuzzy Control

As we can see from the figure 4.1, learning the fuzzy controller's parameters is an extremely time-consuming task. Therefore, the initial parameters of the fuzzy controller were not initialized in the way presented in chapter 4.1.2., but the parameters obtained from the simulation experiments were used (Figure 4.3). These parameters have resulted a performance index of *IAE=5684.1,* which was decreased to *IAE=1407.5* after the 5th run.

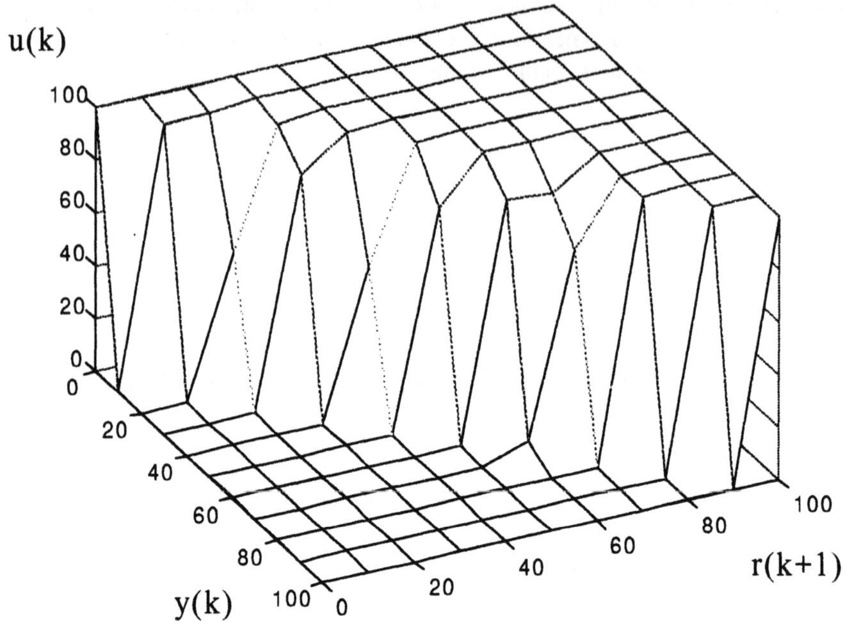

Figure 4.3. The control surface defined by the optimised consequent parameters

Figure 4.4. shows the effect of the fuzzy controller on the performance of the adaptive control system under real-time experiment.

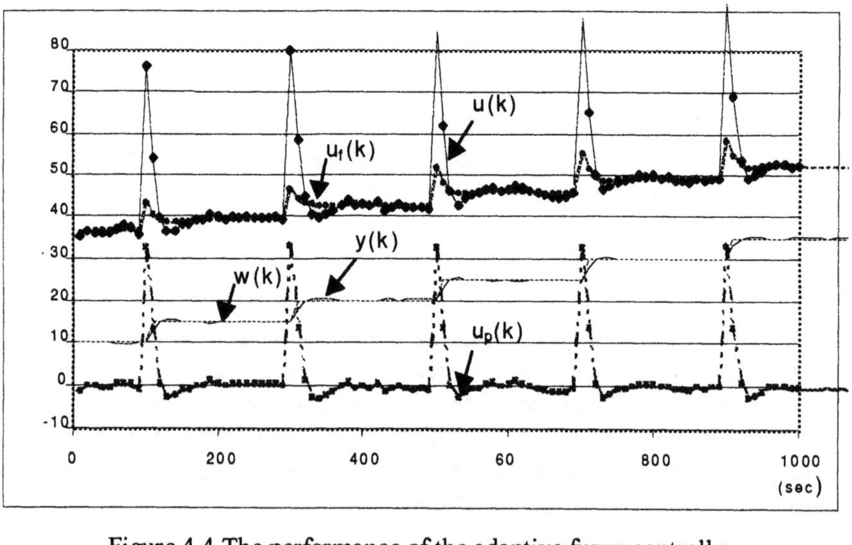

Figure 4.4 The performance of the adaptive fuzzy controller

The result of the simulation and real-time experiments are compared in table 4.2. As can be seen the adaptive fuzzy controller performed better then PI control during both simulation and real-time experiments.

Table 4.2 Comparison of the performance indexes

	Simulation	*Real-time*
PI	1350.8	1917.8
Adaptive Sugeno Fuzzy	751	1407.5

5. Conclusions

This paper presents the capabilities of one-step-ahead adaptive Sugeno fuzzy control. The adaptation algorithm based on the gradient-descent method compensates uncertainty and changes in process dynamics. The proposed adaptive control algorithm was applied to the fuzzy logic control of a laboratory scale liquid level rig. Both in simulation and real-time experiments we have achieved better performance index characterising the operation of the control system using the fuzzy controller than in case of using the conventional optimal PI controller. This type of application can be a proper tool where the already existing linear (PI) control algorithm is to be replaced by a better, more efficient nonlinear adaptive control system.

References

1. Driankov D, Hellendom H 1993 *An Introduction to Fuzzy Control.* Springer-Verlag, London
2. Babuska R, Sousa J, Verbruggen H B 1995 Model-Based Design of Fuzzy Control Systems. *Proceedings of the EUFIT'95* 837-841
3. Jang R R, Sun, C-T 1995 Neuro-Fuzzy Modeling and Control. *The Proceedings of the IEEE May* 83:378-406.
4. Posthlethwaite B 1994 A Model-based Fuzzy Controller. *Trans IChemE* 72(A1):38-46
5. Posthlethwaite B 1997 Building a Model-Based Fuzzy Controller, *Fuzzy Sets and Systems* 79:3-13
6. Posthlethwaite B, Brown M, Sing C H 1997 A New Identification Algorithm for Fuzzy Relational Models and its Applications in Model-based Control. *Trans IChemE* 75(A1):453-458
7. Linkens D A, Kandiah S 1996 Long-Range Predictive control Using Fuzzy Process Models. *Trans IChemE* 74(A1):77-88

8. Graham P, Newel R B 1989 Fuzzy Adaptive Control of a First Order Process. *Fuzzy Sets and Systems* 31:47-65
9. Graham P, Newel R B 1988 Fuzzy Identification and Control of a Liquid Level Rig. *Fuzzy Sets and Systems* 26:255-273
10. Branco C, Dente A 1996 Learning From Examples to Compensate Nonlinearities in Electro Mechanical Drivers: a Fuzzy Logic Approach, The 1st Online Workshop on Soft Computing, http://www.bioele.nuee.nagoya-u.ac.jp/wsc1.
11. Glorennec P Y 1991 Adaptive Fuzzy Control. *Proceedings of the IFSA'91* 33-36.
12. Åström K, Wittenmark B 1989 *Adaptive Control.* Addison-Wesley Publishing Company, New York.
13. Johnson C R, E Tse 1978 Adaptive Implementation of One-Step-Ahead Optimal Control via Input Matching. *IEEE Trans Automatic Control* AC-23:865-872
14. Jang J-S R 1993 ANFIS: Adaptive-Network-based Fuzzy Inference Systems. *IEEE Transactions on Systems, Man and Cybernetics* 23(3):665-685
15. Nomura H, Hayashi I, Wakami N A 1991 Self-Tuning Method of Fuzzy Control by Descent Method. *Proceedings of the IFSA'91* 155-158
16. Guely F, Siarry P 1993 Gradient Descent Method for Optimising Various Fuzzy Rule Bases. *Proc. IEEE-Fuz-93* 1241-1246
17. Jager R R 1995 *Fuzzy Logic in Control.* Ph.D. Thesis, Delft Universitry of Technology, Delft, Netherlands
18. Andersen H C, Lofti A, Tsoi A C 1997 A New Approach to Adaptive Fuzzy Control: The Controller Output Error Method. *IEEE Trans. on Systems, Man, and Cybernetics* 27-B(4)
19. Grace A 1992 *Optimization Toolbox User's Guide,* The Math Works Inc. Massachusetts

An Experimental and Comparative Study of Fuzzy PID Controller Structures

Leandro dos Santos Coelho and Antonio Augusto Rodrigues Coelho

Federal University of Santa Catarina - Department of Automation and Systems
C.P. 476 - 88040.900 - Florianópolis - SC - BRAZIL
Tel: 55 48 3319202 - Fax: 55 48 3319770
E-mail: {lscoelho, aarc}@lcmi.ufsc.br

Abstract

Structures and design issues of fuzzy *PID* (proportional-integral-derivative) controllers (*FLC-PID*'s) are presented and evaluated in this paper. Configuration and basic characteristic of several structures of *FLC-PID* based on models proposed in the literature — *(PD + I)*, (*PI + D* conventional), incremental (*PD + I*), (*PD + PI*) — are here reviewed and implemented. *FLC-PID*'s are assessed on a horizontal balance process, consisting of two propellers driven by two *DC* motors. Such process offers control complexities and can become unstable by using classical controllers. Experimental results, robustness and performance of *FLC-PID*'s are illustrated and discussed.

Keywords

Fuzzy logic control, Fuzzy *PID* Control, Experimental process, Control applications

1 Introduction

PID controllers are widely used in industrial processes due to their dynamic aspects and simplicity of use when there are soft uncertainties in plants. Linear conventional *PID* algorithm, however, may present difficulties in dealing with complex dynamic processes, such as large time-delay and nonlinear characteristics. Thus, many advanced control algorithms have been used and implemented in order to improve the efficiency of the *PID* control under difficult operational conditions in practical activities [1], [2].

Fuzzy logic control implementation has been widely studied in the process control community. Theoretical foundations of the fuzzy sets — proposed by L. A. Zadeh [3] — motivated, initially, the validation of fuzzy logic controllers (*FLC*'s) by Mamdani [4] and, later, spured the emergent commercial development of software and hardware tools based on *FLC* technology, to be used in many fields of knowledge, mainly in Japan [5], [6], [7], [8].

The analysis, discussion of the structures and control design issues of the *FLC-PID*'s are presented in this paper. The different control configurations are: i) *(PD + I)*; ii) (*PI* + *D* conventional); iii) incremental (*PD* + *I*); and iv) (*PD* + *PI*). The control algorithm synthesis are based on the studies of Kwok et al. [9], Malki et al. [10], Ng & Li [11], Li & Gatland [12], and Qin [13]. The approaches by Kwok et al. [9] and Li & Gatland [12] investigated and discussed several *FLC-PID* design structures applied in first- and second-order simulated process. Malki et al. [10] presented the design of a *FLC* type (*PD* + *I*) applied to a robotic manipulator. Li & Ng [11] dealt with the design and the optimisation by genetic algorithms of *FLC-PID* and *FLC-(PI + D)* control applied to a level regulating system composed by two coupled tanks, while Qin [13] presented the design of *FLC-(PI + D* conventional) applied to a simulated process.

In this paper, several *FLC-PID*'s configurations are tested on a horizontal balance process. The complexities present by the process, such as, double integral, instability, nonminimum phase, nonlinearities, and the presence of load disturbance acting on the moment of inertia, call for the evaluation of *PID* control conceptions that utilize the fuzzy logic in a practical environment. Experimental results for setpoint changes, load disturbance, robustness and performance analysis of different configurations are presented. The paper is organized as follows: in section 2, notions of *FLC*'s are shown. Section 3 presents and describes the *FLC-PID*'s issues and design conceptions. In section 4 the experimental process, denominated horizontal balance, is described. *FLC-PID* control experiments of the horizontal balance process are evaluated in section 5. The conclusion and perspectives are shown in section 6.

2 Fuzzy Control Algorithms

Human operators are able to control complex processes by using imprecise information or by being based on operation of such processes. The control strategy adopted by human operators is itself of imprecise nature and, can often be expressed in linguistic terms. The fuzzy sets theory and concepts of the fuzzy logic can be used to translate into mathematical terms the imprecise information expressed by a set of linguistic rules [4], [8].

The *FLC* design philosophy is characterized by the treatment of the knowledge that is usually derived from the acquisition procedure of the knowledge or are automatically synthesized by self-tuning control algorithms. These controllers are typically defined by the nonlinear mapping of system state space for the control space. However, it is possible to identify the result of a *FLC* as a control surface reflecting the operator's (or control engineer's) *prior* knowledge of the process. The control surface is coded into a knowledge base using a compiler to execute rule-base, termsets, and scale factors.

FLC's design presents a data flow that is dealt with a fuzzyfication phase, analysis and execution of rules and, a defuzzyfication phase. The fuzzyfication phase converts numeric variables (real numbers) into linguistic variables (fuzzy numbers), in other words, transforms the crisp data in the corresponding fuzzy set. The analysis and execution of rules (expressed by a set of production rules of the type IF <condition> THEN <conclusion>) are responsible for the evaluation of the rule-base. When a rule is activated, a calculation procedure is realized based on the previous values, before the output rule is attributed. The defuzzyfication phase is the reverse of the fuzzyfication phase, and converts linguistic variable values of the *FLC* output into crisp outputs. The defuzzyfication phase describes the mapping of a fuzzy control action space in non-fuzzy control actions.

3 Fuzzy *PID* Controllers

Several *FLC* configurations have been evaluated and implemented in the process control literature, where the usual *FLC* structure is the *FLC-PI* or the *FLC-PD*. This section presents the structures and design principles of some *FLC-PID*'s, as alternative to the *FLC-PID*'s complete design, which require three inputs, that will substantially expand the rule-base and make the design more difficulty [11], [12].

The emphasis of this paper is to review the design of four *FLC-PID* controllers and to assess the performance on a practical experiment, where a detailed analysis of the theoretical background of *FLC-PID*'s is available in Qiao & Mizomoto [14] and Misir et al. [15]. Next, design characteristics of the *FLC-PID*'s in controlling a horizontal balance process is presented. For the linguistic variables, the attributed values were POsitive (*PO*), NEgative (*NE*), and Zero (*ZE*).

3.1 *FLC* Design — *(PD + I)*

The *FLC-(PD + I)* design was proposed by Ng & Li [11]. *FLC-(PD + I)* is governed by the following control law:

$$u_{PID}(k) = u_{PD}(k) + \phi T_s u_I(k) \tag{1}$$

$$u_{PD}(k) = K_P e(k) + K_D \Delta e(k) \tag{2}$$

$$u_I(k) = u_I(k-1) + u_P(k) \tag{3}$$

$$u_P(k) = K_P e(k) \tag{4}$$

where *e(k)* is the error between setpoint (*sp*) and output, *Δe(k)* is the error variation. The control action, $u_{PID}(k)$, of the *FLC-(PD* + conventional *I)* is constituted by the proportional action plus derivative of the error, $u_{PD}(k)$, and integral action of the error, $u_I(k)$, is obtained from the proportional control of the error, $u_P(k)$. The proportional and derivative gains are attributed by K_p and K_D, respectively; T_s is the sampling period, and ϕ acts as a structural selection parameter, tuning (on or off) the integral action according to the plant nature.

The basic characteristic of the *FLC-(PD + I)* design is the necessity of the configuration of a rule-base of *2-D*, for the fuzzy *PD* control, and a rule-base of *1-D* for fuzzy *I* control. The fuzzy *I* control uses proportional incremental control action of the error, *e(k)*, and it is another relevant characteristic of this design. The basic control diagram and membership functions of *FLC-(PD + I)* are shown in Figures 3.1 and 3.2, respectively.

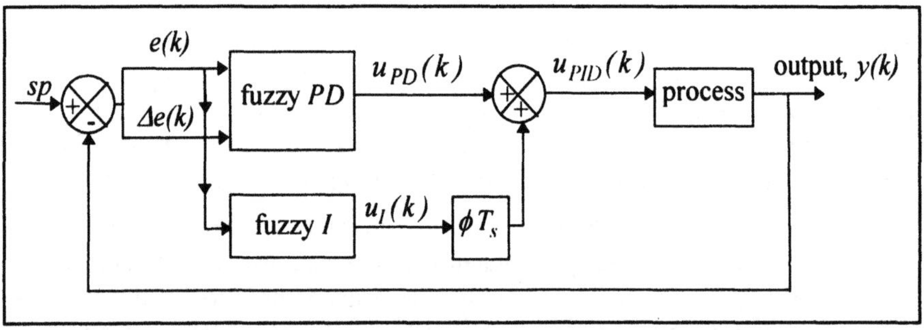

Figure 3.1. Control diagram of the *FLC-(PD + I)*

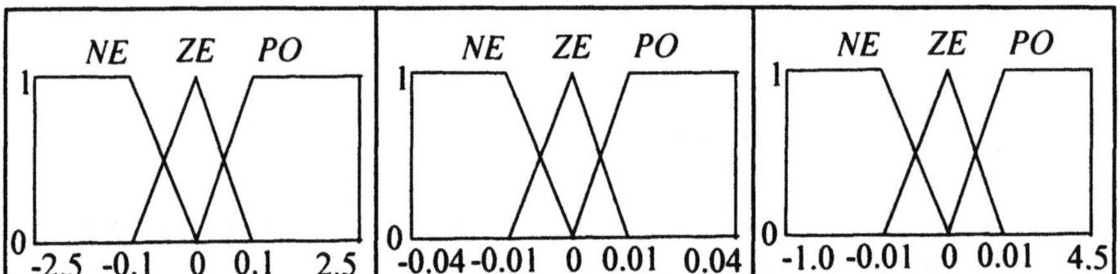

(a) error signal (b) change of the error signal (c) control signal

Figure 3.2. Membership functions of the *FLC-(PD + I)*

3.2 *FLC* Design — *(PI* + Conventional *D)*

The *FLC-(PI* + conventional *D)* design, as proposed by Li & Gatland [12] and Qin [13], consists of a *FLC-PI* plus the derivative control action of the process output, $u_D(k)$, where:

$$u_{PID}(k) = u_{PI}(k) + u_D(k) \tag{5}$$

and the derivative control action has the following transfer function:

$$\frac{Y_D(s)}{Y(s)} = \frac{T_D s + 1}{T_D s / K_D + 1} \tag{6}$$

where T_D and K_D are the derivative time and the derivative gain constants, respectively. The principle of this design is that the fuzzy *PID* control has the benefit of implementing the derivative control on the output, avoiding derivative kicks for step setpoint changes. The basic control diagram and membership functions of *FLC-(PI* + conventional *D)* are shown in Figures 3.3 and 3.4, respectively.

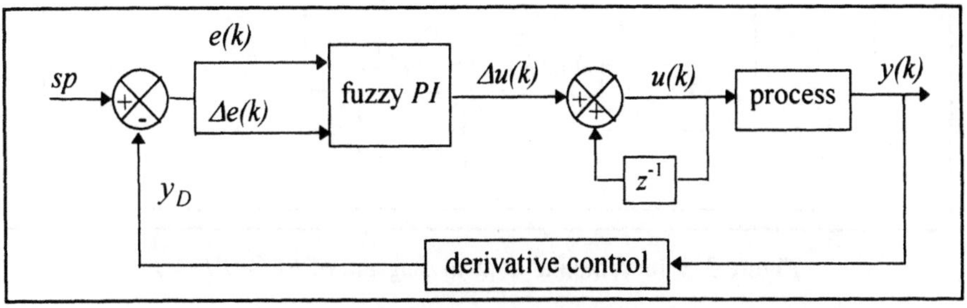

Figure 3.3. Control diagram of *FLC-(PI* + conventional *D)*

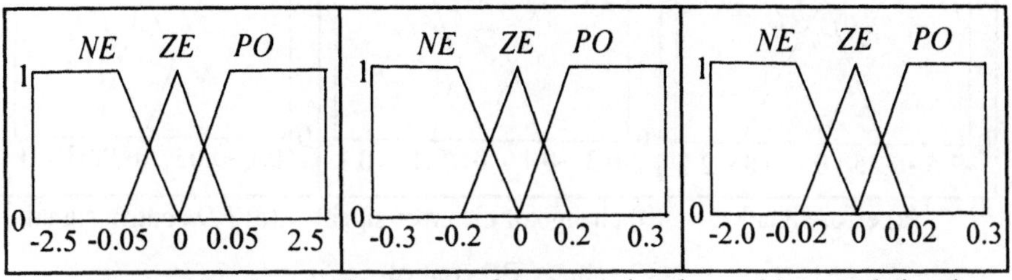

(a) error signal (b) change of the error signal (c) control signal

Figure 3.4. Membership functions of the *FLC-(PI* + conventional *D)*

3.3 *FLC* Design — Incremental (*PD* + *I*)

The incremental *FLC-(PD + I)* design was proposed by Malki et al. [10] and is given by the following control law:

$$u_{PID}(k) = u_{PD}(k) + u_I(k) \tag{7}$$

$$u_{PD}(k) = -u_{PD}(k-1) + Ku_{PD}.\Delta u_{PD}(k) \quad (8)$$

$$u_I(k) = u_I(k-1) + Ku_I.\Delta u_I(k) \quad (9)$$

where Ku_{PD} and Ku_I are the proportional-derivative and integral gains of incremental controls, respectively. The basic control diagram and membership functions of *FLC-(PD + I)* are shown in Figures 3.5 and 3.6, respectively.

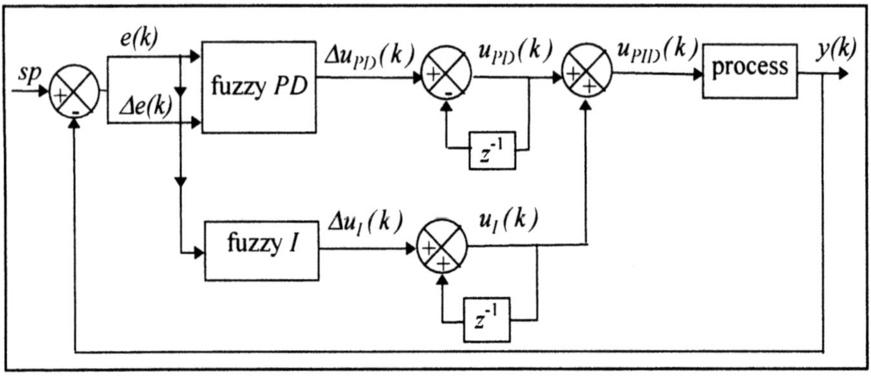

Figure 3.5. Incremental control diagram of *FLC-(PD + I)*

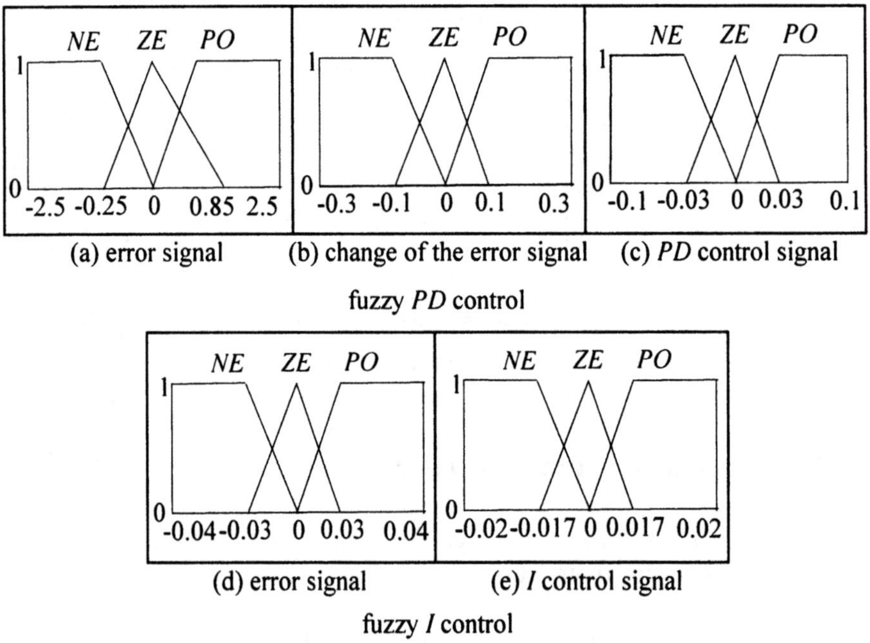

Figure 3.6. Membership functions of the incremental *FLC-(PD + I)*

3.4 *FLC* Design — *(PD + PI)*

The *FLC-(PD + PI)* design consists of a *FLC-PD* in parallel with a *FLC-PI*, being this configuration presented by Kwok [9] and Li & Gatland [14] and the control law is derived by:

$$u_{PID}(k) = u_{PD}(k) + u_{PI}(k) \tag{10}$$

The *FLC-(PD + PI)* design presents a rule-base for the fuzzy *PD* control, another for the fuzzy *PI* control and the implemented design uses the composition of the control actions of fuzzy *PD* and *PI* controls The basic control diagram and membership functions of *FLC-(PD + PI)* are shown in Figures 3.7 and 3.8, respectively.

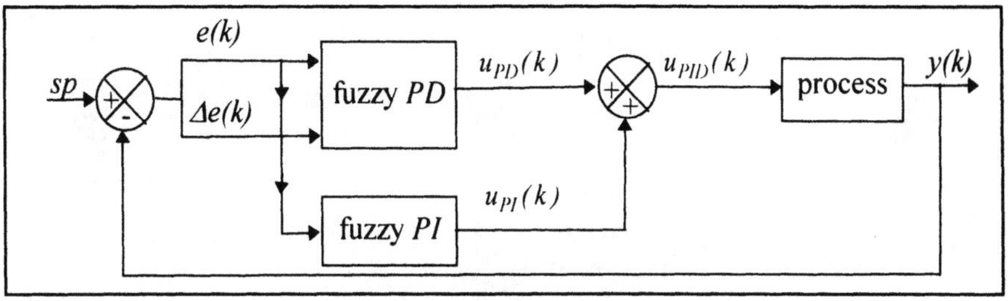

Figure 3.7. Control diagram of *FLC-(PD + PI)*

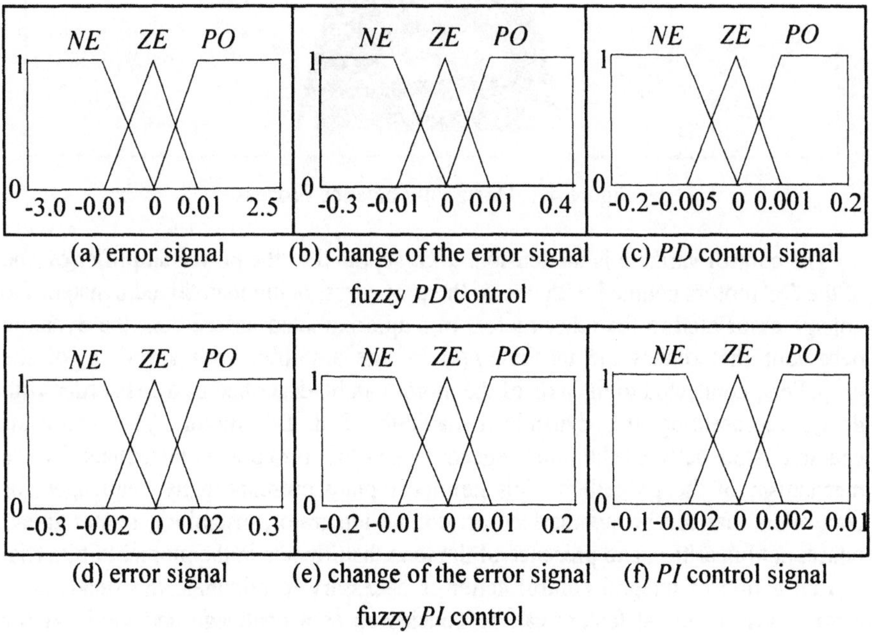

Figure 3.8. Membership functions of the *FLC-(PD + PI)*

4 Description of the Horizontal Balance Process

The process denominated horizontal balance is a practical control system for lab classes and research activities and it was developed in the Department of Automation and Systems at Federal University of Santa Catarina (*DAS/UFSC*). Further information of the process is available in *http://lcp.lcmi.ufsc.br*. The horizontal balance process consists of a beam that horizontally oscillates and it is positioned by two propellers driven by *DC* motors (Figure 4.1).

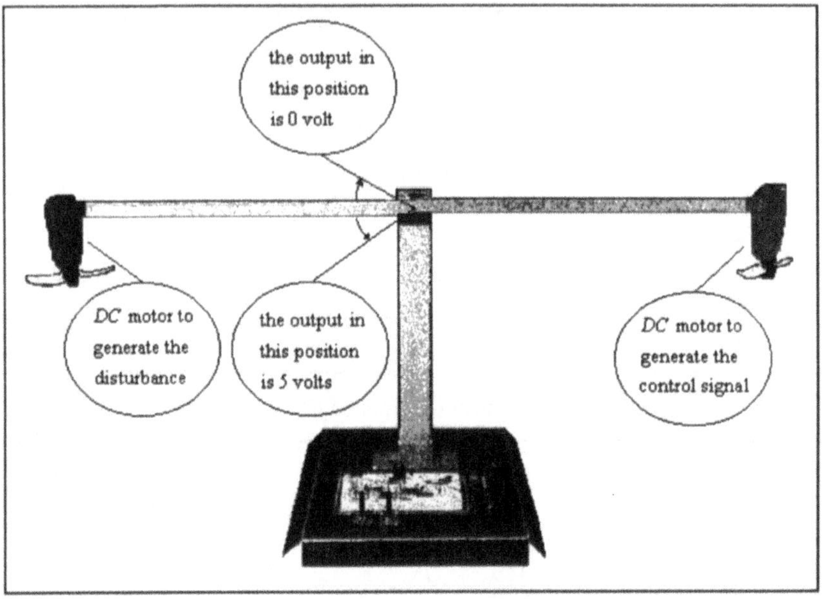

Figure 4.1. Horizontal balance process

The control variable is the tension, *u(k)*, supplied to the power amplifier of one of the *DC* motors coupled with one of the propellers, being maintained constant the voltage associated to the other motor. In a approximated description, the dynamic behaviour that relates the tension, *u(k)*, of the amplifier with speed, *n*, of the propellers, connected to the axis of the motor can be described as a first-order with delay, representing the dynamic behaviour of a *DC* motor. The nonlinear characteristics between the motor speed, *n*, and the torque *M* is introduced by the mechanism of the propellers. This nonlinear plant presents many challenges for controllers that can be summarized as: double integrator, instability, nonminimum phase, nonlinearities, and presence of the load disturbance in the moment of inertia, and an action of integral control action is necessary to eliminate the steady-state error. These practical factors can be considered as a challenge and motivate the implementation of *FLC-PID*'s [16].

5 Servo and Regulatory Control Experiments

Each *FLC-PID* controller was evaluated considering two experiments. The sampling period was 200 msec, and 750 data samples were collected for the experiments (total time of experimentation). The adjustment of the scale factors, rule-base and membership functions of each controller was accomplished by fine tuning and heuristic corrections linked to the knowledge of the process to be controlled. In all experiments the rule-base of *P* controller, in *FLC-(PD + I)* consisted of 3 rules, while the rule-base of *FLC-PD* and *FLC-PI* controllers, consisted of 9 and 5 rules, respectively, as presented in Figure 5.1.

e \ Δe	NE	ZE	PO
NE	PO	PO	ZE
ZE	ZE	ZE	ZE
PO	ZE	NE	NE

(a) *FLC-PD - u(k)* control

e \ Δe	NE	ZE	PO
NE	PO		ZE
ZE		ZE	
PO	ZE		NE

(b) *FLC-PI - $\Delta u(k)$* control

e	NE	ZE	PO
	PO	ZE	NE

(c) *FLC-P - $u_I(k)$* control

Figure 5.1. Rule-base of fuzzy *PD*, *PI* and *P* controllers

In the first experiment the process was submitted to setpoint changes, to evaluate setpoint tracking capability. Initially, the setpoint was specified to 2.5 volts, later modified to 4 volts (samples from 100 to 300), 2 volts (samples from 300 to 450), 3.0 volts (samples from 450 to 600) and, finally 2.5 volts (samples from 600 to 750). Figure 5.2 illustrates the output, setpoint and the control action for the experiment of setpoint changes corresponding to each *FLC-PID* controller.

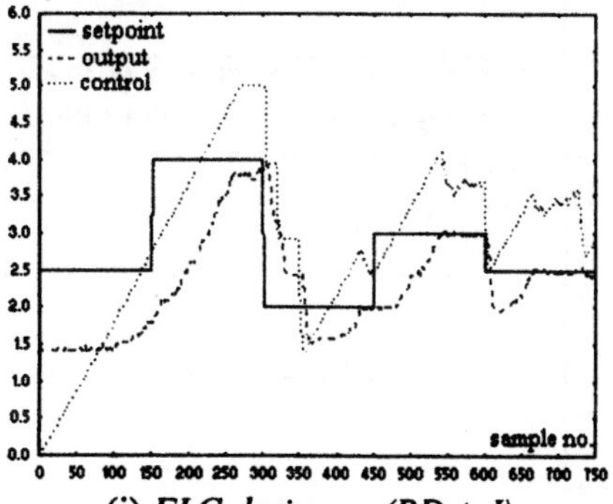

(i) *FLC design — (PD + I)*

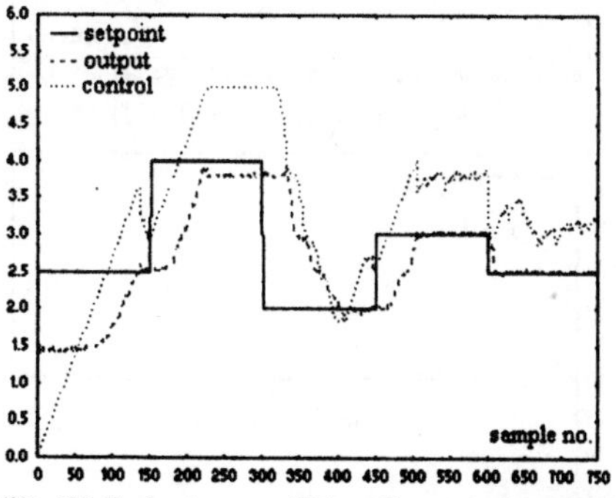

(ii) *FLC design — (PI +* Conventional *D)*

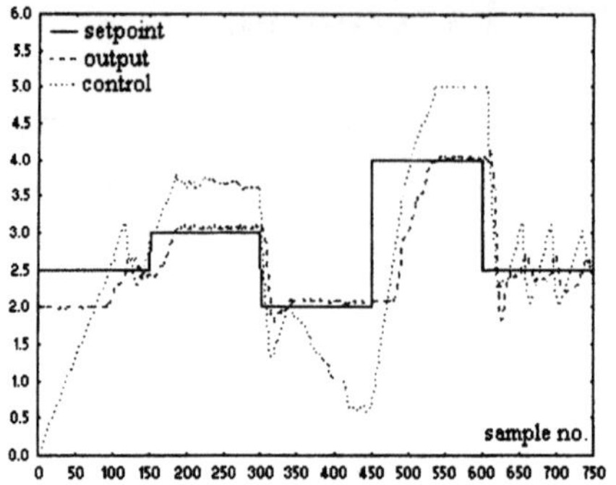

(iii) *FLC design* — Incremental *(PD + I)*

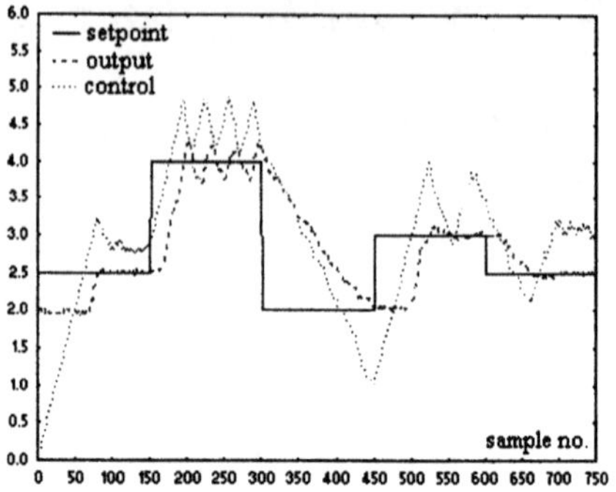

(iv) *FLC design* — *(PD + PI)*

Figure 5.2. Experiment for setpoint changes

In the second experiment a fixed setpoint of 3 volts was used and load disturbances were applied to the process in order to verify the capacity of the control algorithms for disturbance rejection. The disturbance magnitude was +0.3 volts, added from the sample 250 on. Another constant disturbance, of magnitude -0.3 volts was applied from the sample 500 on, and kept until the end of the experiment. The results for the different controllers are shown in Figure 5.3.

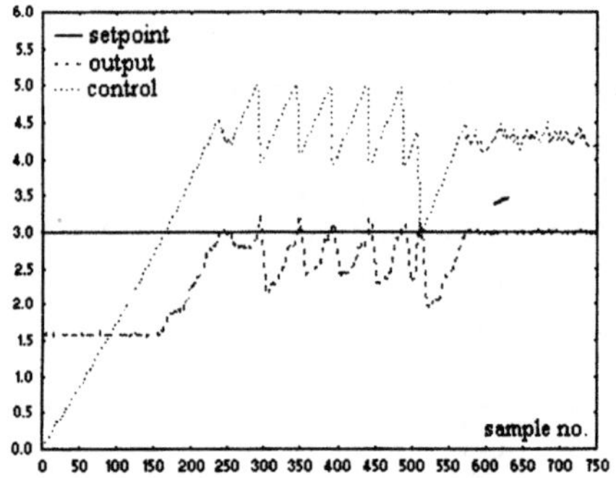

(i) *FLC design* — *(PD + I)*

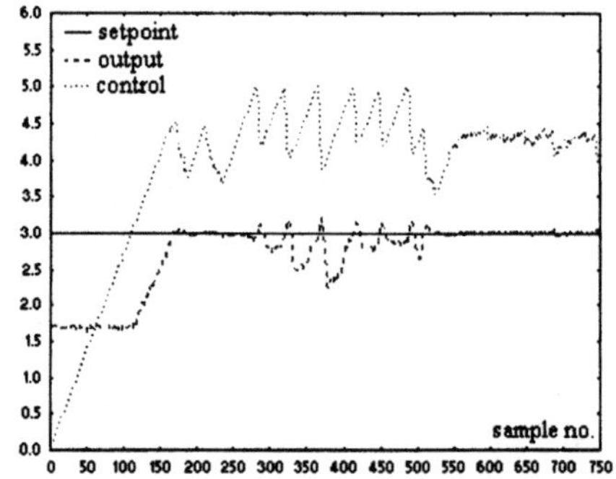

(ii) *FLC design* — *(PI* + Conventional *D)*

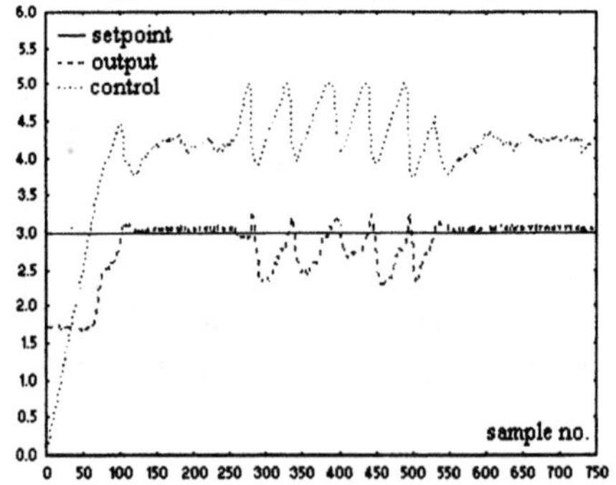

(iii) *FLC design* — Incremental *(PD + I)*

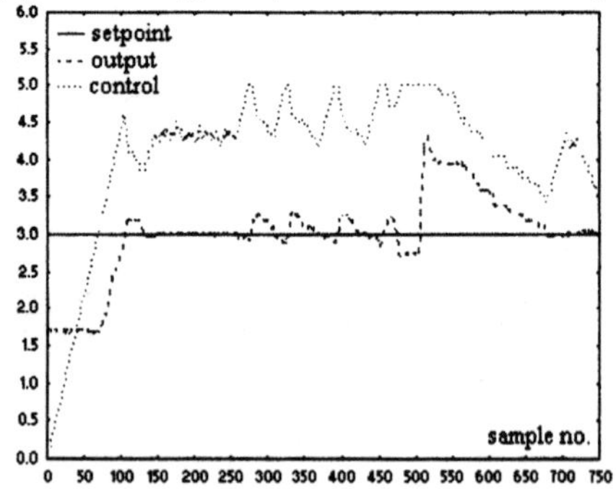

(iv) *FLC design* — *(PD + PI)*

Figure 5.3. Experiments for load disturbance

As shown in these figures, the *FLC-PID*'s were able to overcome practical problems encountered in this nonlinear process and had a good performance in all servo and regulatory conditions tested. The following performance indices were implemented to assess the quantitative measure of the controllers, and were calculated by

(a) absolute error sum, where

$$J_1 = \sum_{k=1}^{N} |e(k)| \tag{11}$$

(b) control variance sum given by

$$J_2 = \frac{1}{N} \sum_{k=1}^{N} \left[u(k) - \bar{u}\right]^2 \tag{12}$$

where $\bar{u}$ is the average of the control signal, and N is the number of collected samples of the experiment. Table 5.1 presents the results of the performance indices for *FLC-PID*'s obtained from setpoint change and load disturbance experiments.

Table 5.1. Results with *FLC-PID*'s in servo and regulatory experiments

experiments	setpoint change		load disturbance	
controller \ criterion	J_1	J_2	J_1	J_2
FLC design — (PD + I)	468.25	1.28	410.82	1.60
FLC design — (PI + Conventional D)	357.35	1.22	239.59	1.16
FLC design — Incremental (PD + I)	249.09	1.67	207.88	0.68
FLC design — (PD + PI)	324.92	1.02	256.69	0.89

It can be observed, from the performance indices, that the controller with the most adequate performance for setpoint change and load disturbance was the incremental *FLC design — (PD + I)*, followed by *FLC design — (PI + PD)* and *FLC design — (PI* + conventional *D)* with similar performance in two experiments.

6 Conclusion

This paper presented several design procedures for the *PID* controller, based on fuzzy control systems. Structures and design principles of *FLC-PID* controllers were described and evaluated. The *FLC-PID*'s implemented were: i) *(PD + I)*; ii)

(*PI* + *D* conventional); iii) incremental (*PD* + *I*); and iv) (*PD* + *PI*). The *FLC-PID* control conceptions were assessed on a horizontal balance experimental process.

The controllers showed appropriate performance for setpoint change and load disturbance tests. However, several factors influenced the performance of the controllers, mainly the adequate choice of the rule-base, the adjustment of membership functions, the determination of the scale factors, the division of state space, the designer's knowledge of the process characteristics, among other factors.

In setpoint change experiments, it was observed that *FLC-PID*'s tracked the setpoint appropriately, except for the first setpoint of 2.5 volts, which can be accounted for by the process inertia at the beginning of the experiment, around 1.8 volts. The experiments for analysis of load disturbances have shown an adequate performance of incremental *FLC-(PD + I)*, with the best performance, if compared to the other controllers.

Advantages of *FLC* over the conventional controllers are: i) it does not need the mathematical model of the process; ii) it can work with imprecise inputs, iii) it can deal with complex processes, with nonlinear characteristics, high-order, time delay, and badly-damped systems, iv) it facilitates the implementation of human specialist knowledge and experience which can be incorporated into linguistic rules (qualitative knowledge) [7], [8].

Perspectives for improving the practical results of the *FLC-PID* controllers, to deal with setpoint change and load disturbance, should addressed a more adequate adjustment of membership functions, with a increase of the rule-base, and a larger number of divisions of the discourse universe. So, it is possible to obtain a more efficient *FLC-PID* controller, which will demand a more difficult adjustment of its parameters and higher computational complexity in the control design if compared with the design methodology adopted in this paper. According to Chen [7], comparisons of *FLC-PID* controllers are now a difficult task, because *FLC-PID* controllers have less than a decade of history, and the *FLC-PID* does not have the maturity of the conventional control structures that have more than 50 years of history. As perspectives for future works, there are the *FLC-PID* optimisation and configuration using advanced hybrid control methodologies, such as the control design by: sliding mode, self-tuning, gain scheduling, robust, adaptive, evolutionary-fuzzy, neuro-fuzzy [17], [18], among others, in order to obtain efficient and robust *PID* controllers.

References

[1] Harris C J, Moore C G, Brown M 1993 *Intelligent Control: Aspects of Fuzzy Logic and Neural Nets*. World Scientific Publishing

[2] Chen C -L, Chang F -Y 1996 Design and Analysis of Neural/Fuzzy Variable Structural PID Control Systems. *IEE Proc-Control Theory Appl*

143(2):200-208

[3] Zadeh L A 1965 Fuzzy Sets. *Information and Control* 8:338-353

[4] Mamdani E, Assilian S 1975 An Experiment in Linguistic Synthesis with a Fuzzy Logic Controller. *Int J on Man Machine Studies* 7(1):1-13

[5] Bartos F J 1996 Fuzzy Logic Reaches Adulthood. *Control Engineering Int* July:50-56

[6] VanDoren V J 1998 Advanced Control Software Goes Beyond PID. *Control Engineering Int* January:57-60

[7] Chen G 1996 Conventional and Fuzzy PID Controllers. *Int J of Intelligent Control and Systems* 1(2):235-246

[8] Raviraj V S C, Sen P C 1997 Comparative Study of Proportional-Integral, Sliding Mode, and Fuzzy Logic Controllers for Power Converters, *IEEE Trans on Industry Applications* 33(2):518-524

[9] Kwok D P, Tan P, Li C K, Wang P 1990 Linguistic PID Controllers. *11th IFAC World Congress, Vol. 7*. Tallinn, Estonia, USSR, pp 192-197

[10] Malki H A, Misir D, Feigenspan D, Chen G 1997 Fuzzy PID Control of a Flexible-Joint Robot Arm with Uncertainties from Time-Varying Loads. *IEEE Trans on Control Systems Technology* 5(3):371-378

[11] Li Y, Ng K C 1997 Reduced Rule-Base and Direct Implementation of Fuzzy Logic Control. *13th IFAC World Congress*. San Francisco, CA, USA, pp 85-90

[12] Li H X, Gatland H B 1995 Enhanced Methods of Fuzzy Logic Control. *Proc of FUZZ-IEEE/IFES'95, Vol. 1*. Yokohama, Japan, pp 331-336

[13] Qin S J 1994 Auto-Tuned Fuzzy Logic Control. *Proc of the American Control Conference*. Baltimore, Maryland, USA, pp 2465-2469

[14] Qiao W Z, Mizumoto M 1996 PID Type Fuzzy Controller and Parameters Adaptive Method. *Fuzzy Sets and Systems* 78:23-35

[15] Misir D, Malki H A, Chen G 1996 Design and Analysis of a Fuzzy Proportional-Integral-Derivative Controller. *Fuzzy Sets and Systems* 79:297-314

[16] Filatov N M, Keuchel U, Unbehauen H 1996 Dual Control for an Unstable Mechanical Plant. *IEEE Control Systems* 16(4):31-37

[17] Coelho L S, Coelho A A R 1998 Computational Intelligence in Process Control: Fuzzy, Evolutionary, Neural, and Hybrid Approaches. *Int J of Knowledge-Based Intelligent Engineering Systems* 2(2):80-94

[18] Ordónez R, Zumberge J, Spooner J T, Passino K M 1997 Adaptive Fuzzy Control: Experiments and Comparative Analyses. *IEEE Trans on Fuzzy Systems* 5(2):167-188

An Accurate COG Defuzzifier Design Using the Coadaptation of Learning and Evolution*

Daijin Kim and HanPyul Lee

Department of Computer Engineering, DongA University,
840, HadanDong, Saha Ku, Pusan, 604 – 714, Korea
Email : dkim@vlsi.donga.ac.kr and hplee@vlsi.donga.ac.kr

Keywords: COG Defuzzifier, FLC, Genetic Algorithms, Lamarckian Co-adaptation

Abstract

This paper proposes an accurate COG (Center of Gravity) defuzzification method that improves both the system's approximation behavior and the control performance of a fuzzy logic controller (FLC). The premise of this paper is that the defuzzifcation is a process of optimal selection and an appropriate selection of the defuzzification method can improve the approximation behavior and the control performance of the FLC greatly. The accuracy of the proposed COG defuzzifier is obtained by representing the output membership functions (MFs) with various design parameters such as the centers, widths, and modifiers of MFs and by adjusting these design parameters with Lamarckian co-adaptation of learning and evolution. This co-adaptation scheme allows to evolve much faster than the nonlearning case and gives a higher possibility of finding an optimal solution due to its wider searching capability. An application to the truck backer-upper control problem of the proposed co-adaptive design method of COG defuzzifier is presented. The approximation behavior and control performance of the proposed COG defuzzifier are compared with those of the conventionally simplified COG defuzzifier in terms of the system error and the average tracing distance, respectively.

1 Introduction

Fuzzy logic controllers (FLCs) have been widely applied to both consumer products and industrial process controls. In particular, FLCs are very effective techniques for complicated and imprecise processes for which either no mathematical model exists or the mathematical model is severely non-linear because FLCs can easily approximate human expert's control behaviors that are working fine under such ill-defined environments [1]. Many researches have been performed for the optimal constructions of the FLCs. However, little attention has been given to the defuzzification method due to its simple operation and the designer's overlook about its role. Some of the reported work in optimal defuzzification method are in what follows. R. Yager and D. Filev [2] introduced a parameterized family of defuzzification operators called the SLIDE (Semi LInear DEfuzzification) method. The determination method was based on a simple linear transformation of the fuzzy output set of the controller. The optimal values of SLIDE's two parameters was determined by the recursive least square solution

* This work was supported in part by the Korea Research Fund grant(1997-001-E00365) and by the DongA University Research Fund grant(1998).

of a set of linear equations based on the concept of Kalman filters. But, due to the nonlinearity of the used algorithm, learning of parameters may converge to some local extremum. A. Bastian [3] proposed a modified COG defuzzification method that could control the nonlinearity at the transition between fuzzy logic rules. The defuzzification weights to the overlapping area of the rule consequences were determined by using the backpropagation learning algorithm in a three-layer feedforward neural network. But, the approach was performed on the specific case where all membership functions of an output variable had the same widths and were located at the same intervals. S. Roy-chowdhury *et. al* [4] used the concept of logical distance and the interaction between neighbors to select an element from a given fuzzy set. But, they simply illustrated a variety number of cooperation modelings among neighbors and did not have any free parameters which could be adapted for realizing an optimal defuzzified value. J. Dickerson and B. Kosko [5] provided the centroidal defuzzification with correlation product whose one of its weight values was the volume of the output membership functions. However, they did not specify how the volume was calculated.

In this paper, we treat the defuzzifcation problem as a process of optimal selection and believe that an appropriate selection of the defuzzification method can improve both the system's approximation behavior and the control performance of the FLC. To realize such an optimization, we introduce a set of free parameters in the defuzzification that are adapted by both evolving and learning them simultaneously. We will show that the parameters are easily determined with the help of Lamarckian co-adaptation model of learning and evolution, where the learning and evolution are taking a local search of design parameters in an individual COG defuzzifier and a global search of design parameters among a population of various COG defuzzifiers, respectively. It is also demonstrated that the Lamarckian co-adaptation scheme expedites the convergence speed to the optimal solution than the evolution-only technique.

This paper is organized as follows. Section 2 explains our proposed COG defuzzifier in detail. Section 3 describes how to determine the optimal design parameters of a COG defuzzifier. Section 4 presents the simulation results of applying the proposed COG design method to the truck backer-upper control problem and the approximation behavior and the control accuracy are compared with those of the conventionally simplified one in terms of the system error and the average tracing distance, respectively. Finally, a conclusion is drawn.

2 A Proposed COG Defuzzifier

Defuzzification is a problem of optimized selection of an element from a fuzzy set. A defuzzification method determines a most appropriate single value to represent a collection of fuzzy elements. The commonly used defuzzification methods are MAX criterion, the Mean of Maximum (MOM), and the Center of Gravity (COG). The Max method gives the rule with the highest possibility as output. The MOM method considers between the possibility values of all the rules and the COG method calculates the center of gravity of the output fuzzy membership functions given by the inference engine. For hardware implementation, the most widely used method is the COG because it produces a better steady-state performance than the other two methods.

The COG defuzzifier generates a crisp value that is the center of area of the fuzzy set. It is actually designed to minimize an objective function that is the membership-weighted mean of the squared distance [6]. Consider a cost function that is weighted by its membership function,

$$J(y_c) = \sum_{i=1}^{n} (y_i - y_c)^2 \mu_Y(y_i), \tag{1}$$

where $y_c \in Y$, and n are the defuzzified crisp value and the number of output fuzzy terms, respectively. By differentiating the $J(y_c)$ to get the minimum of the cost function,

$$\frac{dJ(y_c)}{dy_c} = 2\sum_{i=1}^{n} (y_i - y_c)\{\mu_Y(y_i)\} = 0, \tag{2}$$

we can get a conventionally simplified definition of the COG defuzzifier as

$$y_c = \frac{\sum_{i=1}^{n} y_i \mu_Y(y_i)}{\sum_{i=1}^{n} \mu_Y(y_i)}, \tag{3}$$

where y_i is a singleton value and the membership function $\mu_Y(y_i)$ has the maximum value at $y = y_i$.

However, this conventionally simplified COG defuzzifier causes an inaccurate crisp value because it considers only the output singleton values corresponding to the maximum membership values multiplied by the clipped membership values in calculating the crisp value. Fig. 1 illustrates a fallacy of the conventionally simplified COG defuzzifier. Two membership functions of Fig. 1-(a) have the same size of widths, and two membership functions of Fig. 1-(b) have the different size of widths. But, both cases have the same singleton's values. Since the conventionally simplified COG defuzzifier considers only the membership values at the singletons in calculating a crisp value, it produces the same crisp value in both cases. This is far from our knowledge that the center of gravity is nearer to the membership function having the larger area. In other word, the conventionally simplified COG defuzzifier fails to reflect the different widths of membership functions.

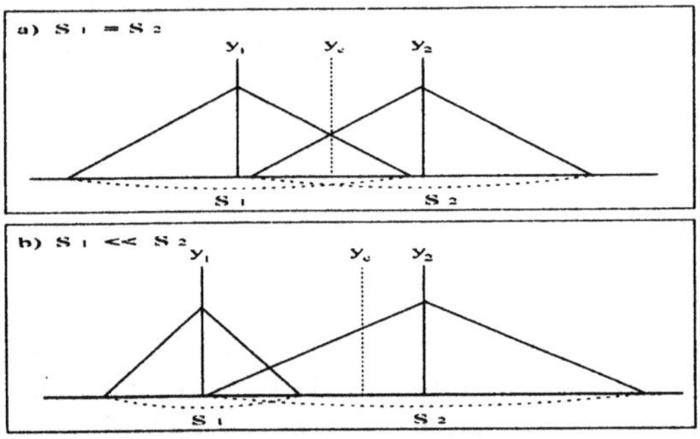

Fig. 1. An illustrative fallacy of the conventionally simplified COG defuzzifer.

In order to correct this fallacy, we suggest a new scheme of the COG defuzzification as follows.

$$y_c = \frac{\sum_{i=1}^{n} A_Y(y_i) \cdot y_i}{\sum_{i=1}^{n} A_Y(y_i)}, \tag{4}$$

where n is the number of discrete fuzzy terms, y_i is the singleton's support value of the ith fuzzy term in the consequence, and $A_Y(y_i)$ is the area under the ith clipped membership value defined as $A_Y(y_i) = \mu_Y(y_i) \cdot s_i$, where s_i is the width of the ith fuzzy term.

Furthermore, we can extend the meaning of each fuzzy term by using a modifier such as "very", "more or less", or "fairly". When we incorporate different modifiers into each fuzzy term, we can extend that the proposed COG defuzzifier calculates the crisp value as

$$y_c = \frac{\sum_{i=1}^{n} A_Y^e(y_i) \cdot y_i}{\sum_{i=1}^{n} A_Y^e(y_i)}, \tag{5}$$

where $A_Y^e(y_i)$ is the effective area under the ith clipped membership value defined as $A_Y^e(y_i) = \mu_Y^{m_i}(y_i) \cdot s_i$, where m_i is the modifier of the ith fuzzy term.

A detailed derivation of an exact equation of the effective area $A_Y^e(y_i)$ in the case of a modified membership function $\mu_Y^{m_i}(y_i)$ is given in [7], where m_i is a modifier that can either concentrate($m_i > 1$) or dilate ($m_i < 1$) the membership function [8]. The derivation result shows that the effective area $A_Y^e(y_i)$ under the clipped membership function of $\mu_Y^{m_i}(y_i)$ is given as

$$A_Y^e(y_i) = \mu_Y(y_i)s_i(1 - \frac{m_i}{m_i+1}\mu_Y^{\frac{1}{m_i}}(y_i)), \tag{6}$$

where m_i, $\mu_Y(y_i)$, and s_i are the modifier of the ith membership function, the membership function of the ith fuzzy term, and the width of the ith membership function, respectively. It is noted that $m_i \to 0$, $A_Y^e(y_i) \to \mu_Y(y_i) \cdot s_i$ and $m_i \to \infty$, $A_Y(y_i) \to 0$.

3 Determination of Design Parameters

3.1 A Gradient-Descent Batch Learning

Assume that a training example set $E = \{e^1, e^2, \cdots, e^N\}$ consisting of N samples has been known in advance. An example $e^t (t = 1, 2, \cdots, N)$ consists of $(x_1^t, x_2^t; y^t)$, where x_1^t and x_2^t are two input values of the tth example and y^t is one output value of the tth example. Fuzzifier replaces input values of an example e^t into a fuzzy term set with the clipped membership values. Fuzzy inference engine performs an approximation reasoning by associating the input fuzzy term set with the precondition parts of the fuzzy rules and produces the consequence's fuzzy term set. Then, the proposed COG defuzzifier computes a crisp value y_c^t about the input values of the example e^t from Eq. (5).

Let the system error of the FLC using the proposed COG defuzzifier (Eq. (6)) over the training example set be E. Then, the system error is computed by

$$\begin{aligned} E &= \frac{1}{2}\sum_{t=1}^{N}(y^t - \frac{\sum_{i=1}^{n} \mu_Y(y_i) \cdot (1 - \frac{m_i}{m_i+1}\mu_Y^{\frac{1}{m_i}}(y_i)) \cdot s_i \cdot y_i}{\sum_{j=1}^{n} \mu_Y(y_j) \cdot (1 - \frac{m_j}{m_j+1}\mu_Y^{\frac{1}{m_j}}(y_j)) \cdot s_j})^2 \\ &= \frac{1}{2}\sum_{t=1}^{N}(y^t - \alpha_i^t \cdot y_i)^2, \end{aligned} \tag{7}$$

where n, N are the number of output fuzzy terms and the number of training examples, respectively, and α_i^t $(= \frac{\mu_Y(y_i) \cdot (1 - \frac{m_i}{m_i+1}\mu_Y^{\frac{1}{m_i}}(y_i)) \cdot s_i}{\sum_{j=1}^{n} \mu_Y(y_j) \cdot (1 - \frac{m_j}{m_j+1}\mu_Y^{\frac{1}{m_j}}(y_j)) \cdot s_j})$ is used to simplify the expression.

From the gradient-descent method, three parameters are adjusted as follows.

$$y_i(T+1) = y_i(T) - \eta_i^y \frac{\partial E}{\partial y_i},$$
$$s_i(T+1) = s_i(T) - \eta_i^s \frac{\partial E}{\partial s_i},$$
$$m_i(T+1) = m_i(T) - \eta_i^m \frac{\partial E}{\partial m_i}, \tag{8}$$

where η_i^y, η_i^s, and η_i^m are the learning rates of the center, width, and modifier of the ith output fuzzy term, respectively.

We compute the gradient values of three parameters by a chain-rule and simple formula of differentiation as follows.

$$\frac{\partial E}{\partial y_i} = \sum_{t=1}^{N} \frac{\partial E^t}{\partial y_c^t} \frac{\partial y_c^t}{\partial y_i} = -\sum_{t=1}^{N} (y^t - y_c^t)\alpha_i^t, \tag{9}$$

where y^t and y_c^t are the desired output value and the actual output value from the proposed COG defuzzifier with respect to the tth training example, respectively. Similarly,

$$\frac{\partial E}{\partial s_i} = \sum_{t=1}^{N} \frac{\partial E^t}{\partial y_c^t} \frac{\partial y_c^t}{\partial s_i}$$
$$= -\sum_{t=1}^{N} (y^t - y_c^t) \left(\frac{\alpha_i^t}{s_i}\right) (y_i - y_c^t). \tag{10}$$

Similarly,

$$\frac{\partial E}{\partial m_i} = \sum_{t=1}^{N} \frac{\partial E^t}{\partial y_c^t} \frac{\partial y_c^t}{\partial m_i} = -\sum_{t=1}^{N} (y^t - y_c^t)\alpha_i^t \left(\frac{(1 - \frac{1+m_i}{m_i} ln(\mu_Y(y_i^t))}{m_i(1+m_i)(1 - \frac{1+m_i}{m_i} \mu_Y^{-\frac{1}{m_i}}(y_i^t))} \right) (y_i - y_c^t) \tag{11}$$

Therefore, learning rules for three parameters are summarized as follows.

$$y_i(T+1) = y_i(T) + \eta_i^y \cdot \sum_{t=1}^{N} \alpha_i^t (y^t - y_c^t),$$
$$s_i(T+1) = s_i(T) + \eta_i^s \cdot \sum_{t=1}^{N} \beta_i^t (y^t - y_c^t)(y_i - y_c^t),$$
$$m_i(T+1) = m_i(T) + \eta_i^m \cdot \sum_{t=1}^{N} \gamma_i^t (y^t - y_c^t)(y_i - y_c^t), \tag{12}$$

where β_i^t and γ_i^t are equal to the terms $\frac{\alpha_i^t}{s_i}$ and $\alpha_i^t \cdot m_i^{eq}$, where m_i^{eq} is defined as $\{\frac{(1 - \frac{1+m_i}{m_i} ln(\mu_Y(y_i^t))}{m_i(1+m_i)(1 - \frac{1+m_i}{m_i} \mu_Y^{-\frac{1}{m_i}}(y_i^t))}\}$, respectively.

3.2 Evolutionary Technique for An Optimal COG Defuzzifer

As explained before, the convergence of the supervised backpropagation learning is dependent on the initial choice of many design and learning parameters and the control performance of the resultant FLC can not guarantee to be optimal. In order to overcome these disadvantages, many researchers suggest to use the evolutionary technique based on genetic algorithms. GAs [9] are any population-based iterative adaptive algorithms that use selection, recombination, and mutation operations based on natural selection and biological genetics. GAs have been proven to be powerful methods in search, optimization and machine learning. They encode a potential solution to a specific problem on a simple chromosome-like data structure and apply recombination operators to these structures to achieve optimization. When GA is applied to find an optimal COG defuzzifier, we need to consider the following five principal problems.

Chromosome Representation Consider a COG defuzzifier whose number of the output fuzzy terms is M. As explained before, each fuzzy term in the output fuzzy variable is specified by the center c_i, the width s_i and the modifier m_i. Therefore, each chromosome that represents an individual COG defuzzifier consists of $3 \times M$ consecutive real numbers, where the first three real numbers represent the center, width, and modifier pairs of the first fuzzy term, and so on. We adopt a real number representation of chromosomes since each component of fuzzy terms is a real number.

Initial Population Generation The initial population is critical for determining the convergence speed of GAs and the control performance of the resultant COG defuzzifier. A chromosome C_r in the initial population is built by randomly generating the output design parameters (c_i, s_i, m_i) pairs P times. They are generated in a way that the centers of the fuzzy terms are located uniformly along the entire universe, the widths of the fuzzy terms are determined to cover the discourse of the output fuzzy variable as evenly as possible, and the modifiers have the values within the range of $[0.5, 2.0]$. P repetitive executions of this generation form the initial population, where P is the population size.

Fitness Function The goodness of the chromosome is evaluated by its fitness function. In our approach, two factors are involved in the fitness function as follows. Firstly, good COG defuzzifier shows good approximation behavior, which means small system error. This requirement can be satisfied by including the system error in the fitness function as $F_e(T) = \frac{1}{1+E(T)}$, where $E(T)$ is the system error defined by Eq. (13). Secondly, good COG defuzzifier shows good control performance. An average tracing distance $d(T)$ over the randomly chosen N_T test examples $\{x_1, x_2, \cdots, x_{N_T}\}$ is defined as $d(T) = \frac{1}{N_T} \sum_{t=1}^{N_T} d(x_t)$, where N_T is the number of test examples. Then, the fitness function $F_p(T)$ with respect to the control performance is represented as $F_p(T) = \frac{1}{1+d(T)}$. When we integrate these two considerations into one fitness function, the fitness function $F(T)$ is represented as

$$F(T) = k_1 \cdot F_e(T) + k_2 \cdot F_p(T), \tag{13}$$

where k_1 and k_2 are appropriate constants that make two fitness components balanced and they satisfy the condition $k_1 + k_2 = 1$.

Genetic Operations The initial population of COG defuzzifiers are then evolved by appropriate genetic operations in order to find an optimal COG defuzzifier for the given control purposes. The detailed explanation about the genetic operations used for an evolutionary COG defuzzifier design is given as follows.

(1) Reproduction We use a mixture of selection methods for reproducing the chromosomes. The first selection method is an elitism that the best chromosome with the highest fitness value is passed in the new population. The second selection method is a modified k-tournament method [10]. In this method, a chromosome having the best fitness value among the k chromosomes selected from the upper class of fitness values randomly is chosen for the reproduction. Two COG defuzzifier chromosomes C' and C'' obtained by repeating the above procedure consecutively create a new chromosome C by applying the crossover and mutation operations explained later. The above reproduction procedure is repeated as many times as $pselect \times |P|$, where $|P|$ is the population size. Finally, the remaining portion of the population set is filled by copying the population set in the order of magnitude of fitness values. Fig. 2 shows a hybrid reproduction method based on a mixture of three different reproduction methods.

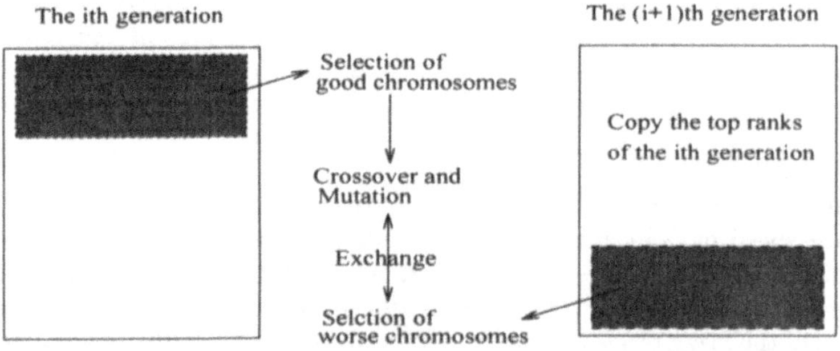

Fig. 2. An illustration of the proposed hybrid reproduction scheme.

(2) Crossover Basically, the crossover operation is based on the fitness-weighted average operation occurred in the frequency of P_c. Let two selected parent chromosomes be C^1 and C^2, and their corresponding fitness values be F_1 and F_2. Since all corresponding design parameters between two selected parent chromosomes are processed in the same way, we consider the specific design parameters of the ith fuzzy term in two selected chromosomes. Let the design parameters of the ith fuzzy terms of two selected chromosomes be (y_i^1, s_i^1, m_i^1) and (y_i^2, s_i^2, m_i^2). Then, the design parameters (y_i', s_i', m_i') of a new chromosome C' created by the crossover operation are obtained by

$$y_i' = \frac{F_1 \cdot y_i^1 + F_2 \cdot y_i^2}{F_1 + F_2}, \quad s_i' = \frac{F_1 \cdot s_i^1 + F_2 \cdot s_i^2}{F_1 + F_2}, \quad m_i' = \frac{F_1 \cdot m_i^1 + F_2 \cdot m_i^2}{F_1 + F_2}. \tag{14}$$

(3) Mutation Mutation over the new chromosome C' is performed as follows. Firstly, a mutation point is randomly selected in the fuzzy terms of the chromosome selected with a probability of P_m. Secondly, the design parameters (y_i'', s_i'', m_i'') of the mutated chromosome C'' are adjusted by

$$y_i'' = y_i' + rand(-\Delta y', \Delta y'),$$
$$s_i'' = s_i' + rand(-\Delta s', \Delta s'),$$

$$m_i'' = \begin{cases} \frac{1}{2} \cdot (1 - m_i) + 1 \text{ if } m_i \geq 1 \\ 2 \cdot (1 - m_i) + 1 \text{ if } m_i < 1, \end{cases} \tag{15}$$

where $\Delta y'$ and $\Delta s'$ are maximum variations of the center position and the width, respectively. The modifier m_i'' turns over the value with respective to 1.0 by the mutation.

3.3 Lamarckian Co-adaptation of the COG Defuzzifier

Recently, the ideas of combining evolution and learning based on the concepts inspired from the biological systems have attracted many increasing attentions. Evolution changes the genotype of organism at the population level while learning changes the behavior of organism at the individual level. Change by evolution is made from reproducing organism selectively and performing genetic operations (recombination and/or mutation) in order to maintain inter-individual variability. The change due to evolution is generationally cumulative in a manner that some changes at a particular generation is superimposed upon other changes in the previous generations. Change by learning is made from interacting with a specific environment during its lifetime and incorporating aspects of the environment through its experience in its internal structure. The change due to learning is individually cumulative in a manner that some changes in an individual at a particular time of its lifetime are influenced by other changes at preceding times.

Fig. 3 shows the proposed Lamarckian co-adaptation scheme [11] for an optimal COG defuzzifier design. Assume that each chromosome has its fitness value $F_i(T)$ at the Tth generation. During its lifetime, each chromosome has been learned by the proposed gradient-descent batch learning presented in Section 3.1. They have been learned differently under their different learning environments due to the choices of different initial parameters. Then, assume that they have the fitness values $F_i'(T)$ after learning at the Tth generation. Some chromosomes have better fitness values after learning (*i.e.* $F_i'(T) > F_i(T)$), but others have worse ones. We take the design parameters having the better fitness values as the learned design parameters (y_i^l, c_i^l, m_i^l) by the following dominant chromosome selection.

$$(y_i^l, c_i^l, m_i^l) = \begin{cases} (y_i, c_i, m_i) \text{ if } F_i(T) > F_i'(T) \\ (y_i', c_i', m_i') \text{ if } F_i'(T) > F_i(T), \end{cases} \tag{16}$$

where (y_i, c_i, m_i) and (y_i', c_i', m_i') are the design parameter pairs before and after learning, respectively. The above equation means that the degraded design parameters by learning are not chosen even though they have been learned.

When we consider only evolution without learning, the design parameters are not changed by learning. So, the learned design parameters (y_i^l, c_i^l, m_i^l) are equal to the design parameters (y_i, c_i, m_i) before learning. After each chromosome has been learned, it is adjusted by the evolutionary technique presented in Section 3.2. This produces a learned and evolved version of design parameters $(y_i^{l+e}, c_i^{l+e}, m_i^{l+e})$ that is just the design parameters at the $(T+1)$th generation. From the above figure, we introduce Lamarckian co-adaptation algorithm that incorporates evolution with learning for an optimal COG defuzzifier design as follows.

Step 1: Initialization.

Prepare a training example set $S = \{e_1, e_2, \cdots, e_N\}$, and set a time index

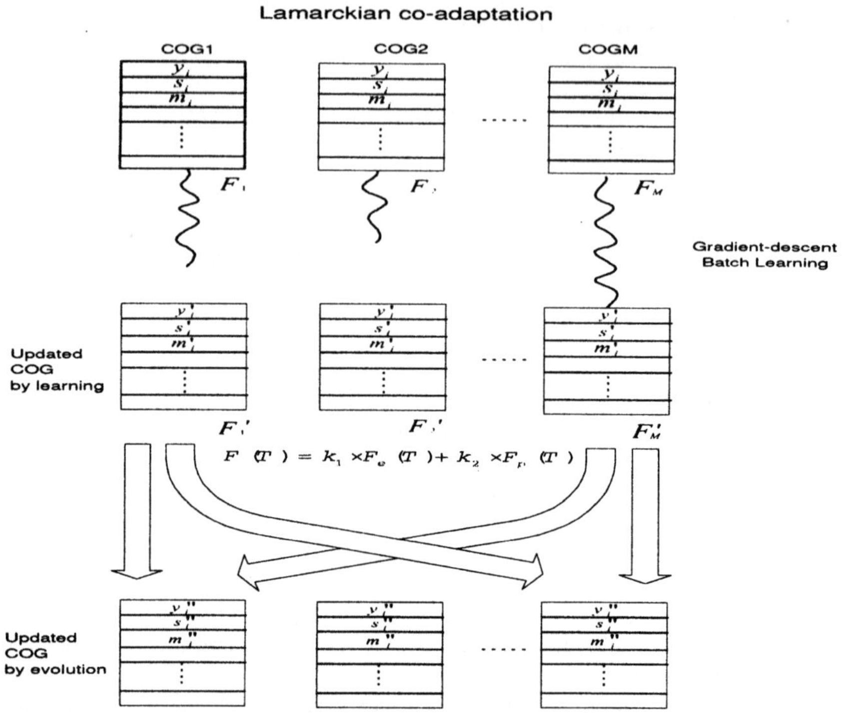

Fig. 3. Lamarckian co-adaptation for an optimal COG defuzzifier.

$T = 0$.
Given a fixed number N, form a pool of P chromosomes $C(T) = \{C_1(T), C_2(T), \cdots, C_P(T)\}$, where each chromosome is corresponding to an individual COG defuzzifier. Since each COG defuzzifier consists of M fuzzy terms and each term consists of (y_i, c_i, m_i), we need to generate $3 \times M$ real numbers.

Step 2: Compute the fitness value $F_i(T)$ of each COG defuzzifier before learning. Use Eq. (13) to evaluate the goodness of each chromosome using (y_i, c_i, m_i).

Step 3: Train each COG defuzzifier. Train each FLC using the gradient-descent batch learning rule (Section 3.1) until the design parameters are not changed any more.

Step 4: Compute the fitness value $F_i'(T)$ of each COG defuzzifier after learning. Use Eq. (13) to evaluate the goodness of each chromosome using (y_i', c_i', m_i').

Step 5: Determine the learned design parameters (y_i^l, c_i^l, m_i^l) using Eq. (16).

Step 6: Update the design parameters using the evolutionary technique (Section 3.2).
Use the proposed hybrid selection method based on the updated fitness value previously.
Recombine and mutate the selected chromosomes using Eq. (14) and (15), respectively.

Step 7: Compute the fitness value $F^{l+e}(T)$ of each COG defuzzifier after learning and evolution.
Use Eq. (13) to evaluate the goodness of each chromosome using $(y_i^{l+e}, c_i^{l+e}, m_i^{l+e})$. Choose the best COG defuzzifier $C_{best}(T)$ whose fitness value is the greatest one $F_{max}(T)$. If $F_{max}(T)$ reaches a predefined value, halt the algorithm with an optimal COG defuzzifier as $C_{best}(T)$.

Step 8: Set $T = T + 1$ and goto Step 2.

Proposed Lamarckian co-adaptive COG defuzzifier design algorithm.

4 Simulation Results and Discussion

The proposed COG defuzzifier is tested on the truck backer-upper control problem and its approximation behavior and control accuracy are compared with those of the conventional COG defuzzifier in terms of the system error and the average tracing distance, respectively. The goal of the truck backer-upper control problem is to back a truck to a loading dock as quickly and precisely as possible. This control problem is a typical non-linear control problem that can not be solved by the conventional control techniques.

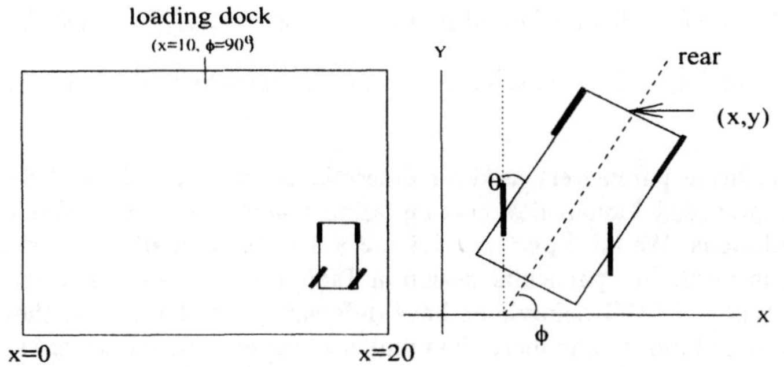

Fig. 4. A model truck and a loading dock used in the truck backer-upper control.

Fig. 4 shows a model truck and a loading dock used in the truck backer-upper control problem. The position of the truck is precisely determined by (x, y, ϕ) where ϕ is an angle between the truck's onward direction and the x axis, and the tracking control of the truck is done by the θ that is an angle between the truck's onward direction and the axis of wheel. The approximate control dynamics of the truck backer-upper control problem is given as (See [12] for details.)

$$x(t+1) = x(t) + cos[\phi(t) + \theta(t)] + sin[\theta(t)]sin[\phi(t)],$$
$$y(t+1) = y(t) + sin[\phi(t) + \theta(t)] - sin[\theta(t)]cos[\phi(t)],$$
$$\phi(t+1) = \phi(t) - sin^{-1}[2sin(\frac{\theta(t)}{b})], \tag{17}$$

where b is the length of truck and $b = 4$ is taken in this work. The design problem of fuzzy controller for the truck backer-upper control problem is reconfigured as to back the truck at a certain position (x_o, ϕ_o) in the interval of $\{0 \leq x \leq 20, -90^\circ \leq \phi \leq 270^\circ\}$ to the loading dock at $x = 10, \phi = 90^\circ$ as fast and precisely as possible.

Even though the Wang's COG defuzzifier [12] is the fixed one, our proposed COG defuzzifier can adjust many design parameters in different ways. Our simulations have been carried out in three different aspects : (1) different combinations of variable design parameters, (2) different types of adjusting design parameters, and (3) different weights of fitness functions. In the first category, we consider three different COG defuzzifiers as

- COG 1 : variants = (y_i), s_i = s, $m_i = 1.0$ (Conventional COG defuzzifier).
- COG 2 : variants = (y_i, s_i), $m_i = 1.0$.
- COG 3 : variants = (y_i, s_i, m_i).

In the second category, we consider two different adjustment techniques as

- Adjustment 1 : Evolution only.
- Adjustment 2 : Lamarckian co-adaptation.

In the third category, we consider three different fitness functions as

- Fitness 1 : $k_1 = 1.0$ and $k_2 = 0.0$ (Consider approximation behavior by system error $E(T)$).
- Fitness 2 : $k_1 = 0.5$ and $k_2 = 0.5$ (Balanced performance between $E(T)$ and $d(T)$).
- Fitness 3 : $k_1 = 0.0$ and $k_2 = 1.0$ (Consider control performance by average tracing distance $d(T)$).

Table 1 shows the execution parameters used for determining the optimal COG defuzzifier through the proposed Lamarckian co-adaptation scheme in many different combinations of simulations. We have performed $3 \times 2 \times 3 = 18$ simulations for the complete experiment in total. One particular notion in Table 1 is the population size. Since three different types of COG defuzzifiers have different types of variants, they have different sizes of populations. The more the number of variants is, the wider the searching space becomes. So, we choose the number of chromosomes in the population as 50, 100, and 150 in the COG1, COG2, and COG3, respectively.

Table 1. A model truck and a loading dock used in the truck backer-upper control.

	COG1	COG2	COG3
Population size	50	100	150
No. of generations	100	100	100
Chromosome size	1×7	2×7	3×7
No. of k-tournaments	3	5	7
pselect	0.5	0.4	0.3
P_c	0.9	0.9	0.9
P_m	0.1	0.15	0.2
η_y	0.1	0.1	0.1
η_s	-	0.1	0.1
η_m	-	-	0.1

Fig. 5 shows the fitness curves of six different combinations of three COG defuzzifiers and two adjustment techniques whose design goal is trying to get a good approximation ability with smaller system error ($k_1 = 1.0$ and $k_2 = 0.0$). In this simulation, the fitness value takes the error performance of COG defuzzifier in the phase of evolution, and the design parameters are adjusted according to the system error of COG defuzzifier in the phase of learning. Here, the solid line and the dotted line represent the fitness curves whose design parameters are adjusted by the evolution only and the Lamarckian co-adaptation method, respectively. From this figure, we note that (1) the Lamarckian co-adaptation adjustment method is more effective than than the evolution-only adjustment method in order to obtain a good approximation ability, (2) the Lamarckian co-adaptation adjustment method reaches the convergent state much faster than the evolution-only adjustment method, and (3) the approximation abilities becomes better in the order of COG defuzzifiers whose design parameter variants are (y, s, m), (y, s), and (y) at the given adjustment method.

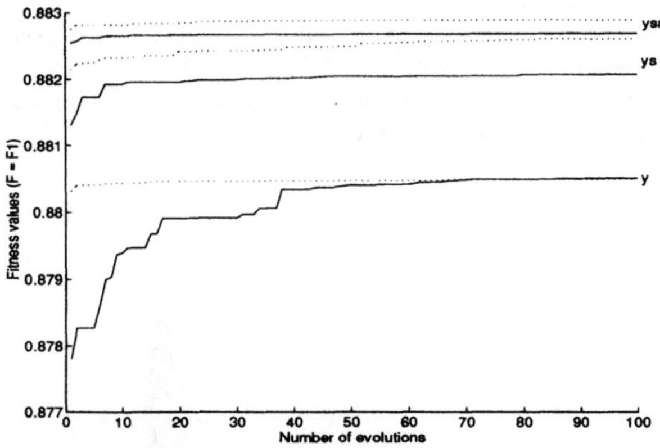

Fig. 5. Fitness for a good approximation.

Fig. 6 shows the approximation errors of seven different types of simulations such as

1. SIM1 : Evolution only with the fitness value $F = F_1$,
2. SIM1-1 : Lamarckian co-adaptation with the fitness value $F = F_1$,
3. SIM2 : Evolution only with the balanced fitness $F = \frac{F_1}{2} + \frac{F_2}{2}$,
4. SIM2-2 : Lamarckian co-adaptation with the balanced fitness value $F = \frac{F_1}{2} + \frac{F_2}{2}$,
5. SIM3 : Evolution only with the fitness value $F = F_2$,
6. WANG(y) : Wang's FLC with the adaptation of the MF's center,
7. WANG(y,s) : Wang's FLC with the adaptation of the MF's center and width.

In these simulations, the approximation error is obtained by computing the system error $E = \sqrt{\frac{1}{236}\sum_{i=1}^{236}(\theta_i - \hat{\theta}_i)^2}$ based on Wang's training sample set [12], where θ_i and $\hat{\theta}_i$ are the ith desired output value and the computed COG defuzzified value, respectively, and the control performance is obtained by computing the average tracing distance $d = \frac{1}{100}\sum_{i=1}^{100}\sum_{j=1}^{n_i}\sqrt{(x_{i,j} - x_{i,j-1})^2 + (y_{i,j} - y_{i,j-1})^2}$ based on the randomly chosen 100 starting points, where $x_{i,j}$ and $y_{i,j}$ are the jth tracing x and y position of the ith starting sample, respectively, and n_i is the number of tracing of the ith starting sample. Here, the solid and dotted line in Fig. 6 represent the approximation errors of the WANG's FLC whose design parameter variant are (y) and (y, s), respectively. From the Fig. 6, we note that (1) the approximation error is the smallest when the design parameters of COG defuzzifier are adapted under the condition of the fitness value $F = F_1$, (2) the approximation error in the case of simulations SIM1 or SIM1-1 becomes smaller in the order of COG defuzzifiers whose design parameter variants are (y, s, m), (y, s), and (y) at the given adjustment method, (3) The Wang's FLC shows a relatively small approximation error among 7 different FLCs and the approximation error is not sensitive to the different types of design parameter variants (y) and (y, s), (4) the approximation ability of SIM3 becomes worse ultimately because the COG defuzzifier of SIM3 is adapted to get a short tracing distance, and (5) the simulations SIM2 and SIM2-2 produce the balanced performances in the approximation ability. From these notions, it is concluded that two design goals take an adverse effect on the fitness function in such a way that smaller approximation error causes longer average tracing distance, vice versa. Nevertheless, our proposed design method provides many flexible COG defuzzifiers that can meet a variety of design goals simply by changing the weight constants k_1 and k_2 of the fitness function.

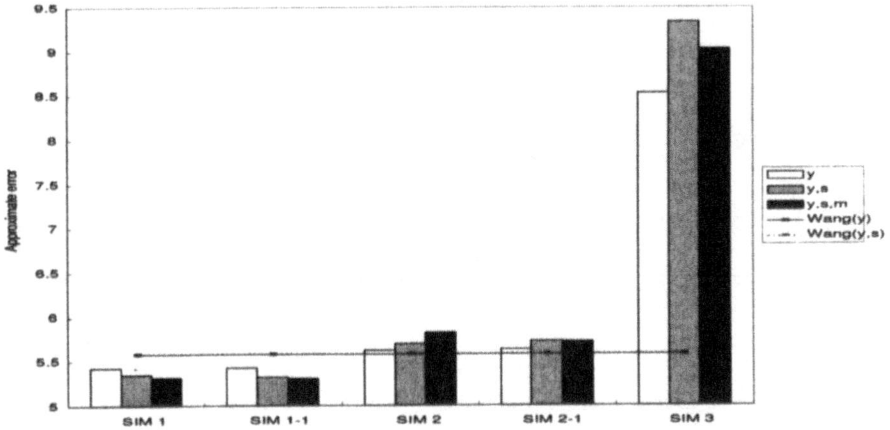

Fig. 6. Approximation errors of different types of simulations.

Fig. 7 shows some tracing trajectories from two randomly selected initial positions $(2.0, 0.0, -81°)$ and $(15.0, 0.0, 265°)$ of two different design goals such as smaller approximation error in the case of simulation SIM1 ($k_1 = 1.0$ and $k_2 = 0.0$) and shorter average tracing distance in the case of simulation of SIM3 ($k_1 = 0.0$ and $k_2 = 1.0$), respectively. Simulation results show that the simulation SIM1 and SIM3 reaches to the goal in the 15 and 13 steps for the left starting position and in the 9 and 14 steps for the right starting position, respectively. This shorter average tracing distance of the simulation SIM3 comes from its design goal of the fitness function $F = F_2$.

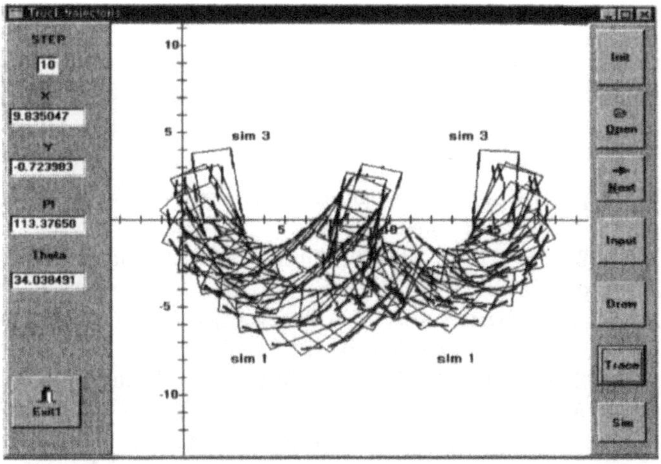

Fig. 7. Different trajectories of two simulations SIM1 and SIM3.

5 Conclusion

This paper proposed a new type of COG defuzzifier whose membership functions are adjusted in their centers, widths, and modifiers. We could attain highly accurate control by adjusting these design parameters as finely as possible. We derived an gradient-descent batch learning algorithm that these design parameters was adjusted simultaneously.

However, simultaneous adjustment of this variety of design parameters for an optimal COG defuzzifier was not an easy task. The conventional adjustment technique based on error backpropagation was not appropriate for determining the design parameters of our proposed COG defuzzifier because the resulting convergence characteristic and control performance were too sensitive to the choice of initial design parameters. The evolutionary technique to be used for avoiding this disadvantage was also not appropriate for our proposed COG defuzzifier because the computation time was too exhaustive due to its enormous searching space.
To overcome these difficulties, we proposed a novel design technique that utilized the biologically inspired concept like Lamarckian co-adaptation. Lamarckian co-adaptation allowed the phenotypic traits acquired by an organism's learning during its lifetime to be passed directly to to the organism's offspring. So, learning could guide the course of evolution directly in a way that learning altered the shape of search space and thereby provided good evolutionary paths towards sets of co-adapted individuals. In the design of an optimal COG defuzzifier, evolution adjusted design parameters inter-COG defuzzifiers through global searching while learning adjusted design parameters intra-COG defuzzifier through local searching. The most important feature of this co-adaptation scheme was to reduce the number of generations for evolution greatly.

We found that two design goals took an adverse effect on the fitness function in such a way that smaller approximation error causes longer average tracing distance, vice versa. Nevertheless, our proposed design method provides many application-specific COG defuzzifiers that can meet a variety of design goals with better approximation ability, better control performance, or balanced performance simply by controlling the weight constants k_1 and k_2 of the fitness function appropriately.

References

1. Mandani E. H. (1974) Application of Fuzzy Algorithms for Control of Simple Dynamic Plant, *IEEE Proc. Control and Science*, **121**(12), 1585-1588
2. Yager R. R., Filev D. P. (1993) SLIDE: A Simple Adaptive Defuzzification Method, *IEEE Transaction on Fuzzy Systems*, **1**(1), 69-78
3. Bastian A. (1995) Handling the Nonlinearity of A Fuzzy Logic Controller at the Transition Between Rules, *Fuzzy Sets and Systems*, **71**(1), 369-387
4. Roychowdhury S., Wang B. H. (1996) Cooperative Neighbors in Defuzzification, *Fuzzy Sets and Systems*, **78**(1), 37-50
5. Dickerson J. A., Kosko B. (1996) Fuzzy Function Approximation with Ellipsoidal Rules, *IEEE Transactions on Systems, Man, and Cybernetics-Part B:Cybernetics*, **26**(4), 542-560
6. Mabuchi S. (1993) A Proposal for Defuzzification Strategy by the Concept of Sensitivity Analysis, *Fuzzy Sets and Systems*, **55**(1), 1-14
7. Kim D., Cho I. H. (1998) An Accurate and Cost-effective COG Defuzzifier without the Multiplier and the Divider, *Fuzzy Sets and Systems*, In print
8. Runkler T. A. and Glesner M. (1993) A Set of Axioms for Defuzzification Strategies Towards a Theory of Rational Defuzzification Operators, *Second IEEE International Conference on Fuzzy Systems*, **2**, 1161-1166
9. Holland J. H. (1975) *Adaptation in Natural and Artificial Systems*, Ann Arbor: The University of Michigan Press
10. Goldberg D. E., Deb K. (1991) A Comparative Analysis of Selection Schemes Used in Genetic Algorithms, In: Rawlins G. (ed) *Fundamental of Genetic Algorithms*, Morgan Kaufmann Publisher, San Mateo, 69-93
11. Ackley D. E., Littman M. L. (1992) A Case for Lamarckian Evolution, In: Langton C. G. (ed) *Artificial Life III*, Addison-Wesley, 3-10
12. Wang L. X., Mendel J. M. (1994) Generating Fuzzy Rules by Learning from Examples, *IEEE Transactions on System, Man, and Cybernetics*, **22**(6), 1414-1427

A Multiagent Intelligent Control System for Glass Industry

Peter Mikulecký [1], Jozef Kelemen [1,3], Róbert Bódi [2], Josef Chmelař [2]

[1] *University of Education, Faculty of Management and Information Technology, 500 03 Hradec Králové, Czech Republic,*
Peter.Mikulecky@vsp.cz
[2] *Glass Service, Ltd., 755 01 Vsetín. Czech Republic,*
bodi@glaserv.cz
[3] *University of Economics,*
852 35 Bratislava, Slovakia
kelemen@euba.sk

Keywords: Expert Systems, Advanced Control, Glass Melting

Abstract

This paper is devoted to an intelligent approach towards the total glass production management in glass furnaces, and to a presentation of recent results. The global approach based on three-layered multiagent expert system architecture is explained, together with some details about various furnace parts control.

1. Introduction

The hardly explainable -- *emergent* -- behaviour of large, complex decentralised systems (e.g. in large computer networks, open information systems, etc.) composed from co-operating or cohabiting, more or less simple, and more or less autonomous agents, as well as some principles of architectures of such type of systems are studied in artificial intelligence and in robotics approximately since 1985. The variety of approaches is based on the paradigm of so called *emergent functionality* instead of the previous traditional paradigm of *deliberation* that dominated the field of AI and robotic researches during the sixties and seventies.

A common property of above mentioned systems from the perspective of the new paradigm is that they are set up from components playing different, more or less autonomous and often quite simple, roles in whole systems. The complicated behaviour of these systems emerges from the individual activities of such agents. *Multiagent systems* may provide a suitable springboard of experimental as well as theoretical studies of the phenomenon of emergent behaviour. Up to now, however, first of all experimental results are at hand.

The field of expert systems has remained up to now not tackled by the paradigm of emergent functionality. The only references we know are those referring on possibilities to build expert systems in multiprocessor architectures.

The success of classical expert systems streams [1] from the fact that these systems are active in the virtual worlds of concepts. The troubles with the expert systems directly connected to real production plants are in that the latter are "situated" in real world, not in the world of their designers' concepts. These systems have to simulate not only experts' "reasoning" on concepts, but also their (perhaps rational) "reactions" in the conditions of the real world. In other words, they must encompass not only what the expert *is telling them to do*, but also what the expert *is really doing*.

This paper is devoted to an intelligent approach towards the total glass production management in glass furnaces, and to a presentation of recent results. The global approach based on three-layered multiagent expert system architecture is explained, together with some details about various furnace parts control.

Some first ideas on the presented matters have been explained in [2], an earlier report describing the achieved results appeared in [3]. A broader context of the presented ideas has been presented in [4]. Some recent results and ideas have been described in [5] and [6].

In this paper we wish to formulate our view concerning the possibility of understanding expertise as an emergent effect of co-operation of a set of specialised agents performing simple symbol-manipulating procedures and generating specific behaviours in order to react to the sensed data. However, the work does not contribute very much to the field of multi-agent control systems, but rather to the application within a particular industry - the glass industry. Some results achieved in this direction in a project of an intelligent control system for glass furnaces will be described.

2. A Motivation for a Multiagent Expert System

In the following, we shall concentrate ourselves to controlled systems behaving continuously under some circumstances and we shall be interested in design of expert systems, which autonomously control the work of such systems. That is, such expert systems which control parameters influencing the "normal" behaviour of the controlled systems in prescribed fashion (power plants or high temperature furnaces are good examples of such systems, for instance). There are several very practical reasons for having multi-agent (decentralised) expert systems instead of the classical ones for such situations. Some of them follow:

- modularity (in design, knowledge acquisition, running, and maintenance as well)
- partial functionality of the system (failure of certain parts of the system does not influence the global functionality)

- respecting the multi-level nature of the expertise in design process as well as in running of the resulting system (desired functionality can be achieved by a co-operation of several, differently specialised experts).

In what follows we shall describe main ideas and features of a recently designed multi-agent expert system for controlling and stabilising the behaviour of an industrial system.

3. Towards the Expert System Architecture Proposal

A rough conceptual model of the proposed system, representing its main functions, is presented in the Figure 1. Let us present some remarks to the rough conceptual model.

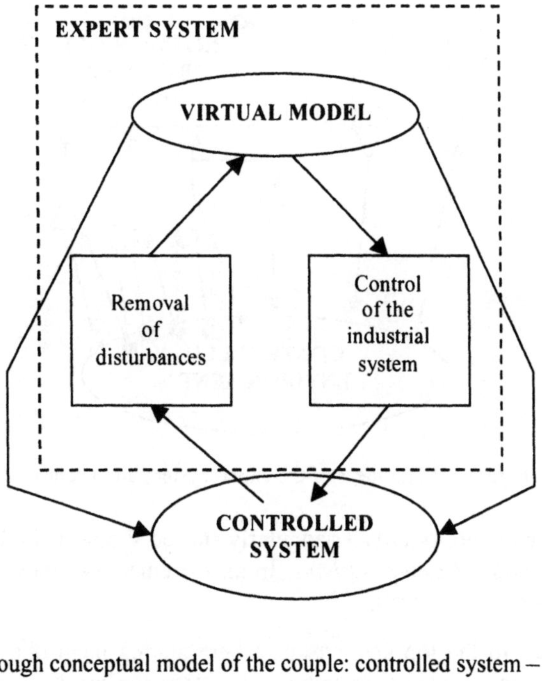

Figure 1: Rough conceptual model of the couple: controlled system – expert system

The expert system consists of two basic functional parts:

- a part focused on the removal of disturbances in measured data (this part of the expert system tests occurrences of possible sensors' failures, displays the failures to the staff and in the time before their recovery it is simulating the missing or incorrect input data on a basis of their possible correlations with other input data)
- a part devoted to the industrial system operation control (this part of the expert system is monitoring states of the controlled system, it proposes and

occasionally also automatically realises control interventions in the environment).

Both parts of the expert system are co-operating with the virtual model of the controlled environment, which is continuously maintained and updated by the expert system using the values provided by sensors.
In what follows, we shall continue the expert system architecture proposal by introducing its main ideas. The ideas are visualised in the Figure 2.

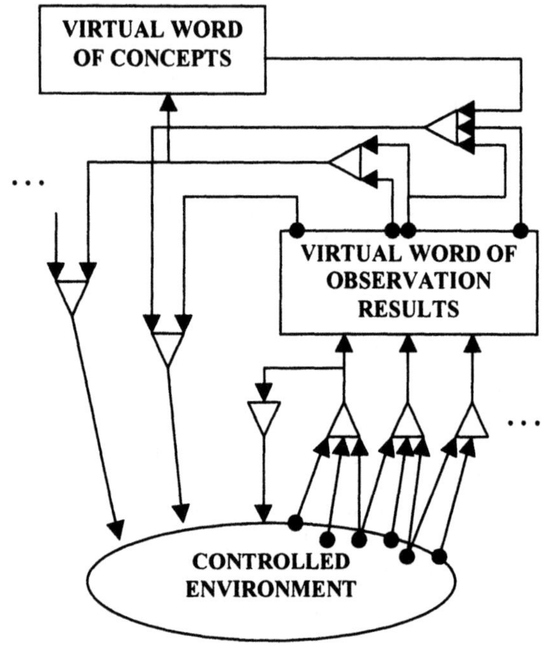

Figure 2: Main ideas of the expert system architecture

Let us explain the picture briefly. Triangularly shaped ∇ objects in the Figure 2 are *small highly specialised expert systems.* In an extreme case, they may be purely reactive software agents, as well.

Triangles oriented to the top are sensors, where among them one is for instance measuring and monitoring a temperature, the other the presence of certain substances, etc. As a result of their mutually connected activities, the VIRTUAL WORLD OF OBSERVATION RESULTS (VWOR) is created.

The VWORs are making use of expert systems of the type ◃ , which are specialised on various types of interventions into the controlled environment. In order to make decisions they are using the VWOR as well as VIRTUAL WORLD OF CONCEPTS (VWC). By another words, the VWC seems to be a *world of advices* given by a human expert (or by another ES contributing with its results to the VWC), and the ◃ may state, in the simplest case, for rules of usual expert systems.

Objects of the type ∇ are expert systems, which *realise control interventions*, recommended by the expert systems of the type ◁. They have a direct feedback with the CONTROLLED ENVIRONMENT, if necessary (e.g., an increase of temperature to a given value need not be realised via Δ →◁→∇, but directly from Δ to ∇).

The advantages of the proposed expert system architecture design are as follows:

- Each proposed module (self-contained expert systems) could be developed and tested independently from the other modules. Their complexity is expected not to be very high.
- Each "path" Δ →◁→∇creates actually a self-contained (and practically usable) expert system. The whole system is a "brick-box" of such modules. This fact enables to create a working prototype of the expert system quite quickly. The prototype can be modularly enlarged yielding thus its further versions up to the final one.
- The proposed modularity could be interesting also from the marketing point of view.
- The proposed architecture yields relatively simple possibility of innovation as well as enlargements.

4. A Knowledge-based Approach to Glass Production

There is a couple of important problems in glass melting which are, in our opinion, issues for a multi-agent expert system approach. Let us mention just a few, but essential problems:

- Melting part of a furnace:
  - glass quality (diagnosing sources of undesirable particles)
  - temperature stability
  - energy efficiency of melting process
  - environmental issues
- Working part of a furnace:
  - temperature stability
- Forehearth:
  - temperature stability
  - temperature homogeneity
  - glass quality
  - conditions for forming and product quality
  - gob weight control

When speaking about a total control of the glass production process, a system aiming to do that has somehow to cope with all of the previously described problems, as all of them can influence the glass quality and efficiency of the production dramatically.

Most work in the process industry is recently devoted to supervisory applications, like alarm analysis, process monitoring, diagnostics, or troubleshooting. The systems developed are mainly used as operator assistants and are exploring the potential of the combinations of distributed control systems with a knowledge-based approach. In a distributed control system, it is quite feasible to let an integrated real-time knowledge-based system, developed using an advanced expert system building tool, to take care of the tuning and monitoring of a number of control loops. In such a case, the expert system is in the function of a supervisor, taking care of all the necessary and controlled parts of the controlled device, e.g., a glass furnace. It is then placed topmost in a hierarchy of the whole furnace managing (controlling) system (see Figure 3), directly communicating with the second level - the information and control system level, where all the advanced process control techniques and means can be used successfully. The hardware on which the control system level is directly implemented creates in this approach the low most, third level.

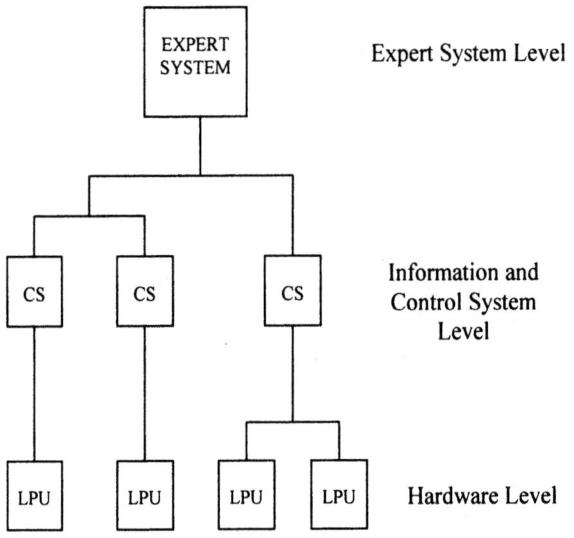

Figure 3: A three-level knowledge-based approach

In a knowledge-based approach towards total intelligent glass furnace managing, the following task should be, in our opinion, addressed:

- monitoring of sensor data, together with their interpretation as inferring situation descriptions from the data; these tasks can be naturally used in early alarming or applying operators' interventions in the most suitable time
- diagnosis should be used, when some malfunctions of the underlying system are observed and remedies from these malfunctions have to be prescribed; diagnosis could be applied to infer failures of most important tech-

nical components of the system, but also to infer problems in technological processes running in these systems

- control and prediction, together with planning are tasks for ensuring the desired underlying system behaviour; together with some advanced controlling techniques, based on a suitable AI technique (fuzzy logic, neural networks, model-based reasoning, case-based reasoning)
- instruction as a tool for operator's training.

In what follows, we shall try to describe briefly our approach to the design and experimental implementation of a system aiming of control a glass-melting furnace intelligently and totally.

5. A Multi-Layered Expert System Architecture

The most suitable architecture for a knowledge-based solution to the intelligent and total glass production management seems to be a multi-layered knowledge-based architecture, consisting of a combination of specialised expert systems mutually interconnected and co-operating in order to share those knowledge which is more generic. This architecture is combined with a number of problem specific advanced control methods, e.g., a model-based predictive control of temperatures in forehearths and feeders, etc. A rough approximation of the prototyped knowledge-based architecture is depicted in the Figure 4.

The proposed architecture reflects the basic requirements: improve level of furnace control to produce the glass of higher quality. In order to achieve this goal, it is necessary to view the whole glass producing process as a number of mutually interconnected and therefore necessary to be co-ordinated tasks. The overall co-ordination and supervision has to be fulfilled by the uppermost (co-ordination) level - the supervisory expert system. At the second level it is supposed to use a number of more specialised dedicated algorithms or expert systems: the expert system for reversal optimisation, the expert system for batch charging, the expert system for stability control of throat temperature and the system for forehearth control.

At a further, lower level an extensive usage of certain advanced control methods is realised. A model-based predictive control approach is used for the efficiency controlling in the forehearths, with some fuzzy neural control application for fuzzy tuning of the controller. A combination of fuzzy control methods with fuzzy linear models is exploited for melting part control. Some more details will be described in what follows.

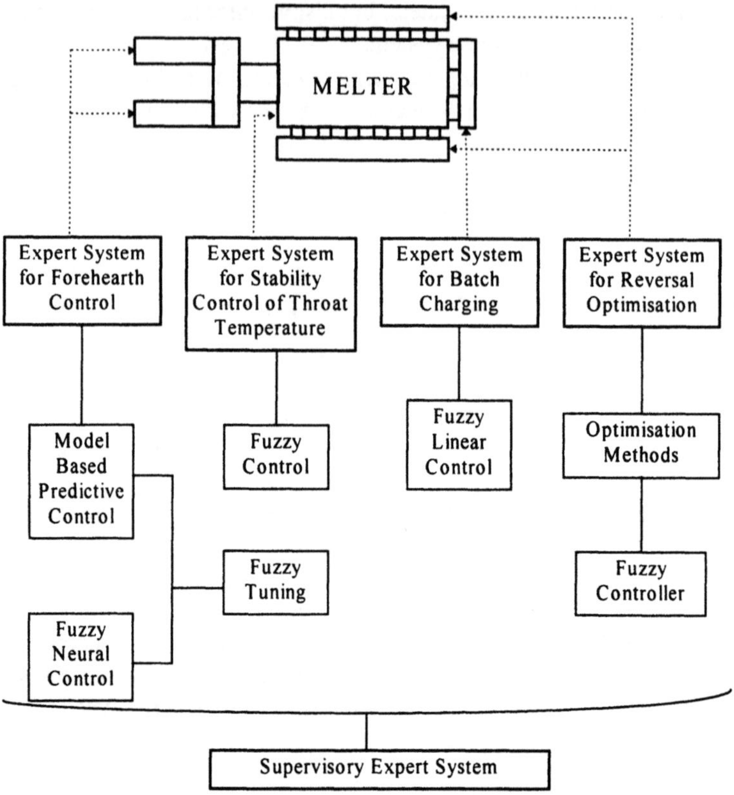

Figure 4: The proposed knowledge-based architecture

6. Implementation and First Experience

The system, implemented using just described architecture, has been recently implemented experimentally, and it has the following basic features:

- it is based on the proposed architecture (Figure 4)
- it has been implemented using a professional real-time expert system development environment
- it is aiming to optimise existing control of the glass furnace acting like an operator
- it has been tested in real production process successfully.

At the **coordination level** of the system, the aim is, under the given constraints, having defined the desired product type:

- to derive automatically the optimal set-point for the gob
- to set up the desired temperature corridors for thermal glass conditioning

- to define the desired throat temperature for the melter

The other desired tasks for the coordination level are, among others, the reversal process optimisation improving thus efficiency of regenerators.

All these aims have to contribute in balancing the ratio between the final product quality and its final price. The quality certainly has to be a compromise reflecting all the constraints. These, and also some other related tasks will be co-ordinated by a top most, rule-based supervisory expert system, co-ordinating all the sub-tasks, performed and controlled by the specialised expert systems dedicated to the control of a specific part of the furnace.

At the **melter controlling level**, in the traditional approach, the melter is controlled manually. It could be possible to use a non-linear PID as well, but the main limitation here is in the fact that a PID is a single-input single- output device. In our approach, we propose to use the fuzzy logic control because it is particularly well suitable for handling non-linear systems. The PID control of a melter is very difficult and needs to be re-tuned quite often. This drawback can be fixed by using a non-linear PID. However, two problems could arise:

- a non-linear PID is even more difficult to tune than a normal PID
- even if a good non-linear PID controller can be set up, we have to remember that the melter temperature controller has to look at more than just the melter temperature before making a change to the gas. It means that this is certainly not a single-input single-output system.

Because of all these factors we have chosen to control the melter using a fuzzy controller approach. A fuzzy controller can be set up and maintained by a person who is even not so experienced, and such a controller can easily be made adaptive.

The goal in controlling the melter is to stabilise the throat temperature. In order to achieve that, it was necessary to divide this task into two subtasks:

- the throat temperature itself
- the crown temperature, influencing the throat.

The manipulated variable here is the total gas for burners in the melter, and its distribution across the burners. Two controllers, working in a cascade have been invented to fulfil the ultimate task of controlling the throat temperature. The whole controlling process is based on a non-linear model, combined with fuzzy linear controllers and rule-based supervisor. The crown temperature control is based on a fuzzy predictive controller with adaptive gain. This approach allows us to cover also certain anomalous states, e.g., a foam appearance, batch coverage, etc.

At the **forehearth controlling level**, there are a number of non-traditional ideas in our approach. First, the heating and cooling are considered to be two independent processes, controlled individually. Second, there is no temperature profile defined, but based on the desired gob temperature a temperature corridor is defined automatically for possible temperature profiles on the feeder. The manipulated vari-

ables in the feeder are gas burners and cooling fans. It is calculated also with the feed-forward variable, reflecting a change of the throat temperature. The task is to find optimal temperatures in the zones in such a way, that the temperatures are within the temperature corridor and the resulting temperature homogeneity at the end of the feeder is fulfilled.
The whole glass conditioning process is controlled using model-based predictive controllers. The disadvantage of this approach is in its dependency on the model quality. In order to achieve this it is necessary to perform a number of tests on the particular furnace and to derive the models preciously. On the other hand, a number of advantages of the used approach can be mentioned:

- no human operator for setting up the temperature profile is needed
- the profile change for production change is derived automatically; the operator says the desired gob temperature and the controller task is to find the optimal temperature profile quickly
- this means a very effective exploitation of the transient time: the production after a change is prepared very rapidly, usually earlier than the machine is pre-set
- the controller's configurability is very good: using a flexible tune possibility the controller's response can be faster or slower, as necessary
- in the case of a thermocouple failure, the controller is re-configured, and the faulty thermocouple is not more involved in the controlling process (this feature is realised in co-operation with a separate fault detection module of the expert system).

7. Conclusions

The techniques used are very much applications of recently developed and broadly exploited methods of soft computing, which can be found, in a number of recent books, e.g., [7] or [8].

The practical experience with using the recent version of the proposed multi – layered expert system is very good, as it is derived from the real production data taken from a glass furnace. The results of a conventional PID control of feeder condition temperatures have been compared thoroughly with the results of an advanced feeder control. The temperature homogeneity of glass is very high (efficiency about 99%) and stable in the second case.

Long term result has been very stable according to production experience. The efficiency is in range 96 - 99% depending on operating parameters and production changes. Similar results are expected in the case of controllers used for the melter control.

The further development of the expert system has to take into account the following aspects or activities:

- assistance in job changes

- glass quality feedback
- information and analysis of process errors
- internal diagnostics and access for system engineer
- an enhanced user friendly operator interface.

The system will eventually supervise the majority of technological actions on the furnace. The major expected benefits of the expert systems be as follows:

- improvement in the production quality by achieving of more stable technology parameters of furnace
- decrease of the human factor influence on furnace control
- reducing the energy consumption
- addressing the environmental issues
- safety improvements using early alarming
- on-line diagnostics and correction of a number of possible furnace instrumentation failures.

The experience and development work on the expert system has been conducted for:

- container glass
- float technology
- fiberglass technology

Although the system is in the development, the first extensive practical tests have brought very promising results. It has been documented that the intelligent and total glass quality management is the right way to the highest quality glass production.

8. References

1. Kelemen J 1996 Towards robots with mind. *Human Affairs 6*: 97-110

2. Chmelař J, Bódi R, Mikulecký P 1997 Towards total glass quality management. In: *Proc. of 4th Int. Seminar on Mathematical Simulation in the Glass Melting, Horní Bečva, Czech Republic,* pp. 172-181

3. Chmelař J, Bódi R, Mikulecký P, Malik S A 1997 Expert systems in intelligent control of continuous glass melting processes. In: *Proceedings from the Workshop on Advanced Control for Industrial Glass Melters, MECC, Maastricht, The Netherlands*

4. Mikulecký P, Ponce D, Hynek J 1997 Knowledge-based projects in industrial applications. In: *Proc. of 4th Int. Seminar on Mathematical Simulation in the Glass Melting, Horní Bečva, Czech Republic,* pp. 191-201

5. Mikulecký P, Bódi R 1997 On emergence of expertise in multiagent real-time systems. FMIT Informatics Technical Report TR97005, Hradec Králové University of Education, Czech Republic

6. Mikulecký P, Bódi R, Chmelař J 1998 Intelligent total glass production control. In: Sarnovský J et al (eds) 1998 *Proc. of the "Artificial Intelligence in Industry" Conf., High Tatras, Slovakia, Tech. University Košice,* pp. 199-206

7. Dubois D, Prade H, Yager R. R (eds.) 1997 *Fuzzy Information Engineering. A Guided Tour of Applications.* J. Wiley, New York

8. Nie J, Linkens D A 1995 *Fuzzy – Neural Control: Principles, Algorithms and Applications.* Prentice Hall, New York

Predictive Control Using Fuzzy Models

Jairo J. Espinosa and Joos Vandewalle

Departement Elektrotechniek -K.U.Leuven
Kardinaal Mercierlaan 94, B-3001 Heverlee-Belgium
jairo.espinosa@esat.kuleuven.ac.be joos.vandewalle@esat.kuleuven.ac.be

Keywords: Predictive Control, Fuzzy Modeling, System Identification, Fuzzy Control

Abstract

This paper presents an algorithm for predictive control using fuzzy models. This strategy is a Predictive Control Strategy for Nonlinear systems. The algorithm takes advantage of the already well known methods for linear systems (like Generalized Predictive Control). Analysis of the modified algorithm is presented and performance comparisons are shown. The performance comparisons are made using a model of a chemical process (Continuous Stirring Tank Reactor) where noise and uncertainties are included.

1 Introduction

In the recent years it had been recognized the important role played by the Model Predictive Control (MPC) strategies specially in the areas of process control [3] [5] [2]. These strategies have been specially attractive because of its reduced number of tuning parameters and high performance [1]. Most of the time spent on tuning predictive controllers corresponds to the identification of the process model used by the controller. Fuzzy modeling had also been recognized as a "practical" strategy for control and for modeling.

Fuzzy models had been very attractive for its capability to capture expert knowledge. In the recent years many algorithms to generate fuzzy models starting from Input-Output data have been developed [11] [9]. Nowadays new algorithms permit the combination of numerical and expert information in the same framework [6]. A framework for non-linear system identification using fuzzy models has been proposed in [14] [15]. The algorithms presented in [15] and [6] guarantee that the obtained fuzzy model will be at least as good as the linear model obtained using classical linear system identification [13].

The wide acceptance of these two strategies motivated the formulation of a control strategy using the theory of Model Predictive Control based on Fuzzy Model. This paper presents such a formulation. Section 2 of the paper presents the fundaments of fuzzy modeling and identification. Section 3 presents the formulation of the predictive control strategy. Section 4 presents an application example of the control strategy compared with a Linear Model Predictive Controller and a simple PID controller. Section 5 presents the conclusions and the future research directions.

2 Fuzzy Modeling and Identification

2.1 Fuzzy modeling

A general fuzzy model can be represented as a non-linear function described by (1). The function f is a weighted sum of L basis functions $w_l(\bar{\mathbf{u}})$.

$$f(\bar{\mathbf{u}}) = \sum_{l=1}^{L} w_l(\bar{\mathbf{u}})\bar{y}_l = \frac{\sum_{l=1}^{L} \mu_l(\bar{\mathbf{u}})\bar{y}_l}{\sum_{l=1}^{L} \mu_l(\bar{\mathbf{u}})} \tag{1}$$

In the fuzzy model L represents the number of rules. The weights $\bar{y}_l$ represent the consequences of the rules. The basis functions $w_l(\bar{\mathbf{u}})$ represent the inference process and the defuzification process. The inference process generate the membership degree of each rule by using the AND operation between the antecedents of the rules. The rules represented by this process have the following form:

$$\begin{array}{c} \cdots \\ \text{IF } u_1 \text{ is } \mu_l^1 \text{ AND } u_2 \text{ is } \mu_l^2 \text{ AND } \ldots \text{AND } u_{N_i} \text{ is } \mu_l^{N_i} \text{ THEN } f = \bar{y}_l \\ \cdots \end{array}$$

where $\bar{\mathbf{u}} = \{u_1, u_2, \ldots, u_{N_i}\}^T$ and μ_l^i represents the fuzzy set corresponding to the i input in the l rule. These membership functions can be triangular, trapezoidal or Gaussian. The AND operator can be a *product* or a *min* operation.

$$\mu_l(\bar{\mathbf{u}}) = \begin{cases} \min\{\mu_l^1(u_1), \mu_l^2(u_2), \ldots, \mu_l^{N_i}(u_{N_i})\} & \text{using MIN operator} \\ \mu_l^1(u_1) \cdot \mu_l^2(u_2) \cdot \ldots \cdot \mu_l^{N_i}(u_{N_i}) & \text{using PRODUCT operator} \end{cases} \tag{2}$$

The "training" of these models or the extraction of the model using input-output data has been treated extensively in [11][9][6]. In the present work the linear structure of the consequences of the rules is exploited to generate a fuzzy model with fixed antecedents where only the consequences are calculated. For N input-output data the inference process can be represented as

$$\underbrace{\begin{bmatrix} y_1 \\ y_2 \\ \vdots \\ y_N \end{bmatrix}}_{\mathbf{Y}} = \underbrace{\begin{bmatrix} w_1^1 & w_2^1 & \dots & w_L^1 \\ w_1^2 & w_2^2 & \dots & w_L^2 \\ \vdots & \vdots & \ddots & \vdots \\ w_1^N & w_2^N & \dots & w_L^N \end{bmatrix}}_{\mathbf{W}} \underbrace{\begin{bmatrix} \bar{y}_1 \\ \bar{y}_2 \\ \vdots \\ \bar{y}_L \end{bmatrix}}_{\boldsymbol{\theta}} + \underbrace{\begin{bmatrix} e_1 \\ e_2 \\ \vdots \\ e_N \end{bmatrix}}_{\mathbf{E}} \tag{3}$$

The aim of the training here is to reduce as much as possible the norm of the vector $\mathbf{E}$. Using quadratic norm:

$$\min_{\boldsymbol{\theta}} \|\mathbf{E}\|_2 = \min_{\boldsymbol{\theta}} \|\mathbf{Y} - \mathbf{W}\boldsymbol{\theta}\|_2 \tag{4}$$

The problem corresponds to a Least-Squares problem. Typically the matrix $\mathbf{W}$ is poorly conditioned for this reason this problem is solved using Recursive Least Squares (RLS). The initial values of the $\boldsymbol{\theta}$ vector can be generated from expert knowledge or using linear regression, this linear regression will be in fact the best linear model you can construct out of these data, so the fuzzy model will have a minimum performance given by the performance of the linear model. Of course the training process will improve the fuzzy model capturing the desired non-linear characteristics of the process.

2.2 Identification using fuzzy models

Several structures can be implemented using fuzzy models as was shown in [15]. But perhaps the most important structures are the NARMAX (Non-linear Auto Regressive Moving Average with eXogenous input)

$$\hat{y}(k+1) = f(y(k), \dots, y(k-m), u(k), \dots, u(k-n)) \tag{5}$$

where $y(k), \dots, y(k-m)$ represent the past outputs of the system and $u(k), \dots, u(k-n)$ represent the past inputs. The NARMAX model is also known as series-parallel model; The other structure is the NOE (Non-linear Output Error)

$$\hat{y}(k+1) = f(\hat{y}(k), \dots, \hat{y}(k-m), u(k), \dots, u(k-n)) \tag{6}$$

where $\hat{y}(k), \dots, \hat{y}(k-m)$ represent the past outputs of the model and $u(k), \dots, u(k-n)$ represent the past inputs. The NOE model is also known as parallel model. The first model is suitable for short term prediction in the presence of noise and the second one is more suitable for simulation and long term prediction.

The correct training of the NOE demands the use of gradient descent techniques with dynamic gradient calculation because of its recurrent nature.

These procedure is very expensive from the computational point of view. A good intermediate solution is to calculate first the NARMAX model and then use its consequences as the initial consequences of the NOE model. A further refinement is obtained by using RLS. Of course the solution will be biased, but most of the time the bias is small.

For the selection of the correct inputs to the model in [15] two methods were proposed. The first method is a free-model method called the ***Lipschitz Quotients***, this method basically evaluates the smoothness of the possible model and the other is the Regularity Criterion in (Group Methods for Data handling) [10]. The first method is sensitive to noise. The second method demands the partition of the training set in two data groups and the calculation of several models which are evaluated using the following cost function

$$RC = \left[\sum_{i=1}^{k_A} (y_i^A - y_i^{AB})^2 / k_A + \sum_{i=1}^{k_B} (y_i^B - y_i^{BA})^2 / k_B \right] / 2 \qquad (7)$$

where

- k_A and k_B are the number of data of the groups A and B.
- y_i^A and y_i^B outputs of the data groups A and B.
- y^{AB} Model output for the group A input estimated using the group B data.
- y^{BA} Model output for the group B input estimated using the group A data.

combined with a search tree where always the branch with the highest RC value is picked. The procedure stops when there is no further decrease in the RC value. This procedure is applied using a reduced model (with only two membership functions on each input). When the sampling rate is too fast the procedure selects only past outputs generating only an autoregressive model not suitable for our control purposes. In such cases the search tree has to be implemented in two trees one to choose the past outputs and the other to choose past inputs.
It is important to remark that the quality of the model will be determined by the characteristics of the exciting input selected for the identification. For non-linear systems the signals should be rich enough not only in frequency but also in amplitude. It means that the use of signals like PRBS (very popular on linear identification) will not be enough because these type of signals are rich in frequencies but not in amplitudes.

3 The Predictive Control Strategy

The Model Predictive Control is an optimal control strategy using the concept of receding horizon. The MPC computes a sequence of manipulated variables to optimize the future behaviour of the plant. The future output is determined for a given horizon N, named the prediction horizon. The predicted output $y(t+k|t)$ for $k = 1, \ldots, N$ depends on the known values of the output and the input up to time t and the future control signals $u(t+k|t), k = 0, \ldots, N_u$ and these are the values to be calculated. N_u is the so-called control horizon. The objective is to keep the output $y(t+k)$ as close as possible to a reference $w(t+k)$. The cost function used to evaluated the "distance" between the reference and the output makes use of a quadratic norm.

$$J = \sum_{k=1}^{N} [w(t+k) - y(t+k|t)]^2 + \sum_{k=0}^{N_u} \rho u(t+k-1|t)^2 \tag{8}$$

Once the "optimal" sequence is calculated only the $u(t|t)$ signal is implemented, for the time $t+1$ a new value of $y(t+1)$ is measured and a new input sequence is calculated.

The present predictive control algorithm uses most of the ideas present in the Generalized Predictive Control (GPC) [1] strategy. The present algorithm assumes the following plant model:

$$y(t) = \hat{y}(t) + n(t) = f(\hat{y}(t-1), \ldots, \hat{y}(t-m), u(t-1), \ldots, u(t-n)) + n(t) \tag{9}$$

where the function $f(.)$ corresponds to the fuzzy model. The noise model is given by

$$n(t) = \frac{C(z^{-1})}{D(z^{-1})} e(t) \tag{10}$$

where $e(t)$ is a *white noise* sequence. A typical choice of the noise model is:

$$n(t) = \frac{C(z^{-1})}{D(z^{-1})} e(t) = \frac{e(t)}{\Delta} \tag{11}$$

where $\Delta = 1 - z^{-1}$. This choice of the noise model will guarantee a zero steady state error for step disturbances and constant references, for details see [1] [2] [3].

The predicted future output is given by:

$$y(t+k|t) = y_{\text{forced}}(t+k|t) + y_{\text{free}}(t+k|t) \tag{12}$$

where $y_{\text{forced}}(t+k|t)$ depends only on the future inputs and $y_{\text{free}}(t+k|t)$ depends only on the past inputs and outputs. In the present case the y_{free} sequence is given by:

$$y_{\text{free}}(t+k|t) = f(\hat{y}(t+k-1),\ldots,\hat{y}(t+k-m),u(t+k-1),\ldots,u(t+k-n))+n(t+k|t) \tag{13}$$

with $u(t) = u(t+1) = \ldots = u(t+k-1) = u(t-1)$ which is equivalent to simulate the system assuming all the future inputs constant and equal to the last applied output $u(t-1)$ and the sequence $n(t+k|t)$ is calculated assuming that $e(t+k|t) = 0$ for $k > 1$. The predicted y_{forced} will be given by

$$y_{\text{forced}}(t+k|t) = \sum_{i=0}^{k-1} g_i \Delta u(t+k-i-1|t) \tag{14}$$

where g_i are the step response coefficients of the plant, calculated on the present operating point by simulating the step response on the model. The optimization problem can be presented in matrix form as:

$$\mathbf{Y} = \mathbf{G}\Delta\mathbf{U} + \mathbf{Y}_{\text{free}} \tag{15}$$

where

$$\mathbf{Y} = \begin{bmatrix} y(t+1|t) \\ y(t+2|t) \\ \vdots \\ y(t+N|t) \end{bmatrix} \qquad \mathbf{G} = \begin{bmatrix} g_0 & 0 & \cdots & 0 \\ g_1 & g_0 & \cdots & 0 \\ \vdots & \vdots & \ddots & \vdots \\ g_{N_u-1} & g_{N_u-2} & \cdots & g_0 \\ \vdots & \vdots & \vdots & \vdots \\ g_{N-1} & g_{N-2} & \cdots & g_{N-N_u-1} \end{bmatrix} \tag{16}$$

$$\Delta\mathbf{U} = \begin{bmatrix} \Delta u(t|t) \\ \Delta u(t+1|t) \\ \vdots \\ \Delta u(t+N_u-1|t) \end{bmatrix}, \mathbf{W} = \begin{bmatrix} w(t) \\ w(t+1) \\ \vdots \\ w(t+N) \end{bmatrix}, \mathbf{Y}_{\text{free}} = \begin{bmatrix} y_{\text{free}}(t+1|t) \\ y_{\text{free}}(t+2|t) \\ \vdots \\ y_{\text{free}}(t+N|t) \end{bmatrix} \tag{17}$$

$$\begin{aligned} y_{\text{free}}(t+i|t) &= f(\hat{y}(t+i-1),\ldots,\hat{y}(t+i-m), \\ &\quad \hat{u}(t+i-1),\ldots,\hat{u}(t+i-n)) + n(t+i|t) \\ \text{with: } \hat{u}(k) &= \begin{cases} u(t-1) & \forall k > t-1 \\ u(k) & \text{otherwise} \end{cases} \end{aligned} \tag{18}$$

Then the cost function can be written as:

$$J = \mathbf{E}^T\mathbf{E} + \rho\Delta\mathbf{U}^T\Delta\mathbf{U} \tag{19}$$

$$\begin{aligned} J &= (\mathbf{W}-\mathbf{Y})^T(\mathbf{W}-\mathbf{Y}) + \rho\Delta\mathbf{U}^T\Delta\mathbf{U} \\ &= (\mathbf{W}-\mathbf{G}\Delta\mathbf{U}+\mathbf{Y}_{\text{free}})^T(\mathbf{W}-\mathbf{G}\Delta\mathbf{U}+\mathbf{Y}_{\text{free}}) + \rho\Delta\mathbf{U}^T\Delta\mathbf{U} \end{aligned} \tag{20}$$

the minimization of the function can be obtained by calculating input sequence $\Delta\mathbf{U}$ such that $\frac{\partial J}{\partial \Delta\mathbf{U}} = 0$

$$\frac{\partial J}{\partial \Delta\mathbf{U}} = 2\mathbf{G}^T(\mathbf{Y}_{\text{free}} - \mathbf{W}) + 2(\mathbf{G}^T\mathbf{G} + \rho I)\Delta\mathbf{U} = 0 \tag{21}$$

then the optimal sequence $\Delta\mathbf{U}$ is

$$\Delta\mathbf{U} = (\mathbf{G}^T\mathbf{G} + \rho I)^{-1}\mathbf{G}^T(\mathbf{W} - \mathbf{Y}_{\text{free}}) \tag{22}$$

The input applied to the plant at time t is:

$$u(t) = u(t-1) + \Delta u(t) \tag{23}$$

where $\Delta u(t)$ is the first element of the vector $\Delta\mathbf{U}$.

3.1 Estimation of the step response

The estimation of the step response is obtained by

$$g(k-1) = \frac{y_{\text{step}}(t+k|t) - y_{\text{free}}(t+k|t) - n(t+k|t)}{du} \tag{24}$$

where du represents the step size and

$$\begin{aligned} y_{\text{step}}(t+k|t) &= f(\hat{y}(t+k-1), \ldots, \hat{y}(t+k-m), \\ &\quad \hat{u}(t+k-1), \ldots, \hat{u}(t+k-n)) \\ \text{with: } \hat{u}(k) &= \begin{cases} u(k) & \forall k \leq t-1 \\ u(t-1) + du & \forall k > t-1 \end{cases} \end{aligned} \tag{25}$$

The estimation of the step response is a very important element on the present algorithm, because the quality of the response of the system is determined by the accuracy of this estimation. The main difficulty to obtain an accurate estimate of the step response around some operating point is to select an appropriate value for the size and the sign of the step du. It is important to remark that the "step response" of a nonlinear system is determined by the operating point, the size and the sign of the input step signal. A good value of du should satisfy some requirements, perhaps the most important is that the value $u(t-1) + du$ should not saturate the "actuators" in the model, because it can lead to an underestimated step response (with smaller gain) and this response will generate a response from the controller larger than it is needed, possibly generating unstable behaviour. Also important is the fact that du should be "close" to the predicted $\Delta u(t)$. Because this value is only known after the optimization a good choice is to select $du = \Delta u(t+1)$ obtained in the previous optimization step $(t-1)$, and it corresponds to the second element of the vector $\Delta\mathbf{U}$. When the system reaches the steady state $\Delta\mathbf{U} \to 0$ it means that if we select du as stated before, the equation (24) will be badly conditioned, it demands the definition of a minimum step size du_{min} to avoid this inconvenient.

4 Predictive Control of a CSTR Using a Fuzzy Model

As example of the application of this control strategy a plant representing a Continuous-Stirred Tank Reactor was chosen. The model of this plant was presented in [16]. The model is described by the following differential equations:

$$\begin{aligned}\dot{C}_a(t) &= \tfrac{q}{v}(C_{a0} - C_a(t)) - k_0 C_a(t) e^{-\frac{E}{RT(t)}} \\ \dot{T}(t) &= \tfrac{q}{v}(T_0 - T(t)) - k_1 C_a(t) e^{-\frac{E}{RT(t)}} \\ &\quad + k_2 q_c(t)\left(1 - e^{-\frac{k_3}{q_c(t)}}\right)(T_{c0} - T(t))\end{aligned} \tag{26}$$

The process describes the reaction of two products, which are mixed and react to generate a compound A with concentration $C_a(t)$, with the temperature of the mixture $T(t)$. This reaction is exothermic. The heat generated acts to slow the reaction. The reaction is controlled by introducing a coolant flow rate $q_c(t)$, which helps to change the temperature and by this means the concentration. C_{a0} is the inlet feed concentration, q is the process flow rate, T_0 and T_{c0} the inlet feed and coolant temperatures, all these values are assumed constant at nominal values. In the same way, k_0, E/R, v, k_1, k_2 and k_3 are thermodynamic and chemical constants. The numerical values of these parameters are given in table 1.

Parameter	Description	Nominal Value
q	process flow-rate	100 l/min
v	reactor volume	100 l
k_0	reaction rate constant	$7.2 \times 10^{10}\text{min}^{-1}$
E/R	activation energy	1×10^4K
T_0	feed temperature	350 K
T_{c0}	inlet coolant temp.	350 K
ΔH	heat of reaction	-2×10^5 cal/mol
C_p, C_{pc}	specific heats	1 cal/g/K
ρ, ρ_c	liquid densities	1×10^3 g/l
h_a	heat transfer coeff.	7×10^5 cal/min/K
C_{a0}	inlet feed concentration	1 mol/l

Table 1: The CSTR Parameters

$$k_1 = -\frac{\Delta H k_0}{\rho C_p} \quad k_2 = \frac{\rho_c C_{pc}}{\rho C_p v} \quad k_3 = \frac{h_a}{\rho_c C_{pc}}$$

The nominal conditions for an product concentration $C_a = 0.1$ mol/l are:

$$T = 438.54\text{K} \qquad q_c = 103.41\text{l/min}$$

4.1 Fuzzy identification of the CSTR

For the identification procedure a sampling time $Ts = 6$ sec. was chosen. A sequence of 7500 samples was generated; the first 5000 samples were used as training set and 2500 as validation set. Figure 1 shows the input and the output signal used for the identification.

The regressors were selected using the regularity criterion for GMDH the structure of the model is:

$$\hat{C}_a(k+1) = f(\hat{C}_a(k), \hat{C}_a(k-1), \hat{C}_a(k-2), q_c(k-1)) \tag{27}$$

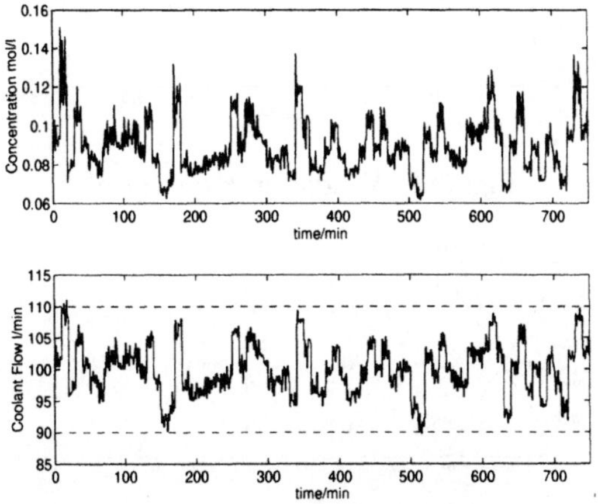

Figure 1: Identification Experiments

The identified model is a NOE model; the fuzzy model has 3 triangular membership functions distributed regularly on the universes of discourse. The number of rules is then $3 \times 3 \times 3 \times 3 = 81$ rules. The membership functions can be observed in figure 3 and the results of the validation in the figure 2. The figure 2 shows the validation error on simulation, the quality of the model is very good and in fact the two signals appear overlapped, the error is plotted in the bottom part of the figure 2 in another scale to make it visible.

4.2 Implementation of the controller and performance results

The predictive controller was implemented using the following parameters: prediction horizon $N = 10$, control horizon $Nu = 2$, $\rho = 0$, $C(z^{-1}) = 1$ and $D(z^{-1}) = 1 - z^{-1}$, $du_{min} = 0.1(q_c^{max} - q_c^{min}) = 2$ l/min. The figure 5 shows

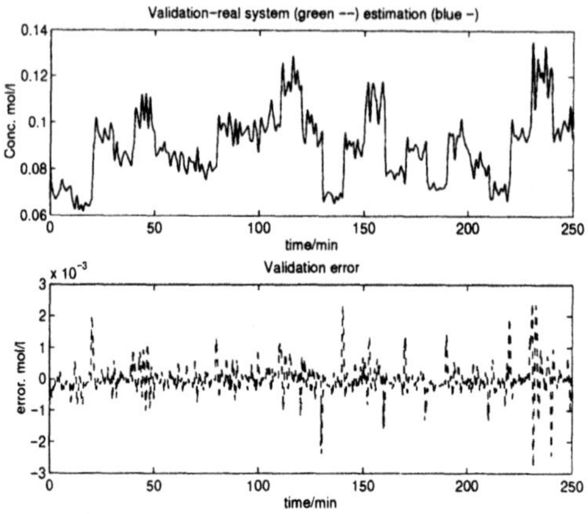

Figure 2: Validation Experiment

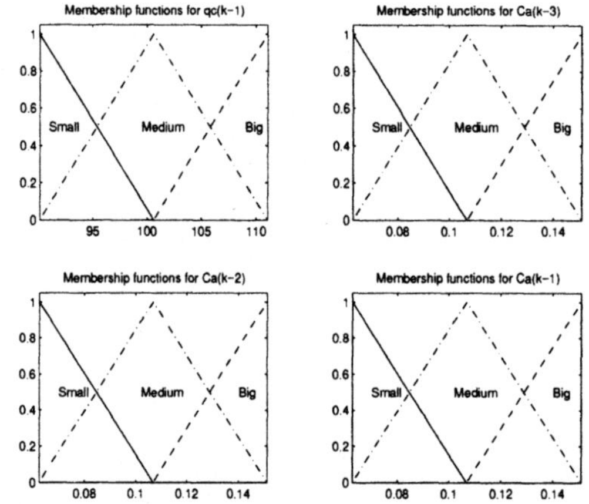

Figure 3: Membership function of the fuzzy model.

the response of the system with a "stair" reference; each step has duration of 10 min and amplitude of 0.005 mol/l. The reference start on 0.08 mol/l and it goes up to 0.12 mol/l. Perturbation steps of 0.005 mol/l are applied in the middle of each step. The reference and the perturbation signal are shown in the figure 4.

To compare the performance of the designed controller two other control strategies were implemented. A PID strategy optimized for the current reference. The parameters of the PID are:

$$K_p = 91.2 \quad T_i = 0.178\text{sec.} \quad T_d = 0.208\text{sec.}$$

The other control strategy was a Generalized Predictive Controller (GPC)

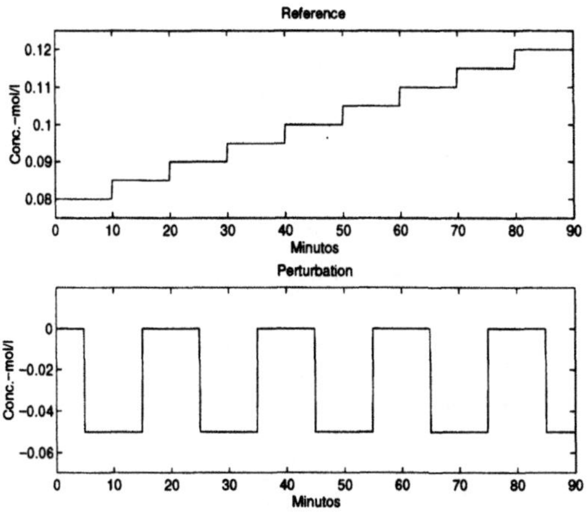

Figure 4: Reference and Perturbation

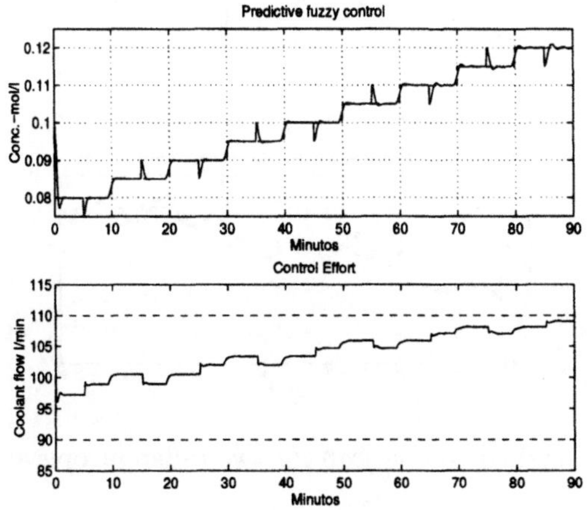

Figure 5: Response of the system

designed with the same parameters as our predictive controller but using an Output Error linear model described as:

$$\hat{C}_a(k+1) = \frac{1.653 \times 10^{-4} z^{-1}}{1 - 2.43z^{-1} + 2.4z^{-2} - 1.189z^{-3} + 0.269z^{-4}} q_c(k) \tag{28}$$

The figure 6 shows details of the performance comparison. It is important to remark the degradation in the performance experimented by the systems with the linear GPC controller and the PID controller when the concentration is near to 0.12 mol/l, observe that the PID controller is not able to stabilize the plant near to this set point. Also the performance of the GPC is very de-gradated.

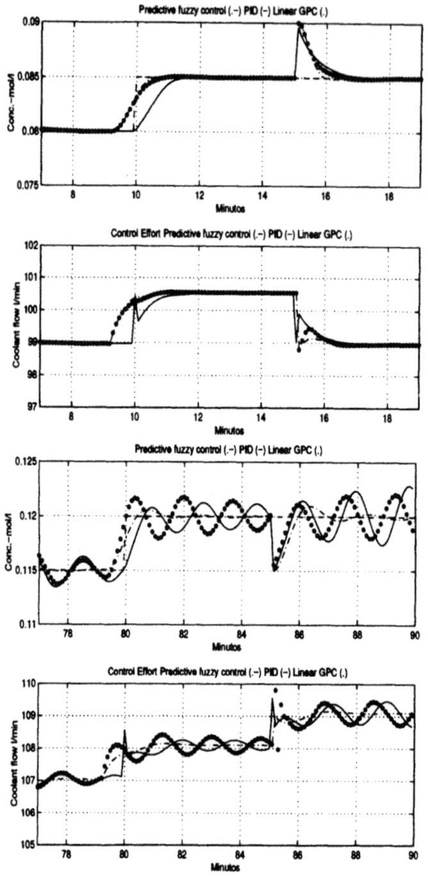

Figure 6: Detailed comparison for two different operating points

5 Conclusions

A Non-linear Model Based Predictive Control strategy based on fuzzy models had been presented. The simplicity of the modeling and the performance of the controller make the design very attractive for the designer. The strategy is a very efficient approach to the problem of Non-linear predictive control. Its efficiency comes from the combination of linear and non-linear strategies into the formulation of the algorithm. The use of fuzzy models had been encourage because of its transparency to the user, this fact makes this modelling technique more powerful and user friendly than neural nets and GMDH that are only data guided and the information given by their parameters is not very clear to the user.

Future work should be done to improve the constraint handling capabilities of the algorithm. Up to now the algorithm took into account the level constraints in the actuator (saturation) to estimate the step response but there is no warranty that the obtained control strategy will respect these constraints.

Acknowledgment

This work is supported by several institutions: the Flemish Government: Concerted Research Action GOA-MIPS (Model-based Information Processing Systems), the FWO (Fund for Scientific Research - Flanders) project G.0262.97 ;Learning and Optimization: an Interdisciplinary Approach, The FWO Research Communities: ICCoS (Identification and Control of Complex Systems) and Advanced Numerical Methods for Mathematical Modeling, The Belgian State, Prime Minister's Office - Federal Office for Scientific, Technical and Cultural Affairs - Interuniversity Poles of Attraction Programme (IUAP P4-02 (1997-2001): Modeling, Identification, Simulation and Control of Complex Systems SCIENCE-ERNSI (European Research Network for System Identification): SC1-CT92-0779.

References

1. Clarke D.W., Mohtadi C., Tuffs P.S., 1987, Generalized Predictive Control-Parts I - II, *Automatica*, **23**, 137-160.

2. Garcia C.E., Prett D.M.,Morari M., 1989, Model Predictive Control: Theory and Practice-a Survey, *Automatica*, **25**, 335-348.

3. Richalet J., 1993, Industrial Applications of Model Based Predictive Control, *Automatica*, **29**, 1251-1274,

4. De Keyser R.M.C., Van de Velde Ph.G.A., Dumortier F.A.G., 1988, A Comparative Study of Self-Adaptive Long-range Predictive, Control Methods, *Automatica*, **24**, 149-163.

5. Camacho E.F.,Bordons C., 1995, *Model Predictive Control in the Process Industry*, Springer Verlag, London.

6. Espinosa J.,Vandewalle J., 1998, *Constructing Fuzzy Models with Linguistic Integrity*, Internal Report,SISTA, Electrothecnik Departement, Katholieke Universiteit Leuven, Belgium.

7. Tan S.,Vandewalle J., 1995, An On-line Structural and Parametric Scheme for Fuzzy Modelling, *Proceeding of the 6th International Fuzzy Systems Association World Congress IFSA-95, Sao Paulo, Brazil, July*, pp.189-192.

8. Pedrycz W., 1994, Why Triangular Membership Functions?, *Fuzzy Sets and Systems*, **64**, 21-30,

9. Jang J.-S.R., 1992,*Neuro-Fuzzy Modeling: Architectures, Analyses and Applications*, Phd. dissertation, University of Berkeley, California, USA.

10. Sugeno M.,Yasukawa T., 1993, A Fuzzy Logic Based Approach to Qualitative Modeling *IEEE Trans. on Fuzzy Systems*, **1**, 7-31.

11. Wang L.X. 1994, *Adaptive Fuzzy Systems and Control*, Prentice Hall, USA.

12. Yager R.R.,Filev D., 1994, *Essentials of fuzzy modeling and control*, John Wiley & Sons, USA.

13. Ljung L., 1987, *System Identification. Theory for the User*, Prentice Hall, USA.

14. Bossley K.M., 1997, *Neurofuzzy Modelling Approaches in System Identification*, Ph.D. Dissertation, University of Southampton, U.K.

15. Espinosa J., Vandewalle J., 1997, Fuzzy Modeling and Identification, A guide for the user, *Proceedings of the IEEE Singapore International Symposium on Control Theory and Applications, Singapore, July*, pp.437-443.

16. Lightbody G., Irwin G.W., 1997, Nonlinear Control Structures Based on Embedded Neural System Models, *IEEE Trans. on Neural Networks*, **8**, 553-567.

Evolutionary Design of a Helicopter Autopilot

Frank Hoffmann, Tak John Koo, Omid Shakernia
Electrical Engineering & Computer Science Department
University of California, Berkeley
email: fhoffman@cs.berkeley.edu

Abstract

This paper presents an evolutionary design method for fuzzy logic controllers, which is based on a self-organizing process that learns the appropriate relationship between control input and output. Our approach employs an evolution strategy that operates on vectors of real numbers which correspond to the gain factors in the conclusion part of fuzzy rules. An incremental learning scheme gradually expands the genome and thereby refines the fuzzy knowledge base that acquires additional fuzzy rules.

The paper describes the fuzzy control autopilot that constitutes the regulation layer of a hybrid flight vehicle management system, which governs the operation of an autonomous model helicopter. The autopilot is composed of four modules that control the longitudinal and lateral motion, altitude and heading. The modules are implemented by a hierarchy of fuzzy controllers. The evolutionary algorithm optimizes the fuzzy rule bases off-line. We compare two design approaches, learning the rule base starting without previous knowledge and tuning an predefined set of fuzzy rules.

1 Introduction

Evolutionary algorithms are search and optimization methods, which are guided by the process of natural evolution. Although they employ different genetic representations, they share the same generic features, a population of competing candidate solutions, random combination and alteration of potentially useful structures to generate new variants and selection of fitter solutions as a means to direct the exploration of the search space.

Fuzzy systems describe knowledge by means of linguistic concepts which require less complexity and precision than mathematical or logical models. Their approximate reasoning scheme exploits the tolerance to imprecision and uncertainty inherent to many real world problems. Fuzzy control provides a flexible means to model the nonlinear relationship between input information and control output. Its robustness with respect to noise and variation of system parameters make fuzzy control a suitable method for intelligent control of unmanned helicopter [13].

Recently, numerous publications propose to integrate the learning capabilities of evolutionary algorithms with the knowledge representation of fuzzy systems [3, 4, 10]. These methods are described by the general term genetic-fuzzy systems. The central idea is to automate the knowledge acquisition step in fuzzy control design by an evolutionary algorithm. In this context learning a fuzzy knowledge base becomes an optimization problem, in which the search space is constituted by the fuzzy membership functions and fuzzy if-then rules. PHILLIPS ET AL. [11] proposed a genetic algorithm algorithm to learn fuzzy logic controllers for helicopter flight.

2 Evolution Strategy

An evolutionary algorithm propagates a population of genetic variants from one generation to the next. A coding scheme transcribes the genetic blue-print of an individual into a potential solution of the optimization problem. A scalar objective function evaluates the quality of solutions in regard to the optimization task. According to Darwin's principle, highly fit individuals obtain a higher chance of reproduction. Selection focuses future search to those regions in which good solutions were already found. Genetic operators such as recombination and mutation generate new offspring which inherit their genetic material from the selected parents. This evolutionary cycle of selection, recombination and mutation gradually improves the quality of solutions that emerge within the population.

In the 1960s RECHENBERG and SCHWEFEL [2] developed evolution strategies, independently from the work of HOLLAND [7] on genetic algorithms at the same time. Both optimization methods are based on similar generic principles, a population of genetic structures, a selection scheme that increases the proportion of better solutions and genetic operators that produce new variants.

The major difference between evolution strategies and genetic algorithms is the way in which the genotype represents a candidate solution. An evolution strategy operates on a vector of real numbers, whereas genetic algorithms manipulate a string of discrete, usually binary, symbols.Evolution strategies are distinguished by self-adaptation of additional strategy parameters, which enables them to optimize the evolutionary process according to the structure of the fitness landscape.

In an evolution strategy the genome of a single individual consists of two sets of real-valued variables. The vector $(x_1, \ldots, x_n)$ contains the object variables, which are tuned by recombination and mutation in order to optimize the objective function. The additional strategy parameters are stored in the vector $(\sigma_1, \ldots, \sigma_n)$ that defines the step-size of mutations applied to the object variables. The mutation operator modifies the corresponding object variable x_i by adding normally distributed noise $N_i(0, \sigma_i^2)$ with variance σ_i^2 and zero mean.

$$x_i' = x_i + N_i(0, \sigma_i^2) \tag{1}$$

The strategy parameters σ_i do not remain constant. They undergo mutation

$$\sigma_i' = \sigma_i e^{\tau' N(0,1) + \tau N_i(0,1)} \tag{2}$$

where τ' is a global mutation factor and τ is a local mutation factor. The random number $N(0, 1)$ is only sampled once for the complete genotype, whereas the mutations $N_i(0, 1)$ are uncorrelated for different genes.

The mutation of strategy parameters enables the evolution strategy to adapt itself to the structure of the fitness landscape. At long sight, selection favors those strategy parameters, which are more likely to generate mutations to object variables that substantially improve their fitness. Due to the mechanism of self-adaptation, an exogenous control of step-sizes, utilized by standard mathematical optimization methods, becomes obsolete.

3 Evolutionary Tuning of TSK Fuzzy Controllers

Among many other applications, fuzzy logic control has been applied to control an intelligent unmanned helicopter [13]. A fuzzy logic controller is a knowledge based system characterized by a set of linguistic variables and fuzzy if-then rules. Fuzzy rules are defined by their antecedents and consequents, which relate an observed input state to a desired control action.

Takagi, Sugeno and Kang (TSK) proposed an inference scheme in which the conclusion of a fuzzy rule is constituted by a weighted linear combination of the crisp inputs [14].

$$\text{if } X_1 = A_1 \text{ and } \cdots X_m = A_m \text{ then } y = c_0 + c_1 x_1 + \cdots + c_m x_m$$

A TSK fuzzy inference system smoothly interpolates among multiple linear controllers defined for specific operating points in the input space. TSK fuzzy systems can be tuned by a learning algorithm, that adapts the continuous gain factors c_i of the linear control rules. For reasons of simplicity we neglect the constant offset term c_0 in the following and only consider the gain factors.

3.1 Incremental Learning Scheme

The number of rules in a fuzzy controller grows rapidly with an increasing number of variables and fuzzy sets. Learning all fuzzy rules at once becomes

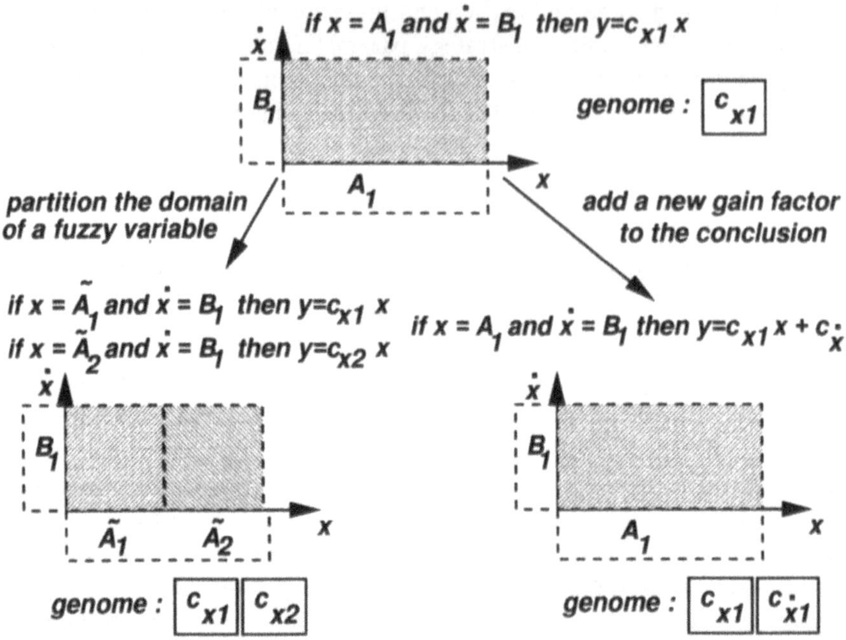

Figure 1: Incremental Refinement of the Fuzzy Rule Base

less and less feasible due to the large number of parameters to be adapted simultaneously. To overcome this problem of increasing complexity, we propose an incremental learning scheme, which partitions the design task into smaller steps.

The evolution strategy first optimizes a fuzzy knowledge base with a small number of rules each of them covering a large region of input space. In later generations the rule base is refined it through superposition of additional rules with finer granulation. Fig. 1 shows an example of the refinement for a fuzzy system with two input variables x and $\dot{x}$. The rule base tuning procedure starts with a single TSK fuzzy rule

$$R_1: \quad \text{if } x = A_1 \text{ and } \cdots \dot{x} = B_1 \text{ then } y = c_{x1} x$$

that covers the entire input space. Its linear control law only depends on the input variable x. The genome is constituted by a single gene that corresponds to the gain factor c_{x1} and its mutation step-size σ_{x1}.[1] This primitive rule R_1 constitutes the point of departure for subsequent refinements of the rule base. The evolution strategy starts to optimize the single gain factor c_{x1}. Within a

[1] In section2 x_i denoted the object variables employed by the evolution strategy. In the following x always relates to the crisp value of input variable x, whereas c_i is used to describe the gain factors encoded by the object variable i.

few generations the population converges to a preliminary solution.

In order to enable a more distinguished control behavior the rule base acquires additional degrees of freedom. The already adapted rule R_1 can be extended in two different ways.

1. Adding a term for the input variable $\dot{x}$ to the consequence part of the rule R_1

   $$R_1': \quad \text{if } x = A_1 \text{ and } \ldots \dot{x} = B_1 \text{ then } y = c_{x1}x + c_{\dot{x}1}\dot{x}$$

   with the result that the control law becomes a linear function of both inputs. The genome is extended by a second object variable for the additional gain factor $c_{\dot{x}1}$ plus its corresponding strategy parameter $\sigma_{\dot{x}1}$.

   Additional genes are incorporated into the genome in a way that minimizes deleterious side-effect off the already adapted genes. Therefore, the new gene $c_{\dot{x}1}$ is initialized with a zero value to guarantee the congruence among the new rule R_1' with its genetic ancestor R_1. The evolution strategy exploits the extra degree of freedom by adjusting the control surface through subsequent mutations applied to the new gain factor.

2. Partition the domain of an input variable, for example x into two partially overlapping fuzzy sets $\tilde{A}_1$ and $\tilde{A}_2$, with the result that the knowledge base now contains two rules R_1', R_2' instead of one.

   $$\begin{aligned} R_1': &\quad \text{if } x = \tilde{A}_1 \text{ and } \ldots \dot{x} = B_1 \text{ then } y = c_{x1}x \\ R_2': &\quad \text{if } x = \tilde{A}_2 \text{ and } \ldots \dot{x} = B_1 \text{ then } y = c_{x2}x \end{aligned}$$

   The genome is expanded by duplicating the gain factor c_{x1} of the original rule R_1 which preserves the previously adapted knowledge expressed by R_1. Afterwards, the genome contains two gain factors c_{x1} and c_{x2} associated to fuzzy rules that cover different regions of input space. In the region where the fuzzy sets $\tilde{A}_1$ and $\tilde{A}_2$ overlap the control output is interpolated among the two fuzzy rules.

The evolution strategy adapts the additional parameters for a few generations before the next rule base expansion takes place. Each expansion step gradually increases the complexity of the fuzzy knowledge base, by either adding a new input variable to the consequence part of all fuzzy rules or by partitioning a fuzzy set of an input variable, resulting in an increased number of rules with finer granulation.

The proper initialization of the newly incorporated genes preserves the input-output mapping already adapted in previous generations. The modularity of fuzzy rules allows it to gradually evolve complex genetic representations starting with a simple genome that only comprises a few parameters.

The cycle of genome expansion and adaptation of the additional genes continues until the rule base accumulates enough parameters to solve the control task. In each expansion step the fuzzy knowledge base can be extended in multiple ways. Either one can add another gain factor to the rule consequences or partition an arbitrary fuzzy set. A simple solution is to hand on the decision to the designer, who generates an expansion schedule based on her knowledge about the structure of the control problem.

Instead of a predefined schedule of expansion, a constructional selection mechanism can be used to regulate the augmentation of the fuzzy knowledge base. It evaluates the functional effect of multiple possible genome expansions selects the one that results in optimal evolvability, or in other words most likely promises the production of fitter offsprings. The idea is to reject gene additions that have a neutral or even negative effect on the performance and to stably embody genes that explore novel degrees of freedom in a way that increases the fitness. The constructional selection mechanism is described in [6] in more detail.

4 Autonomous Helicopter

Unmanned autonomous aerial vehicles have been found indispensable for various applications where human intervention is considered difficult or dangerous. Despite its poor cursing performance, helicopter can be operated in different flight modes, such as vertical take-off/landing, hovering, longitudinal/lateral flight, coordinated turn. Due to its versatile in maneuverability, a helicopter is capable to manœuvre in and out of restricted areas and to hover efficiently for long periods of time. These features make helicopter a valuable tool for power line inspection, terrain surveying, and investigation and clean-up of hazardous waste sites.

The autopilot constitutes the regulation layer of a hybrid flight vehicle management system (FVMS). The FVMS is responsible for planning and controlling the helicopter's operation. Its design is based on a hybrid control approach [1] in which a discrete event systems governs the behavior of a continuous-state plant. The FVMS is responsible for resolving conflicts between air vehicles, planning of the flight path, generating a feasible trajectory and a proper sequence of flight modes, and regulating the helicopter along the trajectory. The proposed system architecture is described in more detail in [9].

4.1 Helicopter Dynamics

Flying a helicopter is a complex control problem including multiple inputs and outputs which in addition are partially coupled. A helicopter is a nonlinear system which is inherently unstable and very sensitive to external disturbances

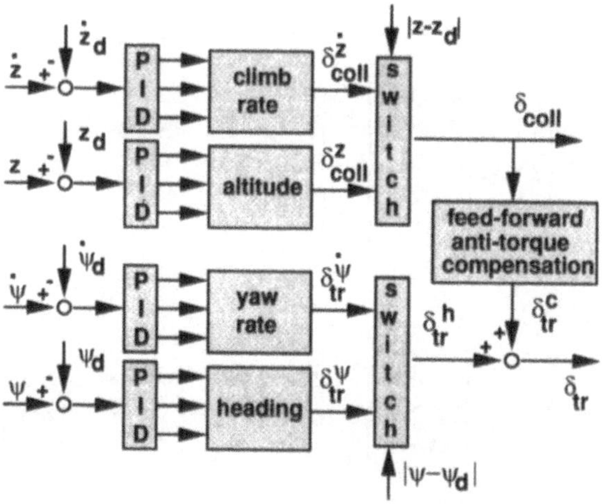

Figure 2: Fuzzy controller architecture for collective and tail rotor pitch

like wind and ground effects. Three rotational and three translational degrees of freedom have to be controlled by four inputs.

The main rotor pitch regulates the vertical thrust which lifts the helicopter. The engine throttle is controlled by an independent engine governor and is not used for modifying the thrust. It maintains a constant main rotor RPM that provides the optimal trim condition for the collective pitch.

The longitudinal and lateral cyclic pitch of the main rotor affect the pitch and roll angle, which on their part determine the forward and side ward motion of the helicopter. Finally, the tail rotor pitch controls the yaw rate and in addition compensates the anti torque generated by the main rotor.

4.2 Controller Architecture

The helicopter autopilot is composed of four separate modules which correspond to the control actuators collective pitch, tail rotor pitch, longitudinal and lateral cyclic pitch. The complexity of the control design can be reduced by carefully considering the helicopter dynamics when subdividing the system into separate modules.

The collective pitch control block attempts to follow a commanded altitude in vertical climb and descent. The heading control block governs the yaw motion during turns and compensates the anti-torque generated by the main rotor in order to maintain a desired heading. The longitudinal and lateral block regulate the horizontal motion and at the same time control the attitude to maintain the helicopter stable. Individual modules contain multiple fuzzy controllers organized in a hierarchical manner that reflects the effect and coupling of different controls. Each block employs a switch which selects the controller that is spe-

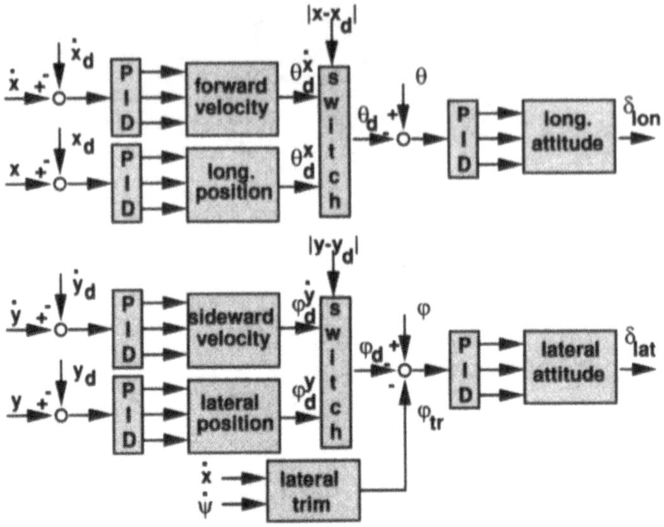

Figure 3: Fuzzy controller architecture for longitudinal and lateral cyclic pitch

cialized for the current flight mode. During a transition between position and velocity control the switch smoothly interpolates among the proposed control actions.

The collective control block is composed of two parallel PID fuzzy controllers, which respectively control the helicopter altitude z and the climb rate $\dot{z}$. If the altitude deviates significantly from the commanded height the climb rate control action passes through the switch. For an intermediate range 1-2m away from the set-point a linear combination of the control outputs is used. When the error in height becomes less than 1m the altitude controller determines the collective pitch δ_{coll}.

The control block for the tail rotor pitch δ_{tr} operates in a similar manner. Two fuzzy PID controllers control the heading ψ and the yaw rate $\dot{psi}$ for hovering and coordinated turns. The main rotor exerts an anti-torque on the helicopter body which has to be compensated by the tail rotor. Feed-forward control trims the tail rotor pitch δ_{tr} in response to a change in collective pitch δ_{coll}.

The control inputs x, x_d, y, y_d are mapped from spatial to body coordinates before they are passed to to the longitudinal and lateral position controllers. This transformation is not shown in Fig. 3. The control inputs $\dot{x}, \dot{x}_d, \dot{y}, \dot{y}_d$ for the velocity controllers are already specified in body coordinates.

The longitudinal control block employs three PID fuzzy controllers to compute the longitudinal cyclic pitch δ_{lon} as a function of the pitch angle θ, longitudinal position error e_x and forward velocity $\dot{x}$. The pitch attitude determines the thrust component in forward direction and therefore the longitudinal motion of the helicopter. The longitudinal position and velocity controllers command a desired pitch angle θ_s to the longitudinal attitude controller. A switch mediates between position and velocity control based on the actual region in state space. The cascaded architecture of the block significantly reduces the number of fuzzy rules compared to a single fuzzy controller that aggregates all three inputs in one step. Connecting the fuzzy controllers in sequence implicitly guarantees a stable pitch attitude while controlling the longitudinal motion.

The lateral control block is based on a similar hierarchical architecture. The lateral position and velocity controllers command the lateral attitude controller to hold a desired roll angle ϕ_s. In forward flight and coordinated turns the lateral velocity $\dot{y}$ should be zero to prevent side-slip. In a coordinated turn roll angle ϕ, yaw rate $\dot{\psi}$ and forward velocity $\dot{x}$ are constrained through the equation

$$\phi = \arctan\left(\dot{\psi}\,\dot{x}/g\right) \tag{3}$$

assuming zero lateral velocity. Therefore, a trim offset ϕ_{tr} is added to the desired roll angle ϕ_s according to Eq.3.

We manually derived fuzzy control rules for each block by a process based on trial and error and some knowledge obtained from a linearized model. At this preliminary design stage each fuzzy controller employed a single PID-type control rule. The control system was able to stabilize the hovering helicopter for small deviations from the nominal operating point.

The performance of the helicopter controllers generated by the evolution strategy is evaluated in a simulation based on a nonlinear model of the helicopter. The model was derived from literature on helicopter dynamics [12]. The aerodynamic equations that describe the forces and moments generated by the main and tail rotor can be simplified assuming that the helicopter operates in hover or low velocity conditions.

5 Evolutionary Tuning of the Autopilot

The genetic-fuzzy system approach allows the designer to specify his performance goals directly by means of a scalar objective function and releases him from a tedious manual tuning and optimization procedure. The method can be used to generate a collection of basic controllers that emphasize on different performance aspects such as robustness, accuracy and speed of execution. This repertoire of basic flight mode controllers constitutes the continuous regulation layer or autopilot of the hybrid flight vehicle management system described in Section ??.

The manually designed autopilot controllers for the altitude, heading, lateral and longitudinal control are tuned in a simulation of typical training situations. The performance of each block is tested in a helicopter manœuvre that commands a sequence of set-points. The fitness function evaluates the time weighted mean square error to the reference signal.

5.1 Lateral and Longitudinal Controller

In the following, the longitudinal and lateral position controller will serve as an example to illustrate the tuning process. The design objective is a fast and accurate system response to a simultaneous longitudinal and lateral position step-input. The simulation lasts 30 seconds and is divided into three sections of 10 seconds each. At time $t = 10$ the helicopter is commanded to move simultaneously 1m forward and side-ward, at time $t = 20$ to return to the origin (see Fig.4).

The objective function is formed by a weighted sum of cumulated state errors. Depending on the way in which the errors are aggregated over time the designer can emphasize different performance aspects such as short reaching time, minimal overshoot or minimal oscillation. The positional error is defined by a two component vector $\vec{e} = [x - x_d, y - y_d]$. The integral of the square error (ISE) which is defined as

$$E = \int_t \vec{e}^T \circ \vec{e}\ dt \tag{4}$$

is often used as a performance index. It achieves a good compromise between fast execution of a manœuvre and sufficient damping.

In our case the helicopter is given the mission to locate and recognize objects on the ground by means of a visual sensor. In order to obtain a stable camera image accurate positioning and fast damping becomes more important than high manœuvrability. Therefore we use a performance index based on the integral of time multiplied by the square error (ITSE)

$$E = \sum_{i=1}^{3} \int_t^{T_i} t * \vec{e}^T \circ \vec{e}\ dt \tag{5}$$

which reduces the effect of large initial errors and penalizes poorly damped oscillations occurring later in the response. Notice, that the index i is used to distinguish the three sections of the simulation.

Fig.4 compares the performance of the original hand designed controllers with the fuzzy rules tuned by the evolution strategy. The ITSE performance criteria was improved by 35% for the longitudinal and 62% for the lateral motion.

5.2 Altitude Controller

The altitude module is optimized in a similar way as the longitudinal and lateral block. At time $t = 10$ the helicopter is commanded to climb to an altitude of $2m$, at time $t = 20$ to descend to its original altitude. The evaluation of the manœuvre is based on the fitness function in Eq.5.

The rule base of the altitude controller is composed of five PID type fuzzy rules of the form

$$R_i^l: \quad \text{if } e_z = A_i \text{ then } \delta_{coll} = c_i^{e_z}\, e_z + c_i^{e_z^D}\, e_z^D + c_i^{e_z^I}\, e_z^I + c_i$$

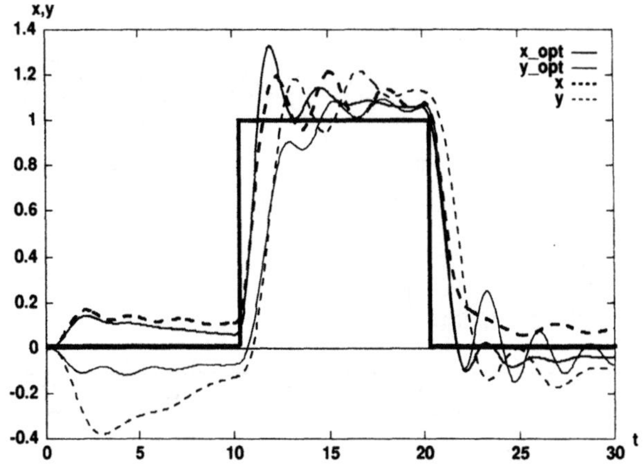

Figure 4: Lateral and longitudinal motion in response to a step-input for original (- - -) and optimized (—) fuzzy controllers.

in which E_z, E_z^D and E_z^I denote the error in altitude, the change in error, the integral of error and δ_{coll} defines the collective pitch of the main rotor blade. The domain of the altitude error is partitioned into the fuzzy sets A_i. The rule consequences are defined by three gain factors and a constant offset c_i, resulting in an overall number of 20 parameters.

In the following we compare the incremental learning scheme with an optimization process that employs a static structure of the complete parameter set. Both methods were tested in two different scenarios, one in which the parameters of the initial population are randomly drawn and the one in which they are initialized with the gain factors of a manually designed PID controller.

In case of the random initialization of fuzzy rules none of the 20 initial controllers with a complete static rule base was able to stabilize the helicopter. Due to the large number of parameters, feasible solutions are sparsely distributed over the vast search space. Therefore, even the chances of finding a second-rate controller through random search are minimal. Aiming for an easier objective by weakening the fitness function presents one way to overcome this problem. Instead of using the ITSE as a criteria the controllers are evaluated according to the time elapsed before the helicopter becomes unstable. The evolutionary algorithm starts using the original performance criteria as soon as the controllers are able to handle the basic stabilization task.

The initialization problem can be also be solved by designing an approximative PID controller manually and use its parameters to initialize the gain factors of the first generation. For the altitude controller we derived an ad hoc PID controller through trial and error. This candidate solution is far from optimal in the sense that a profound controller design based on classical linear control theory will provide a much better performance. This approach constitutes a

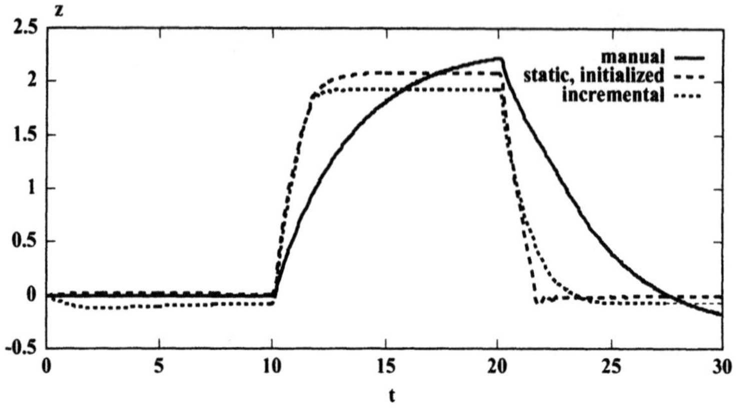

Figure 5: Upward and downward motion in response to a step-input for the manually designed PID (—) controller, the controller designed by a self-tuning process (- - -) and the fuzzy controller designed by the self-organizing process (- - -).

self-tuning process in which the evolutionary algorithm optimizes the performance of an already existing solution. Fig. 5 compares the performance of the manually derived ad hoc PID controller with the TSK fuzzy controller that arises from the evolutionary tuning. The ITSE criteria from Eq.5 is improved by 89%, although most of the progress is due to the poor performance of the initial PID controller.

In the incremental learning case, the optimization becomes a self-organizing process that learns the fuzzy rules starting without any previous knowledge. The genome of the initial population is composed of fewer parameters, which significantly reduces the dimension of search space. Therefore, the likelihood that random initialization produces a stabilizing controller increases. In case of the altitude controller 20% of the initial population manage to keep the helicopter stable during the 30 seconds of the trial. The initial genome is composed of four genes that represent a single fuzzy rule. The number of rules is increased to three after five generations and finally to five after ten generations. The ITSE of the best solution is 34% larger than the fuzzy controller designed by the tuning process (see Fig.5). This result demonstrates that the incremental learning scheme is able to design suitable controllers starting from scratch, which performance is comparable to controllers obtained from a tuning of predesigned fuzzy rules.

6 Summary

This paper proposed a genetic fuzzy system that automates the knowledge acquisition step in fuzzy control design. An evolution strategy adapts a vector of real numbers, which represent the gain factors of TSK-type fuzzy rules. The optimization process gradually expands the structure and number of fuzzy rules by incorporating additional genes into the genome. The rule base design is decomposed into smaller subproblems that become more feasible to evolutionary optimization.

The proposed learning method was applied to design a helicopter autopilot that constitutes the continuous regulation layer of a flight vehicle management system. A collection of controllers for the lateral, longitudinal, altitude and heading control modules are optimized according to the desired control behavior and context of helicopter operation. The visualization tool SmartAerobots is helpful to verify the control design and enables the integration of control and vision algorithms.

The incremental design method was able to learn the rule base of the altitude controller starting without any previous knowledge, whereas the optimization of the full structure controller in one step required the proper initialization of fuzzy rules by means of some previously derived plausible parameters. The results indicate that incorporating existing domain knowledge into the initial population facilitates the evolutionary optimization task. The evolution strategy that starts with a single fuzzy rule circumvents the curse of finding a feasible set of initial parameters in a high-dimensional search space. The altitude controllers designed by the self-organizing process demonstrated comparable performance to the fuzzy controllers obtained from the tuning approach.

Acknowledgment

This work was funded by a research grant (Ho 1790/2-1) from the Deutsche Forschungsgemeinschaft and supported by the Army Research Office under grants DAAH 04-96-1-0341.

References

[1] Antsaklis, P., Kohn, W., Nerode, A., Sastry, S., "Hybrid Systems II", Springer-Verlag, (1995).

[2] Th. Bäck, H. P. Schwefel, "Evolution Strategies I: Variants and their computational implementation", "Evolution Strategies II: Theoretical aspects and implementation", *Genetic Algorithms in Engineering and Computer Science*, Ed. G. Winter et al., pp. 111-140, John Wiley & Sons, (1995).

[3] O. Cordon, F. Herrera, "A General Study on Genetic Fuzzy Systems", *Genetic Algorithms in Engineering and Computer Science*, pp. 33-57, John Wiley & Sons, (1995).

[4] F. Hoffmann, G. Pfister, "Learning of a Fuzzy Control Rule Base Using Messy Genetic Algorithms", *Genetic Algorithms and Soft Computing*, Ed. F. Herrera, J. L. Verdegay, Physica-Verlag, p. 279-305, (1996).

[5] F. Hoffmann, G. Pfister, "Evolutionary Design of a Fuzzy Control Rule Base for a Mobile Robot", *Int. Journal of Approximate Reasoning*, vol. 17, no. 4, (1997).

[6] F. Hoffmann, "Incremental Tuning of Fuzzy Controllers by Means of an Evolution Strategy", Genetic Programming 1998: Proceedings of the Third Annual Conference, Ed. Koza, J., Banzhaf, W. et al. , Madison, Wisconsin, Morgan Kaufmann, (1998).

[7] J. H. Holland, *Adaptation in Natural and Artificial Systems*, MIT Press, Cambridge, MA, (1992).

[8] C. L. Karr, "Design of a Cart-Pole Balancing Fuzzy Logic Controller using a Genetic Algorithm" *SPIE Conf. on Applications of Artificial Intelligence*, Bellingham, WA, (1991).

[9] T.J. Koo, F. Hoffmann , H. Shim, B. Sinopoli and S. Sastry, "Hybrid Control of Model Helicopter", IFAC Workshop on Motion Control, Grenoble,France, (1998).

[10] N.E. Nawa, T. Hashiyama, T. Furuhashi, Y. Uchikawa, "Fuzzy logic controllers generated by pseudo-bacterial genetic algorithm with adaptive operator", Proc. IEEE International Conference on Neural Networks, vol.4, p.2408-2413, Houston, Texas, (1997).

[11] C. Phillips, C.L. Karr, G. Walker, "Helicopter flight control with fuzzy logic and genetic algorithms", Engineering Applications of Artificial Intelligence, vol.9, (no.2), Elsevier, p.175-84, (1996).

[12] Raymond W. Prouty, "Helicopter Performance, Stability, and Control", Krieger Publishing Co., (1995).

[13] M. Sugeno, I. Hirano, S. Nakamura, S. Kotsu, "Development of an intelligent unmanned helicopter", Proceedings the Fourth IEEE International Conference on Fuzzy Systems, p.33-4 vol.5, (1995).

[14] T. Takagi, M. Sugeno, "Fuzzy identification of systems and its applications to modelling and control", *IEEE Transactions on Systems, Man and Cybernetics*, vol. SMC-15, no.1, pp.116-32, (1985).

Decomposition of a Fuzzy Controller Based on the Inference Break-up Method

Marjan Golob

Laboratory for Process Automation
University of Maribor, Faculty of Electrical Engineering
and Computer Science
Smetanova 17, 2000 Maribor, SLOVENIA
Tel.: (+386 62) 220 7161, Fax.: (+386 62) 211 178
mgolob@uni-mb.si

Keywords: Fuzzy PID Controller, Inference Break-up Method, Magnetic Suspension System

Abstract

A concept called the decomposition of multivariable control rules is presented. Fuzzy control is the application of the compositional rule of inference and it is shown how the inference of the rule base with complex rules can be reduced to the inference of a number of rule bases with simple rules. A fuzzy logic based controller is applied to a simple magnetic suspension system. The controller has proportional, integral and derivative separate parts which are tuned independently. This means that all parts have their own rule bases. By testing it was formed out that the fuzzy PID controller gives better performance over a typical operational range then a traditional linear PID controller. The magnetic suspension system and the contact-less optical position measurement system have been developed and applied for the comparative analysis of the real-time conventional PID control and the fuzzy control.

1. Introduction

Fuzzy set theory was introduced by Zadeh [1] in 1965 and has been evolved as a powerful modeling tool that can cope with the uncertainties and non-linearity's of modern control systems. Fuzzy controllers have become popular in recent years because they do not necessarily require a theoretical model of the plant which is to be controlled. Therefore, in order to develop a fuzzy controller, one needs to first have access to a human expert, find quantifiable means to present the expert's experience, and determine a mapping from states of the plant to the fuzzy measures with which the expert's knowledge is quantified. In [2] a supervisory fuzzy logic based controller is applied to the magnetic suspension system. The control architecture consists of two loops; a pole placement controller is utilized in the internal loop, and a supervisory fuzzy controller with a proportional-derivative structure is embedded in the other loop to enhance the transient response of the system. This control design is compared in simulation studies with a classical pole placement controller. Several magnetic suspension systems have been developed and applied for magnetically levitated transit systems by Japanese and American corporations during the last years (Rule and Gilliand [3], Ion at al [4]). In most cases the control system and energy supply requirements to levitate the vehicle have a higher level of complexity. The non-linear nature of the system dynamics coupled with non-linear characteristics of the actuators complicate the controller design. The classical controller development approach relies on a linearization of the system dynamics and on the application of a PID controller to compensate the effects of the non-modeled non-linearity's. By this approach certain system is stabilized close to its nominal operating point. Problems could be existent in the case when the set point is changeable within wide operating range.

A method called inference break-up is presented in this paper and used to model the multivariable fuzzy controller using the decomposition of the control rule base into a set of simple rule bases. We provide a comparative analysis of the conventional PID control and the fuzzy control for the magnetic suspension application. Our main objective is to make an initial assessment of what advantages a fuzzy control approach has over conventional control approaches. The magnetic suspension system is realised by the simple electromagnet and the steel ball. The magnetic force is oposite to gravity force and maintains the suspended steel ball levitated. Details and the matematical model are presented in [5].

The structure of this paper is as follows. A theoretical approach to the fuzzy control based on inference break-up method is described in the next section. The development of the fuzzy controller is presented in Section 3. A real-time laboratory model with contact-less position measurement system is described in the section 4. Comparative analysis of the real-time conventional PID control and the fuzzy control are discussed in Section 5. The concluding remarks are given in Section 6.

2. A Theoretical Approach to the Fuzzy Control Based on the Inference Break-up Method

The control algorithm is represented by fuzzy rules. A multivariable fuzzy system with three inputs and one output is considered. The linguistic description of the process is given by

$$
\begin{array}{l}
\text{IF } X_{1(1)} \text{ AND } X_{2(1)} \text{ AND } X_{3(1)} \text{ THEN } Y_{(1)} \\
\text{AND} \\
\text{IF } X_{1(2)} \text{ AND } X_{2(2)} \text{ AND } X_{3(2)} \text{ THEN } Y_{(2)} \\
\text{AND} \\
\vdots \\
\text{IF } X_{1(i)} \text{ AND } X_{2(i)} \text{ AND } X_{3(i)} \text{ THEN } Y_{(i)} \\
\vdots \\
\text{AND} \\
\text{IF } X_{1(m)} \text{ AND } X_{2(m)} \text{ AND } X_{3(m)} \text{ THEN } Y_{(m)}
\end{array}
\tag{1}
$$

where $X_{k(i)}$ is the fuzzy set of the k-input variable defined in the universe of discourse $\mathcal{X}_k$, k=1,2,3, and $Y_{(i)}$ is the output fuzzy set defined in the universe of discourse $\mathcal{Y}$. m is the number of rules. The fuzzy relation R of the system is expressed as follows

$$
R = \bigvee_{i=1}^{m} \left\{ X_{1(i)} \wedge X_{2(i)} \wedge X_{3(i)} \wedge Y_{(i)} \right\} \tag{2}
$$

where V is the aggregation operator and $\wedge$ is the implication operator. For each rule a fuzzy relation $R_{(i)}$ has to be constructed. To obtain the fuzzy controller relation R, fuzzy relations $R_{(i)}$ are aggregated. Dimension of relation matrix R is $dim[R] = dim[\mathcal{X}_1] \times dim[\mathcal{X}_2] \times dim[\mathcal{X}_3] \times dim[\mathcal{Y}]$. To obtain the new fuzzy output Y', given the current fuzzy inputs X'_1, X'_2 and X'_3 the compositional rule of inference is used:

$$
Y' = X_1' \circ X_2' \circ X_3' \circ R \tag{3}
$$

where $\circ$ is the composition operator of fuzzy relations. Because of the multidimensionality of the fuzzy relation (2) the composition rule of inference (3) is difficult to perform or analytical solutions can usually not be obtained. To overcome this difficulty, it is proposed to breaks up the inference of a multidimensional rule base into the three rule bases of which the inference is easier to perform or has analytical solutions.

Demirli and Turksen [6] showed that rules with two ore more independent variables in their premise can be simplified to a number of inferences of rule bases with simple rules (only one variable in their premise). Their proof is only valid for *sub-min* composition operator and the *min* operation for conjunction in the premise, and is based on the fact that S and R implications are non increasing with respect to their first argument. When the rule:

$$
\text{IF } X_1 \text{ AND } X_2 \text{ AND } X_3 \text{ THEN } Y
$$

is considered and the implication operator *min* is used, the rule can be broken into a set of three rules:

$$(X_1 \wedge X_2 \wedge X_3 \wedge Y) = (X_1 \wedge Y) \vee (X_2 \wedge Y) \vee (X_3 \wedge Y) \tag{4}$$

The proof can be generalized to cases with more then three variables in the rule premise. Normally a rule base consist of more then just one rule. The inference of a rule base with more than one rule also can be broken up into the inference of a number of rule bases with simple rules.

When considering a rule base with two rules, the rule base break up results in:

$$\begin{aligned}
Y' = \left(X_1' \wedge X_2' \wedge X_3'\right) \circ \left\{\left(X_{1(1)} \wedge X_{2(1)} \wedge X_{3(1)} \wedge Y_{(1)}\right) \vee \left(X_{1(2)} \wedge X_{2(2)} \wedge X_{3(2)} \wedge Y_{(2)}\right)\right\} = \\
X_1' \circ \left\{\left(X_{1(1)} \wedge Y_{(1)}\right) \vee \left(X_{1(2)} \wedge Y_{(2)}\right)\right\} \wedge \\
X_2' \circ \left\{\left(X_{2(1)} \wedge Y_{(1)}\right) \vee \left(X_{2(2)} \wedge Y_{(2)}\right)\right\} \wedge \\
X_3' \circ \left\{\left(X_{3(1)} \wedge Y_{(1)}\right) \vee \left(X_{3(2)} \wedge Y_{(2)}\right)\right\} \wedge \\
\left\{\left[X_1' \circ \left(X_{1(1)} \wedge Y_{(1)}\right)\right] \vee \left[X_2' \circ \left(X_{2(2)} \wedge Y_{(2)}\right)\right]\right\} \wedge \\
\left\{\left[X_1' \circ \left(X_{1(2)} \wedge Y_{(2)}\right)\right] \vee \left[X_2' \circ \left(X_{2(1)} \wedge Y_{(1)}\right)\right]\right\} \wedge \\
\left\{\left[X_1' \circ \left(X_{1(1)} \wedge Y_{(1)}\right)\right] \vee \left[X_3' \circ \left(X_{3(2)} \wedge Y_{(2)}\right)\right]\right\} \wedge \\
\left\{\left[X_1' \circ \left(X_{1(2)} \wedge Y_{(2)}\right)\right] \vee \left[X_3' \circ \left(X_{3(1)} \wedge Y_{(1)}\right)\right]\right\} \wedge \\
\left\{\left[X_2' \circ \left(X_{2(1)} \wedge Y_{(1)}\right)\right] \vee \left[X_3' \circ \left(X_{3(2)} \wedge Y_{(2)}\right)\right]\right\} \wedge \\
\left\{\left[X_2' \circ \left(X_{2(2)} \wedge Y_{(2)}\right)\right] \vee \left[X_3' \circ \left(X_{3(1)} \wedge Y_{(1)}\right)\right]\right\} \wedge
\end{aligned} \tag{5}$$

The result of the inference break-up can easily be extended to a large number of rules or more complex rules. The number of parts resulting from the inference break-up can become quite large (in the case of n variables in the premise of the m rules, the number of cross-products is n^m), but many simplifications are possible. First simplification of the inference of a simple rule base can be achieved when the rule base can be divided into a set of simple rule bases which do not interact. The interaction of rule bases means overlap of the rules premises.

When the premises of the rules are assumed to take fuzzy sets which form a fuzzy partition (fuzzy sets are convex and normalized) with no more then two overlapping fuzzy sets, then it can be derived that the number of necessary simple rule bases is 2N-1, where N is the number of fuzzy sets defined on the domain of the output variable. At this point, a crucial simplifying argument is made. From all products over n^m cross-products only three terms are selected. The inference (5) reduces to

$$\begin{aligned} Y_s' = X_1' \circ &\left\{\left(X_{1(1)} \wedge Y_{(1)}\right) \vee \left(X_{1(2)} \wedge Y_{(2)}\right)\right\} \wedge \\ X_2' \circ &\left\{\left(X_{2(1)} \wedge Y_{(1)}\right) \vee \left(X_{2(2)} \wedge Y_{(2)}\right)\right\} \wedge \\ X_3' \circ &\left\{\left(X_{3(1)} \wedge Y_{(1)}\right) \vee \left(X_{3(2)} \wedge Y_{(2)}\right)\right\} \wedge \end{aligned} \tag{6}$$

Using the inclusion operator, the output Y from inference (5) can be written as $Y' \subset Y'_s$. The inclusion sign means that the output fuzzy set Y' is contained in the set obtained by neglecting the terms in the inference (5). In many practical cases this neglected term may not be very significant.

In the Section 3 the development of the simple fuzzy PID controller based on the classic discrete PID controller is presented. The output of the regular fuzzy controller is a nonlinear function of three inputs (error, change-of-error and integral-of-error). The break-up method of regular fuzzy controller inference leads to similar result described by equation (5), with intersections of three basic simple inferences for each input and intersections of many simple cross-inferences. The simple fuzzy PID controller inference is obtained by neglecting all simple cross-inferences and the output Y'_s is a intersection of outputs of three basic simple inferences. Comparing this result with the output of the discrete PID controller, where y_{PID} is the linear algebraic equation, the discrete PID controller can be seen as a subset of the simple fuzzy controller. When designing a simple fuzzy controller and applying specific choices for the shape of membership functions, logical operators and the scaling of inputs and outputs, the simple fuzzy controller can emulate a linear control.

Now define the fuzzy relation R_k from (5) as

$$R_k = \bigvee_{i=1}^{m} \left\{\left(X_{k(i)} \wedge Y_{(i)}\right)\right\} \qquad k = 1,2,3, \qquad m = 2 \tag{7}$$

and the following is obtained from (6):

$$Y_s' = X_1' \circ R_1 \wedge X_2' \circ R_2 \wedge X_3' \circ R_3 \tag{8}$$

To understand better the linguistic description (1) and its mathematical presentation (8), a block-diagram form is proposed in Figure 1.

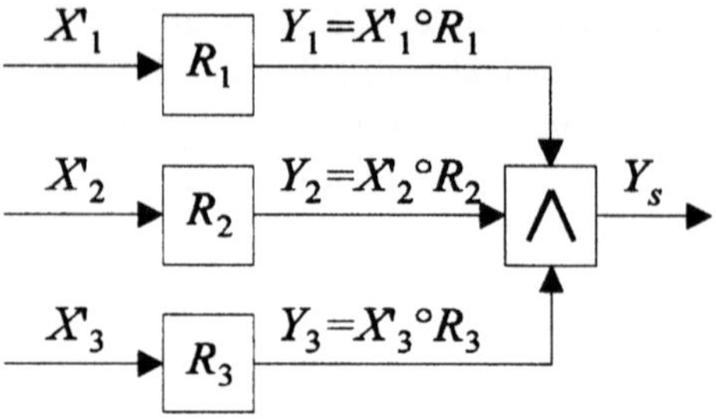

Figure 1: Block diagram of multivariable fuzzy system

The multivariable structure of the fuzzy system which has n inputs contains n functional blocks and an intersection block. Advantages of the block diagram presentation of a multivariable fuzzy system are in the fact that this kind of presentation allows evaluation of the contribution of each component to the overall performance of the system. An interesting possibility concerning the inference break-up is a processor which performs composition of a fuzzy set and a two-dimensional fuzzy relation. A speed-up of the inference can be achieved by using optimized code to perform the composition of a fuzzy set and a fuzzy relation. Because of the inference break-up, the code optimization can be applied to software implementation.

3. Fuzzy PID Controller Design

For a PID-like fuzzy controller the number of rules increases as the third power of the number of membership functions. In this section we would explain how to minimize the number of rules in the case of fuzzy PID controller.

The basic idea of the discrete PID-controller is to choose the control law by considering error $e(kT)$, change-of-error $de(kT) := (e(kT) - e((k-1)T))/T$ and the numerically approximated integral of error $ie(kT) := ie((k-1)T) + Te((k-1)T))$. The PID control law is

$$u_{PID}(kT) = K_P \cdot e(kT) + K_D \cdot de(kT) + K_I \cdot ie(kT) \tag{9}$$

where K_P is a proportional constant, K_D is a differential constant and K_I is an integral constant. For a linear process the control parameters K_P, K_D and K_I are designed in such a way that the closed-loop control is stable. The corresponding analysis can be done by means of the knowledge of process parameters taking into account special performance criteria. In the case of non-linear processes which can be linearized around the operating point, conventional PID-controllers also work successfully. However, the PID-controller with constant parameters in the whole working area is robust but not optimal. In this case, tuning of the PID-parameters has to be performed.

Fuzzy PID controller start from the same assumptions which are decisive for the conventional PID-controller:

- processes either linear or non-linear could be peacewise linearized,
- process could be stabilized taking into account selected performance criteria,
- proportional, differential and integral parameters are non-linear parameters.

The output of the fuzzy controller $u(kT)$ is given by

$$u(kT) = \mathcal{N}(e(kT), de(kT), ie(kT)) \tag{10}$$

where $\mathcal{N}(\cdot)$ is a non-linear function determined by fuzzy parameters.
In the case when we assume the structure of fuzzy PID controller with three input variables and the one output variable with five base fuzzy sets on each fuzzy variable, we get one rule base with maximum 125 rules. To minimize the number of rules the simplification of the fuzzy PID structure is proposed. The basic idea is to decompose of multivariable control rule base into three sets of one dimensional rule bases for each input. The output of the decomposed fuzzy controller $u(kT)$ is given by

$$u(kT) = \frac{\mathcal{N}_P(e(kT)) + \mathcal{N}_D(de(kT)) + \mathcal{N}_I(ie(kT))}{3} \tag{11}$$

where $\mathcal{N}_P(\cdot)$, $\mathcal{N}_D(\cdot)$ and $\mathcal{N}_I(\cdot)$ are non-linear functions determined by three separated rule bases. The structure and principal components of the fuzzy logic based PID controller is shown in Figure 2.

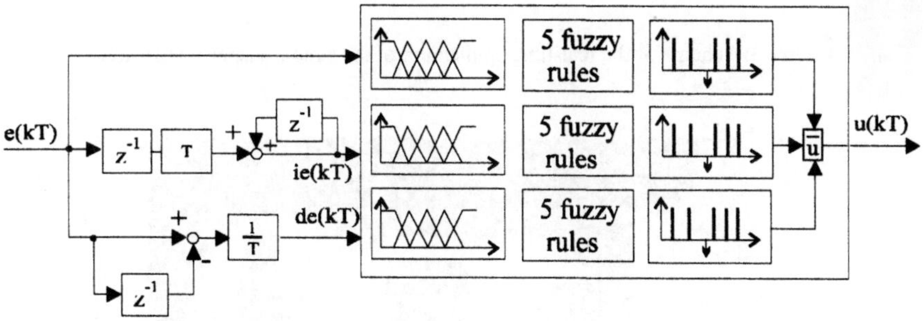

Figure 2: The structure of the fuzzy logic based PID controller

The linguistic description of the knowledge base is given by three inference rules. The output signal is sum of defuzzified proportional, differential and integral action as given in equation (12):

$$u(kT) = \frac{defuzz(U_e) + defuzz(U_{de}) + defuzz(U_{ie})}{3} \tag{12}$$

where $defuzz(\cdot)$ is defuzzification method.

4. The Real-Time Application of the Fuzzy Control for the Magnetic Suspension System

The functional diagram of the real-time application of the simple suspension system is shown in Figure 3. Our fuzzy logic controller based on the personal computer (PC) is extended by the OMRON FB-30AT fuzzy board (with the fuzzy processor FP 3000). An eight channels analog to digital (A/D) converter and two channels digital to analog (D/A) converter with 12 bit resolution is realized on the plug-in PC board. The first cannel of the A/D converter is used to measure input of the control system, namely: ball position which is output of the optical position measurement system. The fuzzy controller software was implemented in ANSI-C. The real-time sampling frequency of 1 kHz was attained. The realization of the model is presented on figure 4.

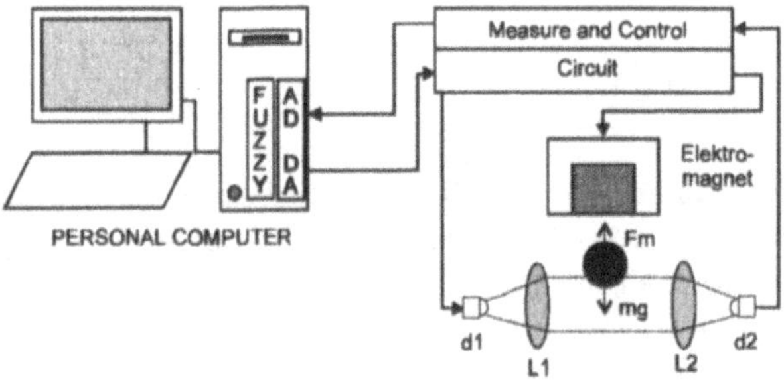

Figure 3: Scheme of the real-time application of the simple suspension system

Figure 4: The realisation of the laboratory model.

The electromagnet used in this study has 1200 coils of copper wire with diameter 1,5 mm. The ferromagnetic core has standard E form. Nominal parameters are:

the coil resistance R is 2,8 Ω, the electromagnet inductance L is 520mH and the maximum air gap D is 25mm. Electrical and mechanical time constants and gains can be calculated from mentioned nominal parameters and obtained from the measured static characteristic $F_m = f(I,X)$.

Nominal parameters, electrical and mechanical time constants and gains calculated with Equation (8) are shown in Table 1.

Table 1. Nominal system parameters

Mass of the steel ball m	Maximum air gap D	Number of coils n	Coil resitance R	Electromagnet inductance L	K_e	T_e	K_m	T_m
0,147 kg	0.025 m	1200	2,8 Ω	520 mH	3,9	0,185 s	5,4	50,5 *ms*

The measurement of the ball position with the simple optical contact-less position measurement system has been realized. The transmitter d1 (LED NLPB-500) with emission angle of 15 degrees is used to generate blue light. The homogenous light wave is provided by optical system of two optical lenses (L1, L2). The light detector (d2) provides an output voltage proportional to the position of the ball in the light wave. The room-light influence is compensated by differential structure of the light detector. By use of the 12 bits A/D converter the error between the set point and the current ball position the integer values from 0 to 4095 are mapped. The same normalization procedure is used for the derivative input and for the integral input. Since ball position measurement is made in the fixed time frame the differential control (determines the rate of error) was implemented by simply subtracting the value of the current error from the previous error in ball position.

All fuzzy inputs are divided into five ranges: Negative Large (NL), Negative Small (NS), Zero (ZE), Positive Small (PS) and Positive Large (PL). Triangular membership functions with 50% overlap are applied to inputs of the fuzzy controller, as shown in Figure 5.

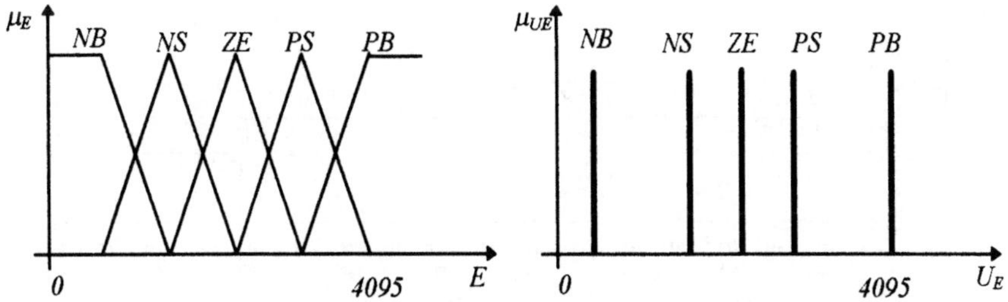

Figure 5: Input and output membership functions

The output of the controller is the output voltage (ranges of 0 to 10 V) from the D/A converter. The final control law is a sum of proportional, differential and integral action. Singleton memberships functions are applied to fuzzy outputs, as shown in Figure 5.

The knowledge base is given by three inference rule bases: the proportional rule base, the differential rule base and the integral rule base. Five rules of the rule base for the proportional part of the fuzzy controller are described in table 2.

Table 2: The rule base of the proportional part of the fuzzy PID controller

Error	Neg. Big	Neg. Small	Zero	Pos. Small	Pos. Big
Output U_e	Neg. Big	Neg. Small	Zero	Pos. Small	Pos. Big

The differential part and the integral part of the fuzzy logic controller are realized with the same type of rule base. The inference is performed using the minimum operator and the composition is done using the maximum operator (Mamdani type of the inference engine). Two defuzzification methods are implemented: Standard methods such as Mean of Maxim and Center of Gravity (COG). The Center of Gravity defuzification method is used in experiments presented in the next Section.

5. Experiments and Results

The real-time implementation of the control for a magnetic suspension system allowed to perform several interesting experiments where the fuzzy controller proved to be very efficient.

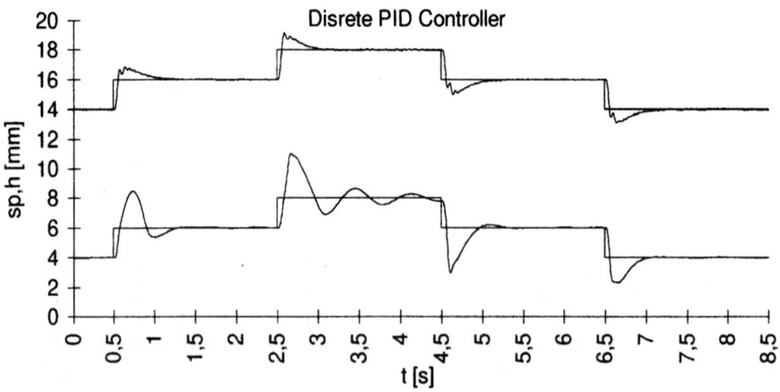

Figure 6: The performance of the discrete PID controller

At first, the ball position control by the discrete PID controller with optimal parameters is considered. The optimization of parameters for the set-point h=16 *mm* was done. The obtain performance (see Figure 6) of PID controller at high set-points (14 - 18 *mm*) was good and highly oscillatory at low set-points (4 - 8 *mm*).

For a more serious comparison of the conventional discrete PID controller and the fuzzy PID controller, a discrete PID controller with gain scheduling was created. The gain scheduling use the set-point to choose suitable parameters which have been optimized at different set-points. The responses of the adaptive controller are shown in Figure 7.

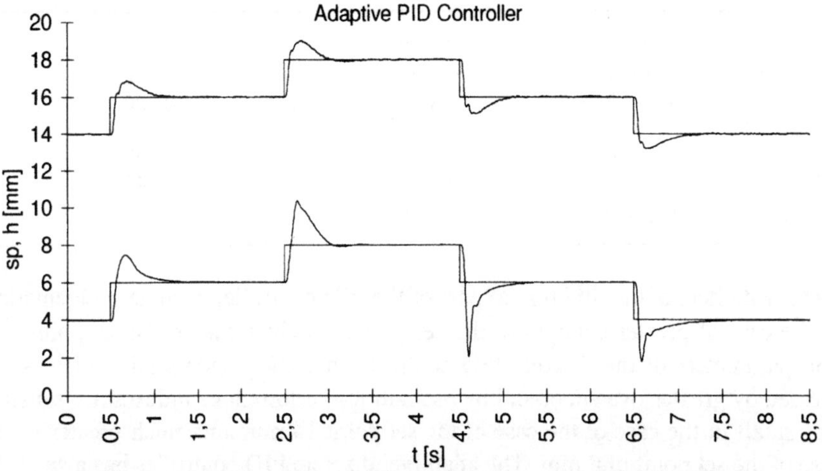

Figure 7: The performance of the discrete PID controller with gain scheduling

Results shown in Figure 8 prove that the fuzzy PID controller is similar to the discrete PID controller with gain scheduling in their performances.

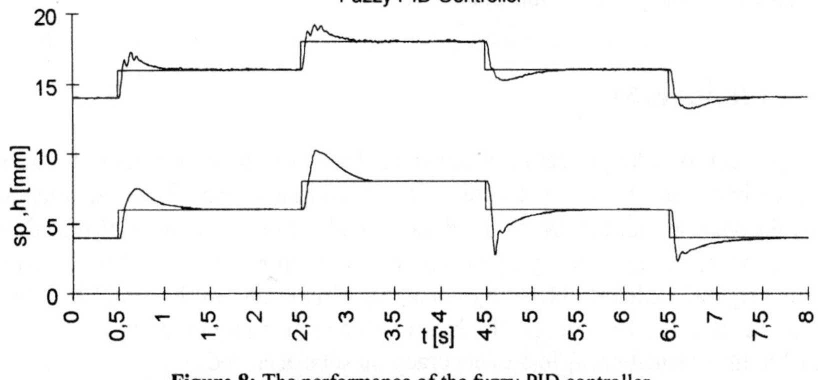

Figure 8: The performance of the fuzzy PID controller

Differences between up-down step responses in Figures 5 to 9 are not evident. It is the reason that results have been compared by use of the performance index IAE and the performance index ISE. Results are presented in Table 3. The discrete PID controller has been optimized for the set point at 14 mm.

Table 3: The IAE performance index and ISE performance index

	Perf. Index	IAE	ISE
	Equation	$\frac{1}{N}\sum_{k=1}^{N}\lvert y_{sp}(k)-y(k)\rvert$	$\frac{1}{N}\sum_{k=1}^{N}\left(y_{sp}(k)-y(k)\right)^2$
Set point 6 mm	PID	0,392	0,613
	Adapt. PID	0,256	0,362
	Fuzzy PID	0,319	0,400
Set point 14 mm	PID	0,163	0,136
	Adapt. PID	0,170	0,142
	Fuzzy PID	0,159	0,127

Increased effect of the differential part of the PID controller become problematical in the case of greater changes of the set point. By changing of the set point at 6 mm parameters of the discrete PID controller become nonoptimal, which is expressed by greater overshoot and by oscillation. Performance indexes are satisfactory small in the case of the case of the set point 14 mm and much greater in the case of the set point of 6 mm. The adaptive discrete PID controller has also at the set point of 6 mm a nonoscilatory response and a great overshoot. IAE and ISE are suitable small. A step response of the fuzzy controller is similar to step response of the adaptive discrete PID controller. It should be mentioned that results of the simple fuzzy controller have been obtained by the basic shape of membership functions. It could be expected that fine tuning of membership functions will results in better performance (smaller overshoot). Fifteen rules in three rule basis have been used in the mentioned test. Note that the rule base can easily be extended with some multivariable control rules.

6. Conclusions

An approach of a fuzzy real-time controller based on the decomposition into the multivariable rule base is presented. The multidimensional fuzzy equation has been decomposed on the set of one-dimensional fuzzy equations. Calculated outputs of the decomposed fuzzy equations are a little more fuzzified then the output of the original multivariable fuzzy equation. This is due to the simplification of the multivariable structure and the lack of a good mapping property of the multivariable fuzzy equation. A loss of accuracy must be accepted.

The model of the magnetic suspension system has been realized by the simple electromagnet and the steel ball. It is evident from test results that the performance of the discrete PID controller depends heavily on the operating parameters of the system. The implementation of the gain scheduling algorithm in the PID structure causes a better performance. The same or better performance has been achieved with the fuzzy PID controller. Advantages of an simplified fuzzy PID structure are:

- decreasing number of rules,
- decomposition of multivariable control rules into three sets of one dimensional rules for each input, simplified the evolution of the rule base,
- a direct relationship between a fuzzy control and a conventional control,
- easy connection between fuzzy parameters and operation of the controller.

An interesting possibility concerning the inference break-up is a processor which performs composition of a fuzzy set and a two-dimensional fuzzy relation. A speed-up of the inference can be achieved by using optimized code to perform the composition of a fuzzy set and a fuzzy relation. Because of the inference break-up, the code optimization can be applied to software implementation. Further researches will be directed to the control of theoretic aspects of simplified fuzzy PID control in vague environments like stability analysis or robustness.

7. References

1. Zadeh, L.A., 1965, Fuzzy sets, *Information and Control* 8, pp.338-353.

2. Tzes, A., Chen, J.C., Peng, P.Y., 1994, Supervisory fuzzy control design for a magnetic suspension system. *Proc. of the 3th IEEE Conf. on Fuzzy Systems*, Vol 1, pp.138-143.

3 Rule, R., Gilliland, R., 1980, Combined magnetic levitation and propulsion: the magtransit concept. *IEEE Trans. On Vehicular Technology*, Vol. VT-29, No. 1. pp.41-49.

4. Ion, B., Trica, A., Papusoiu, G., Nasar, S.A., 1988, Field tests on a maglev with passive guideway linear inductor motor transportation system. *IEEE Trans. On Vehicular Technology*, Vol. VT-38, No. 4, pp.230-236.

5. Golob, M., Tovornik, B., 1998, Fuzzy Ball Position Control in the Magnetic Field, *Proc. of the IIIS World Multiconference on Systems, Cybernetics and Informatics*, Orlando, FL, Vol 4, pp.31-36.

6. Demirli, K., Turksen, I., B., 1992, Rule break up with compositional rule of inference. *Proc. of the 1th IEEE Conf. on Fuzz Systems*, San Diego, pp.949-956.

Part 4: Identification and Modelling

Papers:

Experimental Evaluation of Intelligent Identification Algorithms Applied to a Wind Tunnel Process

Leandro dos Santos Coelho and Antonio Augusto Rodrigues Coelho

Federal University of Santa Catarina - Department of Automation and Systems
C.P. 476 - 88040.900 - Florianópolis - SC - BRAZIL
Tel: 55 48 3319202 - Fax: 55 48 3319770
E-mail: {lscoelho, aarc}@lcmi.ufsc.br

Abstract

This paper evaluates methodologies of the computational intelligence applied to a nonlinear process identification. The different intelligent methodologies are: hybrid genetic algorithm with simulated annealing and modified Elman neural network. Experimental tests of identification are carried out using a laboratory scale wind tunnel plant. Experiments, for the estimation of the mathematical model of the process, are carried out using step function and pseudo-random binary sequence signals as process input. A white noise signal is used in the validation phase. The performance of each technique, in the estimation and validation phases of the process mathematical model, is presented and discussed.

Keywords

Process identification, Genetic algorithms, Neural networks, Experimental application.

1 Introduction

Driven by the investigation in the academic and industrial environments, the technology of intelligent systems, known as computational intelligence (*CI*), is motivating new approaches in control engineering. These new approaches rely on the robustness and possibility of treating nonlinear processes offered by these intelligent tools [1], [2].

Many methodologies of *CI* applied to prediction and identification problems in industrial processes have been proposed in the literature. Mouzoris & Mendel [3] investigated the application of a fuzzy system in a *NARMA* (Nonlinear AutoRegressive Moving Average) system identification. Sjöberg [4] presented a theoretical and practical studies of the application of artificial neural networks in nonlinear process identification. Chipperfield & Fleming [5] presented an overview and discussed application possibilities of the evolutionary computation (*EC*) in control engineering. Wang et al. [6] dealt with neuro-evolutionary systems in problems of process identification. Harris et al. [7] presented design subjects and detailed a study of neuro-fuzzy systems in identification and process control. Karr [8] presented a design of a fuzzy logic controller with learning by genetic algorithms applied to pH practical process.

The methodologies of *CI*, described and analyzed in this paper, are: hybrid genetic algorithm with simulated annealing and modified Elman neural network. The identification experiments are conducted in the identification of a wind tunnel process. This paper is organized as follows. In section 2, the paradigm of the *EC*, known as hybrid genetic algorithm with simulated annealing, is described. Section 3 presents the modified Elman neural network. The description of the wind tunnel process is shown in section 4. The practical experiments in the wind tunnel process identification and the conclusion are presented in sections 5 and 6, respectively.

2 Hybrid Genetic Algorithm with Simulated Annealing

Genetic algorithms (*GA*'s), which were originated from the work of J.H. Holland, thirty years ago, constitute the most studied branch of *EC*. *GA*'s are probabilistic techniques that manipulate a search space, based on natural selection and genetic mechanisms as proposed by C. Darwin and G. Mendel. The basic characteristics that set *GA*'s apart from the traditional optimisation methods are that [9], [10]:

- they work with a coding of a set of parameters, rather than the parameters themselves (when the binary/integer representation of the chromosomes is used);
- they search through a population of points, rather than an isolated point;
- they use only the information of an objective function (fitness);
- they utilize probabilistic transition rules instead of deterministic ones.

GA's have been applied to a wide range of problems dealing with multidimensional and multimodal search. In spite of the effort of many researchers, no formal prove of *GA*'s convergence exist and, additionally, the algorithms do not guarantee an optimal solution [9], [10], [12].

GA's are known to be robust in applications where the "global" search is adequate for treating the problem. However, in "local" searches they do not present an appropriate behaviour. *GA* can become more efficient, if a fine adjustment based on simulated annealing algorithm (*SA*) is utilized for "local" searches.

The *SA* algorithm is a technique of stochastic optimisation for general purpose which has proved to be robust enough in the approximation of a global optimum for many optimisation problems. *SA* is a variation of hill-climbing algorithms where the objective is to minimize an objective function (level of energy) and is based on an analogy with the statistic mechanics of materials under cooling, where physical substances such as metals are taken to high levels of energy and, later, are gradually cooled down until reaching a minimum state of energy.

The annealing procedure is applied to the *GA* at each generation, after the crossover and mutation operations. For fine tune of the *GA* chromosomes values, a low temperature factor and a fast factor of annealing are applied, being this methodology denominated as hybrid *GA* with *SA* (*GASA*) [13], [14].

The basic diagram of the data cycle in *GASA* is composed of selection, crossover and mutation operations, in addition to a genetic thermodynamic operator based on *SA*, according to Figure 2.1. The floating point representation (coding of model parameters to be estimated) is hybridized with *SA* and the *GA*'s operators proposed by Michalewicz [11], while the presence (bit=1) or absence (bit=0) of a term the model to be estimated is represented by 1 bit and, in this case, just the mutation operator is utilized for binary representation proposed by Goldberg [10]. Details of the mathematical models and their terms will be considered in section 5.

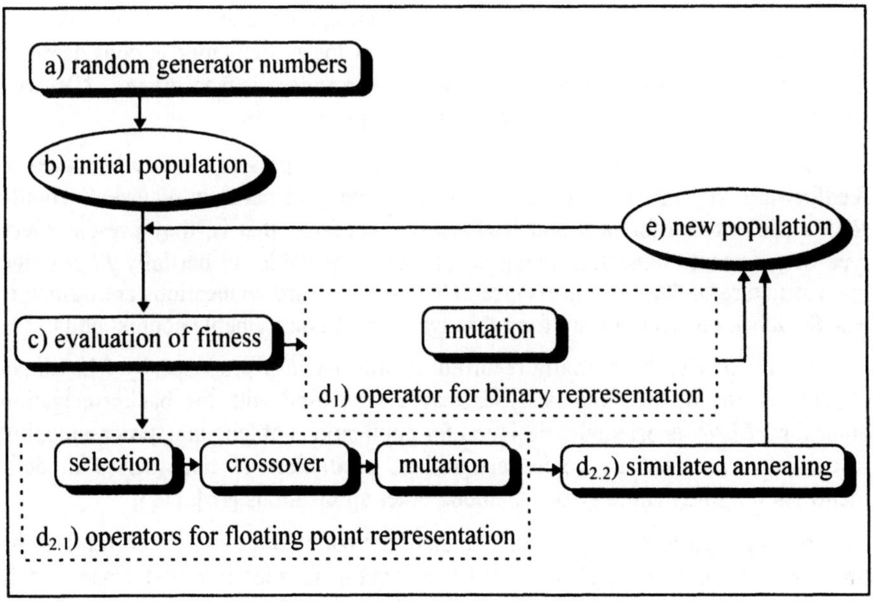

Figure 2.1. Diagram of the data cycle in *GASA*

3 Modified Elman Neural Network

Artificial neural networks (*NN*'s) can be seen as weighted and driven graphs. Artificial neurons represented by nodes and driven arrows (with weights), are connections between the outputs and inputs of the neuron. Based on connection patterns, *NN*'s can be grouped in two categories: feedforward *NN*'s, where the graphs do not have feedback, and recurrent *NN*'s, where there is feedback. Examples of feedforward *NN*'s are multilayer perceptron and radial base *NN*'s. Among feedback *NN*'s there are the competitive networks, Kohonen self-organizing maps, Hopfield *NN* and *ART* models. The main characteristics of the *NN*'s are: parallel processing, associative and distributive memory, and learning, that are evidently similar and inspired by the biological neural networks [1], [4].

NN's can learn through the training measurements and, they are, in fact, a form for approximation of mathematical functions. As they are not based on model, some topologies of *NN*'s are universal approximators of any function with a given arbitrary precision within an appropriate interval [15], which provides a promising tool to be applied in identification, intelligent and adaptive control [1], [2], [4].

Feedforward *NN*'s do not have dynamic memory and the tapped-delay-line method is usually adopted for a *NN* to represent dynamic systems. This method presents some disadvantages such as the learning speed and the high number of neurons in the input layer. These disadvantages are caused, if the order of the system to be identified is unknown. So, in this case, an overestimation is necessary and, consequently, a high number of neurons in the buffer input is considered to deal with high-order systems. This usually provides a feedforward *NN* with structure of high dimension and lower learning speed.

Due to its structure, recurrent *NN*'s (*RNN*'s) do not present the disadvantages of feedforward *NN*'s. *RNN*'s can be classified as totally or partially recurrent. Totally *RNN*'s present feedforward and feedback connections, that is, they present every type of recurrent connection being all of them adjustable. In partially *RNN*'s, the main structure of *RNN* is non-recurrent. The feedforward connections are trainable. The feedback connections are form through a set of non trainable context units.

Among *RNN*'s, the partially recurrent neural network proposed by J.L. Elman (*ELNN*) is one of the simplest types that can be used with the backpropagation algorithm. *ELNN* is originally designed for applications of language processing, but it can also be used in voice recognition, prediction of time series models, identification of dynamic systems, among other applications [16], [17].

Utilizing high learning coefficients, there are oscillations and instabilities in the process of training of *ELNN*. In order to obtain an adequate performance it is necessary to introduce self-connections in the units of context to provide *ELNN* with units of inertia. By doing so the capacity of memorization of the *ELNN* is enhanced to be used for linear and nonlinear activation functions. Thus a modified *ELNN* (*MELNN*) [17], [18] is obtained. A schematic diagram of *MELNN* is presented in Figure 3.1.

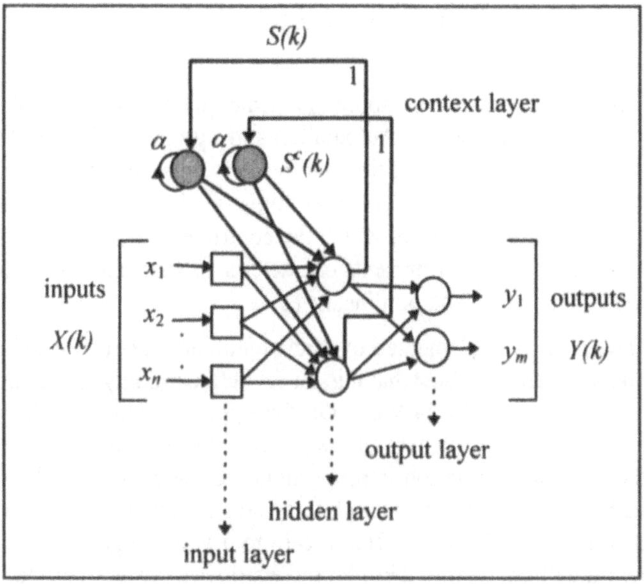

Figure 3.1. Schematic diagram of *MELNN*

In Figure 3.1, the external inputs of *MELNN* are represented by $X(k)$, the outputs of the network by $Y(k)$, the activations of the hidden units by $S(k)$, and the outputs of the context layer are $S^c(k)$. The equations that characterize a *MELNN* are:

$$S(k) = F\{W^{sc}S^c(k),\ W^{sx}X(k-1)\} \tag{1}$$

$$S^c(k) = S(k-1) + \alpha S^c(k-1) \tag{2}$$

$$Y(k) = W^{ys}S(k) \tag{3}$$

where W^{sc}, W^{sx}, W^{ys} are weighting matrices and F is a function vector. In this paper, a linear activation function is adopted. The modification in standard *ELNN* has the characteristic of the j-th unit of context in *MELNN* to be given by

$$s_j^c(k) = s_j(k-1) + \alpha s_j(k-2) + \alpha^2 s_j(k-3) + \ldots \tag{4}$$

The value of α=0.6, is the same for all the self-connections and it is not modified during the training algorithm. Usually α ranges within [0;1], where α nearest to 1 enables the context layer units to aggregate more delayed outputs. The introduction of self-connections in the context units increases the possibility of *ELNN* to model high-order systems [18]. The training algorithm used is the backpropagation with supervised learning [19].

4 Description of the Wind Tunnel Process

The wind tunnel process is a laboratory scale plant (monovariable prototype) developed by the Department of Automation and Systems at Federal University of Santa Catarina (*DAS/UFSC*). Additional information on the process can be found at *http://lcp.lcmi.ufsc.br*. The experimental aims at identifying the wind tunnel process. It is a dynamic system in which the controlled variable is the speed of the air flow inside the duct and it presents characteristics that can also be found in real process such as time delay and nonlinearities.

The wind tunnel is implemented over an aluminium base and consists of a 9 cm wide and 60 cm long air duct having on its left extremity a *DC* motor with a working range from zero to 24 volts. On the opposite extremity, there is another *DC* motor (tachometer) driven by the air flow generated by the first motor (the voltage arising from the rotation is proportional to the air flow). The transmission circuit utilizes several operational amplifiers and adequate the output voltage of the measurement motor (tachometer) from a zero to 5 volts range. The actuator circuit works at the range from zero to 24 volts and employs a power transistor bridge. It is possible to insert a disturbance in the process variable by switching a key which will add a step voltage to the output signal. The wind tunnel control system is shown in Figure 4.1 [20].

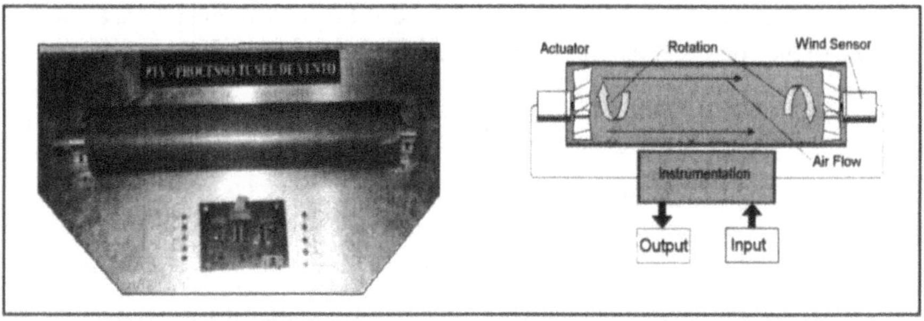

(a) photograph (b) physical setup

Figure 4.1. Characteristic of the wind tunnel plant

Typically wind tunnels are used in aeronautical, automobile industries and in agricultural applications. When subjecting a three dimensional object placed inside the tunnel to an air flow it is desirable that this flow be constant and known. To attend that, it is necessary electronically measure all the variable involved in the process and to have a computer aided controller that can satisfy all the performance specifications (settling time, overshoot).

5 Experiments of Identification

The goal of the system identification is to allow an adjustment of a mathematical model for a dynamic system. This can be obtained by collected measurements, by the adjustment of design parameters and by the adjustment of a mathematical model, until the system output come as close as possible to the measured output samples [21]. The procedure for experimental identification of processes is constituted by four basic stages, according to Figure 5.1.

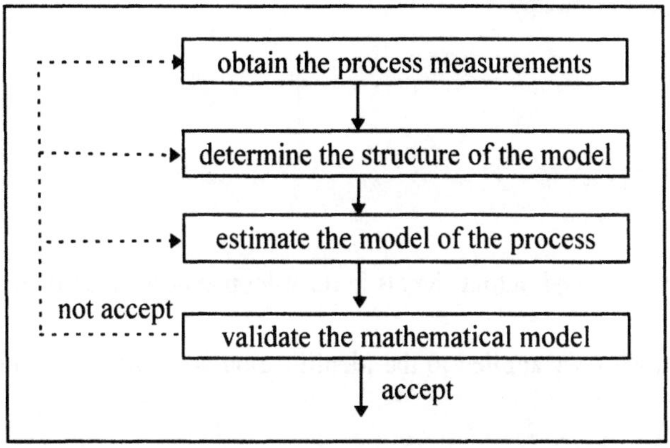

Figure 5.1. Procedure for system identification

The identification of the wind tunnel is suitable if error criterion is in acceptable values (previously defined by the user). The adopted error criterion is based on the minimization of the sum of square of the difference between the values of desired output and the estimated output:

$$F = \frac{1}{2}\sum_{k=1}^{N}[y(k) - \hat{y}(k)]^2 \tag{5}$$

where $y(k)$ is the desired output, $\hat{y}(k)$ is the estimated output, and N is the number of collected samples of the experiment. Experiments for the estimation phase of the process mathematical model are carried out using a step process input signal of magnitude 3, from the samples 0 to 200. Between samples 200 and 667 the process input signal is a *PRBS* (Pseudo-Random Binary Sequence) composed of a step signal of magnitude 3 and a *PRBS* of magnitude ±1.

A sequence of independent random variable (white noise) with zero average and variance one is applied between the samples 667 and 1000, in the validation phase of the mathematical model. The applied input signal (step of amplitude 3 plus

PRBS plus white noise) and the output of the wind tunnel in open-loop case are presented in Figure 5.2.

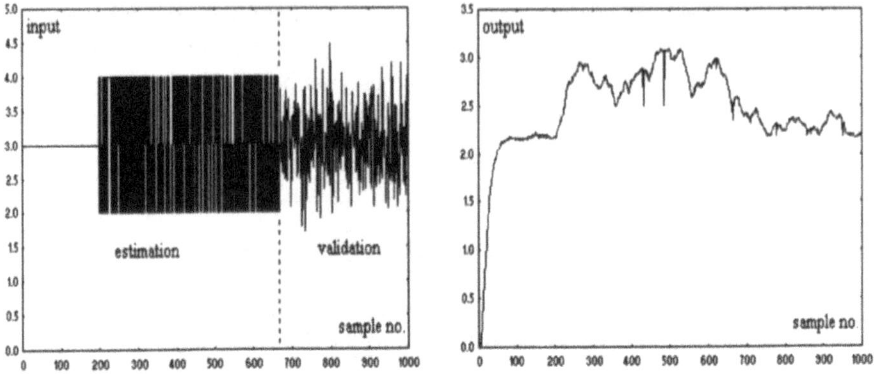

(a) input (b) output

Figure 5.2. Input and output signals in open-loop experimentation of the plant

GASA and *MELNN* applied to the identification are configured as described in Figure 5.3.

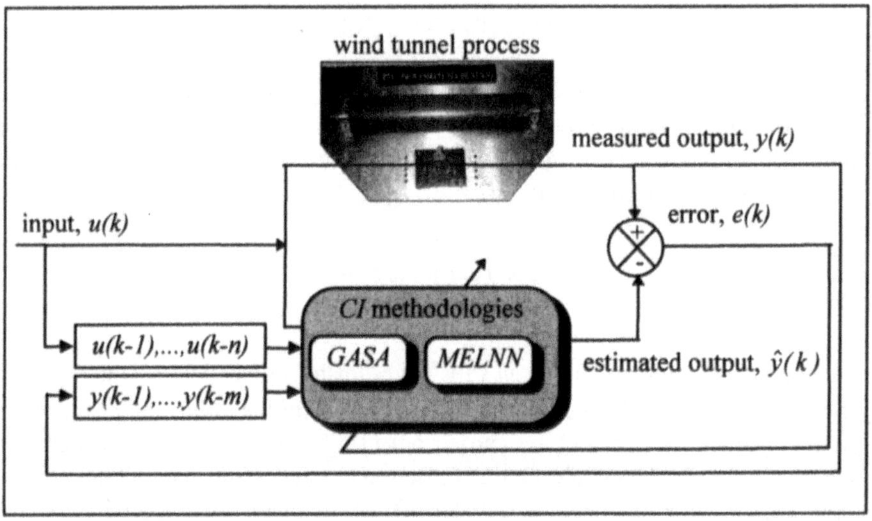

Figure 5.3. Configuration of the wind tunnel identification through *GASA* and *MELNN*

The *GASA(1)-(4)* paradigms estimate the parameters of a mathematical model *NARX* (Nonlinear AutoRegressive model structure with eXogenous inputs) for the wind tunnel. The *NARX* models are characterized by nonlinear difference equations structured by linear parameters. These models are functions of the delayed inputs and outputs and, present nonlinear terms, according to the following equation:

$$\begin{aligned} y(t) = & \sum_{i=1}^{nb} b_i u(k-d-i+1) - \sum_{i=1}^{na} a_i y(k-i) + \\ & \sum_{i=1}^{na} \sum_{j=1}^{nb} \sum_{k=1}^{pep-1} \sum_{m=1}^{pep-1} c_{ijkm} u(k-d-i+1)^k y(k-j)^m + \\ & \sum_{i=1}^{nb} \sum_{j=1}^{nb} \sum_{k=1}^{pep-1} \sum_{m=1}^{pep-1} d_{ijkm} u(k-d-i+1)^k u(k-d-j+1)^m \\ & + \sum_{i=1}^{na} \sum_{j=1}^{na} \sum_{k=1}^{pep-1} \sum_{m=1}^{pep-1} e_{ijkm} y(k-i)^k y(k-j)^m + v(k) \end{aligned} \tag{6}$$

where $y(k)$, $u(k)$, $v(k)$ and d are output, input, noise and time delay, respectively. For the *NARX* model, two coefficients were specified for the polynomial $B(z^{-1})$, $nb=2$, and two coefficients for the polynomial $A(z^{-1})$, $na=2$, and the pow of polynomial expansion $pep=2$, and $d=1$. So, there are $nt = 17$ terms (representation for floating point) + 17 parameters corresponding to the presence or absence of the terms (binary representation) for estimation via *GASA-(1)-(4)* [10], [22].

The techniques called *GASA(5)-(8)*, in a similar way, estimate the parameters of a *ARX* (AutoRegressive model structure with eXogenous inputs) mathematical model for the wind tunnel. The *ARX* linear model adopted is given by

$$y(t) = \sum_{i=1}^{nb} b_i u(k-d-i+1) - \sum_{i=1}^{na} a_i y(k-i) + v(k) \tag{7}$$

For the *ARX* model the specified parameters were $d=1$, $na=3$ and $nb=3$, totalling $nt=6$ terms (floating point representation) + 6 parameters corresponding to presence or absence of the terms (binary representation) for optimisation by *GASA*. The adopted number of generations was 300 for *GASA(5)-(8)*. The search space for *GASA(1)-(8)* of the parameters in the representation for floating point is within the interval [-3.0;3.0].

The *GASA(1)-(8)* present the *SA* algorithm with uniform or Cauchy distribution. Further information about the advantages of acceleration of the evolution given by Cauchy distribution can be found in Ingber [23]. *GASA(1)-(8)* optimisations are carried out with a probability for arithmetic crossover of 0.8, non-uniform mutation with probability of 0.2 (floating point representation) and binary mutation with probability of 0.05 [10], [11]. Other configuration parameters of *GASA(1)-(8)* are shown in Table 5.1.

The adopted configuration parameters of *MELNN(1)-(8)* are shown in Table 5.2. The *MELNN(1)-(8)* are designed with learning coefficient of 0.1, moment term of 0.0001, and α=0.6. An important observation is that the number of neurons in the hidden and context layers are the same.

Table 5.1. Configuration parameters of *GASA(1)-(8)*

technique	*GASA*			*identification*	
	population size	*selection*	*distribution used by SA*	*generations*	*model*
GASA(1)	30	roulette wheel	Cauchy	1000	*NARX*
GASA(2)	30	breeder	Cauchy	1000	*NARX*
GASA(3)	30	roulette wheel	uniform	1000	*NARX*
GASA(4)	30	breeder	uniform	1000	*NARX*
GASA(5)	30	roulette wheel	Cauchy	300	*ARX*
GASA(6)	30	breeder	Cauchy	300	*ARX*
GASA(7)	30	roulette wheel	uniform	300	*ARX*
GASA(8)	30	breeder	uniform	300	*ARX*

Table 5.2. Configuration parameters of *MELNN(1)-(8)*

technique	*neuron(s) in each layer*			*identification*	
	input layer, {u(k-1), y(k-1)}	*hidden layer*	*output layer, y(k)*	*epochs*	*model*
MELNN(1)	2	4	1	1000	black-box
MELNN(2)	2	4	1	1000	black-box
MELNN(3)	2	3	1	1000	black-box
MELNN(4)	2	3	1	1000	black-box
MELNN(5)	2	2	1	5000	black-box
MELNN(6)	2	2	1	5000	black-box
MELNN(7)	2	1	1	5000	black-box
MELNN(8)	2	1	1	5000	black-box

The value obtained for the error criterion F for the estimation and validation phases and, the number of parameters of the obtained mathematical model are presented in Table 5.3 for *GASA* (evaluation of F in 50 experiments) and in Table 5.4 for *MELNN*, respectively.

Table 5.3. Result of the identification of the wind tunnel in open-loop via *GASA* (evaluation of *F* in 50 experiments)

technique	$F_{estimation}$ (0-667 sample)			$F_{validation}$ (668-1000 sample)			*number of terms in the best model*
	best	*average*	*standard deviation*	*best*	*average*	*standard deviation*	
GASA(1)	7.21	24.38	12.01	8.32	38.16	26.25	9
GASA(2)	2.54	9.26	6.72	4.46	16.79	15.42	7
GASA(3)	11.05	29.13	14.86	27.52	57.52	23.98	8
GASA(4)	9.67	28.72	13.05	23.36	32.79	19.98	10
GASA(5)	16.89	21.27	5.24	26.48	26.75	15.06	2
GASA(6)	16.12	19.54	5.19	31.22	24.24	11.76	2
GASA(7)	16.17	20.39	7.39	25.03	55.39	25.30	2
GASA(8)	16.37	22.13	8.05	27.71	33.93	16.44	4

Table 5.4. Result of the identification of the wind tunnel in open-loop via *MELNN*

technique	$F_{estimation}$ (0-667 sample)	$F_{validation}$ (668-1000 sample)	*number of terms in the model*
MELNN(1)	0.67	1.75	36
MELNN(2)	0.72	0.94	36
MELNN(3)	0.74	1.48	24
MELNN(4)	0.74	0.92	24
MELNN(5)	0.74	1.03	14
MELNN(6)	0.75	0.93	14
MELNN(7)	0.74	1.06	6
MELNN(8)	0.74	0.95	6

The convergence of the best individual in *NARX* and *ARX* identification through *GASA* are presented in Figure 5.4. According to these results, the best performance were achieved by *GASA(7)* and *GASA(2)* techniques.

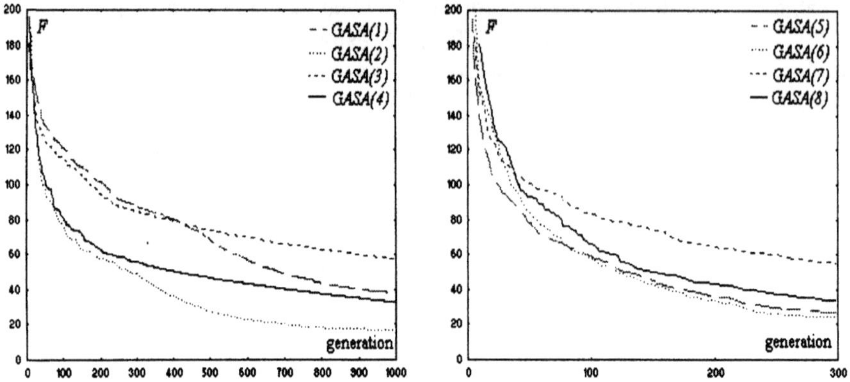

(a) *NARX* model (b) *ARX* model

Figure 5.4. The convergence of the best individual (average in 50 experiments) in identification of the wind tunnel through *GASA* (validation phase)

It was observed, as shown in Table 5.3, that in identification of the *NARX* model the best obtained result was with *GASA(2)* with *F*=2.54, in the identification phase and *F*=4.46, in the validation phase of the model. The mathematical model is given by

$$y(k) = 2.33123u(k-1) + 0.96910y(k-1) - 0.00954u(k-2)y(k-2) + \\ 0.00585u(k-1)^2 - 0.00689u(k-2)u(k-1) + 0.00836u(k-2)^2 + 0.00645y(k-2)^2 \quad (8)$$

In the *ARX* model identification, *GASA(6)* presents the best result in the estimation phase with *F*=16.12, with the same number of terms in the mathematical model of *GASA(7)*, but with a difference in validation phase *F*=31.22. However, the best obtained result was through *GASA(7)* with *F*=16.17 (in the estimation phase) and *F*=25.03 (in the model validation phase). The mathematical model of *GASA(7)* is described by the following equation:

$$y(k) = 0.01812u(k-3) - 0.31733y(k-1) \quad (9)$$

Figure 5.5 shows the estimated response of the wind tunnel for the paradigms with the best result in the estimation and validation phases of parameters for *GASA*.

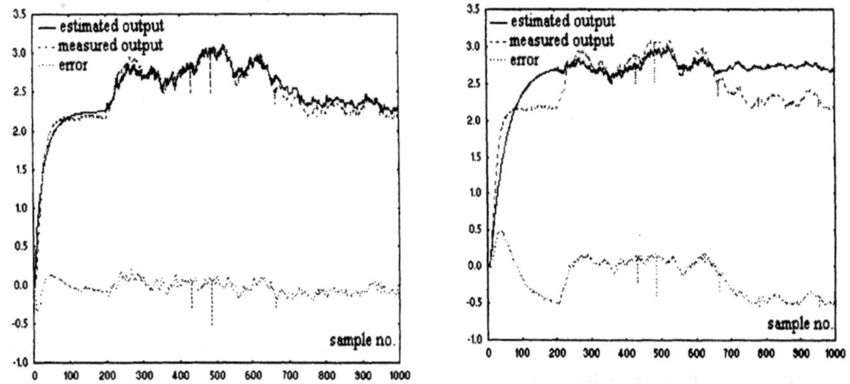

(a) *GASA(2)* (b) *GASA(7)*

Figure 5.5. Estimated response of the wind tunnel in open-loop via *GASA*

All the adopted *MELNN*'s present good results. However, *MELNN(1)* represents the best results, in the estimation phase, of black-box model parameters, as shown in Table 5.4. In the validation phase, the best result was obtained by *MELNN(4)*. In Figure 5.6, the estimated response of the wind tunnel is shown for *MELNN(4)*.

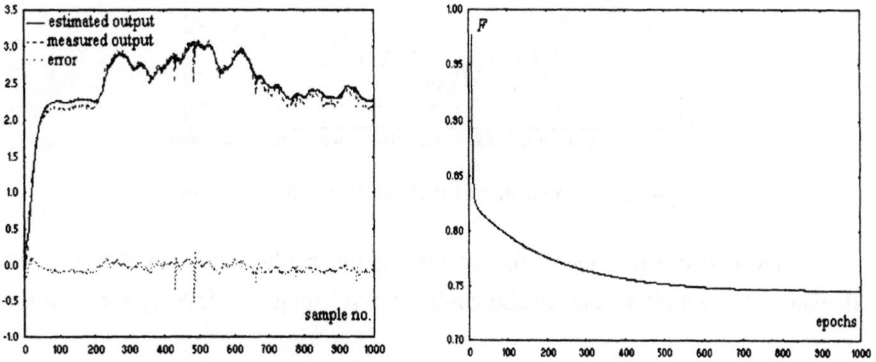

(a) estimated response (b) convergence

Figure 5.6. Estimated response of the wind tunnel in open-loop and convergence via *MELNN*

It can observed from the presented results that *GASA* and *MELNN* provided appropriate experimental mathematical models. *GASA* is linked to the terms of *NARX* and *ARX* models, while *MELNN* presents a black-box model of the wind tunnel, with a more adequate treatment of the nonlinearities of the process due to inherent features of *NN*'s to deal with complex processes. In Figure 5.7, relative to the validation phase, the covariance function of the estimation error, *E(k)*, is evaluated. The best results were obtained by *GASA-NARX* and *MELNN-Black-Box*. The function $E(k)$ is described by the equations:

$$E(k) = \frac{1}{N}\sum_{i=1}^{N-k}(e(i) - \bar{e})(e(i+k) - \bar{e}) \tag{10}$$

$$E(0) = \frac{1}{N}\sum_{i=1}^{N}(e(i) - \bar{e})(e(i) - \bar{e}) \tag{11}$$

where e is the estimation error given by the difference between the measured and the estimated output, $\bar{e}$ is the average of estimation error.

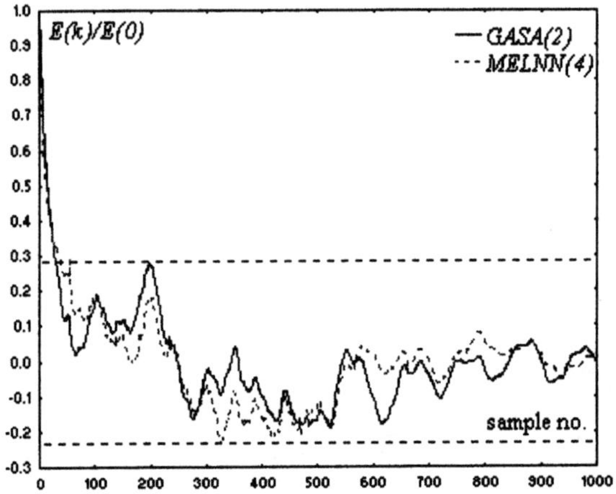

Figure 5.7. Covariance function of estimation error

Another performance index for evaluating the mathematical models can be calculated by the multiple correlation coefficient, R^2, given by following equation:

$$R^2 = 1 - \frac{\sum_{k=1}^{N}(y(k) - \hat{y}(k))^2}{\sum_{k=1}^{N}(y(k) - \bar{y})^2} \tag{12}$$

where $\bar{y}$ is the average of the N samples measured from the process output. The obtained values were R^2=0.95 for *GASA(2)* and R^2=0.98 for *MELNN(4)*. While a value of R^2 close to 1.0 indicates an exact adequacy of the model for measured data, an R^2 between 0.9 and 1.0 may be considered sufficient for many practical applications in control systems.

6 Conclusion

From the experimental results it was observed that two methodologies of *CI*, *GASA* and *MELNN* were adequate to be applied to the identification of the wind tunnel process. *MELNN* provided smaller errors in relation to *GASA*, however it demanded a high computational complexity and higher number of parameters compared to the other techniques. The use of *MELNN* with learning backpropagation was characterized by the obtaining of a structural model for the wind tunnel, but there is the possibility of obtaining a parametric model through the use of Hopfield *NN*'s.

Among the main disadvantages of *MELNN* with learning through backpropagation algorithm are the difficulty in the appropriate selection of the system parameters that are necessary to decrease the learning time of *MELNN*, and the existence of a great number of local minimum in the solution, and, therefore, difficult to determine the global optimal [24], [25].

GASA has been proposed to combine the respective advantages of the two paradigms. *GASA* is a *GA* that uses low-temperature annealing as a local search procedure, however it is not characterized by necessarily find the best solution for the estimation. The *GASA* paradigm, as well as other *EC* methodologies, presents as advantages the possibility of parametric identification of linear processes (or nonlinear) of both discrete or continuous systems (minimum or nonminimum phase); allows to deal with multiobjective and multimodal problems; works with constraints, and it does not depend on derivative terms. The drawback of *GASA* is that for each generation of the algorithm, the fitness of the population of solutions (model parameters) needs to be evaluated.

In spite of the broad, encompassing and multidisciplinary spectrum of applications of several *EC* paradigms, in addition to *GA*'s in control systems — evolutionary programming [26], genetic programming [27], evolution strategies [28], and classifier systems [29] — some issues concerning the evolutionary algorithms are still unresolved, some of these are: determinist rules in the selection of the control parameters of the algorithms, analysis of the convergence properties, in order to improve the efficiency and robustness of *EC* for real-time applications. Future works perspectives concerning to the wind tunnel design include the possibility to assess a comparative study of other tools in identification and control, using intelligent, self-tuning and adaptive methodologies. Perspectives of utilization, theory and practical advances of the *CI* methodologies are paving the way for further technological advancement and increasing integration of the so-called "intelligent machines" in industrial environment.

References

[1] Lu Y -Z 1992 The New Generation of Advanced Process Control. *Control Engineering International*, Mid-March:21-23

[2] RayChaudhuri T, Hamey L G C, Bell R D 1996 From Conventional Control to Autonomous Intelligent Methods. *IEEE Control Systems* 16(5):78-84

[3] Mouzoris G C, Mendel J M 1997 Dynamic Non-Singleton Fuzzy Logic Systems for Nonlinear Modeling *IEEE Trans on Fuzzy Systems* 5(2):199-208

[4] Sjöberg J 1995 Non-Linear Identification with Neural Networks. Dep of Electrical Engineering, Linköping University, Linköping, Sweden

[5] Chipperfield A, Fleming P 1996 Evolutionary Algorithms for Control Engineering. *13th IFAC World Congress*. San Francisco, CA, USA, pp 181-186

[6] Wang Q, Shao H H, Zhang Z J 1997 Genetic Evolved Neural Network for Process Modelling and Optimization. *13th IFAC World Congress*, San Francisco, CA, USA, pp 283-288

[7] Harris C J, Brown M, Bossey K M, Mills D J, Ming F 1996 Advances in Neurofuzzy Algorithms for Real-Time Modelling and Control *Engineering Applications of Artificial Intelligence* 9(1):1-16

[8] Karr C L 1992 Adaptive Process Control Using Fuzzy Logic and Genetic Algorithms. *Proc of 3rd Int Workshop on Neural Networks and Fuzzy Logic*, NASA Conference, *Vol. 1*. Houston, Texas, USA, pp 186-193

[9] Holland J H 1975 *Adaptation in Natural and Artificial Systems*, University of Michigan Press, Ann Arbor

[10] Goldberg D E 1989 *Genetic Algorithms in Search, Optimization, and Machine Learning*. Addison-Wesley, Reading

[11] Michalewicz Z 1992 *Genetic Algorithms + Data Structures = Evolution. Programs*. Springer-Verlag, New York

[12] Uesaka Y 1995 Convergence of Algorithm and the Schema Theorem in Genetic Algorithms. *Proc of the Int Conf on Artificial Neural Networks and Genetic Algorithms*. Alés, France, pp 210-213

[13] Varanelli J M, Cohoon J P 1995 Population-Oriented Simulated Annealing: A Genetic/ Thermodynamic Hybrid Approach to Optimization. *Proc of the 6th Int Conf on Genetic Algorithms*. University of Pittsburgh, Morgan Kaufmann Publishers, San Francisco, CA, USA, pp 174-181

[14] Li Y, Tan K C, Ng K C 1995 Performance Based Linear Control System Design by Genetic Evolution with Simulated Annealing. CSC-95017, Centre for Systems and Control, University of Glasgow, UK

[15] Hornik K, Stinchcombe M, White H 1989 Multilayer Feedforward Networks are Universal Approximators. *Neural Networks* 2:359-366

[16] Elman J L 1990 Finding Structure in Time. *Cognitive Science* 14:179-211

[17] Kremer S C 1995 On the Computational Power of Elman-Style Recurrent Networks *IEEE Trans on Neural Networks* 6(4):1000-1004

[18] Pham D T, Liu X 1992 Dynamic System Modelling Using Partially Recurrent Neural Networks *J of System Engineering* 2(2):90-97

[19] Rumelhart D E, McClelland J L, PDP Research Group 1987 *Parallel Distributed Processing: Explorations in the Microstructure of Cognition, Vol. 1*, MIT Press

[20] Coelho A A R, Bruciapaglia A H, Simas H, Gomes F J 1997 Low-Cost Laboratory Equipment for Analysis and Design of Dynamic Systems. *Proc of 4th Symp on Advances in Control Education*, Istanbul, Turkey, pp 125-130

[21] Ljung L 1995 System Identification. Technical Report 1763, Dep of Electrical Engineering, Linköping University, Sweden.

[22] Li C J, Jeon Y C 1993 Genetic Algorithm in Identifying Non Linear Auto Regressive with Exogenous Input Models for Non Linear Systems. *Proc of American Control Conf.* San Francisco, CA, USA, pp 2305-2309

[23] Ingber L 1989 Very Fast Simulated Re-Annealing *Math Comp Modelling* 12:967-973

[24] Pal C, Kayaba N, Morishita S, Hagiwara I 1995 Dynamic System Identification by Neural Networks *JSME Int J*, Series C, 38(4):686-692

[25] Jain A K, Mao J 1996 Artificial Neural Networks: A Tutorial *IEEE Computer* 29(3):31-44

[26] Ma J T, Wu Q H 1995 Generator Parameter Identification using Evolutionary Programming, *Electrical Power & Energy Systems* 17(6):417-423

[27] Bettenhausen K D, Marenbach P, Freyer S, Rettenmaier H, Nieken U, 1995 Self-Organizing Structured Modelling of a Biotechnological Fed-Batch Fermentation by Means of Genetic Programming. *Proc of IEE/IEEE GALESIA'95.* England, UK, pp 481-486

[28] Coelho L S, Coelho A A R 1998 Genetic Algorithms and Evolution Strategies Applied in Identification and Control: Case Study. In: Chawdhry P K, Roy, R, Pant, R K, *Soft Computing in Engineering and Manufacturing.* Springer-Verlag, London, pp 430-438

[29] Huang R, Fogarty, T C 1991 Adaptive Classification and Control-Rule Optimisation via a Learning Algorithm for Controlling Dynamic System. *Proc of the 30th Conf on Decision and Control.* Brighton, UK, pp 867-868

Improvement of Membership Function Identification Method in Usability and Precision

Ayumi Yoshikawa

Kyoto Institute of Technology, Kyoto 606-8585 JAPAN, ayumi@sip.dj.kit.ac.jp

Keywords: Trapezoidal membership function, Computer-aided identification of membership function, Boundary asymptotic estimation method, Computerized fuzzy graphic rating scale method

Abstract

Fuzzy sets are used in various fields. For those who have little knowledge on fuzzy set theory we find that improving usability of membership function identification method is really important. This paper aims to propose this new method improved in usability and precision, and then to verify the method through experiments. First, we propose the method to identify a trapezoidal membership function, BASE method (boundary asymptotic estimation method). The features of this method are ternary rating of membership grades, asymptotic estimation of boundaries of 1-level set and support set, and effective and recursive selection of elements using computers. Results from psychological experiments are showing that the BASE method is superior to the computerized fuzzy graphic rating scale both in usability for inexperienced users of fuzzy set theory and precision.

1. Introduction

Wrong interpretation of settling membership functions: As already known, we cannot define unique membership function (MF in short) for each fuzzy concept because of vagueness and diversity of data from individual differences. Many researchers therefore consider that they can settle the functions arbitrarily without adequate procedures. Some adequate procedures however must be applied to handle individual differences of their shapes, such as in subjective evaluation [1].

Previous works and their problems: Due to the above necessity several researchers proposed some identification methods and verified them through experiments [2-8]. Most of the methods attach weight to precision of the identified MFs, namely, ability to describe the shapes. However, as fuzzy sets are used in various fields, improving usability of the methods is really important for those who have little knowledge on fuzzy set theory. In above sense the methods proposed in the past have rooms for improvement.

Aims of this work: Aims of this paper are to propose a usable identification method for a trapezoidal MF and to validate the method experimentally. First, we propose a new MF identification method (MFIM in short) with considering its usability for "*novice*" users of fuzzy set theory in the section 2. Next in the section 3 three different user interfaces of the proposed method, BASE, are evaluated upon its implementation onto computers. Then in the section 4 the proposed method is compared with the computerized fuzzy graphic rating scale [8] both in usability and precision aspects. In this paper we deal with identification methods for a trapezoidal MF, i.e., unimodal distribution, since most fuzzy concepts and fuzzy numbers used in applications are unimodal ones and even the identification methods for unimodal ones are not well developed.

2. Theoretical Explanations of BASE Method

2.1 Aims of This Section

In this section, we first review two major previous MFIMs as background of this work and describe problems concerning usability. Then we propose a new MFIM, BASE method, and explain its algorithm and the features.

2.2 Previous MFIMs and Some Problems

Categorization of MFIMs: Many researchers have already proposed many kinds of MFIMs, and some researchers have compared those methods through experiments in aspect of the obtained MF [2-8]. **Figure 2.1** shows a categorization of major MFIMs by Nakamura [9]. The methods are divided into two major categories. One category includes methods that a subject rates membership grades for elements presented by either experimenters or systems. We call this type of the methods RM: rating membership grades for presented elements. The other includes methods that a subject answers elements or intervals that correspond to membership grade(s) presented by either experimenters or systems. We name this type of the methods AE: answering elements corresponding to membership grade(s). As seen from the figure 2.1, the RM has a remarkable variety of estimation of membership grades. In the variations, DR: direct rating is often used.

Schema of two popular methods: Both the DR and the AE are popular methods for identifying MFs in some real applications. As preparation of our new method proposed later, we explain their strategies and problems that spoil their usability. **Figure 2.2** shows schemes of how a MF is identified using the DR and the AE.

In the DR the discrete elements to be rate are chosen in an equal interval. A subject or responder rates to what extent the elements belong to the fuzzy concept to be identified. The subject imagines only partial shape of the MF around the presented element through "*scope for rating*." However, there are some problems to lower usability. First, the method forces subjects to describe the membership grade precisely and rigidly. Second, the more

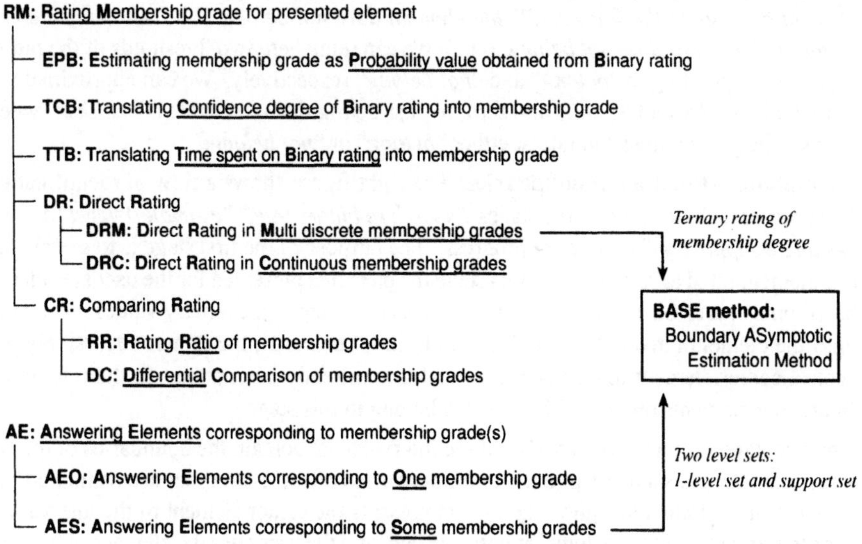

Figure 2.1. Categorization of MFIMs based on ways of subjective judgments and position of the BASE method in the categories.

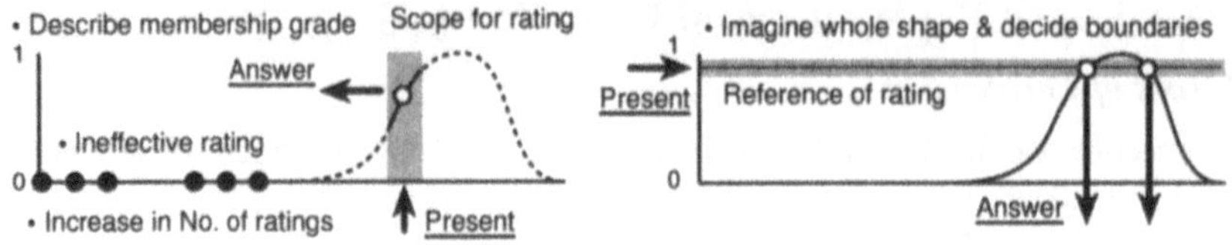

(a) The DR (b) The AE

Figure 2.2. Schemes of two major methods: DR and AE, and their problems

precise MF we obtain, the more elements the subjects have to rate. Third, it forces them to rate ineffective elements not necessary to, if their rating values were well deduced from the values of other elements.

In the AE a subject answers a set of elements or an interval that corresponds to a given membership grade. An advantage of this method is that it can complete the identification by answering only small number of boundaries. There are problems in this method however that spoil its usability. Namely, the subject is forced to imagine the whole shape of the fuzzy set before identification and to decide the boundaries directly comparing the fuzzy set with the presented membership grade as "*reference of rating.*"

2.3 Principle of BASE Method

Algorithm of the BASE method: As mentioned above, both of the two popular methods often used have some problems that lower their usability. To reduce some of the problems, we propose a new method, named BASE method: *boundary asymptotic estimation method* [10]. This method estimates *indirectly* two level sets for *a trapezoidal MF*, i.e., a unity level set and a support set. That is, the obtained fuzzy set is comparable to one obtained from the AE. This method adopts *ternary rating* of the membership grade for the element selected *effectively and recursively.*

Ternary rating: Figure 2.3 shows an algorithm of the BASE method and illustrates its process of identification of a MF. As shown in the upper left figure, a user of this method rates a membership degree of the element presented by the system at ternary values: "*the element belongs to the fuzzy set,*" "*the element does not belong to it,*" and "*the element neither belongs nor does not belong to it.*" We can think here two thresholds of the rating: λ_1, λ_0, corresponding to "*belong*" and "*not belong*" respectively. We can approximate λ_1, λ_0 to the ideal thresholds, *unity* and *zero*, by instructing the user to rate at "*neither*" whenever she/he gets confused to rate at either "*belong*" or "*not belong.*"

Asymptotic estimation of boundaries: The right figure shows a flow of identification. The process is divided into three stages: "*search in binary tree,*" "*estimate 1-level set,*" and "*estimate support set*" in order of progress. The purpose of the first stage is to search only one element rated at "*belong.*" At this stage the elements presented for the user are selected based on a unique binary tree. The user rates the element at the ternary values repetitively until the element rated at "*belong*" is found. In other words, the system recursively presents a center element of two elements that have already been rated. In the middle left figure, the elements numbered from 1 to 5 belong to this stage.

Another stages aim to narrow width of the intervals that contain the boundaries of the two level sets. At the second stage the two boundaries of the 1-level set are estimated. As shown in the middle left figure, the system presents the center element of the interval that has the element rated at "*belong*" at only one end, and the user rates the element. These two states are repeated until all the upcoming intervals that should be rated become smaller than the given minimal width (the elements numbered from 6 to 11). At the last stage the

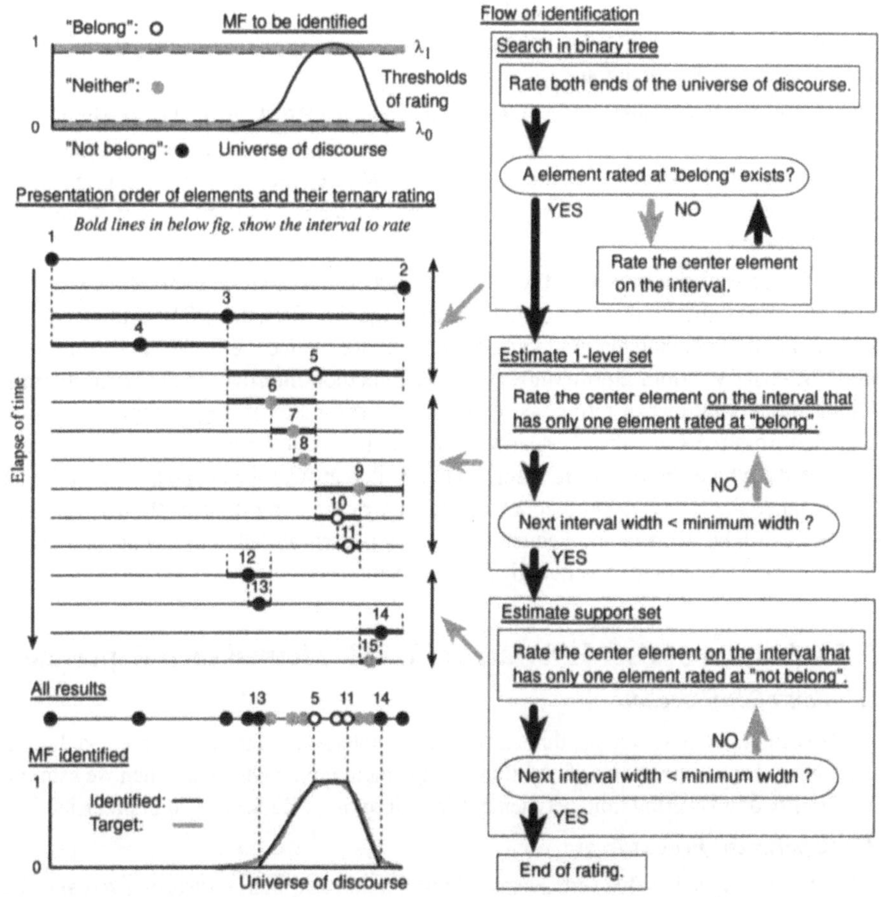

Figure 2.3. A schema of how a MF is identified by the BASE method.

two boundaries of the support set are estimated. The same way as the second stage is used except for process that the anchor element is changed from "*belong*" into "*not belong*." In the figure the elements numbered from 12 to 15 belong to this stage.

Finally, as shown in the bottom left figure, the trapezoidal fuzzy set is constructed. The lower and upper elements rated at "*belong*," i.e., the element of 5 and 11 in the figure, are assigned to the lower and upper boundaries of the 1-level set respectively. Meanwhile the upper element rated at "*not belong*" in all the elements located in the left side of the lower boundary of the 1-level set, i.e., the element of 13 is assigned to the lower boundary of the support set. Also the lower element rated at "*not belong*" in all the elements located in right side of the upper boundary of the 1-level set, i.e., the element of 14 is assigned to the upper boundary of the support set.

Verification of the algorithm through computer simulation: We executed a computer simulation experiment to confirm that the algorithm works completely. We divided the universe of discourse into 32 intervals. 58905 possible trapezoidal and triangular MFs were constructed from either two, three or four of arbitrary different elements in the 33 elements. Then we checked whether or not all the MFs are identified by the proposed algorithm. As the results, no error is detected. So it confirms that this algorithm works to estimate an arbitrary normal convex fuzzy set as a trapezoidal MF completely. The result also shows expected mean value and SD of the number of ratings per fuzzy set are 14.6 and 2.5 under the above condition.

Position of the BASE method: As mentioned above, the BASE method equips both of the features of the DR and the AE. Therefore, there is no category that includes the BASE method exactly. As drawn in figure 2.1, it locates between the DRM: direct rating in multi discrete membership grades and the AES: answering elements corresponding to some membership grades.

Good features of the BASE method: We can reduce and improve some of the problems concerning usability of the DR and the AE using the BASE method. The problem of precise rating of a membership grade in the DR is reduced since the BASE method requires only the ternary rating. Furthermore the BASE method improves the other problem of increasing the number of rating, as seen from the results of averaged number of rating obtained from the simulation. Meanwhile the problems in the AE that user is forced to imagine the whole shape prior to identification and to answer the boundaries directly disappear since the BASE method adopts the ternary rating and estimates these boundaries indirectly. Above all, an improvement in usability is exemplified in the section 4 comparing with the AES through psychological experiments.

3. Evaluation of User Interfaces of BASE Method thru Experiment

3.1 Aims of This Section

In this section, we first evaluate three different user interfaces of the BASE method through experiment from a viewpoint of usability to implement it on computers. Then we examine availability of answering time per element for improving MFs identified in precision.

3.2 Experiment Procedures

Stimuli: We adopted twelve verbal expressions of tallness, which consists of a verbal hedge and "tall" such as "very tall," as fuzzy concept to be identified. The stimuli were divided into three groups, and the groups were assigned to three different feed back methods explained below. The first stimulus in each group is a dummy one. So these three stimuli are not used in analysis.

Systems: Figure 3.1 shows a user interface of the BASE method. In this experiment, we used Macintosh computer with a 13 *inches* display. The interface consists of a fuzzy concept to be identified, a graphic scale with a moving pointer, a feed back area of histories of rating, and four buttons for answering. The graphic rating scale has two verbal anchors and three ticks. It is 256 *pixels* in length, corresponding to around 95 *mm* on the 13 *inches* display. The left and right extremities of the scale mean "*absolutely short*" and "*absolutely tall*" respectively in this experiment. The pointer located under the scale indicates a current element to be rated. We adopted eight *pixels* as the minimal interval. The subjects are asked to rate membership degree of the presented element as an extent that two degrees of tallness between the element and the fuzzy concept agree with. So we adopted {"*agree*", "*not agree*", "*neither*"} as the ternary rating values instead of {"*belong*", "*not belong*", "*neither*"} mentioned in the section 2.

Expressing histories of rating: To examine influence of expressing histories of rating on its usability, we compared three different feed back methods. One is that neither positions of elements nor their rating values are displayed. Another one is that only the positions are displayed in feed back area. Last one is that both the positions and their values are dis-

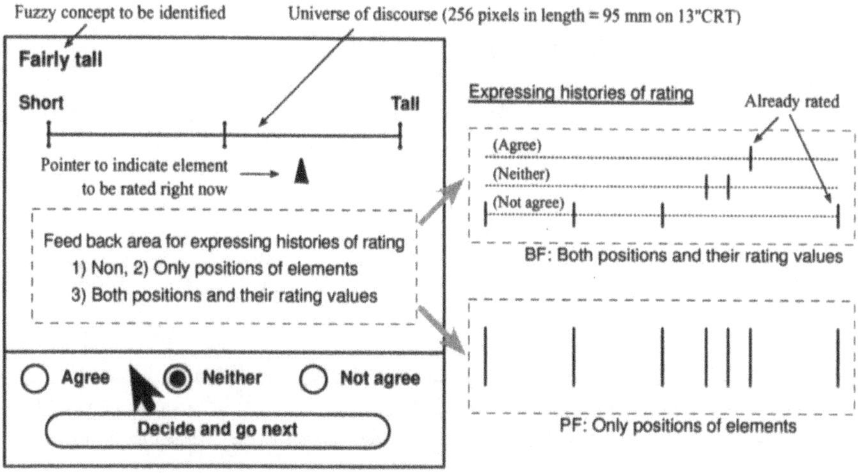

Figure 3.1. Examples of the BASE method implemented on computer and illustrations of different feed back methods.

played in the area. We call these conditions "*NF*," "*PF*" and "*BF*" in order.

Measuring items: We measured answering time per element in a unit of 1/60 *s* as well as the ternary value for each element. Moreover, we asked the subjects to rank their usabilities between the three different feed back methods using the pair comparison method.

Subjects: 36 subjects participated in this experiment as volunteer. They were graduate students and undergraduates at Kyoto Institute of Technology. All the subjects had little knowledge on fuzzy set theory. All of them were native Japanese speakers. Thus all materials for the experiment were displayed in Japanese. These subjects were divided into six groups. The groups were assigned to different order of the three feed back methods to counterbalance influence of experiment order on the answering time.

3.3 Results and Discussions

3.3.1 Feed Back Method Used in The BASE Method

Subjective evaluation of the three feed back methods: Figure 3.2 shows the results obtained from the pair comparison of their usability between the different feed back methods. As seen from these figures, BF is significantly usable than NF and PF at significance level of 5 and 0.1 % in the pair test, where the frequencies rated at "*even*" are not used in this test. Meanwhile, there is no significant difference of the frequencies between NF and PF.

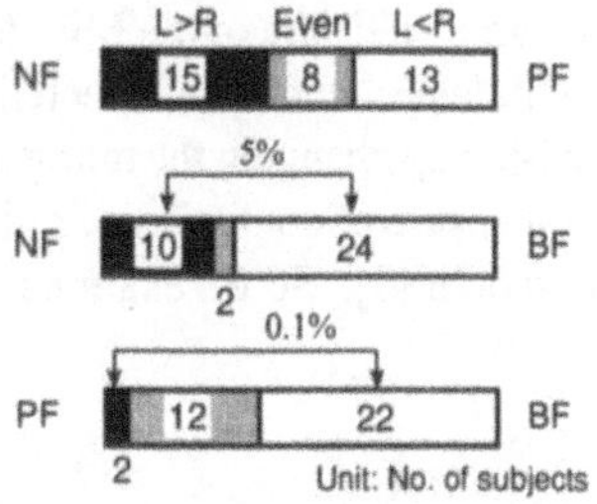

Figure 3.2. Results of pair comparison of usability between the different feed back methods.

Objective evaluation of the three methods: Figure 3.3 shows examples of identified MFs and results of their answering time. We then averaged the number of ratings per hedge, answering time per hedge, and width of the 1-level set and the support set for each feed back method. **Figure 3.4** reveals the mean values and their SDs for the three feed back methods along with the results from the t-test of difference between two mean values in two-tail. As seen from the figures, there is no significant difference among three different feed back methods except for the pair of NF

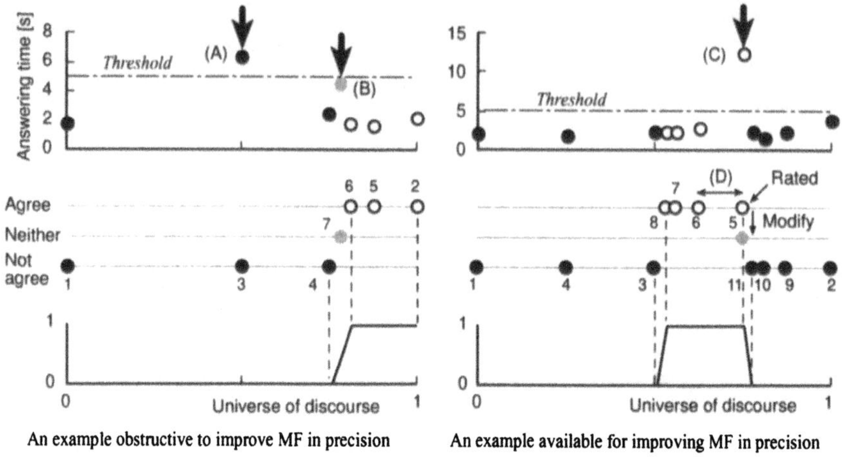

Figure 3.3. Examples of obtained results and availability of answering time for improvement in precision of MF identified

and BF in the averaged answering time and the pair of NF and PF in the width of the support set at significance level of 5, 10% respectively.

Recommended feed back method: In summary the obtained results of objective evaluation reveal that the differences of the feed back methods do not influence the objective values, i.e., the MFs. Contrarily the results of subjective evaluation clearly show that the subjects prefer "*expressing positions of elements and their ratings*" to the other feed back methods. Consequently, we recommend expressing both the elements and the ratings as the feed back method of the BASE method based on the subjective evaluation.

3.3.2 Availability of Answering Time toward Improvement of MF in Precision

Problem of forced binary rating in the BASE method: One of the features of the BASE method is that the element hard to rate at either "*belong*" or "*not belong*" is simply rated at "*neither.*" This feature is useful for improving not only in usability, but also in precision, mentioned in 2.3. However, some users prefer forced binary rating to ternary rating. The forced binary rating causes two thresholds: λ_1, λ_0, to bring near 0.5 in figure 2.3. So it decreases precision of the MF identified. If an element that is forcibly rated at "*belong*" or "*not belong*" would be detected on identification anyhow, we could improve the MF in precision by changing the rating into adequate one, i.e., "*neither.*" Answering time is known as a measure of confusion of rating. Oda has already tried to estimate membership degree based on it [5]. So we examine availability of answering time per element.

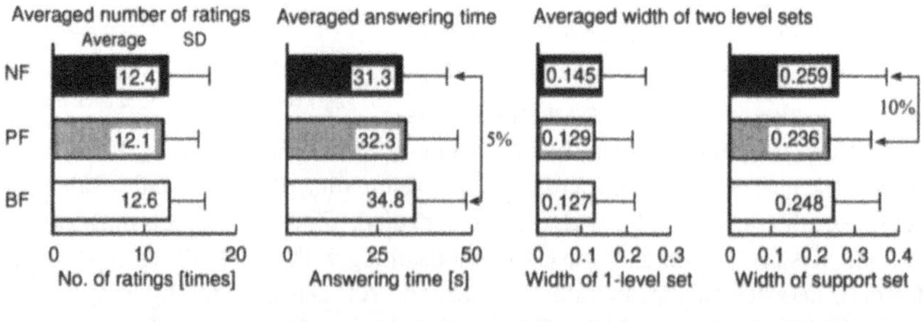

Figure 3.4. Comparison of the obtained results among three different feed back methods.

Availability of answering time: If we could use the answering time for improving, the three conditions must be satisfied. First, there is an element that has remarkably long delay of the answering time. Second, the delayed element is at an end of level sets. Third, the element is detected on identification. We check these conditions from the measured answering time in order.

For the first condition, there are the elements that have longer answering time than the others, marked with an arrow in the figure 3.3. However the second condition is not satisfied. For example, the element of (A) in the left figure belongs to "*not agree*" obviously, but it takes long time to be rated. These kinds of delay are caused by the other sources, such as lowness of arousal level, but not confidence of rating. Contrary like as the element of (C), the answering time could be used for improving. Finally for the third condition we could adequately discriminate these long delayed answers *after the identification*. However, it is hard to settle the threshold for detection on identification, since a range of answering time cannot be estimated prior. From these results, we conclude as follows. It is difficult to improve the MF identified in precision by changing the subject's rating values based on detecting long delay of the answering time for each element on identification. However, the long delay is effective to check the inadequate rating, i.e., forced binary rating with confusion, after end of all the identification.

4. Comparative Studies on BASE Method

4.1 Aims of This Section

In this section, we compare the BASE method with the computerized fuzzy graphic rating scale (FGRS) [8], which is one of the popular methods often used in the applications. Then we evaluate usability of these two methods for "*novice*" users of fuzzy set theory and their precision with repetitive identification.

4.2 Experiment Procedures

Stimuli: We adopted seven verbal expressions of tallness, like as those used in section 3. The first stimulus was used for dummy, thus the total six verbal expressions except for the first were analyzed.

Experiment design: This experiment consists of two sub experiments: "*non repetition experiment*" and "*repetition experiment*." In the non repetition experiment 29 subjects identified the seven stimuli using both the BASE method and the FGRS. We adopted the expression of both of the elements and their ratings as its user interface of the BASE method referring to the result obtained in the section 3. Meanwhile, we adopted a method for answering each two ends of the 1-level set and the support set as the FGRS [11]. These four boundary elements in the FGRS are measured in one pixel. We obtained the answering time and the ternary rating value per element in the BASE method, and the answering time and four boundary elements per stimulus in both. To examine influence of experiment order of both methods on results, we divided the subjects into two groups, and assigned the different order to each group. After identifying both of them the subjects were asked to compare ease of rate between the two methods using the pair comparison.

Meanwhile, in the repetition experiment ten subjects identified the same stimuli as in the non repetition experiment at four times. Five subjects from each group in the non repetition experiment participated. All the subjects identified in order from the BASE method to the FGRS at twice, and in opposite order at twice by turns. The results at the first repetition were diverted from the results in the non repetition experiment. There is no change of experiment design from the non repetition experiment except for that subjective evaluation of usability was executed after only first and forth repetition.

Subjects: All the subjects were graduate students and undergraduates at Kyoto Institute of Technology, and participated as volunteer. They were native Japanese speakers. So all the materials for experiment were displayed in Japanese. All the subjects had little knowledge on fuzzy set theory.

4.3 Results and Discussions

4.3.1 Usability of Both of the Methods: FGRS and BASE

Subjective evaluation of usability of FGRS and BASE: Figure 4.1 shows the results obtained from the pair comparison between both methods. As seen from the results of the non repetition experiment, those who have little knowledge on fuzzy set theory tend to prefer the BASE method to the FGRS in subjective usability. However, the difference of the frequencies between both methods (BASE : FGRS = 19 : 8) is not significant at 5% level (in two-tail) in the pair test. Meanwhile, in the repetition experiment the ratio of the BASE method versus the FGRS changed from 8 : 2 after the first repetition into 5 : 5 after the last repetition. As a reason of this change, we can think three subjects who change their evaluation come to imagine the whole shapes of the MFs easily, thanks to the repetition. So they regard the criterion of usability as "*smallness of effort spent on identification*" instead of "*easy to rate.*"

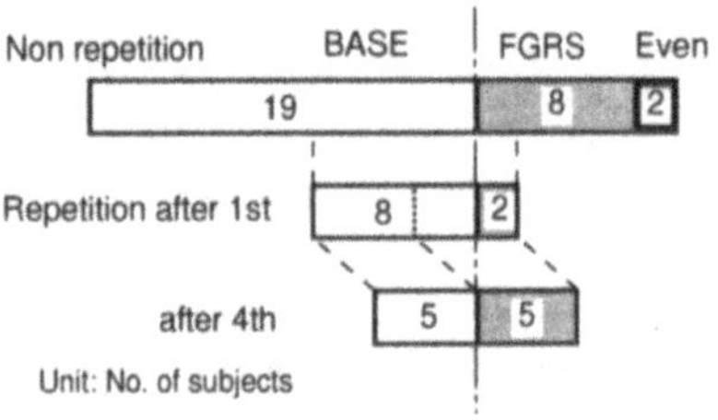

Figure 4.1. Results of pair comparison between the FGRS and the BASE method.

Usability based on answering time: Figure 4.2 shows the averaged answering time per hedge in the FGRS and the averaged answering time per element and the total number of ratings in the BASE method. In the non repetition experiment the averaged answering time per hedge in the BASE method and the FGRS took 42.15 *s* and 16.67 *s* respectively. The averaged answering time from the BASE method is about 2.5 times larger than that from the FGRS. Meanwhile, in the repetition experiment the answering time per hedge in the FGRS shortened as the experiment progressed as seen from the left figure in the figure 4.2. There are significant differences of the averaged answering time between the fourth repetition and the others at significance level of 0.5, 5 and 10% to the first, second and third repetition, in order, in the t-test of difference between mean values. It can be considered

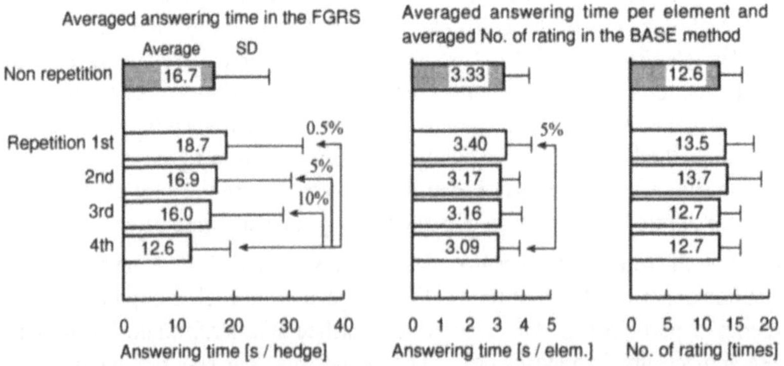

Figure 4.2. Results related to answering time from comparison between the FGRS and the Base method.

that the shortening the answering time origins from so-called "*learning effect*" related to shape of the MFs. Contrary in the BASE method the repetition of identification does not influence the averaged answering time per element, except for the pair of the first and fourth repetition (significance at 5% level). These results clearly show that the BASE method allows users to identify MFs without preceding training.

Conclusion of usability: In summary the results obtained are as follows; from subjective evaluation it is clear that those who have little previous knowledge on fuzzy set theory prefer the BASE method to the FGRS. Similarly the results of the answering time show that the BASE method does not require its user to learn the shape of MF. Consequently, it is concluded that the BASE method is satisfied with the aim of proposing the usable method without the knowledge on fuzzy set theory.

4.3.2 MFs Obtained from Both FGRS and BASE Methods

Evaluation measure: Now we discuss the MFs obtained from the BASE method and the FGRS. To compare the MF, we brought along following measures: *matching measure*, *similarity measure*, *position index*, *shape index*, *inclusion measure*. We adopted the measures related to the logical product, the logical sum and the cardinality as the matching, similarity and inclusion measures [12]. Formula (1) to (3) represents these measures.

Matching measure: $\text{hgt}\,(A \cap B)$ (1)

Similarity measure: $|A \cap B| / |A \cup B|$ (2)

Inclusion measure ($A \supset B$): $|B| / |A \cup B|$ (3)

where A and B mean a fuzzy subset on a universe of discourse: X. $A \cap B$ and $A \cup B$ mean the logical product set and logical sum set between A and B, and $|A|$ denotes the cardinality of the fuzzy subset A. hgt (A) means maximum grade of the fuzzy subset A. The position index is the distance between two gravity centers of fuzzy sets. The shape index is the similarity measure of the two fuzzy sets that are shifted to meet their gravity centers. These indices can adequately describe a disagreement between two fuzzy sets in each meaning than the similarity measure [13].

Preparation of analysis: First, we calculated these measures between the MFs obtained from the BASE method and the FGRS for each same hedge and each subject, and then averaged them for each group. There were 70 and 68 effective data averaged in the group of order from the BASE method to the FGRS: B to F, and from the FGRS to the BASE

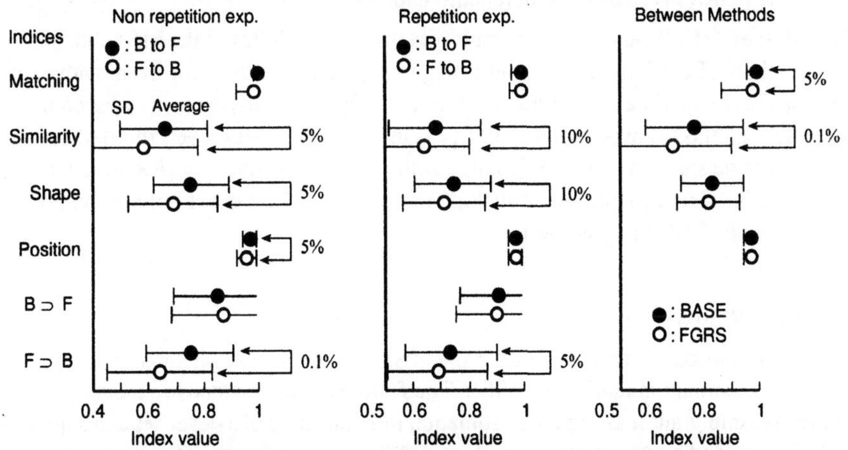

Figure 4.3. Results related to MF obtained from the FGRS and the BASE method

method: F to B respectively in the non repetition experiment. Meanwhile, in the repetition experiment 120 effective data were averaged for each group. **Figure 4.3** shows the results.

Difference between methods: The mean values for all subjects are 0.724 (SD: 0.153), 0.963 (0.032), 0.728 (0.144), and 0.971 (0.026) in order of the shape and the position index in the non-repetition, and the shape and the position index in the repetition. These values reveal the MFs obtained from both methods do not completely agree with each other. However, we must notice the difference of the minimum measurable interval between both methods. The interval in the FGRS is one pixel, while that in the BASE method is eight pixels. To estimate an influence of the difference, we think one trapezoidal fuzzy set averaged for all the sets obtained from this experiment. The widths of the 1-level set and the support set for the averaged fuzzy set are 0.138 and 0.259, normalized length of universe of discourse as unity. Meanwhile, the minimal interval in the BASE method corresponds to 0.031. Under these conditions, the position index calculated between the averaged fuzzy set and that shifted to 0.031 becomes 0.969. Similarly the shape index calculated between the original and that expanded each boundary of the original at 0.031 becomes 0.759. These values approximate to the obtained values, 0.963 and 0.724 respectively. Consequently the difference of the minimal measurable interval can decrease values of the measures, thus it is hard to remove the influence of the difference under these experimental conditions.

Influence of experiment order: As seen from the figure 4.3, the mean values for the "B to F" are higher than those from the "F to B" in both the non-repetition and the repetition experiment. Especially, for the similarity measure and the shape index, the differences between the averages are significant at 5% in the non-repetition experiment and 10% in the repetition experiment. These results suggest that identification using the BASE method prior to the FGRS helps the subjects to image the adequate shapes of the MFs and answer easily. That is the evidence of the BASE method being usable.

Stability of MF with repetition: We calculated the matching and similarity measure and the shape and position index of all possible combinations of four repetitions for each hedge and each subject, and averaged them for each method. The right figure in the figure 4.3 shows the results. As seen from the figure, for all four indices the values of the BASE method are higher than those of the FGRS. Moreover, there are significant differences of the matching and similarity measure at significance level of 5 and 0.1% in two-tail in the t-test of difference between averages. So these results reveal that the BASE method is superior to the FGRS in precision with repetitive identification.

Precision of MF: In summary the results obtained are as follows; the MFs obtained from both methods, FGRS and BASE, indicate good agreement except for irremovable difference on these experiment conditions. From the influence of experiment order on the MFs the BASE method shows higher usability than the FGRS. Moreover, the BASE method turns out to be superior to the FGRS in stability in the experiment of repetitive identification of MFs. Consequently, we conclude the BASE method is not only usable, but also superior to the FGRS in precision.

5. Conclusions

We have proposed the new method for identifying a trapezoidal MF, named the boundary asymptotic estimation method (BASE). The features of the method are the ternary rating of membership grades, asymptotic estimation of boundaries of 1-level set and support set, and effective and recursive selection of elements using computers. The results of verification experiment have shown that the BASE method is superior to the computerized fuzzy graphic rating scale both in usability for "*novice*" users of fuzzy set theory and precision.

We have already rewritten the program of the BASE method in the JAVA language. The BASE method therefore is available for use with internet browsers and works along with other systems on any operating system. In fact, Yoshikawa has already used the BASE method for obtaining individual meaning of verbal hedges in computer aided design of categorical rating scales [14].

However, the problem of forced binary rating has remained. To fix and reduce it, I would like to find and test other measures adequate for detecting confusion of rating during identification.

Acknowledgments

This study was supported by Tateishi Science and Technology Foundation in 1994.

References

[1] Yoshikawa A 1998 Fuzzy Rating and Membership Function Identification Method (Jpn). *J Japan Society for Fuzzy Sets Theory and Systems* 10:184-192

[2] Chameau J, Santamarina J C 1987 Membership Functions I: Comparing Method of Measurement. *Int J Approximate Reasoning* 1:287-301

[3] Turksen I B 1991 Measurement of membership functions and their acquisition. *Fuzzy Sets and Systems* 40:5-38

[4] Hersh H M, Caramazza A 1976 A Fuzzy Set Approach to Modifiers and Vagueness in Natural Language. *J Experimental Psychology: General* 105:254-276

[5] Oda T 1992 Pilot study on measurement of membership grade: comparison between response time method and rating scale method (Jpn). In: *Proc 2nd workshop on fuzzy system for non-engineering* 68-71

[6] Wallsten T S, Budescu D V, Rapoport A, Zwick R, Forsyth B 1986 Measuring the Vague Meaning of Probability Terms. *J Experimental Psychology: General* 115:348-365

[7] Takeuchi H 1991 Measuring the Meaning of Linguistic Hedges by Fuzzy Rating Scale Method (Jpn). *Mathematical Linguistics* 17:365-376

[8] Hesketh B, Pryor R, Hesketh T 1987 An application of a computerized Fuzzy Graphic Rating Scale to the psychological measurement of individual differences. *Int J Man-Machine Studies* 29:21-35

[9] Nakamura K 1994 Fuzzy theory and Behavior science (Jpn). In: Japan Society for Fuzzy Sets Theory and System (eds) *Fuzzy theory ,and humanity and social science*, Nikkan-kougyo-shinbun, Tokyo 194-198

[10] Yoshikawa A, Fujimoto D, Nishimura T 1994 Improvement on method for identifying membership functions using computers (Jpn). In:*Proc 4th FAN Symposium* 7-12

[11] Yoshikawa A, Fujimoto D, Nishimura T 1995 Objective and Subjective Comparison between Two Methods for Identifying Membership Functions: Computerized Fuzzy Graphic Rating Scale and Boundary Asymptotic Estimation Method (Jpn). In:*Proc11th Fuzzy system symposium* 405-408

[12] Dubois D, Prade H 1980 *Fuzzy Sets and Systems: Theory and Applications*, 9-35, Academic Press, San Diego

[13] Yoshikawa A 1997 Independence of Functions of Verbal Hedges from Modificands. *Biomedical Fuzzy and Human Sciences* 3:77-82

[14] Yoshikawa A Fuzzy theoretic approach to computer aided desing of categorical rating scales (Jpn). now submitting

General Parameter Radial Basis Function Neural Network Based Adaptive Fuzzy Systems

Daouren F. Akhmetov and Yasuhiko Dote

Department of Computer Science & Systems Engineering
Muroran Institute of Technology
27-1, Mizumoto-cho, Muroran 050, JAPAN
Phone: +81-143-47-3267; FAX: +81-143-47-3374
E-Mail: dauren@oyna.cc.muroran-it.ac.jp

Keywords: Adaptive Fuzzy System, Rule Generation, Neural Network, Learning Algorithm

Abstract

In this paper an automatic fuzzy rule generation problem through the artificial neural network (ANN) approach is considered. The unknown fuzzy relation reconstruction problem is treated as an optimization of the structure and parameters of the neural network. The functional equivalence between some classes of fuzzy systems and radial basis function networks (RBFNs), namely, their localized sensitivity to input value, is a background of the proposed approach. The improved structure and advanced learning feature RBFN is developed based on General Parameter (GP) method of complex system identification. The criterion of the GP RBFN (General Parameter Radial Basis Function Network) structure optimality is derived using the GP steady state statistics. The derived criterion is used then for the development of the GP RBFN structure self-organization procedure. As a result, an Adaptive Fuzzy System (AFS) with capability to extract fuzzy If-Then rules from input and output sample data is proposed. Simulation examples are given.

1. Introduction

A capability of fuzzy systems to solve a wide class of mapping problem has been approved by control theorists and practical engineers. One of the most important design issues of fuzzy systems is the construction of appropriate fuzzy rules set. There are two common approaches - manual rule generation and automatic rule generation. The first one is very time consuming, and its applicability may be problematic for high dimensional fuzzy rule set case or poor domain knowledge data base. Therefore, many researchers have been involved into automatic fuzzy rule generation (extraction) problem [1-5].

The basic idea behind these approaches is to estimate fuzzy rules through learning from input to output sample data. For example, Kosko [2] has developed fuzzy associative memory (FAM), a class of neural networks, in order to store a set of fuzzy rules and to process them in parallel. The fuzzy associative memory learns fuzzy rules by clustering the sample data in the product space or the linguistic space. In this approach each cluster represents a fuzzy rule. The problem here is to overcome the possible conflicts between the generated rules [3]. Wang and Mendel [5] have proposed a method to resolve the conflict problem, while assigning a degree to each rule.

The direct usage of artificial neural networks (ANN) to identify fuzzy rules and to tune their membership functions has been also reported [6]. But inherent intractability of neural network approach complicates the understanding of the identified fuzzy rules due to distributed way of information processing with ANN . To resolve the difficulty, a fuzzy multilayer perceptron have been proposed in order to generate a number of fuzzy rules which can be obtained by the connection weights computed in the training phase [7]. Fuzzy rules make it possible to infer the output class membership values of an unknown input and estimate a measure of certainty of that decision in the testing phase. Recently, a class of radial basis function networks (RBFNs) are also extensively used to explicitly extract fuzzy rules from the given data. The RBFN proposed by Moody and Darken [8] is a type of neural network which utilizes local receptive fields, called basis functions, for function mapping. The RBFNs have demonstrated their usefulness in a variety of applications including classification, prediction, and system modeling.

The functional equivalence between fuzzy systems and the RBFNs have been shown, and the adaptive network based fuzzy inference system, which is trained by the backpropagation algorithm, have been developed by Jang et al. [9,10]. Following the idea that fuzzy IF-THEN rules are naturally related to the basis functions, Wang and Mendel [11] have represented fuzzy systems as series expansions of fuzzy basis functions.

It is important to note that the above architecture can only realize the inference schemes of either the Sugeno type (the consequence is linear combination of inputs) or a special case of the Sugeno type schemes (the consequence is a constant) [4]. To extend other types of fuzzy reasoning than the Sugeno type, Horikawa et al. [1] have proposed fuzzy neural networks which implement the process of fuzzy reasoning whose consequence part has fuzzy variables. However, the mentioned fuzzy neural networks require proper initial rule basis which can be created using the expert knowledge. It is easy to see that creating the initial rule basis may be difficult as the number of inputs becomes high. Furthermore, in both of the above methods [1,10], it is necessary to determine the number of fuzzy rules to be identified before training since the networks are trained by the backpropagation algorithm.

Fuzzy systems with adaptive capability to extract fuzzy implicative rules from sample data by a learning algorithm were already proposed [12,13]. Three different architectures of the RBF based AFS are developed in order to accomodate not only the Sugeno type but also a more general fuzzy inference scheme, i.e., when the consequence is fuzzy. The proposed RBF based AFS has several advantages against other methods trained by the backpropagation algorithm. It does not require the initial rule base. Fuzzy rules are generated by recruiting the basis function units until desired accuracy is achieved, therefore, it is not necessary to specify the number of fuzzy rules to be identified. Along with these, the RBF based AFS is compatible with existing fuzzy systems.

The paper presented deals with modified RBFN application to automatic fuzzy rule generation. The proposed adaptive fuzzy system is based on the General Parameter RBFNs (GP RBFNs), which has advanced learning features and various applications reported in preceding papers [14-20]. The statistics characteristics (expectation and variance) of the GP at the steady stage of the learning procedure provide an information about model accuracy. This feature has been used, for example, for determination of a linear model order and (or) time lag [14], time series analysis and fault diagnosis problem[15,17,19]. As the research results have shown, it is very promising to apply the derived method to the wide class of nonlinear mapping problem through neural network approach. For example, a fuzzy procedure for GP RBFN structure self-organization has been designed [16].

The rest of the paper is organized as following. First, the problem of a fuzzy rule set generation through neural network representation is formulated, and some already elaborated methods are discussed. Then, the theoretical backgrounds of GP approach are presented including its convergence (stability) analysis. Next, GP RBFN structure and learning procedure are explained, and the GP RBFN based adaptive fuzzy system (AFS) is described. In conclusion, simulations are presented and research results are discussed.

2. Fuzzy Systems Through Radial Basis Function Neural Networks

The fuzzy system maps an input fuzzy set X into an output fuzzy set Y in accordance with fuzzy relation R describing the fuzzy rule set (1).

$$Y = X \circ R \tag{1}$$

where $\circ$ is a compositional rule of inference [21].

A fuzzy rule generation problem may be considered as a problem of fuzzy relation R determination from numerical data [e.g.,1,3,5]. Approximation abilities of neural networks [22] make it possible to reformulate the fuzzy mapping problem (1) and to state a neural network based mapping problem instead, which is of the form (2).

$$\hat{Y} = F(X) \tag{2}$$

where $\hat{Y}$: estimate of fuzzy system (1) output set Y; F: input-output mapping operator of the neural network. Therefore, the initial problem of the reconstruction of the fuzzy relation R has become a problem of a neural network structure St and parameter Par optimization in order to minimize a norm (3).

$$\left\| Y - \hat{Y} \right\| \to \min_{(St,Par)} \tag{3}$$

Under the following conditions the functional equivalence between a radial basis function network and a fuzzy inference system can be obtained [9]:

- the number of hidden nodes in the RBFN is equal to the number of fuzzy if-then rules;
- the membership function of each rule is Gaussian function;
- the t-norm operator used for computing the firing strength of each rule is multiplication;
- the output of each fuzzy if-then rule is composed of a constant under consideration using the same method (defuzzification method) to derive output.

The output x of RBFN is described by equation

$$x = \sum_{i=1}^{K} w_i R_i(\tilde{u}) + w_0 \tag{4}$$

where K: number of basis functions; w_i: weights of the network; $R_i(\tilde{u})$: radial basis functions of network input $\tilde{u}$; $\tilde{u} = [u_1, u_2, ..., u_N]^T$: input vector of dimensionality N; w_0: bias term. For example, Gaussian functions

$$R_i(\tilde{u}) = \exp[-\|\tilde{u} - \tilde{c}_i\|^2 / \tilde{\delta}_i^2], \quad i = \overline{1, K} \tag{5}$$

where $\tilde{c}_i = [c_{i1} c_{i2} ... c_{iN}]^T$ is the center of i-th function in the input signal space; $\tilde{\delta}_i^2 = [\delta_{i1}^2 \delta_{i2}^2 .. \delta_{iN}^2]^T$ is a receiptive field width vector.

If the above mentioned conditions of the RBFN and the fuzzy inference system are satisfied, any RBFN learning rule can be applied to adapt fuzzy systems [12]. Some of authors using RBFN for approximation problems propose network training procedures with weights $w_i (i = \overline{0, K})$ adjusting while centers $\tilde{c}_i$ and widths $\tilde{\delta}_i^2$ are fixed. Alternative methods are based on simultaneous corrections of all the network parameters. The latter allows to reduce number of units and to simplify network structure resulting in the same approximating abilities. But learning procedure complexity significantly increases.

There are well-known methods of RBFN parameter determination, both supervised and unsupervised. For example, K-means clustering algorithm is used to search the cluster centres. Appropriate widths are calculated using nearest neighbor heuristics

[8]. The gradient descent method was also utilized for this problem [23]. Although RBFN are characterized by substantially faster training than, for example, backpropagation networks, there is a problem to decrease RBFN learning time for high input space dimensionality case. The general parameter approach is one of the possible ways to improve RBFN's training features and to simplify their estimation accuracy analysis [14-20].

3. General Parameter Method for System Identification

There is a trade-off between network approximation ability and learning procedure convergence rate [24]. Recurrent learning procedures are based on adjusting model schemes.

Assume that the plant is nonlinear on input variables:

$$x(t) = \tilde{a}^T \tilde{f}(t) + \eta(t) \tag{6}$$

where $\tilde{a}$: unknown parameter vector of dimensionality N; $\tilde{f}(t) = [\tilde{f}_1(t),...,\tilde{f}_m(t)]^T$: vector function of input variables; $\tilde{f}_i(t) = [f_{i1}(\tilde{u}),...,f_{ip_i}(\tilde{u})]^T$: subvector function of input variables; $\tilde{u}(t)$: input vector; $\sum_{i=1}^{m} p_i = N$, $1 \le m \le N$. GP-model of the plant (6) is of the form

$$\hat{x}(t) = (\hat{a}(0) + Q\tilde{\beta}(t-1))^T \tilde{f}(t) \tag{7}$$

where $\hat{x}(t)$: model output; $\hat{a}(0) = [\hat{a}_1(0),...,\hat{a}_m(0)]^T$, $\hat{a}_i(0) = [\hat{a}_{i1}(0),...,\hat{a}_{ip_i}(0)]^T$: initial model parameter vector and its i-th subvector; $Q = diag(q_1\tilde{I}_1,...,q_m\tilde{I}_m)$: (N x m)-matrix; $\tilde{I}_i = [11..1]^T$: $(p_i \times 1)$ – vector of units; q_i : weight coefficients; $\tilde{\beta}(t) = [\beta_1(t),...,\beta_m(t)]^T$ general parameter vector adjusted with algorithm

$$\tilde{\beta}(t) = \tilde{\beta}(t-1) + (x(t) - \hat{x}(t))\Gamma(t)Q^T \tilde{f}(t) \tag{8}$$

where $\Gamma(t)$: (m x m)-diagonal matrix of gains. The number of the model (7) parameters to be adjusted, i.e. the number m $(1 \le m \le N)$ of general parameter vector $\tilde{\beta}(t) = [\beta_1(t),...,\beta_m(t)]^T$ components, is reduced to increase the learning convergence rate. The following theorem gives the algorithm's (8) convergence conditions of the common form.

Theorem 1: *For plant (6) and model (7) the convergence conditions of the algorithm (8) are of the form (9), where* $\Gamma_{\bullet}(t)$ - diagonal matrix with elements equal to corresponding diagonal elements of matrix $Q^T E\{\tilde{f}(t)\tilde{f}^T(t)\}Q$; $E\{\cdot\}$ - expectation symbol.

$$0 < \Gamma(t) < \Gamma_{\bullet}^{-1}(t) \tag{9}$$

It was shown [e.g.,18] that, for fixed values of signal/noise ratio, the quadratic form $S(t) = E\{\Delta\tilde{\beta}^T(t)Q^T Q\Delta\tilde{\beta}(t)\}$ of the general parameter error vector $\Delta\tilde{\beta}(t) = \tilde{\beta}(t) - C$, where C is general parameter vector expectation at steady state (asymptotical value), may be considered as an estimation accuracy measure. Besides that, the convergence rate (the rate of the value $S(t)/S(0)$ decrease) depends on general parameter vector dimensionality m, not on model dimensionality N.

The general parameter approach is one of the possible ways to improve ANN's training features. The main characteristics of the GP approach are the possibility of estimation accuracy evaluation during training stage and improved convergence rate due to artificial reduction of adjusting model parameter space dimensionality, especially useful for initial steps when approximation error is large.

4. General Parameter Radial Basis Function Neural Networks

First, let consider the GP approach to RBFN weights adjusting. As soon as RBFN is linear on its weights, the GP method may be implemented in a straightforward manner. The equation describing GP-RBFN for a single output network is

$$x = \sum_{j=1}^{K}(w_j^0 + \beta)\exp[-\sum_{i=1}^{N}(u_i - c_j)^2/\delta_j^2] + (w_o^o + \beta) \tag{10}$$

where w_j^0 : fixed initial values of network weights; β : scalar general parameter to be adjusted with the following algorithm

$$\beta(t) = \beta(t-1) + \gamma_\beta(x^* - x)\{\sum_{j=1}^{K}\exp[-\sum_{i=1}^{N}(u_i - c_j)^2/\delta_j^2] + 1\} \tag{11}$$

where γ_β: adjusting gain; x^*: desired network output. The simplest GP-RBFN structure is shown in Fig.1.

Simulations were performed to investigate convergence characteristics of algorithm (11). Two schemes were modelled. In both cases, the parallel model identification approach was realized with "sample" RBFN as a plant. The structure of the "plant" was assumed fixed and known while plant's parameters ("sample" RBFN's weights) were to be estimated by "model", i.e. RBFN of the same structure as a plant.

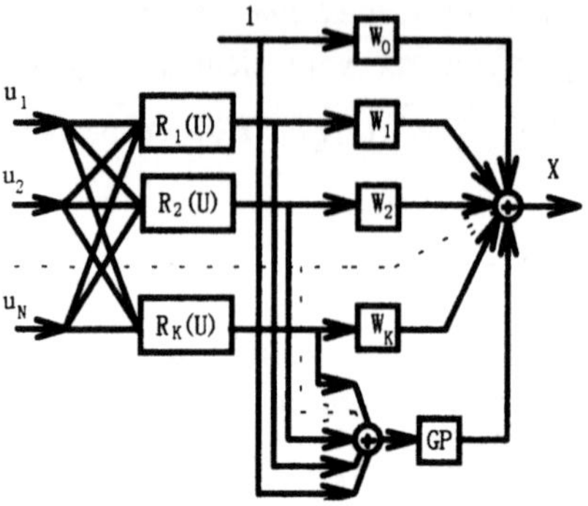

Figure 1. The simplest GP-RBFN.

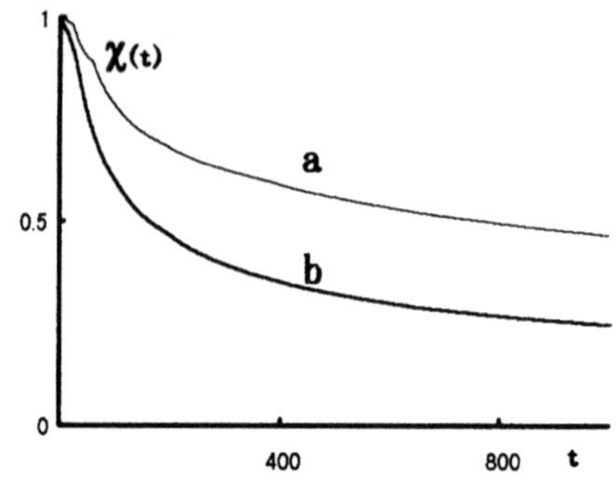

Figure 2. Learning algorithm convergence: a) conventional RBFN; b) GP-RBFN.

Parameters of the model ("model" RBFN's weights) were adjusted independently by two algorithms starting from the same initial conditions. In the first case, a conventional RBFN was used with individual adjusting of its weights. In the second case, GP-RBFN was simulated with learning algorithm (11). The efficiency of both methods were compared by using the following measure of convergence speed:

$$\chi(t) = \frac{E\{\|\tilde{r}(t)\|^2\}}{\|\tilde{r}(0)\|^2} \tag{12}$$

where $\tilde{r}(t) = \tilde{w} - \tilde{w}(t)$: parameter error vector at moment t; $\tilde{w}$: vector of "true" parameters (vector of the "plant" RBFN weights); $\tilde{w}(t)$: estimate of the vector $\tilde{w}$ at moment t (vector of the "model" RBFN weights); E - expectation symbol; $\|\cdot\|$: Euclidean norm. Measure (12) is an average estimate of a learning algorithm convergence speed in the sense of relational reduction of the square norm of the parameter error vector.

Simulation results under the same initial conditions for both methods are presented in Fig.2. Due to the artificial reduction of the adjusting parameter space dimensionality learning speed of GP-RBFN has increased relatively conventional RBFN.

5. GP RBFN Based Adaptive Fuzzy Systems (AFS)

RBFN to be used in adaptive fuzzy system (AFS), in common case, is assumed to be trained by means of the minimum necessary number of rules (hidden unit number) determination and adjusting of the mean and variance vectors of individual hidden nodes as well as their weights. In this paper, the simplest GP RBFN based adaptive fuzzy system for automatic fuzzy rule number determination is proposed (Fig.3).

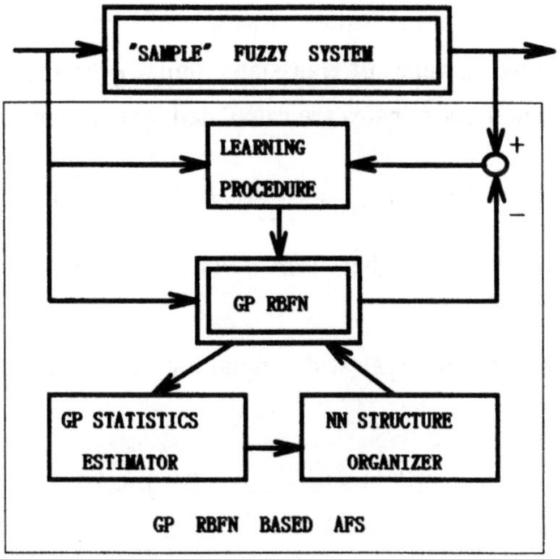

Figure 3. GP RBFN adaptive fuzzy system.

Only the network weights have been assumed to be adjusted by the GP algorithm while the centres and widths of unit sensitive zones were completely determinedwith the network input signal range and unit number during each training epoch. A "sample" fuzzy system has been presented by RBFN with the "unknown" number of hidden units (i.e., fuzzy rules). Starting from the single-unit-GP-RBFN, the network learning has been performed by the scalar general parameter adjusting in the Learning Procedure block. The steady state general parameter expectation $E\{\beta\}$ and variance $D\{\beta\}$ have been calculated by GP Statistics Estimator. The approximation quality criterion (13) was evaluated for the current GP RBFN structure, and decision on changing of network structure parameter adjusting in the Learning Procedure block. The steady state general parameter expectation $E\{\beta\}$ and variance $D\{\beta\}$ have been calculated by GP Statistics Estimator. The approximation quality criterion (13) was evaluated for the current GP RBFN structure, and decision on changing of network structure (additional unit recruiting) was made by NN Structure Organizer.

$$Q = \frac{D\{\beta\}}{E\{\beta\}} \tag{13}$$

Therefore, the GP RBFN AFS determines the "true" fuzzy rule number by incrementally recruiting the radial basis function units and continuous estimation of the approximation quality through criteria (13) evaluation for each fixed GP RBFN structure. The network to be determined is the network with least value of Q, and its unit number is assumed to be equal to the fuzzy rule number of the "sample" fuzzy system.

The proposed procedure is explained below for the simplest case of the GP RBFN AFS with scalar input signal.

1) For a given maximum value u_{max} of the centered input signal $u\ (E\{u\}=0)$ and known number of Gaussian units q (for the first stage, $q=1$) the sensitive zone center coordinates are calculated by relationship (14).

$$C_i^q = -u_{max} + \frac{u_{max}}{q}(2i-1), \quad i = \overline{1,q} \tag{14}$$

where i is a current unit number. For $q=1$ and $i=1$, for example, one can receive $C_1^1 = 0$.

2) The initial (basic) sensitive zone width equal for all network units is calculated as (15).

$$\delta_0^q = \frac{u_{max}}{q\sqrt{-\ln\rho}}, \quad 0<\rho<1 \tag{15}$$

where ρ is the Gaussian function value at the boundary point of the sensitive zone. Variations of ρ provide different degree of conjugent sensitive zone (membership function) overlapping. This results in the system fuzziness variations. The meaning of the above mentioned parameters is explained by Fig.4 for RBFN with 3 units ($q=3$).

3) For network structure determined during two above steps the scalar general parameter learning procedure is performed based on input-output sample data accordingly to the algorithm (11). Simultaneously the general parameter expectation $E\{\beta\}$ and variance $D\{\beta\}$ are estimated with some conventional method, for example, by the moving average calculation.

4) After GP statistics stabilization the criteria (13) is evaluated and memorized.

5) If $q=1$ then GOTO 8).

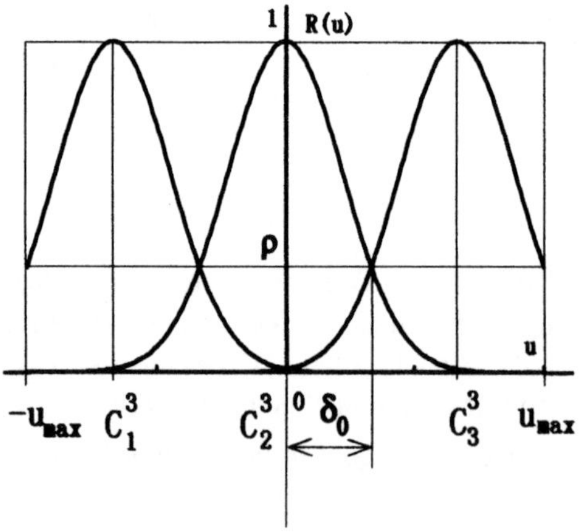

Figure 4. Definition of GP RBFN basic parameters.

6) The modified system criteria value Q^{q+1} is compared with its preceding value Q^q: If $Q^{q+1} < Q^q$ then GOTO 8) else $q = q - 1$.

7) Stop.

8) The structure of GP RBFN is modified by one more Gaussian unit recruiting: $q = q + 1$. The steps 1) - 6) are repeated.

The result of the algorithm 1) - 8) implementation is a number of the Gaussian function units in GP RBFN which provides the best approximating accuracy. In the terms of fuzzy system theory it means the fuzzy rule number determination problem solution.

6. Simulation Results

The proposed algorithm of fuzzy rule number determination by means of the GP RBFN AFS usage has been studied through computer simulation.

Both, the number of "sample" network units and number of its inputs have been chosen equal to 5. All the "sample" network weights were of the same value 1 to indicate that the corresponding basis functions are involved into the output value calculation. In terms of fuzzy approach it means that the membership function related to nonzeroed weight is actually playing its role in the input-output mapping. The GP RBFN weight initial values $w_j^0, j = \overline{1, q_m}$ have been assigned equal to 0 (no *a priori* information about "valid" membership functions available). The number of GP RBFN units q_m has been changed during the learning procedure correspondingly to the Q-criteria variation trend. However, the GP RBFN input number was fixed and equal to that of "sample" network.

Centres of the Gaussian units sensitive fields were assumed to be uniformly distributed in the input signal space with latter known for AFS. The sensitive widths of all units were chosen equal. The different sensitive zone overlapping degree cases were studied. During the simulations the value of ρ was constant: $\rho = 0.01$ ($\delta_0^q = 0.093$), i.e., the basic level of overlapping was 1%. The real sensitive zone width values δ^q have been calculated as follows:

$$\delta^q = \alpha \delta_0^q \tag{16}$$

where α : the overlapping coefficient. In Fig.5 the examples of different overlapping degree Gaussian functions are given (network with 3 units).

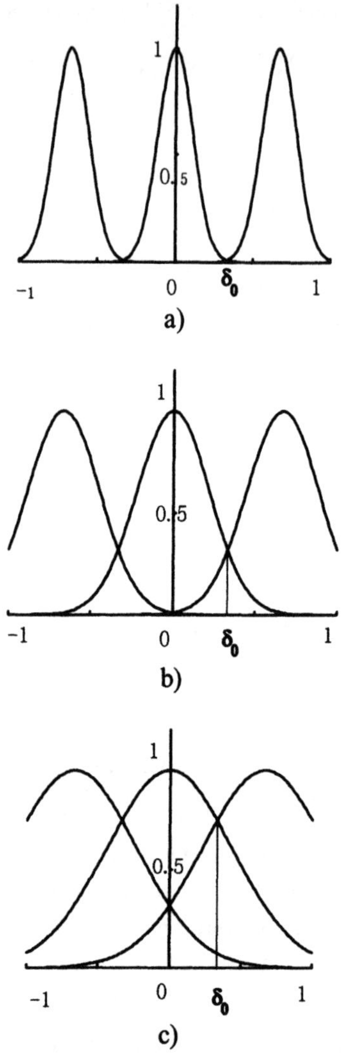

Fig.5. Different Overlapping Degree Gaussian Functions (Number of Units is 3):
a) α =1; b) α =2; c) α =4.

Two different type input signals were used for network training. In the first simulation series it was a sine function. In the second series the pseudo-random sequence with uniform distribution in the region of [-1,1] was used. In Fig.6 the GP RBFN learning curves are shown for sinusoidal inputs and different number of hidden units. The steady state (asimptotical value) error is 0 only for unit number of 5 (the same as for "sample" network) (Fig.6c). The analogous results have been gained for the random input signals (Fig.7) with the best approximation accuracy for 5-unit GP RBFN (Fig.7c).

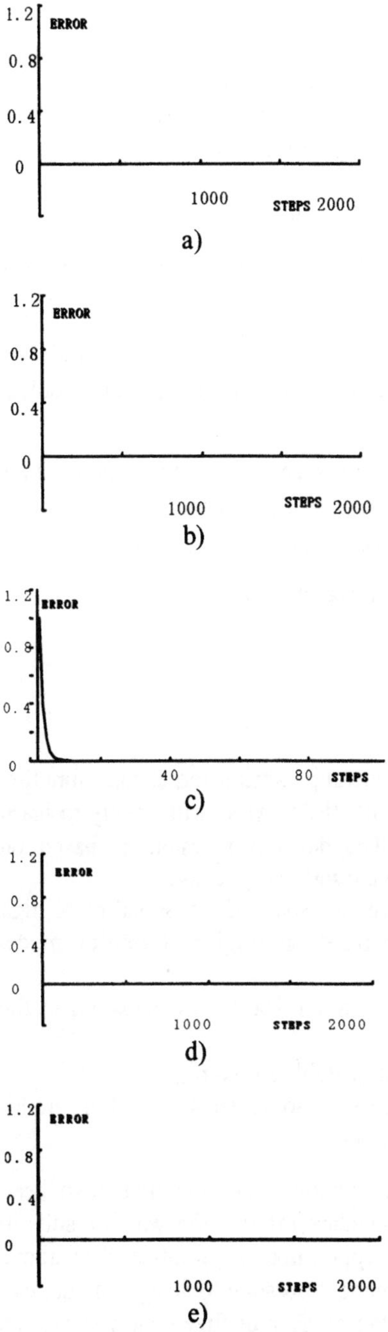

Figure 6. GP RBFN learning with sinusoidal input:
a) q_m=1; b) q_m=4; c) q_m=5; d) q_m=6; e) q_m=9.

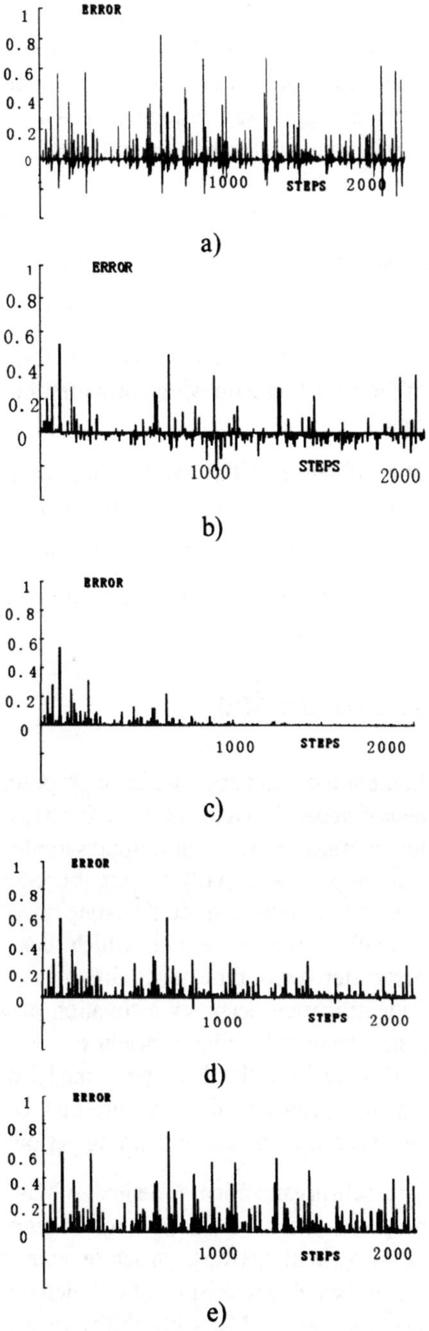

Figure 7. GP RBFN learning with random input:
a) q_m=1; b) q_m=4; c) q_m=5; d) q_m=6; e) q_m=9.

The optimal unit number judgment abilities (the optimal number of fuzzy rules) of the GP RBFN based AFS are demonstrated in Fig.8. The proposed system has been analyzed under different learning conditions. Two factors have been determined to have a significant influence to GP RBFN AFS performance. There are the overlapping coefficient α and the GP adjusting algorithm gain γ_β.

The upper Q-criteria curves in Fig.8 correspond to the case when γ_β=0.6. The sensitivity of the proposed system has increased with adjusting gain γ_β increase (γ_β=0.1...0.6), but the latter is bounded with convergence conditions of GP algorithm (Theorem 1). Therefore, this theorem gives the upper boundary estimate of the adjusting gain which provides the applicability of GP approach to the problem stated.

The influence of the overlapping coefficient α is seen from experimental curves plotted for δ_0=0.093 (ρ=0.01) and α=1..10 (Fig.8a-8f correspondingly). For the higher values of α the dependence of the system performance from γ_β value reduced and the "true" rule number determination was more clear.

7. Conclusion

In the presented paper, we have proposed the general parameter radial basis function neural network based adaptive fuzzy system (GP RBFN AFS) with ability to learn fuzzy rules from input-output sample data. The developed system is based on general parameter (GP) training procedure with the following features:
- artificial reduction of adjusting model parameter space dimensionality, - high (initial) convergence rate which does not depend on original (sample) model parameter vector dimensionality,
- identification accuracy estimation possibility by GP statistics calculation in the steady state of learning procedure,
- GP-model is a flexible type of model with an adjustable structure,
- quazi-optimal model structure may be derived with some decision making multi-step procedure based on learning period information.

The self-organization method of GP RBFN structure determination have been proposed based on steady stage of learning procedure information which results in quasi-optimal network structure with desired approximation abilities. The above mentioned characteristics of GP method have made it possible to design an adaptive fuzzy system (AFS) with ability to optimize the number of fuzzy rules generated during multi-step learning procedure. The method developed may be easily implemented into various type artificial neural network structures to solve a wide class of signal processing, control and related problems.

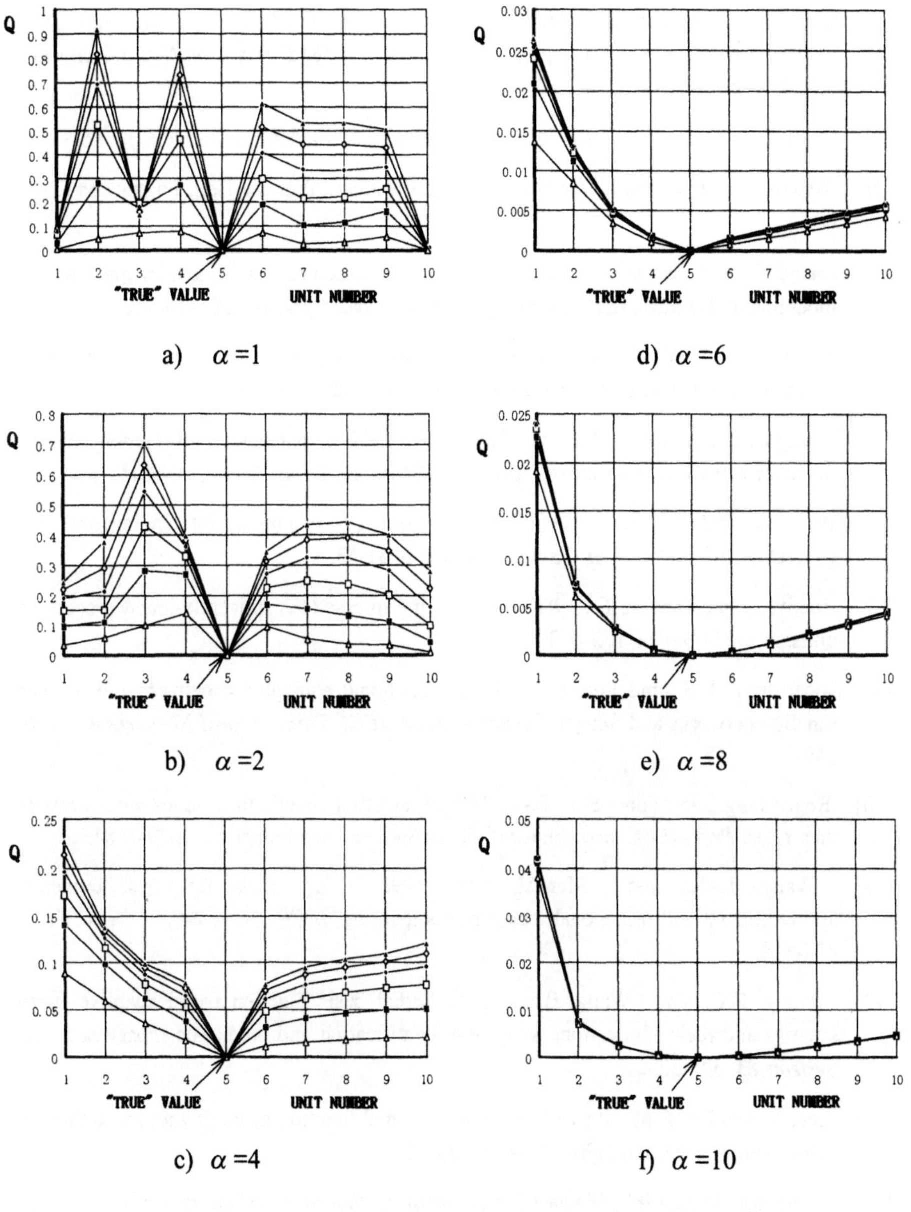

Figure 8. Q - criteria curves for α =1..10, γ_{β} =0.1...0.6

(upper curves - for γ_{β} =0.6, lower - for γ_{β} =0.1)

8. References

[1] Horikawa, S., Furuhashi, T., and Uchikawa,Y., 1992, On fuzzy modeling using fuzzy neural networks with the backpropagation algorithm, *IEEE Trans. on Neural Networks,* **3**, 801- 806.

[2] Kosko, B., 1987, Fuzzy associative memories, in: A.Kandel, Ed., *Fuzzy Expert Systems.*

[3] Kosko, B., 1992, *Neural Networks and Fuzzy Systems*, Prentice-Hall, Englewood Cliffs, NJ.

[4] Takagi, T., and Sugeno, M., 1985, Fuzzy identification of systems and its applcation to modeling and control, *IEEE Trans. Systems, Man and Cybernet.* **15**, 116-132.

[5] Wang, L.-X., and Mendel, J.M., 1992, Generating fuzzy rules by learning from examples, *IEEE Trans. Systems, Man and Cybernet.* **22**, 1414-1427.

[6] Yamaguchi, T., Imasaki, N., and Haruki, K., 1990, A reasoning and learning method for fuzzy rules with associative memory, *IEE Japan Trans.* **110-C**, 207-215.

[7] Mitra, S., and Pal, S.K., 1995, Fuzzy multi-layer perceptron, inferencing and rule generating, *IEEE Trans. on Neural Networks* **6**, 51-63.

[8] Moody, J., and Darken, C., 1989, Fast learning in networks of locally-tuned processing units, *Neural Comput.* **1**, 281-294.

[9] Roger Jang, J.-S., and Sun, C.-T., 1993, Functional equivalence between radial basis function networks and fuzzy inference systems, *IEEE Trans. Neural Networks* **4,** 156-159.

[10] Roger Jang, J.-S., and Sun, C.-T., 1993, Predicting chaotic time series with fuzzy if-then rules, *Proc. IEEE Internat. Conf. Fuzzy Systems*, San Francisco, 1079-1084.

[11] Wang, L.-X., and Mendel, J.M., 1992, Fuzzy basis functions, universal approximation and orthogonal least squares learning, *IEEE Trans. Neural Networks* **3**, 807-814.

[12] Kwang, B.C., and Wang, B.H., 1996, Radial basis function based adaptive fuzzy systems and their applications to system identification and prediction, *Fuzzy Sets and Systems* **83**, 325-339.

[13] Lee, S., and Kil, R.M., 1991, A Gaussian potential function network with hierarchically self-organizing learning, *Neural Networks* **4**, 207-224.

[14] Akhmetov, D.F., 1989, *Elaboration of strip width forming adaptive control system in rough stand group of hot rolling mill*, Candidate of Technical Science Degree Dissertation, Kazakh Polytechnical Institute, Almaty, Kazakhstan.

[15] Akhmetov, D.F., and Dote, Y., 1995, Novel "on-line" identification procedure using artificial neural network, *Proc.of the IFAC Workshop on On-Line Fault Detection and Supervision in the Chemical Process Industries*, Newcastle, U.K., June, 90-97.

[16] Akhmetov, D.F., and Dote, Y., 1996, General parameter neural networks with fuzzy self-organization, *Proc. of the Artificial Neural Networks in Engineering (ANNIE'96)*, St. Louis, Missouri, U.S.A., Nov., 191-196.

[17] Akhmetov, D.F., and Dote, Y., 1997, Fault diagnosis with general parameter radial basis function neural network, *Proc. of the IEEE/ASME International Conference on Advanced Intelligent Mechatronics (AIM'97)*, Tokyo, Japan, June, 83.

[18] Akhmetov, D.F., Dote, Y., and Shaikh, M.S., 1997, System identification by the general parameter neural networks with fuzzy self-organization, *Prep. of the 11th IFAC Symposium on System Identification (SYSID'97)*, Kitakyushu, Japan, July, **Vol. 2,** 829-834.

[19] Akhmetov, D.F., and Dote, Y., 1997, General parameter time series analysis method for fault diagnosis, *Proc. of the 2nd Asian Control Conference (ASCC'97)*, Seoul, Korea, July, **Vol.3,** 87-90.

[20] Dote, Y., Akhmetov, D.F., and Shaikh, M.S., 1997, Control and diagnosis for power electronics systems using soft computing, *Proc. of the 2nd Asian Control Conference (ASCC'97)*, Seoul, Korea, July, **Vol.3,** 491- 494.

[21] Zadeh, L.A., 1978, Fuzzy sets as a basis for a theory of possibility, *Fuzzy Sets and Systems* **1,** 3-28.

[22] Hornik, K., Stinchcombe, M., and White, H., 1989, Multilayer feedforward networks are universal approximators, *Neural Networks* **2,** 359-366.

[23] Sorsa, T., and Koivo, H.N., 1993, Application of artificial neural networks in process fault diagnosis. *Automatica,* **Vol.29, No 4**, pp.843-849.

[24] Poggio, T., and Girosi, F., 1990, *Extensions of a theory of networks for approximation and learning: dimensionality reduction and clustering*, Artificial Intelligence Lab. Memo. **No** 1167. MIT, Cambridge, MA.

Uneven Division of Input Spaces for Hierarchical Fuzzy Modeling

Kanta Tachibana Takeshi Furuhashi

Bio-Electronics Laboratory, Department of Information Electronics,
Graduate School of Engineering, Nagoya University, Japan
Tel: +81–52–789–2793, Fax: +81–52–789–3166
Email: kanta@bioele.nuee.nagoya-u.ac.jp,
furuhashi@bioele.nuee.nagoya-u.ac.jp

Keywords: Fuzzy Modeling, Genetic Algorithm, Fuzzy Neural Network

Abstract

Fuzzy modeling is a promising technique to describe input-output relationships of nonlinear systems with fuzzy if-then rules. It is easy for us to understand the linguistically expressed nonlinear relationships. This paper presents a new dividing method of input spaces for a hierarchical fuzzy modeling method using Fuzzy Neural Network (FNN) and Genetic Algorithm (GA). Uneven division of input space for each submodel in the hierarchical fuzzy model can be achieved with the proposed method. The obtained hierarchical fuzzy model are probable to be more concise and more precise than those identified with the conventional methods.

1. Introduction

Fuzzy modeling[1,2] is a method to describe the characteristics of nonlinear systems using fuzzy rules. The method has a distinguishing feature in that it can express complex nonlinear input-output relationships linguistically. For automatic acquisition of fuzzy rules, combinations of fuzzy logic and neural networks have been studied[3–8]. The Fuzzy Neural Network (FNN) in [5] has a capability to identify fuzzy rules and tune the membership functions through the back propagation learning algorithm. This FNN has been applied to the fuzzy modeling of nonlinear systems[9,10]. Sufficient data, which cover whole input space, is hard to obtain from the actual plant with many input variables. Hierarchical fuzzy modeling method using multiple FNNs was proposed by Nakayama, Furuhashi and Uchikawa[11] to make the fuzzy model have a high generality. Each sub model in the hierarchical fuzzy model has a smaller number of input variables, and it does not need many data to describe the input-output relationships in the sub space. The studies on fuzzy modeling has been extended to clustering approach for the antecedent structure identification[12] of Takagi-Sugeno-Kang (TSK) model[1,2], and to reduction of model complexity[12–16] of given fuzzy models.

Karr, Freeman, Meredith[17] proposed a combination of fuzzy logic and Genetic Algorithm (GA). Unlike the neural network learning, GA does not need the teaching signal to obtain the fuzzy controller. GA finds fuzzy rules using the payoff for the success/failure of its control actions. GA was applied to identification of hierarchical structure of fuzzy model[18] from given input-output pairs of data. The authors[19] have applied GA to selection of input variables of FNN hierarchical model. This method is very effective in the case where the target system has a strong nonlinearity. GA can find appropriate sets of input variables and a proper number of membership functions for each selected input variable for each sub-model from many candidates. The authors[20] also proposed an efficient method for rough search of sub-models. The method in [19] uses the FNN for evaluation of the sets of input variables and the membership functions determined by chromosomes in the GA operations, and the learning of the FNN takes much computation time. The proposed method in [20] replaces the FNN with a quick fuzzy rule identification method to roughly search sub-models. The FNN is used to fine tune the roughly selected models.

This paper presents a new dividing method of input spaces for the hierarchical fuzzy modeling method using the FNN and GA. The proposed method can find fuzzy sub-models with unevenly divided input spaces. The chromosome determines a combination of input variables, and the uneven division method generates an FNN sub-model with unequally divided input space. The effectiveness of the proposed method is significant in finding more concise and more precise hierarchical fuzzy model than those by the conventional methods. It becomes easier for us to grasp the nonlinear input-output relationships from the concise model.

Numerical experiment is done to demonstrate that the proposed method can find a simpler and more precise sub-model with unevenly divided input space by comparing the results to those obtained by the conventional method.

2. Hierarchical Fuzzy Modeling Using FNN

2.1 Fuzzy neural network

The FNN presented by the authors is a multi-layered back-propagation (BP) model with a specially designed structure for easy extraction of fuzzy rules from the trained NN. This paper uses Type I of the FNNs in [9]. Fig. 1 shows an example of the FNN.

This is a case where the FNN has two inputs x_1 and x_2, one output y and three membership functions for each input. The Back Propagation (BP) learning algorithm can be applied to modify the connection weights w_c,w_g, and w_b. From this network, the following simplified fuzzy inference can be extracted:

$$\begin{array}{c} R^i : \mathrm{if} x_1 \mathrm{is} A_{i,1} \mathrm{and} x_2 \mathrm{is} A_{i,2} \\ \mathrm{then} y = b_i \\ (i = 1, 2, \cdots, n) \end{array} \tag{1}$$

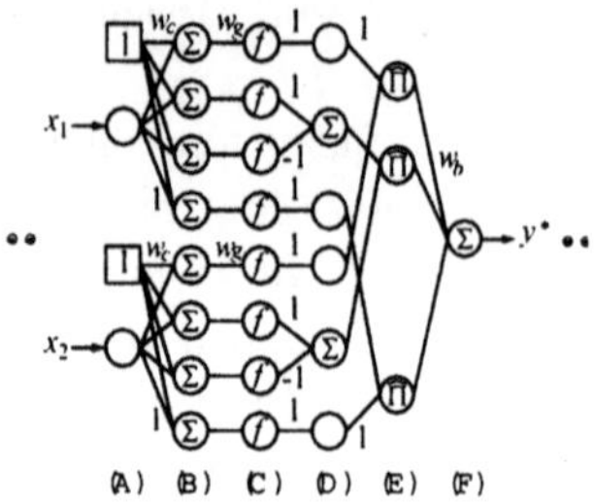

Figure 2.1: Fuzzy Neural Network

$$\mu_i = A_{i,1}(x_1)\, A_{i,2}(x_2) \tag{2}$$

$$\widehat{\mu}_i = \frac{\mu_i}{\sum_k \mu_k} \tag{3}$$

$$y^* = \sum_{i=1}^{n} \widehat{\mu}_i b_i \tag{4}$$

where R^i is the i-th fuzzy rule. $A_{i,1}$, $A_{i,2}$ are labels of membership functions, b_i is a singleton in consequence, n is the number of fuzzy rules, μ_i is the truth value of R^i, $\widehat{\mu}_i$ is the normalized truth value, and y^* is the inferred value.

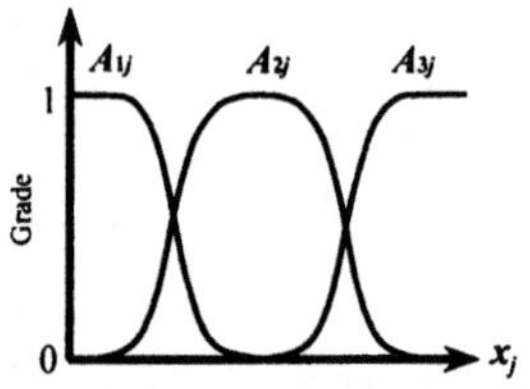

Figure 2.2: Membership functions

Fig. 2 shows an example of membership functions in the antecedent formed in (A)-(D)-layers. The connection weights w_c, w_g determine the positions and slopes of the sigmoid functions f in the units in (C)-layer, respectively. Each membership function consists of one or two sigmoid functions. The outputs of the units in (D)-layer are the values of membership functions. The products of these values are the inputs to the units in (E)-layer and the outputs of the units are the normalized truth value in the antecedent $\widehat{\mu}_i$ in eq.(3). The output of the unit in (F)-layer is the sum of the products of the connection weights w_b and the normalized truth values. The connection weights w_b correspond to the singletons in the consequence b_i in eq.(1). The output in (F)-layer is, therefore, the inferred value y^* in eq.(4).

Since the center-of-gravity method is used in (E)-layer, the updating method of connection weights, i.e. BP algorithm, needs some modifications. The learning algorithm for the FNN is well described in [9].

The feature of this FNN is that fuzzy rules can be extracted easily from the trained FNN. Three layered neural network can identify the input-output relationships. However, it is hard to extract rules from the three layered neural network.

2.2 Procedure of Hierarchical Fuzzy Modeling

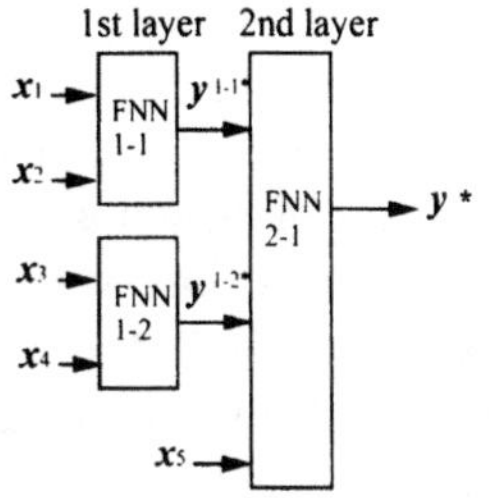

Figure 2.3: A hierarchical fuzzy model using FNNs

Fig. 3 shows an example of hierarchical fuzzy model which consists of FNN sub-models. The figure shows a case where the object has 5 inputs $(x_1, x_2, x_3, x_4, x_5)$ and one output (y) and the fuzzy model has a two level hierarchical structure. In Fig. 3, y^{1-1^*}, y^{1-2^*}, y^* are the inferred values of the fuzzy sub-models. In the 1st layer, the fuzzy model with the inputs x_1 and x_2, and the model with x_3 and x_4 are lined in parallel. The outputs of these models are y^{1-1^*} and y^{1-2^*}, respectively. These two fuzzy sub-models in the 1st layer greatly contribute to the input-output relationships of the system. The input variable x_5 is used for a small adjustment of the model. This fuzzy model reduces the number of divisions of each input space by constructing sub-models in a hierarchical manner. This reduction of divisions prevents the model from over-fitting. The obtained fuzzy model, therefore, has the generalities.

The authors have proposed a hierarchical fuzzy modeling method using the FNNs and GA[11, 14]. Each sub-model was built by the FNN and a proper set of input variables and sets of membership functions for the sub-model were selected by GA. A hierarchical structure was constructed by finding proper sub-models one by one.

This paper proposes a new hierarchical fuzzy modeling method which realizes uneven division of input space. GA is utilized to find an appropriate set of input variables. A model having the selected variables is generated by the proposed uneven division method of input space and it is tuned by the FNN learning. The procedure of this hierarchical fuzzy modeling is as follows:

1. The input-output data is divided into two groups A, B whose statistical characteristics are nearly the same. The group A constitutes training data, and the group B is used as a test set. The model identified from the data of group A is called model A. The number of layer h is initially set at 1.

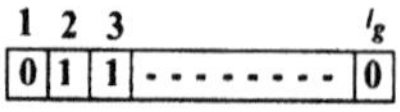

Figure 2.4: An example of chromosome

2. Using GA and the FNN, a sub-model is identified. A combination of input variables is encoded into a chromosome as shown in Fig. 4. The number of genes l_g is the same as that of the whole candidates of input variables. If the value of a gene is 1, the corresponding input variable is used for the model. If it is 0, the input variable is not selected. The evolution of individuals is carried out by the following procedure:

   (a) Chromosomes are initialized to have 0 in each gene. The number of chromosomes is n_g. The binary number in each gene is flipped to 1 with the probability of p_i.

   (b) Chromosomes are evaluated. The chromosome determines a combination of input variables to be used for the sub-model. The division process of the input space to be described in subsection 2.3 is carried out. In this process, the identification of singletons in the consequent parts of FNN is done with the data of group A. The performance index F used in this fuzzy modeling process is given by

$$F = n_B \log E + cS, \qquad E = \frac{\sum_{i=1}^{n_B} \left(y_i^{BA} - y_i^{B}\right)^2}{n_B} \tag{5}$$

   where n_B is the number of the data of group B, y_i^B is the output data of group B, y_i^{BA} is the inferred value of model A with data B, and S is the number of subspaces. The first term evaluates the generality of the identified model, and the second term is a criterion for the conciseness of the model. Coefficient c adjust the weights on the generality and the conciseness.

   An appropriate division of input space for each input variable is obtained.

   (c) Individuals are ranked with this performance index. The worst n_w chromosomes are replaced with copies of elite chromosome.

(d) Crossover and mutation operations are applied to the population. Crossover is applied to the whole population except for one elite at the rate of p_c. Parents are randomly selected, and one point crossover with randomly selected crossover point is applied. Mutations are applied to each gene of all the chromosomes except for that of the elite chromosome at the rate of p_m.

(e) Stop if the performance of the elite chromosome does not improve during m_{end} generations. If else, go to step (b).

The next step is fine tuning of the acquired model by the back propagation learning of the FNN[9]. The membership functions in the antecedent as well the singletons in the consequence of the best model found in the above process are modified to obtain a better model. For the stopping condition of this learning, the following criterion is used:

$$\begin{aligned} C &= \sqrt{P_A + P_B} + \sqrt{G_A + G_B} \\ P_A &= \sum_{i=1}^{n_A} \left(y_i^A - y_i^{AA}\right)^2 \\ P_B &= \sum_{i=1}^{n_B} \left(y_i^B - y_i^{BB}\right)^2 \\ G_A &= \sum_{i=1}^{n_A} \left(y_i^{AB} - y_i^{AA}\right)^2 \\ G_B &= \sum_{i=1}^{n_B} \left(y_i^{BA} - y_i^{BB}\right)^2 \end{aligned} \tag{6}$$

where n_A and n_B are the numbers of data groups A, B, respectively; y_i^A and y_i^B are the outputs of data A and data B, respectively; y_i^{AA} and y_i^{BB} are the inferred value of model A with data A and that of model B with data B, respectively; y_i^{AB} and y_i^{BA} are the inferred value of model B with group A and that of model A with group B, respectively. The first term on the right hand side in eq.(6) is the precision of the model, and the second term is the criterion for evaluation of the generalities of the model.

This identified model is called model h-1 and its output is denoted by y^{h-1^*}. The model h-k means that it is the k-th model in the h-th layer.

3. If the number of remaining input variables not used in the model h-1 is more than two, another model to compensate the error of model h-1 is identified using GA. This error is used as the teaching signal for the next model. Some of the remaining input variables would be selected. The fine tuning of the acquired model is, then, carried out. The model is denoted by model h-2. If more variables remain, model h-3 for the compensation of the error of model h-2 will then be identified. This modeling is repeated until the number of remaining input variables becomes less than two. The model h-1 will be used for the identification of the models in the succeeding layer. The outputs of the models h-2, -3, $\cdots$ will be the candidates for the input variables of the models in the next layer.

4. Fuzzy modeling in the next layer is done using the sub-models identified in the previous layer. The output of model h-1 is always used. A combination of this output y^{h-1^*} and some of the outputs of models h-2, -3,

··· as well as the input variables not used in model h-1 is chosen by GA. The acquired model is denoted by $(h+1)$-1.

5. The evaluation criterion C of model h-1 and model $(h+1)$-1 are compared. If the latter value is less, repeat the procedure from (3). If not, stop. Since the input variables which are used in the previous layer are not used in the succeeding layers, the acquired structure is simple.

2.3 Unequal Division of Input Space

Unequal division of input space for the fuzzy modeling is described in this section. The input space is divided so that the variations of outputs of data across the subspaces are minimized. The procedure of input space division is as follows:

1. When the given data have I input variables, I dimensional input space is to be divided. The input space is initially has no division, i.e. the number of subspaces S equals 1. Figure 5 shows a case where two inputs x_k, x_l are given. The top figure shows the initial division of input space. A fuzzy model with one rule is made with an FNN, trained, and evaluated under the criterion F in eq.(5). The training of the FNN here is only to adjust the singleton in the consequence for efficient training.

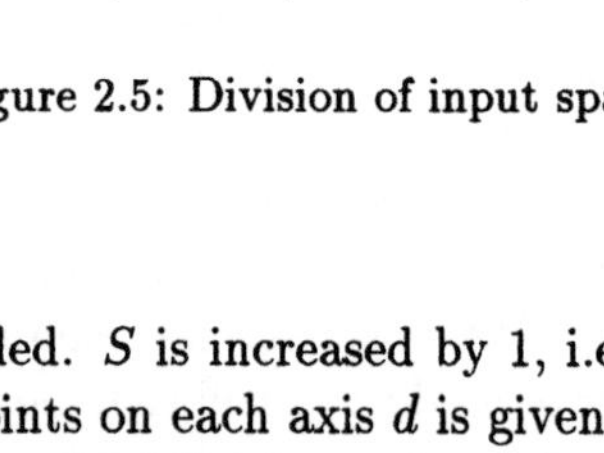

Figure 2.5: Division of input space

2. Input space is divided. S is increased by 1, i.e. $S = S + 1$. Number of possible division points on each axis d is given *a priori*. Figure 5 shows that it has three possible division points, $d = 3$. There are $I \times S \times d$ possible division points in this case. Each division is evaluated one by one as shown in the figure with the criterion given by

$$V = \sum_{s}^{S} v_s \tag{7}$$

V is the accumulation of variation of data outputs in each sub space. The variation in a sub space s is expressed as:

$$v_s = \frac{\sum_j^{n_s} (y_j - \overline{y}_s)^2}{n_s}, \qquad (8)$$
$$\overline{y}_s = \frac{\sum_j^{n_s} y_j}{n_s}$$

where n_s is the number of data in the sub space s, y_j and $\overline{y}_s$ are the data output and the average of the data outputs in this sub space, respectively. The division with the smallest V value is selected. An FNN with S rules is generated, trained, and evaluated. The training of the FNN is only to adjust the singletons in the consequence for efficient training.

3. If F is not improving, stop. The model obtained before the above step is selected as the final fuzzy model. If it is improving, step 2 is repeated. Figure 6 shows an example of unequal division of the input space x_k, x_l, and the membership functions constructed by the FNN.

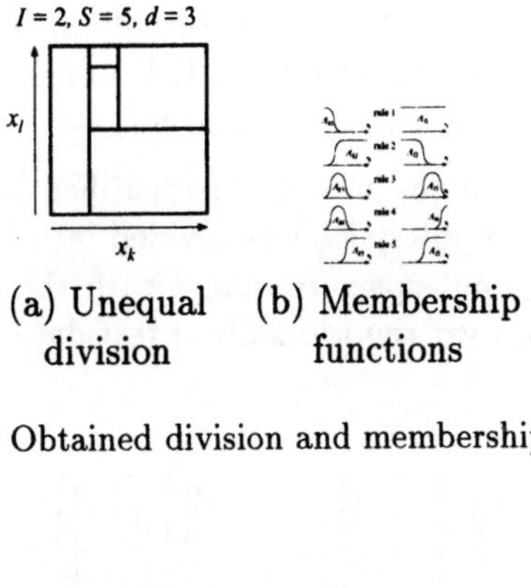

(a) Unequal division (b) Membership functions

Figure 2.6: Obtained division and membership functions

3. Numerical Experiment

A numerical experiment using the proposed modeling method was done in this section. Model 1-1 was identified by this experiment. The object nonlinear system was given by the following equation:

$$y = (-4 + x_0^{0.5} + x_1^{-1})^2 + 5\sin(x_2 + x_3) + \exp(1 + x_4 + x_5) \qquad (9)$$

$$\begin{aligned} x_6 &= x_0^{0.5}, & x_7 &= x_1^{-1}, \\ x_8 &= x_1^{-2}, & x_9 &= x_0^{0.5} x_1^{-1}, \\ x_{10} &= x_0^{0.5} + x_1^{-1}, & x_{11} &= x_2 + x_3, \\ x_{12} &= x_4 + x_5 \end{aligned} \qquad (10)$$

where x_0 to x_5 were input variables. x_6 to x_{12} were arranged input variables expressed by eq.(10). x_{13} was used as a dummy variable, that had no relationship with y. The number of candidates of inputs l_g was 14. The ranges of input variables are shown in Table 1. The ranges were decided so that the input variables influenced the output almost equally.

Table 3.1: Range of variable

variable	range
x_0	$\{0, 1, \cdots, 20\}$
x_1	$\{0.5, 0.7, \cdots, 1.5\}$
x_2, x_3	$\{-1.0, -0.9, \cdots, 5.0\}$
x_4	$\{0.0, 0.1, \cdots, 0.5\}$
x_5	$\{-1.5, -1.4, \cdots, 0.5\}$
x_{13}	$\{0, 1, \cdots, 17\}$

Two sets of eighty pairs of input-output data each for A and B group were generated. The input-output data were normalized within the range $[0, 1)$ for the fuzzy modeling. The parameters of GA were set as follows: $n_g = 20$, $n_w = 2$, $p_i = 0.1$, $p_c = 0.85$, $p_m = 0.01$ and $m_{end} = 10$.

The input variables of the model 1-1 identified by the proposed method were x_{10} and x_{11}. The input space was divided into 7 subspaces. Figure 1(a) shows the division of input space obtained with the proposed method. The root mean square error over the normalized test data was 0.141.

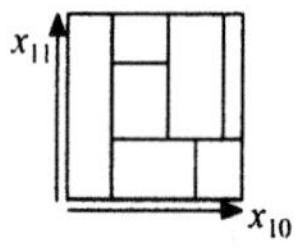

(a) proposed method

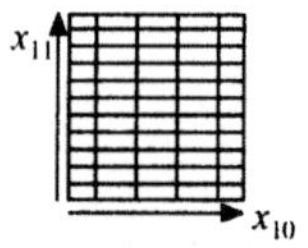

(b) conventional method

Figure 3.1: Division of Input Space

The model 1-1 identified with the conventional method in [20] also had the same combination of input variables, x_{10} and x_{11}, and the number of membership functions were 5 for x_{10} and 11 for x_{11}. The number of subspaces was 55. Figure 1(b) shows the division of input space with the conventional method. The root mean square error over the normalized test data was 0.114. Obtained models are summarized in Table 2.

Table 3.2: Comparison of the models 1-1

	proposed method	conventional method
selected variables	x_{10}, x_{11}	x_{10}, x_{11}
rooted mean square error	0.141	0.114
number of rules	7	55

4. Conclusions

FNN has an distinguishing feature in that it can describe input-output relationship linguistically. Hierarchical fuzzy model uses multiple FNNs as sub-models. Hierarchical fuzzy model reduces total number of fuzzy rules in the case where there are many input variables. However, the conventional hierarchical fuzzy modeling method tends to generate many fuzzy rules because membership functions are allocated evenly for each input variable and the number of fuzzy rules is the product of the number of membership functions. That does not utilize the advantages of fuzzy modeling: comprehensibity of the model.

This paper presented a new method of input space division for the hierarchical fuzzy modeling. Set of input variables for each sub-model is determined by GA. Uneven division of input space of each sub-model is realized with the proposed method. The significant feature of the new method is that it can find a more concise and precise fuzzy model than those identified with the conventional methods. A numerical experiment was done to demonstrate the feature of the proposed method. A more understandable model was identified with the new method.

5. References

1. Takagi T, Sugeno M 1985 "Fuzzy Identification of Systems and Its Applications to Modeling and Control", IEEE Trans. on Syst., Man, and Cybern., Vol.15, No.1, pp.116–132

2. Kang G, Sugeno M 1987 "Fuzzy Modeling", Trans. of the Society of Instrument and Control Engineers, Vol.23, No.6, pp.650–652

3. Takagi H, Hayashi I 1988 "Artificial-Neural-Network Driven Fuzzy Reasoning", IIZUKA-88, pp.183–184

4. Furuya T, Kokubu A, Sakamoto T 1988 "NFS: Neuro Fuzzy Inference System", IIZUKA-88, pp.219–230

5. Horikawa S, Furuhashi T et al. 1990“A Fuzzy Controller Using a Neural Network and its Capability to Learn Expert's Control Rules”, Proc. of the Int'l. Conf. on Fuzzy Logic & Neural Networks (IIZUKA-90), pp.103–106

6. Yamaguchi T, Imasaki N, Haruki K 1990 “Fuzzy Rule Realization on Associative Memory System”, IJCNN-90-WASH-DC, Vol.II, pp.720–723

7. Yoshida K, Hayashi Y, Imura A, Shimada N 1990 “Fuzzy Neural Expert System for Diagnosing Hepatobiliary Disorders”, Proc. of the Int'l. Conf. on Fuzzy Logic & Neural Networks (IIZUKA-90), pp.539–543

8. Sanchez E 1990 “Fuzzy Connectionist Expert Systems”, Proc. of the Int'l. Conf. on Fuzzy Logic & Neural Networks (IIZUKA-90), pp.31–35

9. Horikawa S, Furuhashi T, Uchikawa Y 1992 “On Fuzzy Modeling Using Fuzzy Neural Networks with the Back-Propagation Algorithm”, IEEE Trans. on Neural Networks, Vol.3, No.5, 801–806

10. Hasegawa T, Horikawa S, Furuhashi T et al. 1992 “A Study on Fuzzy Modeling of BOF Using a Fuzzy Neural Network” Proc. of the 2nd Int'l Conf. on Fuzzy Logic & Neural Networks(IIZUKA'92), pp.1061–1064

11. Nakayama S, Furuhashi T, Uchikawa Y 1993 “A Proposal of Hierarchical Fuzzy Modeling Method”, Journal of Japan Society of Fuzzy Theory and Systems, vol.5, no.5, pp.1155–1168

12. Yager R R, Filev D P 1993 “United Structure and Parameter Identification of Fuzzy Models”, IEEE Trans. on Systems, Man, and Cybernetics, Vol.23, No.4, pp1198–1205

13. Song B G, Marks II R J, Oh S, et al. 1993 “Adaptive Membership Function Fusion and Annihilation in Fuzzy If-Then Rules”, Proc. of Second IEEE Int'l Conf. on Fuzzy Systems, pp.961–967

14. Chao C T, Chen Y J, Teng C C 1996 “Simplification of Fuzzy-Neural Systems Using Similarity Analysis”, IEEE Trans. of Syst., Man, and Cybernetics, Vol. 26, No.2, pp.344–354

15. Babuska R, Setnes M, Kaymak U, van Nauta Lemke H R 1996 “Rule Base Simplification with Similarity Measures”, Proc. of Fifth Int'l Conf. on Fuzzy Systems, pp.1642–1647

16. Yen J, Wang L 1996 “An SVD-based Fuzzy Model Reduction Strategy”, Proc. of Fifth Int'l Conf. on Fuzzy Systems, pp.835–841

17. Karr C L, Freeman L, Meredith D 1989 “Improved Fuzzy Process Control of Spacecraft Autonomous Rendezvous Using a Genetic Algorithm”, SPIE Conf. on Intelligent Control and Adaptive Systems, pp.274–283

18. Shimojima K, Fukuda T, Hasegawa Y 1995 "Self-tuning Fuzzy modeling with Adaptive Membership Function, Rules, and Hierarchical Structure Based on Genetic Algorithm", Fuzzy Sets and Systems, Vol.71, No.3, pp.295–309

19. Matsushita S, Kuromiya A, Yamaoka M, Furuhashi T, Uchikawa Y 1996 "Determination of Antecedent Structure of Fuzzy Modeling Using Genetic Algorithm", Proc. of 1996 IEEE Int'l Conf. on Evolutionary Computation (ICEC'96), pp.235–238

20. Furuhashi T, Matsushita S, Tsutsui H 1997 "Evolutionary Fuzzy Modeling Using Fuzzy Neural Networks and Genetic Algorithm", proc. of 1997 IEEE Int'l Conf. on Evolutionary Computation (ICEC'97), pp.623–627

Ensembles of Evolutionary created Artificial Neural Networks and Nearest Neighbour Classifiers

Christoph M. Friedrich

University of Witten/Herdecke; Institute for applied Biotechnology and Systems Analysis; Alfred-Herrhausen Str. 50; D-58448 Witten; E-mail: chris@uni-wh.de; http://www.tussy.uni-wh.de/˜chris

Abstract. Recent developments have shown the usefulness of combinations of classifiers. This paper presents a 3-Level Stacked Generalizer consisting of evolutionary created Artificial Neural Networks with different architectures and weight settings and a nearest neighbour classifier. The components of the Stacked Generalizer have been obtained by an Evolutionary Algorithm to optimize the architecture of an Artificial Neural Network.
Several combination and new selection methods are discussed and their usefulness is shown in practical examples. The use of Ensembles allows it, to obtain better classifiers in earlier generations of the evolutionary optimization process, reducing the necessary computing time. It can be shown that it is more efficient to use information obtained from the complete population, enriched by classical classifiers, rather than taking the best individual.

1 Introduction

One way to improve the quality of a classifier is to optimize the underlying algorithm or to search for new algorithms. A simpler approach is to combine the results of several different classifiers by a combination algorithm. Many scientists have presented methods to obtain useful Ensembles of classifiers. The main objective is to create Ensembles of highly accurate members, which are diverse (there is no gain in combining identical classifiers) and make their errors independent of each other. The central question in this context is, what kind of classifiers should be combined and how.

This paper, like [8] and [16], focuses on the combination of evolutionary created Artificial Neural Networks (ANN's). These Ensembles will be expanded by a k-NN classifier obtaining a heterogenous Ensemble. At first, a system to create evolutionary designed ANN's is described in brief. Then the used nearest neighbour approach will be presented and the methodology of Ensembles of ANN's will be shown in detail and references to recent publications will be given. This

* An extended online version of this paper, with additional links to further online information is available at: http://www.tussy.uni-wh.de/˜chris/publications/wsc3

will be followed by the description of a new way to combine Ensembles of networks called a 3-Level Stacked Generalizer. The description will be supplemented by an explanation of Ensemble member selection and combination algorithms. Finally the usefulness of the proposed method will be experimentally proven for real world tasks.

2 Evolutionary designed Artificial Neural Networks

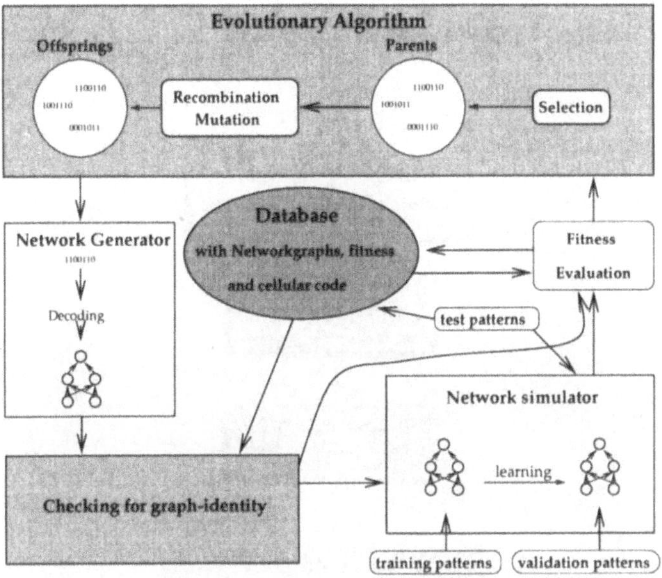

Fig. 1. Evolutionary algorithm to optimize the architecture of an Artificial Neural Network

The design of the architecture of an ANN is a challenging task. The architecture influences the learning capability and generalization ability. Beside of constructive and pruning techniques, Evolutionary Algorithms have been suggested by many authors to find good performing architectures for ANN's. For a summary of recent results, see the paper of Branke [2] or consult the homepages of researchers in this area in appendix A. In this paper an Evolutionary Algorithm with a modified cellular encoding [7] technique has been used. Figure 1 shows the principle working structure of this algorithm. For detailed information about the used algorithm the paper [5] is suggested.

3 Nearest Neighbour Algorithms

Nearest neighbour algorithms are a famous and simple form of classifier. Given a set of stored training examples (prototypes), an unseen test point x will be classified by computing a distance in input space from x to each of the training prototypes. In the case of a k-NN, the k nearest prototypes will be selected to calculate the class for the test point x by majority decision of the k nearest neighbours. For the calculation of the distance, several metrics have been suggested, one of the simplest is the euclidean distance. For detailed descriptions of nearest neighbour classifiers see [11] or [14, 3] for Ensembles of nearest neighbour classifiers. In this paper, a k-NN classifier based on euclidean distance metric with k selected by the leave-one-out method, has been used.

4 Ensembles of Neural Networks

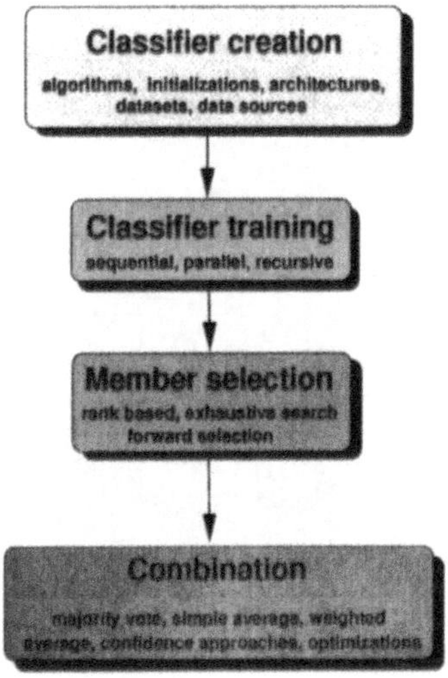

Fig. 2. Necessary steps and possible choices for Ensembles of classifiers

Combinations of ANN's, sometimes called Ensembles of Networks [13] are an alternative to the tradition of generating a population of networks, by changing the initial weights or modifying the architecture and choosing the best performing population member. Figure 2 shows the necessary steps using Ensembles of classifiers. Several methods for the creation of Ensemble members and their combination have been suggested. Among them are:

- Variation of the initialization of weights
- Variation of the network architectures

- Variation of the classification algorithms
- Variation of the training datasets
- Variation of the data sources

After determining, which method should be used for the creation of classifiers, the learning method for the Ensemble has to be chosen. In general it is possible to use parallel, sequential or recursive methods for learning. In sequential methods like boosting or its modern form, arcing [4], the classifiers are learned sequentially and the dataset for the next classifier is created using the misclassification information of the former classifier. Finally the learned networks are assembled by a combination algorithm. The recursive form of learning, Stacked Generalization, [15] combines the learned classifiers with another classifier and several of those outputs may be combined again. The parallel learning method is the special case of a 2-Level Stacked Generalizer.

Another important aspect is the selection of a subset of classifiers that should be combined. In [1] it is argued, that it is only useful to combine classifiers that are negatively correlated. Otherwise Ensembles will be used that have accurate and diverse members, but the gain of combining is only a loss in Ensemble accuracy. In section 5.1 several new member selection methods will be proposed.

The last step is the choice of a combination algorithm. In the context of classfication with ANN's, majority voting is a common method. Other techniques like simple averaging or weighted averaging based on the error on the validation set for the weights have been applied. In [16] a recursive least square (RLS) algorithm has been used to find optimal values for the weights. For further information on this topic consult the homepages of researchers in this area at appendix B

5 3-Level Stacked Generalization

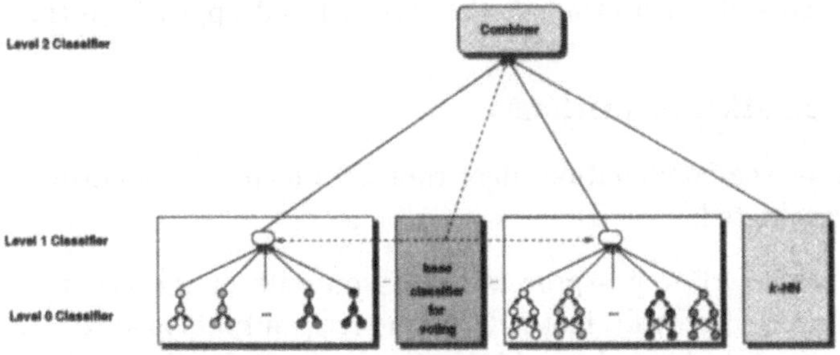

Fig. 3. A 3-Level Stacked Generalizer

In this paper a 3-Level Stacked Generalizer is presented. Figure 3 shows the structure of this Ensemble. The Level-0 classifiers are ANN's created by an Evolutionary Algorithm to optimize the architecture of ANN's. During the evolutionary cycle some of them are evaluated several times for different initial weight settings and stored in the used graph-database (see [5] for details). The Level-1 classifier combines the output of some selected Level-0 classifiers, with the same architecture from the graph-database and feeds the outputs to the Level-2 classifier. At the Level-2, members are selected from the population created by the Evolutionary Algorithm, with different architectures expanded by a nearest neighbour classifier. These are combined and supply the result of this Ensemble. All the networks are evaluated during the evolutionary optimization process and it is not necessary to relearn them again, it is therefore an extra gain to combine the networks during the evolutionary process.

5.1 Ensemble member selection

For a population size of n, the space of possible non empty subsets is of size $(2^n - 1)$. Therefore it is impractical for larger populations to search for an optimal subset. In this work, smaller Ensemble subsets (The Level-0 classifiers) are nevertheless selected by exhaustive search with the Ensemble error on the validation set as optimization criterion. For the complete population (Combination of Level-1 classifiers) a different approach has to be found. First a fitness-rank based list of classifiers is created. Then the following new selection methods have been used by the author:

top down: Start with the best classifier and try to add one classifier from the rank-based-list of classifiers in the direction best to worst. If there is no gain in Ensemble performance the classifier is dropped from the Ensemble.

bottom up: Start with the worst classifier and try to add one classifier from the rank-based-list of classifiers in the direction worst to best. If there is no gain in Ensemble performance the classifier is dropped from the Ensemble.

middle out: Start with the median member of the rank-based-list and try to add in an alternating way better or worse members from the list. If there is no gain in Ensemble performance the classifier is dropped from the Ensemble.

5.2 Combination algorithms

The choice of the combination algorithm is crucial. The following algorithms have been evaluated:

Majority vote: After postprocessing the outputs of the classifiers with the WTA (winner takes all) method, getting crisp outputs in an 1 of M encoding, a majority vote is used to obtain the result of the Ensemble. Ties, in the case of even Ensemble size, are broken arbitrarily using the base class. The base class is the class with the highest *a priori* probability in the learning set.

Unisono vote: After postprocessing with WTA, an unisono vote is used. In the case of undecidedness the rejection is solved arbitrary using the base class information.

Simple Average:

$$o_{c,\{ens\}} = \frac{1}{n}\sum_{i=1}^{n} o_{c,i} \tag{1}$$

Weighted Average:

$$o_{c,\{ens\}} = \sum_{i=1}^{n} w_i o_{c,i} \tag{2}$$

with w_i based on the Error of the validation and learning set, n denotes the size of the Ensemble and c is the index for the class output node in an 1 of M encoding.

Entropy Approach: In [10] it is shown, that ANN classifiers with a 1 of M encoding of the output units summing to one, estimate Bayesian *a posteriori* probabilities, when a sum of squared errors or a cross-entropy error function is used. This allows to see the outputs of a neural network not only as crisp classification results like in the majority vote algorithm, but to use additional probability information of the outputs. The *a posteriori* probabilities for the k-NN classifier can be directly calculated from the class distribution of the k nearest neighbours. In [6] a combination algorithm, called the Entropy approach, based on this information is presented. For every sample, the function $H(c)$ is calculated and the class with $\max(H(c))$ is selected.

$$H(c) = \frac{1}{n}\sum_{i=1}^{n} \frac{o_{c,i}}{-\sum_c o_{c,i} \ln o_{c,i}} \tag{3}$$

5.3 Related Works

Opitz and Shavlik [8] describe a similar system. In their algorithm, the fitness function of the evolutionary algorithm is changed to actively search for accurate and diverse population members. On the other hand they use an unusual representation and special genetic operators, more suited for the design of their presented knowledge-based Neural Networks than for standard ANN's. As an Ensemble member selection method, they use the n best networks and combine them solely by weighted average.

Yao and Liu [16] use their Evolutionary Algorithm EPNet to create a Population of good performing ANN's and combine them by majority voting, weighted average, and a recursive least squares algorithm. Beside of using the complete population, they propose a Genetic Algorithm to find better subsets of the population to be combined.

Breiman [3] used an Ensemble of nearest neighbour algorithms with different training datasets, chosen by bootstrapping. He proves that it is not useful to combine stable classifiers like nearest neighbour algorithms. In contrast Skalak [14] successfully used nearest neighbour classifiers that are based on a smaller set of prototypes (learning instances). He explains this success by the increase of instability in his algorithm.

All of them only used homogenous Ensembles, consisting of ANN's or nearest neighbour classifiers. In contrast, this article describes a heterogenous Ensemble of networks and a k-NN classifier.

6 Experimental Results

The presented 3-Level Stacked Generalizer has been empirically tested on three problems. The first was the diabetes problem from the benchmark suite of L. Prechelt [9]. The task is to classify from diagnostic data (e.g. blood pressure, result of glucose tolerance test etc.), whether a female Pima Indian is diabetes positive or not. This case with 8 inputs and 2 output parameters is complex, because some values for the input patterns are not available and set to zero. It decepts the classification algorithm, but is realistic for real world tasks. Furthermore the results can be compared with data available from literature and from previous works on evolutionary created ANN's by the author. Table 1 shows the classification error on the test set for different classifiers. The tests for the diabetes problem have been divided into two subtests. The first test was investigating, if Ensembles of the same network architecture, learned with the same training patterns but with different weight initializations are better than the best member of the Ensemble. These networks will be later used as the Level-0 classifiers. The graph-database obtained by the experiments undertaken in [5] was used and Ensembles (maximum size 5) have been created by exhaustive search from about 500 different networks of this database. In all cases the best Ensemble showed better or equal performance than the mean of all networks and even outperforms the best network, it should be noted that for comparison the classification error on the validation set and not the test set was used. Otherwise there would have been an unjustified bias towards the test set. The equality with the best network is a property of the exhaustive selection method. In table 2 the results for different combination algorithms are presented. It should be noted that in about 30% of the cases the classification error on the test set is increased. This phenomenon could be observed in all experiments and is a sign of overfitting the learn and validation set. It can be avoided using a second validation set for the selection of the Ensemble members, but in the observed problems the amount of available patterns was to small to use a second validation set.

Problem	Base	k-NN	LVQ	standard ANN	evolved ANN	3-Level SG + k-NN
Diabetes	38.5	28.1	-	24.8	23.1	22.9*
CERA (2)	30	15	15	15	10	10
CERA (4)	45	35	35	40	35	30

Table 1. Classification error in percent on the test set for different algorithms. *=the algorithm dropped the k-NN from the Ensemble

The second subtest was made with the full 3-Level Stacked Generalizer enriched with a k-NN classifier, that means during the evolutionary process of

Problem	Unisono	Majority	simple	weighted	entropy
Diabetes	72.4	85.6	90.8	93.3	92.2
CERA(4)	56.3	83.2	92.3	94.1	97.0
CERA(2)	42.4	58.9	61.8	65.6	66.3

Table 2. Percent of best Ensembles (without k-NN) better (inequal) than the best network

creating ANN's the complete population was used to select useful Ensemble members. At this level, with a population size of 100, exhaustive search was impossible and the three selection methods have been used. For the diabetes problem only 6% of the full Ensembles used the k-NN, this can be explained by the poor performance of the k-NN, in contrast to the evolved networks for this problem. The mean size of the Level-2 Ensembles was 4.3 with clear differences between the different member selection schemes. It was not surprising, that for all made tests, the Ensemble size for the bottom up selection, resulted in greater Ensembles than the middle out and the top down algorithm. Compared to Ensembles of all classifiers of a population, the selection methods gave better results. The best found Ensemble resulted in an classification error on the test set of 22.91% which is slightly better than the mean of five runs of the best network found by the Evolutionary Algorithm (23.1% classification error). But Ensembles with this classification error were found in early generations reducing the computation time drastically.

The second Test with the 3-Level Stacked Generalizer was made with the Cortical evoked response audiometry (CERA) data for the objetive evaluation of auditive perception in children showing central auditory processing disorders or language based learning disabilities. For detailed descriptions of this dataset and previous evaluations with Kohonen type networks see [12]. The dataset consists of only 112 patterns, with 18 inputs and 4 or 2 classes. It was divided into a learning set with 76 patterns, a validation set with 16 patterns, and a never seen test set with 20 patterns. In the 2 class scenario the classifier should only discriminate between a person having language based learning disabilities or not. In the 4 class scenario it should additionally be discriminated between adults, infants and different levels of learning disabilities. In Table 1 the results are summarized.

The subtests were done with equal settings like in the diabetes tests and results are similar. The k-NN was used in about 32%/38% (CERA(4)/CERA(2)) of the full Ensembles and the mean size was 3.4/2.8. The problem of overfitting the learning and validation set was not as important as in the diabetes dataset.

The different combination algorithms show performances comparable to the results of table 2. The entropy approach was the clear winner and the unisono algorithm gave only poor results. For the different selection algorithms only a difference in Ensemble size was found, but no significant gain in classifier performance.

It could be clearly demonstrated, that the usage of Ensembles of evolutionary created networks and a k-NN classifier, combined by different combination algorithms yields comparable or better results with less computing time, than using the single best network. The usage of classical classifiers like the k-NN algorithm helped to find good Ensembles earlier in the process of optimization. The proposed member selection methods showed better results than using the complete population. The usage of *a posteriori* probabilities with the Entropy combination algorithm gave the best combination results.

7 Conclusion and Outlook

A 3-Level Stacked Generalizer consisting of evolutionary created ANN's and a nearest neighbour classifier has been presented. Several combination algorithms and Ensemble member selection methods have been presented and experimentally evaluated. It could be shown, that it is more efficient to take Ensembles rather than using the single best network. Additionally it can be concluded, that the choice of Ensemble members and combination algorithm is of central importance. The use of *a posteriori* probabilities in the Entropy approach showed the most gain in combining. The 3-Level Stacked Generalizer allowed, to obtain good results for different weight initializations and architectures in early generations of the Evolutionary Algorithm reducing the necessary computing time. Furthermore the use of nearest neighbour algorithms to enrich the Ensembles showed better results in earlier generations. In ongoing research the combinations of evolutionary created ANN's and other classical classifiers, like decision trees should be investigated. This could lead to a toolbox chosing automatically the necessary classifiers that bring the best classification performance for a given problem. The fitness function of the Evolutionary Algorithm should be changed to lead the search not to optimal single networks, but to find optimal members of an Ensemble. The greatest problem using this approach is the need for a large number of training and validation patterns, making its usefulness problematic for real world tasks with small training sets.

Acknowledgements:

I like to thank Dr. Rainer Schönweiler of the Medizinische Hochschule Hannover for the CERA dataset and Prof. Claudio Moraga for useful comments.

References

1. K. M. Ali. *Learning Probabilistic Relational Concept Descriptions.* PhD thesis, University of California, Irvine, 1996. http://www.ics.uci.edu/~ali/phd/thesis.ps.Z.
2. J. Branke. Evolutionary algorithms for neural network design and training. In *Proceedings of the 1st Nordic Workshop on Genetic Algorithms and its Applications*, 1995. ftp://ftp.aifb.uni-karlsruhe.de/pub/jbr/Vaasa.ps.
3. L. Breiman. Bagging predictors. Technical Report TR 421, University of California, Berkeley, 1996. ftp://ftp.stat.berkeley.edu/pub/users/breiman/bagging.ps.Z.

4. L. Breiman. Bias, variance, and arcing classifiers. Technical Report TR 460, University of California, Berkeley, 1996. ftp://ftp.stat.berkeley.edu/pub/users/breiman/arcall.ps.Z.
5. C. M. Friedrich and C. Moraga. Using genetic engineering to find modular structures and activation functions for architectures of artificial neural networks. In B. Reusch, editor, *Computational Intelligence; Theory and Applications (LNCS1226); Proceedings of the 5th Fuzzy Days*, pages 150–161, Dortmund, Germany, 1997. Springer. ftp://ftp.cis.ohio-state.edu/pub/neuroprose/friedrich.nnarch-activation.ps.Z.
6. J. Ghosh, S. Beck, and C.-C. Chu. Evidence combination techniques for robust classification of short-duration oceanic signals. In *SPIE Conf. on Science of Artificial Neural Networks (SPIE Vol. 1710)*, pages 266–276, 1992. http://www.lans.ece.utexas.edu.
7. F. Gruau. Cellular encoding of genetic neural networks. Technical Report 92-21, Ecole Normale Superieure de Lyon, Institut IMAG, 1992. ftp://lip.ens-lyon.fr/pub/Rapports/RR/RR92/RR92-21.ps.Z.
8. D. W. Opitz and J. W. Shavlik. Actively searching for an effective neural network ensemble. *Connection Science*, 8(3&4):337–354, 1996.
9. Lutz Prechelt. PROBEN1 — A set of benchmarks and benchmarking rules for neural network training algorithms. Technical Report 21/94, Fakultät für Informatik, Universität Karlsruhe, D-76128 Karlsruhe, Germany, September 1994. ftp://ftp.ira.uka.de/pub/papers/techreports/1994/1994-21.ps.gz.
10. M. D. Richard and R. P. Lippmann. Neural network classifiers estimate bayesian a posteriori probabilities. *Neural Computation*, 3, 1991.
11. B. D. Ripley. *Pattern Recognition and Neural Networks.* Cambridge University Press, 1996.
12. R. Schönweiler and M. Ptok. Kohonen-Netze zur Klassifizierung später akustisch evozierter Potentiale. In A. Grauel, W. Becker, and F. Belli, editors, *Proceedings of the 4.th International Workshop on Fuzzy-Neuro-Systeme '97 (FNS'97) - Computational Intelligence*, pages 146–153, Soest, Germany, 1997.
13. A. J. C. Sharkey. On combining artificial neural nets. *Connection Science*, 8(3&4):299–314, 1996.
14. D. B. Skalak. *Prototype Selection for Composite Nearest Neighbor Classifiers.* PhD thesis, University of Massachusetts, Amherst, 1996. ftp://ftp.cs.umass.edu/pub/techrept/techreport/1996/UM-CS-1996-089.ps.
15. D. H. Wolpert. Stacked generalization. *Neural Networks*, 5:241–259, 1992.
16. X. Yao and Y. Liu. Making use of population information in evolutionary artificial neural networks. *IEEE Transactions on Systems, Man and Cybernetics*, 28B(2), 1998. ftp://www.cs.adfa.oz.au/pub/xin/smc096-05-0503.ps.gz.

A GANN Researchers Homepages and Bibliography

http://www.tussy.uni-wh.de/~chris/gann/gann.html

B Ensemble Researchers Homepages and Bibliography

http://www.tussy.uni-wh.de/~chris/ensemble/ensemble.html

Part 5: Data Mining

Papers:

Application of Multi-dimensional Fuzzy Analysis to Decision Making
Alexandr A. Savinov

Information-Theoretic Fuzzy Approach to Knowledge Discovery in Databases
Oded Maimon, Abe Kandel and Mark Last

Intelligent Electronic Catalogs for Sales Support – Introducing Case-Based Reasoning Techniques to on-line Product Selection Applications
Ivo Vollrath, Wolfgang Wilke and Ralph Bergmann

A Genetic Algorithm for Generalized Rule Induction
Alex A. Freitas

Application of Multi-dimensional Fuzzy Analysis to Decision Making

Alexandr A. Savinov

Laboratory of AI Systems
Institute of Mathematics, Moldavian Academy of Sciences
str. Academiei 5, MD-2028 Kishinev, Moldavia
Phone: 3732-73-81-30, Fax: 3732-73-80-27
E-mail: savinov@math.md
http://www.geocities.com/ResearchTriangle/7220/

Key words: Multi-dimensional fuzzy analysis, Fuzzy decision support systems, Fuzzy resolution, Rule induction, Fuzzy logical inference.

Abstract

The goal of multi-dimensional fuzzy analysis consists in discovering different properties in multi-dimensional fuzzy distributions represented either extensionally (database) or intensionally (knowledge base). In this paper we show how this approach can be applied to such problems as decision making and knowledge discovery in databases. For uniform and efficient representation of fuzzy knowledge and data we propose a technique of sectioned matrices. For carrying out logical inference we use a new operation of fuzzy resolution which is a generalization of the conventional resolution. With the help of this operation we find fuzzy prime disjunctions which later are used for making decisions in concrete situations. For discovering hidden dependencies in data a new fuzzy covering method is used.

1. Introduction

Let us consider the following general problem. Given n variables $x_1,x_2,\ldots,x_n$ each of which takes its values from the sets $X_1,X_2,\ldots,X_n$, respectively. The universe of discourse O is equal to the Cartesian product of all sets of the values

$$O = X_1 \times X_2 \times \ldots \times X_n$$

and consists of n-tuples

$$o = \langle x_1,x_2,\ldots,x_n \rangle$$

There is some distribution over the universe of discourse. The general goal of multi-dimensional analysis consists in finding different properties of distributions and their transformations to other forms.

Depending on the character (modality) of the distribution and the form of its representation multi-dimensional analysis can be divided into subdomains. For example, traditional multi-dimensional statistical analysis aims at discovering properties of probabilistic distributions, e.g., mean values or correlation, represented extensionally in the form of a table of data. (Such an important case as the analysis of intensionally represented multi-dimensional probabilistic distributions, e.g., by means of rules, is not studied in this subdomain.)

Note that the key point in the whole approach is *multi-dimensionality*, i.e., it is precisely multi-dimensionality that is responsible for non-triviality of problems and difficulty of procedures. In particular, such a well known problem as finding satisfiability can be easily reformulated as a problem of multi-dimensional analysis of two-valued distributions over n-dimensional hyper-cube which are represented by means of CNF or DNF. Although multi-dimensional analysis can be used for extensionally represented distributions, e.g., to analyze tables of data, the most interesting and difficult problems arise in analyzing intensionally represented distributions. In the former case the distribution is represented pointwisely, i.e., each statement represents the value of the distribution in one point of the universe of discourse. In the latter case we use elementary distributions over the values of individual variables and some operations to represent multi-dimensional distributions. Thus *intensionality* of representation is the second key point of multi-dimensional analysis.

In this paper we consider multi-dimensional analysis of fuzzy finite distributions [13]. We suppose that the sets $X_1,X_2,...,X_n$ called domains are finite and consist of $n_1,n_2,...,n_n$ members, respectively. If the variable x_i has only two values (they are supposed to be 0 and 1, i.e., $X_i=\{0,1\}$ and $n_i=2$), then it is said to be boolean variable. In finite case the power of the universe of discourse is equal to $n_1 \times n_2 \times ... \times n_n$. In the case of all boolean variables the universe of discourse is equal to n-dimensional hyper-cube with the power 2^n.

The second our supposition concerns the character of the distributions being studied. Namely, we suppose that the distributions take their values from the interval [0,1]. If the distribution is crisp, i.e., takes only values 0 and 1, it is said to be boolean. If only boolean variables and boolean distributions are used then we obtain just the case well studied in such fields as boolean algebra, propositional calculus, switching theory etc. Multi-dimensional analysis differs from these and other similar directions in that it has its own goal — finding properties of distributions, although many of these approaches often have very similar and even identical main notions and procedures.

The third main supposition is that the operations of minimum and maximum are used to combine fuzzy distributions. In particular, this property distinguishes fuzzy case from the probabilistic one.

According to these suppositions we assume that any problem domain being considered is described in terms of attributes and their values. If such a suitable set of attributes and values exists then it is referred to as an *attribute model* of the problem domain [1]. All combinations of the values of attributes represent possible objects or states of the problem domain. By knowledge we mean some fuzzy constraints on possible states of the problem domain represented by the corresponding multi-dimensional fuzzy distribution. If there is not any knowledge about the problem domain then any combination of attribute values is possible (meaningful) and is assigned value 1. Otherwise some states are fuzzily disabled, i.e., they are assigned values which are less than 1 (0 means that the state or object is completely impossible, meaningless).

The first problem is how to represent such knowledge in attribute model by means of multi-dimensional fuzzy distributions. In section 2 we propose a method of sectioned matrices which allows us to represent knowledge and data uniformly and efficiently. The second problem is how to carry out logical inference and make conclusions proceeding from some knowledge and data. It is described in section 3. The third problem considered in the paper is how to generate knowledge from available data. In section 4 we propose a method to transform a sectioned matrix of data into a sectioned matrix of knowledge.

2. Knowledge and Data Representation

In this section we describe a sectioned matrix technique [2, 3] for representing multi-dimensional fuzzy distributions. An elementary distribution is a distribution over one domain and represents the corresponding proposition about the attribute. For example, if there is an attribute ALCOHOLISM STAGE with the values from the set {First, Second, Third}, then the elementary proposition

ALCOHOLISM STAGE = {First : 1, Second : 0.2, Third : 0}

means that the corresponding fuzzy distribution has value 1 in the point First, value 0.2 in the point Second and value 0 in the point Third.

Elementary propositions can be represented only extensionally since they are 1-dimensional, i.e., they can be represented only by enumerating their values in all points of the domain of definition. We will write elementary propositions in the form of one *section*, i.e., simply enumerating all its values, e.g., {1, 0.2, 0}. One number in the section is said to be a *component*. Thus the section {1, 0.2, 0} consists of three components 1, 0.2 and 0.

A sectioned *vector* is defined as a set of sections for all attributes. It can be interpreted as disjunction or conjunction. If the vector is interpreted as disjunction then all its sections are supposed to be combined with the operation of maximum. (From the logical point of view its propositions are combined with the connective OR.) If the vector is interpreted as conjunction then all its sections are supposed to be combined with the operation of minimum. For example, the following constructions are fuzzy sectioned vectors for an attribute model consisting of 3 attributes:

$$\{1, 0.2, 0\}, \{0, 1\}, \{0, 0, 0\}$$
$$\{0, 1, 0.5\}, \{0.5, 0.7\}, \{1, 1, 1\}$$

Depending on its interpretation, i.e., on the connective used for combining sections, each sectioned vector represents one or another multi-dimensional distribution over the universe of discourse O. If the sectioned vector $\mathbf{d}$ is interpreted as disjunction then it defines the following multi-dimensional fuzzy distribution over the space O:

$$\mathbf{d}(o) = \mathbf{d}(\langle x_1, x_2, \ldots, x_n \rangle) = \max[\mathbf{d}_1(x_1), \mathbf{d}_2(x_2), \ldots, \mathbf{d}_n(x_n)]$$

where $\mathbf{d}(o)$ is equal to the value of the distribution in the point o and $\mathbf{d}_i$ are sections of the vector. In other words, to find the value $\mathbf{d}(o)$ we have to choose the maximal of n components (one component from each section) corresponding to the values of variables. Note that the section consisting of all 0 is equivalent to its absence.

If the sectioned vector $\mathbf{k}$ is interpreted as conjunction then it defines a multi-dimensional fuzzy distribution over the space O in the dual way:

$$\mathbf{k}(o) = \mathbf{k}(\langle x_1, x_2, \ldots, x_n \rangle) = \min[\mathbf{k}_1(x_1), \mathbf{k}_2(x_2), \ldots, \mathbf{k}_n(x_n)]$$

To compute its value in some point we have to take minimal of n components corresponding to the values of variables. If some section consists of all 1 then it is equivalent to its absence.

With the help of disjunctions we can represent general dependencies between attributes. A number of such disjunctions combined with the operation of minimum represents a knowledge base.

Let us consider an example from the field of screening alcoholism [4] where the goal is to determine a patient alcoholism stage (the methods of treatment depend on this diagnosis). The problem domain can be described by the following three attributes:

DEPENDENCE = {Psychical, Psychico-physical, Physical}
LOSS OF SELF-CONTROL = {Quantitative, Situational}
ALCOHOLISM STAGE = {First, Second, Third}

If there are not known dependencies between these attributes then any state of the problem domain is possible, i.e., any combination of their values is considered meaningful, and the knowledge base represents the multi-dimensional distribution which is equal to 1 in all points of the universe of discourse.

One way to express our knowledge about the problem domain consists in indicating a state (or a fuzzy interval of states) which is disabled. Then the corresponding combinations of the attribute values are meaningless. For example, if nobody has ever seen anything like the object

⟨Psychical, Situational, Third⟩

(i.e., a patient at the third stage of alcoholism with the psychical dependence and situational loss of self-control) then such a knowledge can be written as the disjunction

$$\mathbf{1} = \{0, 1, 1\},\{1, 0\},\{1, 1, 0\}$$

which pricks a whole in the unit level distribution (Fig. 1).

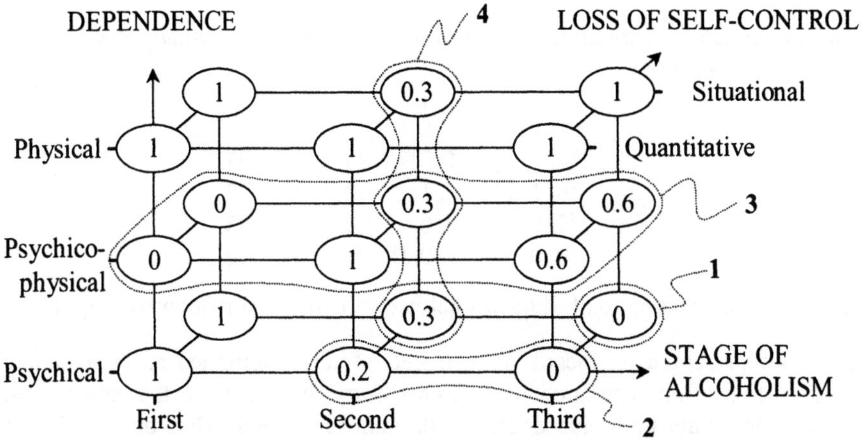

Figure. 1. Fuzzy distribution over the universe of discourse

Another way to represent knowledge is by means of conventional implications. Each implication can be easily transformed into disjunction by negating its left part. For example, the following two implications

DEPENDENCE = {Psychical : 1, Psychico-physical : 0, Physical : 0} AND
LOSS OF SELF-CONTROL = {Quantitative : 1, Situational : 0} →
ALCOHOLISM STAGE = {First : 1, Second : 0.2, Third : 0}

DEPENDENCE = {Psychical : 0, Psychico-physical : 1, Physical : 0} →
ALCOHOLISM STAGE = {First : 0, Second : 1, Third : 0.6}

can be written as the following two disjunctions

$$\mathbf{2} = \{0, 1, 1\}\{0, 1\}\{1, 0.2, 0\}$$
$$\mathbf{3} = \{1, 0, 1\}\{0, 0\}\{0, 1, 0.6\}$$

Rules represented in the form of implication usually describe dependence of the goal from the rest of the attributes. Sometimes it is convenient to express our knowledge in the form of a backward dependency where we describe properties (symptoms, situations) which are typical of the concrete conclusion (decision). In other words, we suppose that a concrete decision has been made and then ask the question what other conditions it is characterized by. For example, let us suppose that we know one more dependency where situational loss of self-control is usually (degree 0.3) not typical of the second stage of alcoholism:

$$\text{ALCOHOLISM STAGE} = \{\text{First} : 0, \text{Second} : 1, \text{Third} : 0\} \rightarrow$$
$$\text{LOSS OF SELF-CONTROL} = \{\text{Quantitative} : 1, \text{Situational} : 0.3\}$$

Then it can be written in the form of disjunction

$$\mathbf{4} = \{0, 0, 0\}, \{1, 0.3\}, \{1, 0, 1\}$$

The whole matrix representing our knowledge consists of 4 disjunctions **1**, **2**, **3**, **4** and looks as follows

$$\mathbf{D} = \begin{array}{|lll|l} x_1 & x_2 & x_3 & \\ \{0, 1, 1\} & \{1, 0\} & \{1, 1, 0\} & \mathbf{1} \\ \{0, 1, 1\} & \{0, 1\} & \{1, 0.2, 0\} & \mathbf{2} \\ \{1, 0, 1\} & \{0, 0\} & \{0, 1, 0.6\} & \mathbf{3} \\ \{0, 0, 0\} & \{1, 0.3\} & \{1, 0, 1\} & \mathbf{4} \end{array}$$

and the corresponding multi-dimensional fuzzy distribution is shown in Fig. 1.

Data is represented by means of sectioned vectors interpreted as conjunctions. Each section of one conjunction represents fuzzy constraints on the values of the corresponding attribute independent of the other attributes. Thus observable data used as an input for logical inference and making decision can be represented as one conjunction each section of which contains information about one attribute. If there is not any observable data then such vector of data contains all 1 (all values are possible). If we have known that some attribute cannot take certain value then the corresponding component is decreased in proportion to our degree of confidence. For example, if it is known that the patient does not have physical dependence than it can be written as the sectioned vector

$$\{1, 1, 0\}, \{1, 1\}, \{1, 1, 1\}$$

In addition, if we suppose with the degree 0.5 that he does not have quantitative loss of self-control then we obtain the vector

$$\{1, 1, 0\}, \{0.5, 1\}, \{1, 1, 1\}$$

Thus as we obtain and enter new information, components of the vector of data (and the whole distribution values) are decreased. Obviously, each section must contain at least one 1, otherwise our data is inconsistent.

If all components of conjunction are equal to 0 except for the one component in each section, for example,

$$\{1, 0, 0\}, \{1, 0\}, \{1, 0, 0\}$$
$$\{0.7, 0, 0\}, \{0.7, 0\}, \{0.7, 0, 0\}$$

then such a vector represents the distribution which is equal to 0 in all except for the one points of the universe of discourse where it is equal to the minimal of n non-0 components in each section. Thus by means of such vector we can represent one element of data, e.g., one row of a database, along with the corresponding degree of membership, and with the help of a set of such conjunctions we can represent any multi-dimensional distribution.

3. Multi-attribute Decision Making

The problem of fuzzy decision making in concrete situation for an attribute model of the problem domain can be formulated as follows [5, 6]. There is some data on the observable situation represented in the form of fuzzy constraints on the possible attribute values and written as a vector-conjunction **f**. In addition, there is a knowledge about fuzzy dependencies between attributes which is represented by means of a sectioned matrix **D**. The problem is to find fuzzy constraints on the values of a goal attribute(s) which are equal to the projection of the joint multi-dimensional distribution including data and knowledge onto the goal variable(s). Thus logical inference is reduced to the problem of finding projection of some multi-dimensional distribution. In other words, if we know that some *combinations* of attribute values are disabled (knowledge about the problem domain) and, in addition, certain values of *individual* attributes are disabled (data on the observable situation) then we have to find disabled values of the goal attribute(s).

Once we have built our knowledge base we are interested in carrying out consultations, i.e., obtaining concrete conclusions on necessary attributes in concrete situations. For example, we might be interested in obtaining possible stages of alcoholism for a patient for whom we know that he does not have physical dependence, i.e., the situation is described by the conjunction

$$\{1, 1, 0\}, \{1, 1\}, \{1, 1, 1\}$$

In principle, we could solve this problem extensionally using combination and projection principle [7, 8]. For this it is necessary to calculate the values of 4 distributions corresponding to 4 disjunctions from the knowledge base and the values of 1 distribution corresponding to 1 conjunction representing data in all points from the universe of discourse. Then we have to combine all 5 distributions with the help of minimum operation and to find the ultimate distribution representing both knowledge and data. And finally we have to find projection of this distribution onto the third variable by finding maximal values in 3 subspaces of the universe of discourse each of which corresponds to one value of the goal attribute. In our example we can easily infer (Fig. 1) that the projection on the goal attribute is equal to $\{1, 1, 0.6\}$, i.e., in this situation the third stage of alcoholism is disabled with the degree 0.6. Obviously, this method is inapplicable to real examples since the universe of discourse is very large space and it is not possible to compute the distribution values in all its points. Some usefulness of this procedure is that we deeper understand the essence of multi-dimensional logical inference at the level of one point ("pointwise" logical inference). Thus we have to use intensional approaches to logical inference which would explicitly use disjunctions to make conclusions.

One such method also uses combination and projection principle but in other form. In this method we try to make conclusions from individual disjunctions and then combine all results by finding conjunction of all projections. After the first pass when we found all projections of individual disjunctions we should take the conclusions along with the data and use it as new data for the next pass. Then, at the next pass we make new conclusions from individual disjunctions and again add them to the previous data. The process is stopped when no new conclusions can be obtained.

For this method we need a procedure of finding projection of one disjunction and one conjunction. It can be proved that this projection is computed in two steps. First, we have to find componentwise conjunction (minimum) of two sectioned vectors. Then we have to take the sectioned vector interpreted as disjunction resulted from this step and find its projection onto the goal attribute. Projection of one disjunction is equal to its goal section to which we add with the help of maximum operation the maximal component in all the rest of the sections. (For the sake of simplicity in this paper we do not take into account the issue of handling disjunction constant which is connected with the notion of reduced disjunctions [3, 5, 6, 9, 13].)

In our example, first, we find componentwise conjunction of the vector of data with each disjunction and obtain the matrix

$$\mathbf{D} = \begin{array}{|lll|l} \multicolumn{1}{c}{x_1} & \multicolumn{1}{c}{x_2} & \multicolumn{1}{c}{x_3} & \\ \{0, 1, 0\} & \{1, 0\} & \{1, 1, 0\} & \mathbf{1} \\ \{0, 1, 0\} & \{0, 1\} & \{1, 0.2, 0\} & \mathbf{2} \\ \{1, 0, 0\} & \{0, 0\} & \{0, 1, 0.6\} & \mathbf{3} \\ \{0, 0, 0\} & \{1, 0.3\} & \{1, 0, 1\} & \mathbf{4} \end{array}$$

(In this matrix 3rd components in 1st section are equal to 0.) Then we have to find 4 projections of the individual disjunctions onto 3rd attribute. Since all 4 disjunctions involve 1 either in 1st or in 2nd section (non-goal sections), we obtain that all their projections are equal to {1, 1, 1}, i.e., they do not have non-trivial projections onto the goal attribute (addition to any section of the value 1 with the help of maximum operation results in the section consisting of all 1). We have not inferred any conclusion and therefore the procedure is stopped. Comparing this result with that of the previous procedure (pointwise logical inference) we see that this procedure (it could be called "disjunctionwise" logical inference) does not guarantee obtaining exact result. Note that the method of inference on individual disjunctions (rules) followed by combination of conclusions dominates on the application to inference engines. It is caused by its simplicity and that exactness of inference is not the most important requirement in many systems — frequently it is sufficient if rules work independently.

Let us consider a method of logical inference which guarantees that the conclusion, i.e., projection obtained, will be exact. This method requires that the matrix of fuzzy disjunctions be equivalently transformed into the form where it contains all *prime* disjunctions [3, 5, 6, 9, 13]. Informally prime disjunctions are the most strong propositions about the problem domain provided that all of them are consequences of the initial matrix, i.e., if any component of prime disjunction is decreased then it is not already a consequence of the matrix. Thus before we can carry out logical inference we have to generate all prime disjunctions for our knowledge base.

One approach to building prime disjunctions is based on a new original operation of fuzzy resolution [3, 5, 6, 9-11, 13, 14] which generalizes traditional resolution in that it can be used (i) in the case of multi-valued variables (proposed in [2]), and (ii) for fuzzy distributions (proposed in [3]). If the resolution is applied to k-th section of two disjunctions **u** and **v** then the resolvent $\mathbf{w}=\mathbf{u}\langle x_k\rangle\mathbf{v}$ resulted from this operation is computed as follows:

1. k-th section $\mathbf{w}_k$ is equal to conjunction of two corresponding sections $\mathbf{u}_k$ and $\mathbf{v}_k$, i.e., each component in $\mathbf{w}_k$ is equal to the minimum of two corresponding components in $\mathbf{u}_k$ and $\mathbf{v}_k$.

2. non-k-th sections $\mathbf{w}_i$ are equal to disjunction of two corresponding sections $\mathbf{u}_i$ and $\mathbf{v}_i$, i.e., each component in $\mathbf{w}_i$ is equal to the maximum of two corresponding components in $\mathbf{u}_i$ and $\mathbf{v}_i$.

The whole procedure for generating prime disjunctions consists in applying the resolution operation to all disjunction pairs on all variables and adding new disjunctions to the matrix. Of course, it is natural to check if new disjunction is a consequence of one of those which are already in the matrix. When all disjunctions have been generated the matrix has to be simplified by removing (absorbing) disjunctions which follow from others. Then new pass has to be carried out. The procedure is stopped when any new resolvent is absorbed by disjunctions in the matrix, i.e., we are not able to generate new disjunctions.

After the first pass of this procedure we obtain the matrix

$$
\mathbf{D}_1 = \begin{array}{|lll|l}
x_1 & x_2 & x_3 & \\
\{0,1,1\} & \{0,1\} & \{1,0.2,0\} & \mathbf{2} \\
\{1,0,1\} & \{0,0\} & \{0,1,0.6\} & \mathbf{3} \\
\{0,0,0\} & \{1,0.3\} & \{1,0,1\} & \mathbf{4} \\
\{0,1,1\} & \{0,0\} & \{1,1,0\} & \mathbf{5}=\mathbf{1}\langle x_2\rangle\mathbf{2} \models \mathbf{1} \\
\{0,0,1\} & \{1,0\} & \{1,1,0.6\} & \mathbf{6}=\mathbf{1}\langle x_1\rangle\mathbf{3} \\
\{0,1,1\} & \{1,0\} & \{1,0.3,0\} & \mathbf{7}=\mathbf{1}\langle x_3\rangle\mathbf{4} \\
\{0,0,1\} & \{0,1\} & \{1,1,0.6\} & \mathbf{8}=\mathbf{2}\langle x_1\rangle\mathbf{3} \\
\{0,1,1\} & \{0,0.3\} & \{1,0.2,1\} & \mathbf{9}=\mathbf{2}\langle x_2\rangle\mathbf{4} \\
\{1,0,1\} & \{1,0.3\} & \{0,0,0.6\} & \mathbf{10}=\mathbf{3}\langle x_3\rangle\mathbf{4}
\end{array}
$$

Note that the disjunction **1** has been absorbed by the resolvent $\mathbf{5}=\mathbf{1}\langle x_2\rangle\mathbf{2}$, i.e., resolvent absorbed one of its parents. The matrix $\mathbf{D}_1$ consists of 9 disjunctions and no one of them is a consequence of another. At the second pass we have to find all resolvents of the disjunctions in matrix $\mathbf{D}_1$. After absorbing disjunctions **5**, **7**, **6**, **8**, **9** we obtain the matrix

$$\mathbf{D}_2 = \begin{array}{|lll|l} x_1 & x_2 & x_3 & \\ \{0,1,1\} & \{0,1\} & \{1,0.2,0\} & \mathbf{2} \\ \{1,0,1\} & \{0,0\} & \{0,1,0.6\} & \mathbf{3} \\ \{0,0,0\} & \{1,0.3\} & \{1,0,1\} & \mathbf{4} \\ \{1,0,1\} & \{1,0.3\} & \{0,0,0.6\} & \mathbf{10{=}3}\langle x_3\rangle\mathbf{4} \\ \{0,1,1\} & \{0,0\} & \{1,0.3,0\} & \mathbf{11{=}2}\langle x_2\rangle\mathbf{7} \models \mathbf{5,7} \\ \{0,0,1\} & \{0,0\} & \{1,1,0.6\} & \mathbf{12{=}3}\langle x_1\rangle\mathbf{5} \models \mathbf{6,8} \\ \{0,0,1\} & \{1,0\} & \{1,0.3,0.6\} & \mathbf{13{=}4}\langle x_3\rangle\mathbf{6} \models \mathbf{6} \\ \{0,1,1\} & \{0,0.3\} & \{1,0.2,0\} & \mathbf{14{=}5}\langle x_3\rangle\mathbf{9} \models \mathbf{9} \end{array}$$

At the next pass we generate only one disjunction which absorbs disjunction **10**

$$\mathbf{D}_3 = \begin{array}{|lll|l} x_1 & x_2 & x_3 & \\ \{0,1,1\} & \{0,1\} & \{1,0.2,0\} & \mathbf{2} \\ \{1,0,1\} & \{0,0\} & \{0,1,0.6\} & \mathbf{3} \\ \{0,0,0\} & \{1,0.3\} & \{1,0,1\} & \mathbf{4} \\ \{0,1,1\} & \{0,0\} & \{1,0.3,0\} & \mathbf{11{=}2}\langle x_2\rangle\mathbf{7} \models \mathbf{5,7} \\ \{0,0,1\} & \{0,0\} & \{1,1,0.6\} & \mathbf{12{=}3}\langle x_1\rangle\mathbf{5} \models \mathbf{6,8} \\ \{0,0,1\} & \{1,0\} & \{1,0.3,0.6\} & \mathbf{13{=}4}\langle x_3\rangle\mathbf{6} \models \mathbf{6} \\ \{0,1,1\} & \{0,0.3\} & \{1,0.2,0\} & \mathbf{14{=}5}\langle x_3\rangle\mathbf{9} \models \mathbf{9} \\ \{1,0,1\} & \{1,0\} & \{0,0.3,0.6\} & \mathbf{15{=}3}\langle x_3\rangle\mathbf{13} \models \mathbf{10} \end{array}$$

This is the final matrix of all prime disjunctions since we are not able to generate stronger disjunctions — all new resolvents are absorbed.

The fact that the matrix contains all prime disjunctions guarantees that projection on any variable obtained as a conjunction of projections of individual disjunctions is exactly equal to the real projection of the corresponding multi-dimensional distribution. Thus to carry out logical inference we impose our data represented as the conjunction **f** onto each prime disjunction and then combine all their projections onto the goal attribute. For example, in our case we have to change all 3-rd components in 1-st section (i.e., 3-rd column) onto 0 and then it becomes obvious that only disjunction **12** has non-trivial projection onto the attribute x_3 which is equal to the exact value {1, 1, 0.6}. If, in addition, we suppose that the patient does not have quantitative loss of self-control, i.e., the whole situation is described by the conjunction

$$\{1,1,0\},\{0,1\},\{1,1,1\}$$

then two more disjunctions fire. Disjunction **4** has projection {1, 0.3, 1}, and disjunction **13** — {1, 0.3, 0.6}. The final conclusion is characterized by the projection {1, 0.3, 0.6}. We also can find additional information by finding projections on other attributes. In the above situation we see that disjunction **15** gives projections {1, 0.6, 1} onto the attribute x_1, i.e., we can conclude that the patient does not have psychico-physical dependence (2-nd value of the 2-nd attribute is disabled with the degree 0.6). By means of the matrix of all prime disjunctions we can also analyze knowledge base. In particular, we can find hidden dependencies, i.e., dependencies which follow from explicitly formulated disjunctions, or find symptoms which can influence one or another attribute.

It is obvious that the method of inference based on finding all prime disjunctions is inapplicable to real problem domains since their number is extremely high. However, it is not necessary to build all prime disjunctions — for decision support systems it is enough to carry out only several passes especially on early stages of debugging (more disjunctions can be generated when the knowledge base has been created and is ready for use). To compensate the loss of exactness multiple passes of inference can be conducted, i.e., conclusions obtained at earlier passes are used as data along with the old data at the next pass.

This approach to fuzzy knowledge representation and logical inference has been implemented in the fuzzy expert system shell EDIP for MS-Windows 3.x [12, 15].

4. Knowledge Discovery in Databases

Let us suppose that a database is represented by means of the matrix of conjunctions **K** which represents some distribution over the universe of discourse *O*. The problem is to express this distribution in the form of dependencies between attributes represented by disjunctions. Thus it is necessary to transform the matrix of conjunctions **K** representing data into the matrix of disjunctions **D** representing knowledge. Such a possibility is provided by a generalized covering method. This method just like the above described fuzzy resolution operation generalizes the conventional covering method in two directions: (i) it allows using multi-valued variables instead of only boolean ones, and (ii) it can be used with fuzzy matrices representing fuzzy distributions.

General covering method is intended for finding prime disjunctions given a matrix of conjunctions (DNF). Prime disjunctions are built gradually by means of searching the space of all possible disjunctions starting from the initial disjunction consisting of all 0 and then increasing its different components. As components increase they influence corresponding columns of the matrix and some its lines are *covered*. When all lines of the matrix are covered we say that the disjunction obtained is prime one (formally it has to be guarantied by the procedure). An order in which the procedure searches the space of disjunctions is determined by the order in which we increase the components. A line of the matrix is said to be covered if it is a consequence of the current disjunction being built.

An important property of this procedure is that it allows us to search only a part of the whole space of disjunctions which contains required dependencies between attributes. For example, if we are searching for dependencies between the first and the third attributes then we increase components in the corresponding sections only.

Let us suppose that we do not have an expert knowledge about stages of alcoholism but there is a database of case histories which can be transformed to the sectioned matrix (fuzzy DNF) consisting of 14 conjunctions

$$\mathbf{K} = \begin{array}{c} x_1 \quad\quad x_2 \quad\quad x_3 \\ \left|\begin{array}{lll} \{1, 0, 0\} & \{1, 0\} & \{1, 0, 0\} \\ \{0.2, 0, 0\} & \{0.2, 0\} & \{0, 0.2, 0\} \\ \{1, 0, 0\} & \{0, 1\} & \{1, 0, 0\} \\ \{0.3, 0, 0\} & \{0, 0.3\} & \{0, 0.3, 0\} \\ \{0, 1, 0\} & \{1, 0\} & \{0, 1, 0\} \\ \{0, 0.6, 0\} & \{0.6, 0\} & \{0, 0, 0.6\} \\ \{0, 0.3, 0\} & \{0, 0.3\} & \{0, 0.3, 0\} \\ \{0, 1, 0\} & \{0, 0.6\} & \{0, 0, 0.6\} \\ \{0, 0, 1\} & \{1, 0\} & \{1, 0, 0\} \\ \{0, 0, 1\} & \{1, 0\} & \{0, 1, 0\} \\ \{0, 0, 1\} & \{1, 0\} & \{0, 0, 1\} \\ \{0, 0, 1\} & \{0, 1\} & \{1, 0, 0\} \\ \{0, 0, 0.3\} & \{0, 0.3\} & \{0, 0.3, 0\} \\ \{0, 0, 1\} & \{0, 1\} & \{0, 0, 1\} \end{array}\right| \begin{array}{l} \mathbf{1} \\ \mathbf{2} \\ \mathbf{3} \\ \mathbf{4} \\ \mathbf{5} \\ \mathbf{6} \\ \mathbf{7} \\ \mathbf{8} \\ \mathbf{9} \\ \mathbf{10} \\ \mathbf{11} \\ \mathbf{12} \\ \mathbf{13} \\ \mathbf{14} \end{array} \end{array}$$

(This matrix is semantically equivalent to the above matrix of disjunctions **D**, i.e., they represent the same 3-dimensional fuzzy distribution.) It is required to discover a dependence between two attributes x_1 and x_3.

Disjunctions representing dependencies of two attributes contain components in the corresponding sections only (more exactly, they can contain components which are less than 1 also in other sections but for the sake of simplicity we will not touch this issue). Thus we have to gradually increase components in sections 1 and 3. It can be noticed that the third column (3-rd value of the 1-st attribute) contains the maximal number of non-0 components therefore it is natural to cover the maximum number of lines by increasing the corresponding component. Let us set 3-rd component of the disjunction to 1 and cover 6 last conjunctions **9**, **10**, **11**, **12**, **13**, **14** of the matrix which after removing them is equal to

$$\mathbf{K}_1 = \begin{array}{c} x_1 \quad\quad x_2 \quad\quad x_3 \\ \left|\begin{array}{lll} \{1, 0, 0\} & \{1, 0\} & \{1, 0, 0\} \\ \{0.2, 0, 0\} & \{0.2, 0\} & \{0, 0.2, 0\} \\ \{1, 0, 0\} & \{0, 1\} & \{1, 0, 0\} \\ \{0.3, 0, 0\} & \{0, 0.3\} & \{0, 0.3, 0\} \\ \{0, 1, 0\} & \{1, 0\} & \{0, 1, 0\} \\ \{0, 0.6, 0\} & \{0.6, 0\} & \{0, 0, 0.6\} \\ \{0, 0.3, 0\} & \{0, 0.3\} & \{0, 0.3, 0\} \\ \{0, 1, 0\} & \{0, 0.6\} & \{0, 0, 0.6\} \end{array}\right| \begin{array}{l} \mathbf{1} \\ \mathbf{2} \\ \mathbf{3} \\ \mathbf{4} \\ \mathbf{5} \\ \mathbf{6} \\ \mathbf{7} \\ \mathbf{8} \end{array} \end{array}$$

Now we see that the first two columns both have 4 non-0 components. We can choose any of them (and we must do it in the case of searching for all dependencies) but let us choose the second column since it has greater sum of the components. Then the disjunction $\{0, 1, 1\}, \{0, 0\}, \{0, 0, 0\}$ covers 4 last lines **5**, **6**, **7**, **8** in the matrix $\mathbf{K}_1$ and we obtain

$$\mathbf{K}_2 = \begin{array}{c} \\ \end{array}\begin{array}{|lll|l} x_1 & x_2 & x_3 & \\ \{1, 0, 0\} & \{1, 0\} & \{1, 0, 0\} & \mathbf{1} \\ \{0.2, 0, 0\} & \{0.2, 0\} & \{0, 0.2, 0\} & \mathbf{2} \\ \{1, 0, 0\} & \{0, 1\} & \{1, 0, 0\} & \mathbf{3} \\ \{0.3, 0, 0\} & \{0, 0.3\} & \{0, 0.3, 0\} & \mathbf{4} \end{array}$$

Now we see that this matrix can be covered by the third section {1, 0.3, 0} (the component 1 covers disjunctions **1**, **3**, and the component 0.3 — **2**, **4**). Thus we obtain that the disjunction

$$\{0, 1, 1\}, \{0, 0\}, \{1, 0.3, 0\}$$

covers the whole matrix of conjunctions and as such it represents a fuzzy dependence between attributes x_1 and x_3. Notice that it is equal to the prime disjunction **11** from the matrix $\mathbf{D}_3$.

5. Conclusion

In the paper it has been described how new original results from the field of multi-dimensional fuzzy analysis can be applied to the problem of decision making and rule induction. In particular, a uniform and efficient method of sectioned matrices has been proposed for extensional and intensional representation of multi-dimensional fuzzy distributions. If knowledge is represented by means of such a matrix interpreted as fuzzy CNF and data is represented by means of a sectioned vector interpreted as conjunction, then logical inference consists in finding projection of the whole fuzzy distribution on the goal attribute. There has been described a procedure for solving this problem based on the operation of fuzzy resolution. If data is represented by a sectioned matrix interpreted as fuzzy DNF, then the rule induction consists in finding the most general disjunctions which cover the fuzzy distribution it represents. We described a generalized fuzzy covering method to find such rules. The main problem of multi-dimensional fuzzy analysis consists in computational difficulty. For two algorithms described in the paper we have proposed an approximate variant which can be applied to real tasks.

6. References

[1] Levchenko, V.I., 1990, Diagnostic system based on finite predicates, Preprint, Institute of Mathematics and CC AS Moldavia, Kishinev, Shtiintsa (in Russian).

[2] Zakrevsky, A.D., 1989, Logical inference in finite predicates, Preprint No.6, Institute of Technical Cybernetics, AS Belorussia, Minsk (in Russian).

[3] Savinov, A.A., 1991, Matrix representation of fuzzy knowledge in attribute models, Preprint, Institute of Mathematics and CC, AS Moldavia, Kishinev, Shtiintsa (in Russian).

[4] Levchenko, V.I., and Savinov, A.A., 1994, Diagnosis by fuzzy constraints in attribute model, 2nd Eur. Congr. on Intelligent Techniques and Soft Computing—

EUFIT'94, Aachen, Germany, September 20-23, 382-385.

[5] Levchenko, V.I., and Savinov, A.A., 1993, The matrix representation of fuzzy knowledge and its application to the expert systems design, Computer Sci. J. of Moldova **1**(1), 62-84.

[6] Levchenko, V.I., and Savinov, A.A., 1993, Matrix representation of fuzzy predicates and its application in expert systems, Izvestia RAN, Tehnicheskaia kibernetika No.5, 1993, 126-140 (in Russian).

[7] Zadeh, L.A., 1975, The concept of a linguistic variable and its application to approximate reasoning — Part I. — Information Sciences **8**, pp. 301-357.

[8] Zadeh, L.A., 1979, A theory of approximate reasoning, In: Machine Intelligence **9** (Hayes J.E., Michie D. and Mikulich L.I., Eds.). — New York: Elsevier, pp. 149-194.

[9] Savinov, A.A., 1993, Fuzzy propositional logic, Fuzzy Sets and Systems **60**(1), 9-17.

[10] Savinov, A.A., 1993, Fuzzy propositional logic for the knowledge representation, First European Congress on Fuzzy and Intelligent Technologies—EUFIT'93, Aachen, Germany, September 7-10.

[11] Savinov, A.A., 1996, Some properties of new resolution rule in the logic of possibility distributions, 4th European Congress on Intelligent Techniques and Soft Computing—EUFIT'96, Aachen, Germany, September 2-5, 178-182.

[12] Levchenko, V.I., and Savinov, A.A., 1992, The representation of fuzzy knowledge in the diagnostic expert system shell EDIP, Proc. 2nd Int. Conf. on Fuzzy Logic and Neural Networks—IIZUKA'92, Iizuka, Japan, July 17-22.

Online references

[13] Savinov, A.A., 1997, Fuzzy Multi-dimensional Analysis, online paper, http://www.geocities.com/ResearchTriangle/7220/fmda_0.html.

[14] Savinov, A.A., 1997, New Fuzzy Resolution Operation, online paper, http://www.geocities.com/ResearchTriangle/7220/fresol.html.

[15] Fuzzy expert system shell EDIP, http://www.geocities.com/ResearchTriangle/7220/edip.zip

Information-Theoretic Fuzzy Approach to Knowledge Discovery in Databases

Oded Maimon[1], Abe Kandel[2], Mark Last[3]

[1] *Department of Industrial Engineering, Tel-Aviv University, Tel-Aviv 69978, Israel, e-mail: maimon@eng.tau.ac.il*

[2] *Department of Computer Science and Engineering, University of South Florida, Tampa, Florida 33620, USA, e-mail: kandel@babbage.csee.usf.edu*

[3] *Department of Industrial Engineering, Tel-Aviv University, Tel-Aviv 69978, Israel, e-mail: last@eng.tau.ac.il*

Keywords: Knowledge discovery in databases, Data mining, Information theory, Fuzzy databases, Dimensionality reduction

Abstract

We suggest a novel, unified approach to automating the entire process of Knowledge Discovery in Databases (KDD). The approach builds upon Shannon's Information Theory, statistical estimation methods, and Fuzzy Logic. The KDD stages to be automated include: dimensionality reduction, discovering informative rules (patterns), predicting values of unknown attributes, and cleaning a dataset from lowly reliable data.

The relational database model is extended by partitioning the relation scheme into a subset of candidate input (predictive) attributes and a subset of target (classification) attributes. A multilevel information-theoretic connectionist network is constructed to evaluate the mutual information between input and target attributes. The optimal network structure is found by a stepwise gradient procedure. The network connection weights are used to extract informative rules and to calculate reliability degrees of target attributes.

The approach is applied to a real-world manufacturing database containing typical data on production batches at a semiconductor factory.

1. Introduction

Fayyad et al. [1] define the process of Knowledge Discovery in Databases (KDD) as "the nontrivial process of identifying valid, novel, potentially useful, and ultimately understandable patterns in data". The KDD process involves, usually, the following stages: data selection, pre-processing and transformation, data mining (which is the core of the KDD process), and interpretation of results.

Data selection is aimed at choosing the dataset to be analyzed. Due to storage and processing constraints, only a part of the real-world database is usually selected. The selection parameters may define both the records and the attributes of interest to a data miner. The existing methods of reducing the number of attributes (dimensionality reduction) are summarized in sub-section 1.1 below.

Data pre-processing includes data cleaning, handling missing and unknown values, data transformation, and other activities improving the quality of the dataset. Some data mining algorithms can be applied at this stage to the problem of discovering and removing (or correcting) "outliers" in data. As indicated by [2], there is a danger of removing valuable patterns together with really garbage ones. This means that the notion of garbage (erroneous, unreliable, etc.) data should be clearly defined as a part of the cleaning process, but such a common unified definition is still missing in KDD literature. Sub-section 1.2 presents current approaches to data quality and data cleaning. The fuzzy logic model of managing imperfect data is separately covered in sub-section 1.3.

Fayyad et al. [1] distinguish between two main types of *Data Mining*: *verification-oriented* (the system verifies user's hypothesis) and *discovery-oriented* (the system finds new rules and patterns autonomously). The discovery-oriented methods can be further partitioned into *descriptive* (e.g., visualization) and *predictive* (like regression and classification). The most common classification methods (aimed at assigning a record to one of a specified number of disjoint classes) are covered in sub-section 1.4 below.

Interpretation of results is another important aspect of the KDD process. The discovered patterns should be presented to a user in an understandable manner (e.g., as *if... then* ... rules). When data is subject to noise (which is the case in most manufacturing databases), a statistical measure of accuracy can be associated with each rule. Presenting numerous patterns, without any scoring, is very inefficient, and thus, some measures of importance, relevance, interestingness, etc. are required.

As shown below, the existing methods of dimensionality reduction, data cleaning and data mining can only be applied to individual stages of the KDD process. Most of these methods are also limited by restricting assumptions on the sample size, attribute type, data distribution, dependency relations, etc. The integration of different methods into a single discovery process is usually a costly manual procedure. A global automated approach, applicable to the entire KDD process and independent of the dataset nature, is still missing and will be presented here.

1.1 Dimensionality Reduction

Dimensionality (i.e., the number of database attributes or groups of attributes) constitutes a serious obstacle for efficiency of most data mining algorithms. Though the methods of *feature selection* are aimed at selecting a minimum subset of relevant attributes, they are not readily applicable to dimensionality reduction in a conventional database, since their assumptions are characteristic mostly to the pattern recognition problems. Thus, the Chi2 algorithm of [4] applies only to ordinal attributes with first-order dependency relations, Almuallim and Dietterich [5] work on the problem of finding the smallest set of Boolean, noise-free features, and Caruana and Freitag [6] assume that there is an unlimited source of data to test the generalization performance of the learning algorithm.

1.2 Data Quality and Data Cleaning

As indicated by Wang et al. [7] data reliability is one of *data quality dimensions.* Other data quality dimensions include [7-9]: accuracy, timeliness, relevance, completeness, consistency, precision, etc. Various definitions of these and other dimensions can be found in [8]. An attribute-based approach to data quality [8] is based on the entity-relationship (ER) model (described in [10]). According to [9], the reliability "indicates whether the data can be counted on to convey the right information". Unreliable (deficient) data represents an inconformity between the state of the information system and the state of the real-world system. The paper [9] follows a "Boolean" approach to data reliability: the information system states are assumed either correct, or incorrect. No "medium" degree of reliability is provided.

An information theoretic approach to data cleaning is presented in [2]. The paper assumes that erroneous ("garbage") data has a high information gain. This means that the most "surprising" patterns (having the lowest probability to be predicted correctly) are suspicious to be unreliable.

1.3 Fuzzy Databases

Kandel et al. [11] mention unreliable information as one of sources of data uncertainty, other sources including fuzziness of human concepts, incomplete data, contradicting sources of information, and partial matching between facts and events. According to [11], the main drawback of the probabilistic approaches to uncertainty (e.g., the Bayesian approach) is their limited ability to represent human reasoning, since humans are not Bayesian when reasoning under uncertainty.

Since, in a general case, data reliability is a *linguistic variable* (the data can be considered "very reliable", "not so reliable", "quite unreliable", etc.), the models of fuzzy databases seem to be helpful for treating reliability of database attributes. As indicated by [12], the main problem of fuzzy databases is to propagate the level of uncertainty associated with the data (reliability degree in our case) to the level of uncertainty associated with answers or conclusions based on the data. Klir and Yuan [13] suggest a powerful tool for analyzing and summarizing values of fuzzy attributes - *ordered weighted averaging operations (OWA).*

Zemankova and Kandel [12] and Kandel [11] propose a Fuzzy Relational Data-Base (FRDB) model which enables to evaluate fuzzy queries from relational databases.

The attribute values in the FRDB can represent membership or possibility distributions defined on the unit interval [0,1]. According to this model, a single value of a membership distribution can be used as a value of a fuzzy attribute.

The Fuzzy Data model developed by Takahashi [14] assumes that some nonkey attributes may have values defined by fuzzy predicates. All key attributes and some other attributes are assumed to have nonfuzzy values only. Any tuple in Takahashi data model has a *truth value* z defined over the unit interval [0, 1]. The value of z is interpreted as a degree to which the tuple is true, with two special cases: $z = 0$ when the tuple is completely false and $z=1$ when the tuple is completely true. A similar idea of associating a single truth value (a *weight)* with each tuple is described by Petri [15].

The concept of reliability degree associated with every column in a fuzzy spreadsheet table is used by Nakajima and Senoh [16]. According to their definition, the degree of reliability can take any continuous value between zero and one, but no explicit interpretation of this variable is provided.

1.4 Data Mining Methods

Many classification techniques of Data Mining are based on Neural Networks, which are remarkable for their learning efficiency, but strongly limited in their interpretation capabilities [17]. According to [18], the drawbacks of applying Neural Networks to Data Mining include: long learning time, difficulty in extracting classification rules from the network structure, and, also, difficulty in incorporating prior knowledge about the application domain in a neural network.

An approach to extracting association rules of the form "*if X, then Y*" was developed by Srikant and Agrawal [19]. An estimated conditional probability P (Y/X) is termed by Srikant and Agrawal a "rule confidence", while an estimated joint probability P (X, Y) is termed a "rule support". As opposed to the Neural Network methods, Srikant and Agrawal suggest an understandable interpretation of each extracted rule (in terms of its confidence and support), but they do not study the interaction between multiple (possibly redundant or contradicting) association rules, valid in the same record.

The conditional dependency and independence relations between multiple variables, or causes, are directly treated by the Bayesian learning methods, also known as Bayesian belief networks (see [20]). Each node in a Bayesian network stands for a variable (or event). The *parents* (immediate predecessors) of a node represent the causes of the appropriate event. According to [21], an event is "overdetermined", if it has more than one set of sufficient causes.

If the dependency relations between predictive attributes and the class attribute can be represented in a simple, disjunctive form, one can use a decision tree algorithm, like the ID3 algorithm of Quinlan [22]. The classification is performed separately for each leaf representing a conjunction of attribute values in a rule. The longer the path from the tree root to the leaf, the greater the number of rule attributes. There are many methods of pruning decision trees at the sub-tree level (e.g., see [23]). However, the pruning of one path does not necessarily mean that the same attribute is omitted at all other paths, thus decreasing the total number of input attributes.

The information theoretic connectionist learning methods [24-27] suggest a clear advantage for discovering information patterns in large sets of imperfect data, since they use meaningful network structure, based on the Information Theory. The main elements of a single-layer information-theoretic connectionist network (developed in [25] for a speech recognition application) include: a layer of input nodes representing *events*, an output layer representing *actions*, and an activation function associated with each output action. A connection weight is given an explicit meaning by defining it as a mutual information between a single event (word) and a particular action.

The speech recognition method of [25] has been enhanced by us for discovering informative rules of any order in a relational database. In [27] we use a single-layer connectionist network (assuming independence of all input attributes). Here we are extending the approach presented by us in [27] to the problem of finding the minimal number of network layers representing a minimum set of conditionally dependent input attributes.

1.5 Paper Organization

This paper is organized as follows. In section 2 we present an extended relational database model, an information-theoretic connectionist approach to knowledge discovery, and an algorithm for reducing the number of predicting attributes by simplifying the network structure. The extraction of informative rules from the information-theoretic network is also demonstrated.

A novel, fuzzy approach to calculating reliability degrees of target attributes is described in section 3. We also suggest some ways of improving the quality of data by using degrees of reliability as fuzzy attributes.

In section 4 we present the results of applying the proposed approach to knowledge discovery in a real-world manufacturing database.

Section 5 concludes the paper with summing-up the main contributions of the method and outlining some directions for the future research.

2. Information-Theoretic Connectionist Network

2.1 Extended Database Model

To build an information-theoretic connectionist network, we make the following partition of the relation scheme [10]:

A subset $C \subset R$ of *candidate input* attributes ($|C| \geq 1$). This is a subset of attributes that *can be* used to predict the values of *target* attributes (see next).

A subset $O \subset R$ of *target* ("classification") attributes ($|O| \geq 1$). We assume that an attribute cannot be both a candidate input and a target ($C \cap O = \varnothing$). However, some attributes (e.g., the key ones) may be out of both subsets ($C \cup O \subseteq R$).

I_m -a subset of m input attributes selected by the dimensionality reduction procedure (see sub-section 2.3 below) from a subset of *candidate input* attributes ($I_m \subseteq C$).

2.2 Connectionist Network Structure

Without loss of generality, we consider a multi-layer network of a single target attribute. The connectionist network has the following components:

m - total number of hidden layers in a network. Each hidden layer is associated with an input attribute (and vice versa) by representing the interaction of that attribute and the input attributes of the previous layers.

z – a hidden node. Each hidden node at a hidden layer No. *l* represents a conjunction of values of input attributes associated with the first *l* hidden layers.

L_l - a subset of hidden nodes at a hidden layer No. *l*

K – a subset of distinct target nodes V_j for each value j in the domain of the target attribute.

w_z^j- a connection weight between a hidden node z and a target node V_j. Only the weights exiting from *unsplitted* nodes (see next-subsection) are used for predicting the values of the target attribute

The learning algorithm for building a multi-layer network having a minimum number of layers is presented in the next sub-section.

2.3 Dimensionality Reduction Procedure

The dimensionality reduction procedure is aimed at finding a minimum subset of input attributes explaining the target attribute. The procedure includes the following steps:

Step1 - Given a relation scheme, define every attribute as a *candidate input* attribute, a *target* attribute, or neither.

Step 2 - Enter a minimum confidence level for splitting a network node.

Step 3 - Initialize the information-theoretic network (one root node associated with all tuples, no input attributes, one target layer for values of the target attribute).

Step 4 - Repeat for the maximum number of hidden layers (total number of candidate input attributes):

Step 4.1 - Repeat for every candidate input attribute i':

Step 4.1.1 - Calculate the conditional mutual information [28] of the candidate input attribute i' and the target attribute i, given the final layer of hidden nodes (MI ($A_{i'}$;

A_i / I_m)) . A node is splitted if it provides a significant decrease in the conditional entropy of a target attribute (see [29]).

Step 4.1.2 – Go to next candidate input attribute.

Step 4.2 - Find a candidate input attribute maximizing the conditional mutual information ("the best candidate attribute").

Step 4.3 - If the maximum conditional mutual information is greater than zero, make the best candidate attribute an input attribute and define a new final layer of hidden nodes, else **stop** the procedure.

Step 4.4 - Add the conditional mutual information of the best candidate attribute to the total mutual information of the target attribute and the input attributes. Go to next hidden layer.

Step 5 – **End** the procedure. Output the list of input attributes, total mutual information, and the list of hidden nodes.

2.4 Extracting Informative Rules

Disjunctive informative rules are extracted from the information theoretic connection weights linking the unsplitted and the final layer nodes to the target layer nodes. Each connection weight is calculated by:

$$w_z^j = \mathrm{P}(\mathrm{V_j}; z) \bullet \log \frac{P(V_j / z)}{P(V_j)}$$

Where

$P(V_j;z)$ - an estimated joint probability of the target value V_j and the node z.

P (V_j/ z) - an estimated conditional (*a posteriori*) probability the target value V_j, given the node z.

$P(V_j)$ - an estimated unconditional (*a priori*) probability of the target value V_j.

The most informative rules are found by sorting the information-theoretic connection weights (w_z^j) in descending order. According to the well-known Pareto principle (which is confirmed experimentally in section 4 below), a small number of informative rules is expected to explain a major part of the total mutual information. The *predicted* value of a target attribute in a tuple can be calculated by a simple *maximum a posteriori* (MAP) rule: select the value j maximizing the conditional probability, given the node (P(Vj / z)).

3. Fuzzy Approach to Data Reliability

3.1 Definition of Data Reliability

We propose the following definition of data reliability, based on the fuzzy logic approach:

Definition. *Degree of Reliability of an attribute A in a tuple k is defined on a unit interval [0,1] as the degree of certainty that the value of attribute A stored in a tuple k is correct from the user's point of view.*

This definition is consistent with the definition of fuzzy measures in Klir and Yuan [13], since a set of correct attribute values can be viewed as a "crisp" set, and we are concerned with the certainty that a particular attribute belongs to that set. The data reliability is defined here as a subjective, user-dependent measure. Two special cases are: degree of reliability = 0 (the data is clearly erroneous) and degree of reliability = 1 (the data is completely reliable, as assumed in most regular relational databases).

3.2 Calculating Degrees of Reliability

We compute the degree of reliability of the actual value of a target attribute No. i in tuple k by the following formula:

$$t_k[\mathrm{R}_i] = \frac{2}{1 + e^{\beta \bullet d_{ik}}}$$

Where

β - The exponent coefficient representing a subjective attitude of a particular user to the reliability of "unexpected" data. The default value of β is one. Higher values of β represent a "conservative" approach: any value different from the predicted one has almost zero reliability.

$d_{ik} = \max_j\{\log P(V_{ij}/z)\} - \log P(V_{ij^*}/z)$ - The distance between the predicted value and the actual value of a target attribute i in tuple k (bits). If both values are the same, the distance is zero and, consequently, the reliability is one.

j* - actual value of a target attribute i in tuple k.

4. Application Example

4.1 Database Description

The approach has been applied to a real-world manufacturing database containing typical data on Work-in-Process (WIP) at a microelectronics plant. The database is a part of the plant ERP (Enterprise Resource Planning) system. A simplified version of the same database is used by us in [27]. The dataset contains 1,042 tuples, each tuple referring to a production batch, identified by a key attribute "**Batch ID**". The batch size is given by the attribute **NW** (Number of Wafers).

Manufacturing process of a batch (often called "routing") includes a number of stages (operations). The main purpose of WIP database is to provide the Production Control department with the exact information on location of different batches. When an operation is completed, it is reported to the system and a batch is transferred automatically to the next operation on its routing list (assuming a transfer batch quantity to be equal to a process batch quantity).

The attribute "**Time**" (flow time) calculates automatically the number of days passed since a batch entered the current operation (completed the preceding stage). Batches with higher **Priority** are supposed to spend fewer days at a stage than batches with lower priority. Experimental batches (indicated by the attribute **Experimental**) are expected to stay more time at the same stage than the regular batches. The number of days at a stage may depend also on **Product Type**, **Chip Size** and **Status**.

Discovering attributes and rules predicting the attribute "Time" may help the production control manager to shorten the overall flow time of manufacturing batches by finding the main factors increasing and decreasing this parameter. The attribute "Time" may also be incorrect in some tuples, since some operations are not reported properly to the system. This causes the number of days at stage to be higher or lower than the real one. Thus, a prediction model may also serve as a tool of improving the quality of work-in-process data.

During the pre-processing stage, the qualitative attributes ("Experimental", "Product Type", "Chip Size", "Priority", and "Status") have been categorized to numeric values and the quantitative attributes ("Number of Wafers" and "Time") have been discretized according to pre-determined intervals. Meaningful missing values of input attributes have been encoded to special values.

4.2 Dimensionality Reduction

Only four candidate input attributes (out of six) have been selected as input attributes by the dimensionality reduction procedure. The input attributes are (in descending order of informativeness): Status, Product Type, Priority, and Chip Size. Thus, the first important result is that the batch priority (imposed by the Production Control) affects the flow time much less than the batch status and the product type.

The operational conclusion may be to change the prioritization policy for the production floor.

The prediction accuracy of the resulting four-layer information-theoretic network is 0.511 (as compared to the *a priori* prediction accuracy of 0.372 only). The relatively low prediction accuracy of the final network can be explained by an extremely "noisy" nature of this real-world database. Thus, if a multiple linear regression is applied to the same data, the prediction accuracy obtained is only 0.458.

Eight out of 69 extracted rules were found responsible for more than 90% of the conditional mutual information between the target attribute and the set of significant input attributes. The most informative rules (having the highest connection weights) are particularly actionable. For example, one rule says that discrepancy batches wait more than a week for the same operation which may be a symptom of a prolonged treatment by the Material Requirement Board (MRB). The corrective action may be to define a shortened procedure for most common discrepancy problems. The algorithm has also revealed that certain product types have longer flow times, probably indicating of an engineering problem in their manufacturing process.

4.3 Data Cleaning

Since we have extended a regular relational database to a fuzzy relational database by adding fuzzy attributes expressing reliability degrees of nonfuzzy attributes, any operation of a fuzzy relational algebra (see [11,13-15]) can be applied to the extended database. The original database has been extended to a fuzzy relational database by adding a fuzzy attribute: reliability degree of the target attribute "Time". Seven lowly reliable tuples have been identified by applying a *selection* operator of the form "What are the tuples with the reliability degree lower than 5%?" An *aggregation* operator has been used to calculate the reliability of work-in-process data at each production stage. The weight of reliability degree of every batch has been taken as the ratio between the number of wafers in that batch and the total number of wafers at the stage.

5. Conclusions

In this paper, we have presented a novel, information-theoretic fuzzy approach to Knowledge Discovery in manufacturing databases. The original contributions of the proposed approach include:

1) The method is applicable to data sets of mixed nature (comprising continuous, ordinal, nominal, and binary-valued attributes).
2) The dimensionality reduction of the data set is attained by a greedy procedure, minimizing the number of predicting attributes.

3) The multi-layer search procedure explicitly reveals the significant dependency relations of any order between attributes, resulting in the removal of redundant and irrelevant features.

4) Discovered rules are ranked in terms of their contribution to the mutual information between the input attributes and the target attribute.

5) A novel, fuzzy approach to data reliability enables to combine the "objective" information of data probability with a "subjective" user attitude towards the reliability degree of unexpected (lowly probable) data.

The issues for the future research in this direction include: utilizing prior knowledge on attribute dependencies and probabilities, optimizing partitions of input and target attributes, integrating the approach with other data mining methods, and discovering rules in non-relational databases.

References

1. Fayyad U, Piatetsky-Shapiro G, and Smyth P 1996 From Data Mining to Knowledge Discovery: an Overview. In: Fayyad U, Piatetsky-Shapiro G, and Smyth P (eds) 1996 *Advances in Knowledge Discovery and Data Mining*, AAAI Press, Menlo Park, pp 1-30

2. Guyon I, Matic N, and Vapnik V 1996 Discovering Informative Patterns and Data Cleaning. In: Fayyad U, Piatetsky-Shapiro G, and Smyth P (eds) 1996 *Advances in Knowledge Discovery and Data Mining*, AAAI Press, Menlo Park, pp 181-203

3. Elder IV J F, and Pregibon D A 1996 Statistical Perspective on Knowledge Discovery in Databases. In: Fayyad U, Piatetsky-Shapiro G, and Smyth P (eds) 1996 *Advances in Knowledge Discovery and Data Mining*, AAAI Press, Menlo Park, CA, pp 83-113

4. Liu H, and Sutiono R 1997 Feature Selection via Discretization. *IEEE Trans Knowledge and Data Eng* 9, 4: 642-645

5. Almuallim H, and Dietterich T G 1992 Efficient Algorithms for Identifying Relevant Features. In: *Proc. of the 9th Canadian Conf. on AI*, Vancouver, Canada, pp 38-45

6. Caruana R, and Freitag D 1994 Greedy Attribute Selection. In: *Proc. of the 11th Conf. on Machine Learning*, New Brunswick, NJ, pp 28-36

7. Wang R Y, Reddy M P, and Kon H B 1995 Toward Quality Data: an Attribute-Based Approach. *Decision Support Sys* 13: 349-372

8. Wang R Y, Storey V C, and Firth C P 1995 A Framework for Analysis of Data Quality Research. *IEEE Trans Knowledge and Data Eng* 7, 4: 623-639

9. Wand Y, and Wang R Y 1996 Anchoring Data Quality Dimensions in Ontological Foundations. *Comm ACM* 39, 11: 86-95

10. Korth H F, and Silberschatz A 1991 *Database System Concepts*. McGraw-Hill, New York

11. Kandel A, Pacheco R, Martins A, and Khator S 1996, The Foundations of Rule-Based Computations in Fuzzy Models. In: W. Pedrycz (ed) 1996 *Fuzzy Modelling, Paradigms*

and Practice, Kluwer, Boston, MA, pp 231-263

12. Zemankova-Leech M, and Kandel A 1984 *Fuzzy Relational Databases - a Key to Expert Systems*. Verlag TUV, Koln
13. Klir G J, and Yuan B 1995 *Fuzzy Sets and Fuzzy Logic: Theory and Applications*. Prentice-Hall Inc, Upper Saddle River, NJ
14. Takahashi Y 1993 Fuzzy Database Query Languages and Their Relational Completeness Theorem. *IEEE Trans Knowledge and Data Eng* 5, 1: 122-125
15. Petry F E 1996 *Fuzzy Databases, Principles and Applications*. Kluwer, Boston, MA
16. Nakajima H, and Senoh Y 1996 A Spreadsheet-Based Fuzzy Retrieval System. *Int J Intelligent Sys* 11: 661-670
17. Mitchell T M 1997 *Machine Learning*. McGraw-Hill, New York
18. Lu H, Setiono R, and Liu H 1996 Effective Data Mining Using Neural Networks. *IEEE Trans Knowledge and Data Eng* 8, 6: 957-961
19. Srikant R, and Agrawal R 1998 Mining Quantitative Association Rules in Large Relational Tables. To appear in: *IEEE Trans Knowledge and Data Eng*
20. Pearl J 1988 *Probabilistic Reasoning in Intelligent Systems: Networks of Plausible Inference*. Morgan-Kaufmann, San Mateo
21. Spirtes P, Glymour C, and Scheines R 1993 *Causation, Prediction, and Search*. Springer Verlag, New York
22. Quinlan J R 1986 Induction of Decision Trees. *Machine Learning* 1: 81-106
23. Quinlan J R 1987 Simplifying Decision Trees. *Int J Man-Machine Stud* 27: 221-234
24. Goodman R M, Miller J W, and Smyth P 1988 An Information Theoretic Approach to Rule-Based Connectionist Expert Systems. In: *Proc. of Neural Information Processing Systems*, pp 256-263
25. Gorin A L, Levinson S E, Gertner A N, and Goldman E 1991 Adaptive Acquisition of Language. *Computer Speech and Language* 5, 2: 101-132
26. Gorin A L, Levinson S E, and Sankar A 1994 An Experiment in Spoken Language Acquisition. *IEEE Trans Speech and Audio Processing* 2, 1: 224-239
27. Maimon O, Kandel A, and Last M 1998 Fuzzy Approach to Data Reliability. To appear in: Pons O, Kacprzyk J, and Vila A (eds) *Knowledge Management in Fuzzy Databases*, Springer-Verlag.
28. Cover T M 1991 *Elements of Information Theory*. Wiley, New York.
29. Attneave F 1959 *Applications of Information Theory to Psychology*. Holt, Rinehart and Winston, New York

Intelligent Electronic Catalogs for Sales Support

Introducing Case-Based Reasoning Techniques to On-Line Product Selection Applications

Ivo Vollrath, Wolfgang Wilke, and Ralph Bergmann

Department of Computer Science
University of Kaiserslautern, PO-Box 3049
67653 Kaiserslautern, Germany
http://wwwagr.informatik.uni-kl.de/

Keywords: Electronic Catalogs, Sales Support, Case-Based Reasoning, Electronic Engineering, Databases

Abstract

The number of electronic catalogs has grown rapidly during the past few years. Most of these catalogs use standard databases for storing and retrieving product information. Using ordinary databases for product catalogs, however, has the major drawback that it is often very difficult to find the products desired: very often, the database does not return a matching product at all or it returns many products that have to be examined manually. To overcome this problem, we propose the use of Case-Based Reasoning (CBR) techniques as an approach to requirement-oriented retrieval of products. CBR incorporates product knowledge into the database by means of a similarity measure. *Recently, a number of commercial electronic catalogs based on CBR have been realized. We take a closer look at Analog Devices' on-line catalog of operational amplifiers, which helps an engineer to find a suitable amplifier for her specific requirements.*

1 Introduction

The number of electronic catalogs – especially on CD-ROMs and on the Internet – has grown rapidly during the past few years. In certain areas they even have become as important as their printed counterparts. This is especially true for electrical engineering applications where many vendors offer product information over the Internet and most of their customers use this service frequently. Most of these catalogs employ standard database approaches for storage and retrieval of product information. While today, databases are well understood and many off-the-shelf solutions exist, we claim that they can only be used efficiently by advanced users – users that are already very familiar with the contents of the database. This is especially a problem for electronic catalogs meant to be used directly by end consumers or engineers who, in many cases, are not experts in the family of products they are looking for, or at least do not have a good overview over all the products and their specialties inside the catalog. Imagine an electronics engineer looking for a device to integrate into her new circuit that pushes the limits of the most advanced devices from her catalog. How can this person select the device *best fitting* her needs if she is no expert in the area of devices she is looking for? A standard database offers no support for her.

In this paper we give examples of how difficult it is often to find desired product information in electronic catalogs. We introduce the basic idea of Case-Based Reasoning (CBR) and discuss the possibilities that this technique offers to overcome many of these problems. A number of commercial applications using part of the proposed techniques recently have been realized[1]. Information brokerage in high precision manufacturing and electronic parts catalogs are covered by these applications. We provide a case study of a commercial application. Analog Devices' on-line catalog of operational amplifiers is presented in detail and we close with a short discussion.

2 The Problem with Ordinary Databases

So why should a customer searching for product information on the web not be satisfied with a huge database of products and a database search engine on top of it? The most important answer to this question is: "Because this system entirely fails to provide *sales support.*" We illustrate this problem with the following scenario: A customer enters a shop that sells electronic parts. Imagine the shop assistant acts like a database search engine:

Customer: Hello, I need this transistor 'POC-4711'.
Assistant: Sorry, we currently do not have this transistor in stock. *(No further explanation, no alternatives.)*
C: Er, well, do you have the military version 'POC-4711-m', then?
A: Sorry, we do not have this one in stock either. *(The Customer probably leaves the shop now, but let's see what happens if she doesn't.)*
C: Well, do you have any transistor coming close to the POC-4711's amplification?
A: Can you please exactly state what you mean by saying 'close'? Can you specify a range?
C: (Sigh!), let's say 30 to 35.
A: Sorry, we do not have any ...
C: *(Leaving frustrated, looking for another shop.)*

Now let's see how a real shop assistant would have managed the situation:

C: Hello, I need this transistor 'POC-4711'.
A: Sorry, the 'POC-4711' is out. But I can offer you the 'BAC-42'. It behaves quite similar.
C: What is it's amplification?

[1] At http://wwwagr.informatik.uni-kl.de/~lsa/CBR/wwwcbrindex.html you find a collection of links to CBR based online product catalogs.

A: 37.
C: *(Will probably buy the 'BAC-42', but let's see what happens if she doesn't:)* Hm, I don't know... I might get into trouble with that.
A: Okay, why don't you connect two 'YH-12'? They will do the job.
C: *(Satisfied, buys two YH's.)*

Finding the desired information or product in an online catalog can be rather frustrating given the standard database implementations that are commonly used today. With standard databases it is very difficult for the occasional user to "balance" a database query in order to not produce too many results – or just no answer at all. A customer using some kind of search facility to find a product in an online catalog must fulfill certain requirements in order to have a chance to get usable results: If she knows the product's part number in advance she will certainly get the desired product information. If the customer is not looking for a specific part number but for some product that *has certain qualities,* she is in a much more difficult situation. In that case she must be very familiar with the contents of the database or otherwise she will get either no answer to her query or a whole bunch of answers that might not even be sorted in any usable way and therefore do not help at all.

By incorporating product knowledge into the database CBR is much more flexible and powerful than by just offering menu driven product selection or a searchable database. In this paper we present a scenario of how *similarity based retrieval* can partly emulate a shop assistant and offer real *sales support* to the end consumer. A CBR system's product knowledge enables it to judge which products optimally fit the customer's needs. CBR systems with configuration capabilities can even *adapt* a product to the user's demands which otherwise would not satisfy her requirements.

The fact that today's standard online catalogs cannot offer intelligent sales support requires of the customer that she is a semi-expert in the field of products she is interested in, in order to solve this problem. It is very important to put the knowledge of the shop assistant into the retrieval component of the online catalog. The system must have enough *domain knowledge* about the products to be able to aid the customer in her search. This would not only satisfy the customer, but it would also help the manufacturer or broker in selling her products. In Sect. 4 we will see how Case-Based Reasoning can help to overcome these deficiencies.

3 Extended Search Approaches

Currently there exist a number of systems that try to help the user searching for certain information or a certain product on the web in an intelligent way. Well known types of systems are meta search engines as well as search agents.

But these systems do not offer the kind of sales support this article is focused on. This is because they have no knowledge to differentiate between *similar* products of the same type and what their difference of value is to the user. Especially, query routing systems mainly rely on the intelligence of the engines the query is routed to and have only little additional knowledge of their own.

Often a customer needs support in selecting a product that best meets her requirements from a catalog of many similar products. This kind of sales support (or *decision support*) requires knowledge on a much lower level of abstraction than search agents can provide. A conventional printed product catalog offers only little sales support to the customer but sometimes there is extensive information about how one can find the best product for her needs. If you ever tried to select a mountain bike from a catalog you know what we are talking about: *How do I determine the right frame size for me? What kind of wheels do I need? What are break boosters good for? Do I need them?* etc. It usually takes a lot of time to read all the additional information and to decide for a certain product. Interactive technologies offer much more possibilities to overcome these problems and prepare for a much better substitute for a shop assistant than printed media do. With a good interactive catalog with sales support capabilities the customer should not need to become an expert in the type of product she is looking for.

In general, there are two opposing ways to design interfaces for such catalogs. The *product-based* approach offers support for guided navigation through the product space. This can be compared to finding a certain product by searching through a highly structured table of contents of a printed catalog. With this approach the customer step-by-step is lead through the decision process. To stay with our mountain bike example, she might first be asked for her size and weight, then for the intended use of the bicycle (off-road, city, etc.) and so on, until the system finally suggests a small number of bikes suitable for her. In more complex applications the fixed order of decisions can become a major problem. Questions the user cannot answer at a given time are also difficult to handle. After all, it is a common experience of people doing after sales support using such a system that they find it very tiring having to go through all the presented questions and selections.

The *requirements-based* approach allows the user to specify all or part of her requirements by using some (more or less pre-structured) query form. The simplest facet of this kind of systems are searchable databases where the user specifies the demanded attributes of a product. This allows for very flexible queries but as we have seen before it is only useful for expert users who are familiar with the contents of the database.

Case-Based Reasoning offers solutions for requirement-based query interfaces that preserve the flexibility of searchable databases but integrate the advantages of product-based approaches and make them useful for both first-time users and experts. By incorporating product knowledge into the database it is much more flexible and powerful than just offering *both* menu driven product selection *and* a searchable database.

4 The Concept of Case-Based Reasoning

The basic idea of Case-Based Reasoning (a good overview is given in [2]) is to solve new problems by comparing them to old problems that already have been solved in the past (Fig. 1). The key assumption is that if two problems look *similar,* then the solutions to these problems are often similar as well. However, CBR systems must know how to decide whether two cases are similar or not.

Old problems and their solutions are stored in a database of *cases,* called the *case base.* Often the cases are stored as collections of *attribute-value pairs* but sometimes it is useful to explicitly represent the hierarchical structure of the cases by describing them as structured objects, using inheritance, object decomposition and possibly other relations between the object's parts.

When a new problem has to be solved, the CBR system searches for the most similar old problem. The solution of this old problem may optionally be *adapted* to better meet the requirements of the new problem. CBR systems have been successfully used for various applications such as fault diagnosis of complex machines, classification, decision support, support for planing, configuration and design tasks, legal reasoning and various other application areas [3]. Another commercially interesting field where CBR is widely used and has proved its qualities are help desk applications. For further links to information on CBR applications and its theory see the CBR homepage at http://wwwagr.informatik.uni-kl.de/~lsa/CBR/.

In our application scenario the cases are descriptions of the products to be sold. The problem description is a specification of a single product and possibly demands that are satisfied by it. The solution to the problem is an unambiguous reference to the product. For configurable products such as workstations, automobiles, complex machines, etc. the solution is not only the part number, but possibly the entire configuration. When a customer enters a query (perhaps into a query form), this query is regarded as a new problem and the CBR system tries to solve this problem by comparing it to the cases in the case base.

Imagine an online catalog of holiday accommodations: When someone wants to spend her holidays on the beautiful island Pariku in a 3-star hotel, her preferences are interpreted as a problem description. The CBR system will then find one or more similar cases which best match the query. There might be no 3-star hotel on Pariku, so the system might suggest a 2-star hotel on Pariku and alternatively a 3-star hotel on the neighbor island Pung-Chitu.

We described the important roles of the similarity measure and the retrieval and adaptation components of a CBR system. We will now take a closer look at these aspects.

4.1 Similarity Measures

A frequently asked question is "What makes Case-Based Reasoning different from ordinary database systems?". Perhaps the most important answer to this question is: "CBR

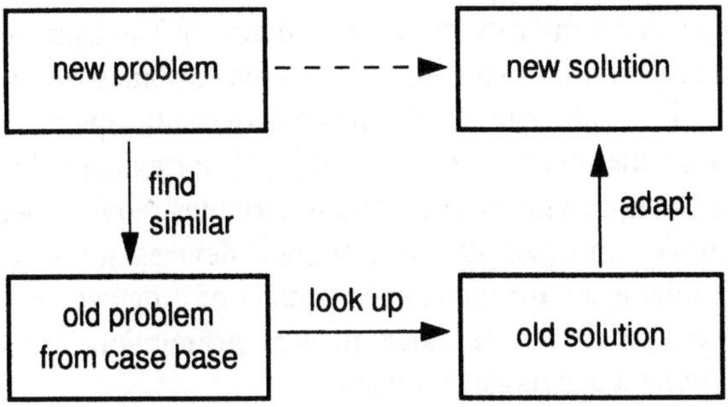

Fig. 1. The idea of CBR is to solve new problems by reusing the solutions of old ones.

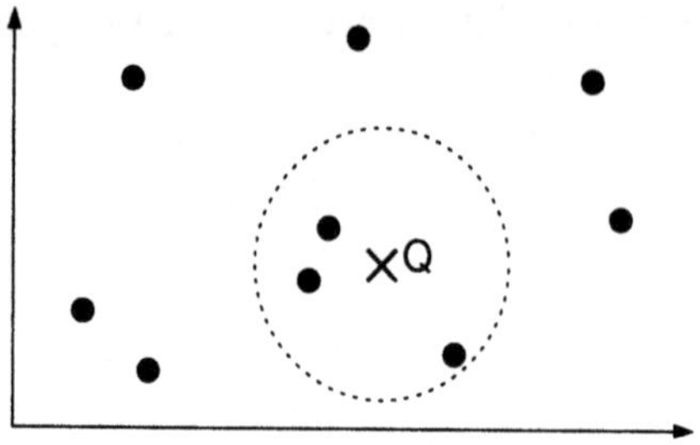

Fig. 2. Usage of similarity in CBR: The cases are distributed over an n-dimensional problem space. The circle around the query Q indicates a similarity threshold for the cases to be retrieved.

systems have additional domain knowledge built into them." The main part of this additional knowledge is implemented in a *similarity measure* which is a function that assesses the similarity of a given query to the cases in the case-base. The similarity values are ordinal values that are often normalized to the interval $[0, 1]$. A value of 0 means "does not satisfy the query at all" and a value of 1 says "that's exactly what you asked for".

To understand how such a similarity measure is used to find the best solutions for a given problem consider the cases being represented as a fixed length vector of n attributes. These attributes may have numerical values or their values can be arranged to reflect some kind of order. Part of the description of an electronic device, for example, might be:

```
case1 = ( 5,       supply voltage [Volt]
          0.25,    max. power consumption [Watt]
          32,      number of pins
          ... )
```

The problem space can now be seen as an n-dimensional space where similar problems are placed 'close' together (Fig. 2). The term 'close' is defined by the similarity measure. When a new problem is presented to the system, the similarities to all the cases in the case base are calculated (the number of visited cases may possibly be optimized by suitable indexing techniques). The cases within a similarity range of a certain threshold (or sometimes a fixed number of similar cases) are then presented to the user.

But how can such a similarity measure be defined? The usual approach is to start with individual (local) similarity measures for all the attributes. For numerical attributes a natural approach is to calculate the difference between the query value and the case's value and normalize the result to the interval $[0, 1]$, for example by applying the formula $1 - \frac{d}{d+1}$ to the difference d. But certain attributes require specialized functions. Imagine, for example, our description of electronic devices. It is often (but not always) the case that the tolerances for the supply voltage of a device are much tighter than the user's tolerances for the device's max. power consumption. The similarity function must reflect this domain dependent knowledge.

In real-world applications not all attribute values are numerical values. A case description very often contains boolean or symbolic attributes. In such cases, the local similarity measure can be described by a table which defines the similarities for all possible pairs of attribute values (Fig. 3). Even more complex attribute types such as taxonomy types or complex objects sometimes are required.

After the local similarity functions have been defined, the global similarity of two cases must be derived from the local similarities (and possibly from additional information like constellations of certain attributes). The usual way to do this is to apply a weighted sum to all the local similarities. Consider a query q is described by the attributes $q_1, \ldots, q_n$ and a case c is described by $c_1, \ldots, c_n$ where attribute values with corresponding indices belong to the same attribute. The similarity σ between q and c can be calculated from the local similarities σ_i as follows:

$$\sigma(q, c) = \sum_{i=1}^{n} w_i \sigma_i(q_i, c_i) \qquad \text{where} \quad \sum_{i=1}^{n} w_i = 1 \quad \text{and} \quad w_i \geq 0 \text{ for all } i$$

The weights w_i have to be assessed by an expert but sometimes it is useful to give the user the possibility to manipulate them in order to express individual preferences. This general method of defining a similarity measure as a weighted sum can be modified and refined in many ways. In general the computation of the similarity measure depends very much on the data model and on the type of application.

Finding a suitable similarity measure often is the most critical part during the design and implementation of a CBR system. Once a good similarity measure has been found and implemented, maintaining a CBR system is rather easy. The similarity measure typically doesn't need to be changed for a long time compared to the knowledge bases of other types of expert systems. Often the only part that undergoes significant changes over time is the case base, whereas the domain-model and similarity measure rarely need to be changed.

4.2 Retrieval Techniques

We said before that, in order to find the cases that best satisfy the query, the similarities of the query to all cases from the case base have to be calculated. But it is not always necessary to do a linear search which, of course, would be a time-consuming process. To significantly reduce the retrieval times of CBR-Systems a number of indexing techniques have been developed. Generally it is difficult to adapt standard database indexing algorithms to similarity based systems. This is because these database approaches do not take into account soft similarity values, but only support hard selection criteria.

One example of efficient indexing structures are *k-d trees* [4] which are modified versions of decision trees that allow to significantly reduce the number of visited cases while soft similarity values are still taken into account. Another approach to fast case retrieval are *case retrieval nets* [5]. The idea is to represent the cases and their attributes as a network of interconnected *information entities.* Starting with the query's information entities activated, a spreading activation algorithm is used to retrieve the best matching cases. Case retrieval nets are successfully used in some existing on-line applications for travel agencies.

Both of these indexing approaches significantly speed up the retrieval times by *compiling* the similarity measures into indexing structures. They become inefficient if the case-base needs to be updated very frequently, i. e., if the number of updates is in the order of the number of read accesses.

4.3 Adaptation

Another point that makes CBR different from standard database approaches is its ability to not only retrieve existing cases, but also *adapt* their (old) solution (here: the product that is to be sold) to the newly presented problem and thus create new solutions that were not represented in the case base [6]. This requires, of course, that the solution is configurable. Let's look at a simple example: Imagine a car dealer who runs a CBR-based catalog for her stock of used cars. A customer who is looking for a used car enters her preferences into the retrieval system. One feature that the car should have is a sliding roof but the dealer's stock does not contain any car with a sliding roof. A CBR system with adaptation capabilities will know that a sliding roof can be built into a car later. So the system will automatically adapt the descriptions of the best matching cars to contain an entry saying that there *is* a sliding roof. The price for the sliding roof and the wage for its installation is automatically added to the 'price'-attribute. This happens totally transparent to the user.

The knowledge needed to perform such kinds of adaptation must be represented in some suitable form. One possible representation is a set of *rules* that perform certain actions given that the required preconditions for these actions are valid. A rule for our sliding roof example could be written as follows:

```
if  query.slidingroof == true and
    case.slidingroof == false and
    case.cabriolet == false
do: case.slidingroof := true;
    case.price := case.price + $200
```

Depending on the application domain, the adaptation process can be more or less complicated. To our knowledge there is currently no commercial CBR-based product catalog that supports adaptation. The reason for this probably is that new technologies are always first applied to simple problems and only later to more difficult tasks. Also, there are many products that cannot be adapted in any way because they have no modifiable structure. On the other hand, adaptation is indispensable if the system should be able to perform some kind of configuration task, like interactive configuration of personal computers. The WEBSELL project described in Sect. 6 works on adaptation techniques for CBR-based on-line applications. [7] gives an overview of configuration technologies based on CBR.

4.4 Advantages

Let us summarize some of the advantages of CBR-based product databases over standard database techniques:

- Near misses are avoided: If the customer asks for features that cannot be fully satisfied by products from the database, she is still offered something *close* to her request.

- The fact that there is an answer to almost any query greatly reduces frustration on the customer's side and helps the vendor in selling *nearly perfect* solutions.
- The solutions are ranked by their similarity to the query. This is essential when the system finds dozens or hundreds of solutions. In opposition to other retrieval systems (such as some Internet search engines) this ranking is not only based on some statistical frequencies but it is based on *knowledge* that has been acquired from real experts.
- The adaptation process can even *create* solutions not contained in the original database. Thus CBR offers support for interactive or automatic product configuration.

5 Existing CBR Applications

The number of online catalogs with built-in CBR technology is yet rather small. Nevertheless, the quality of their search results are very promising. We encourage the reader to check out the online versions of the products described at:
http://wwwagr.informatik.uni-kl.de/~lsa/CBR/wwwcbrindex.html

5.1 A Case Study: Analog Devices

We will now have a close look at Analog Devices' online catalog of operational amplifiers. Some background information: Analog Devices is one of the major manufacturers and sellers of electronic devices in the U.S. According to the company's annual report [8], the company has direct sales offices in 17 countries. Many of Analog Devices' customers are small electronics firms or electronics departments of larger firms. Analog Devices currently employs a number of engineers to provide central applications support for their customers. There is also an online catalog with certain standard search capabilities and selection trees. Many people calling the hot-line do so because they could not find a suitable match for their problem in the printed or online product catalog. Roughly half of the calls to the hot-line support relate to product selection. It is obvious that the number of calls to the selection support service could possibly reduced significantly, if only the online catalog was able to provide intelligent sales support. So it has been decided to offer CBR-based search facilities in addition to the standard database interface. In a first stage the application domain for this system was chosen to be Analog Devices' product family of operational amplifiers, but the system is currently being extended to cover at least three more product families. Their product catalog contains about 130 operational amplifiers suitable for a wide range of applications. In many cases the new CBR system is able to support the customers in selecting a device that satisfies their requirements. This reduces the number of calls to Analog Devices' hotline and enables some engineers to care about more advanced support problems that cannot be solved automatically.

Case Representation

The data sheet for an operational amplifier consists of about 40 parameters which can be logically structured into categories of parameters, such as electrical input and output specifications, functionality, dimensions, etc. Most of the parameter values are high exponent real numbers, but there are also symbolic and string parameters. No object decomposition or hierarchical representation has been used for this application. The retrieval interface provides hyper-links to explanations of each of the 40 attributes.

The Similarity Measure

To assess the similarity of two operational amplifiers, the similarities for all corresponding parameter values are calculated by applying local similarity functions to each pair of corresponding parameters. These local parameter similarities are then used to calculate an overall similarity value for the two devices. The overall similarity is computed as the weighted average of the individual similarities. The original weight factors have been suggested by experts in operational amplifiers but they can be influenced by the customer according to her priorities. The local similarities for discrete and continuous values are calculated in different ways.

Discrete similarity measures are defined by a table which explicitly lists the similarity values for all possible attribute combinations (Fig. 3). Note that the table is asymmetric: it makes a difference whether the query's attribute value is x and the case's value is y or vice versa. If, for example, the user asks for a device with industrial standard temperature specifications, she will be satisfied by a part that fulfills military requirements (provided that no other attributes, such as the price, stand against it). On the other hand, the request for military standards can not fully be satisfied by a part that only has industrial specifications. This situation is reflected by the asymmetry of the similarity table.

This principle also holds for most of the *continuous* attributes of an operational amplifier. The only difference is that their similarity measures cannot be as easily represented in a table, but must be formulated as a function. An example of one such similarity function is shown in Fig. 4. This function assumes that there is a minimum and a maximum attribute value. If the case's attribute value is less than or equal to the query attribute's value their similarity is 1. Otherwise it is a value somewhere between 0

query \ case	com.	industr.	military	space
com.	1	1	1	1
industr.	0.6	1	1	1
military	0.4	0.6	1	1
space	0.1	0.4	0.6	1

Fig. 3. An example similarity function for discrete attribute values taken from Analog Devices' operational amplifier application. This is the (simplified) similarity function for the parameter "maximum temperature range" of an operational amplifier. The symbols 'commercial', 'industry', 'military' and 'space' describe temperature range standards of electronic devices. If, for example, the query specifies 'industrial' standards, the CBR system regards 'military' and 'space' to fulfill the requirements, but 'commercial' has a similarity of only 0.6.

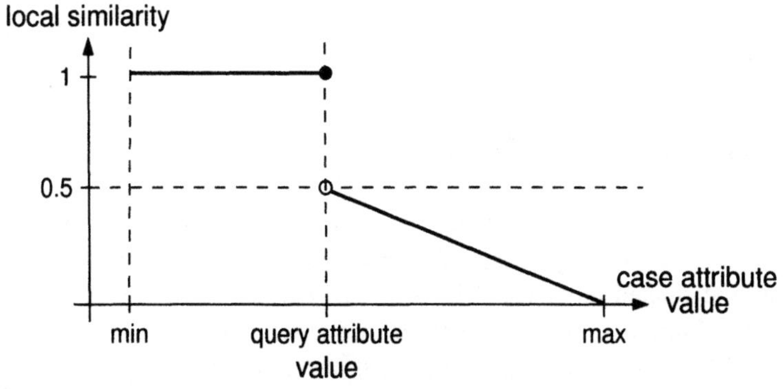

Fig. 4. Another example similarity function but for continuous attribute values. The similarity function for the parameter "current noise density" of an operational amplifier is shown. The corresponding attribute of an old case has similarity 1 if its value is less than or equal to the query's value. Otherwise the similarity is significantly smaller.

and 0.5. For certain other attributes the function must be reversed to return 1 for values greater than or equal to the query's value and 0 to 0.5 for smaller values.

This application does not use adaptation. This is because operational amplifiers are unchangeable parts that cannot be reconfigured to the customer's individual demands.

The Benefits: Sales Support

A typical sales support session goes as follows: The customer enters the parameter values she needs into the query form. The system will then retrieve the ten best matches to the request. If the results do not exactly fit the customer's needs, she will usually increase the priorities of the parameters that are most important for her. Again, the system displays the ten best matches to the refined query. If the results still do not satisfy the customer, she might fill more parameter slots that she left empty so far, thus further improving the quality of the returned results. When finally a suitable device has been found, the customer can link directly to its detailed data sheet.

This is very similar to a real shop situation when a customer consults the shop assistant: The assistant step by step learns about the customer's exact demands until finally she finds the product that best fits the customer's needs. A CBR system has part of the knowledge of a real shop assistant built into it. This knowledge is used to interpret the user's query and greatly enhances the quality of the retrieved data. Even the first request immediately produces usable results. Near misses are avoided. If the system returns more than one answer, the results are ranked by their similarity to the query. This makes it easy for the user to refine her request until she is satisfied with the results. This is why we claim such a system being capable of intelligent sales support.

But not only the customer gains profit from this intelligent sales supporting catalog. As this approach will be pursued further, Analog Devices in the future will be able to allow some of their engineers to spend more time addressing more complex customer support requirements. Encouraged by the first success of the new CBR catalog, Analog Devices have decided to further extend the application of CBR technology to other product families. Work on these follow-up projects has already started by time of writing of this article.

6 Current and Future Work

The research project WEBSELL which has been started recently has the primary goal to explore techniques to support this sales process especially by employing Case-Based Reasoning and related methods. WEBSELL is a research and development project funded by the Commission of the European Communities (ESPRIT contract P27068, the WEBSELL project: Intelligent Sales Assistants for the World Wide Web) with the partners: TecInno (Germany, prime contractor), Adwired (Switzerland), EuroWEB (Germany), Irish Medical Systems (Ireland), Trinity College Dublin (Ireland) and the University of Kaiserslautern (Germany).

Another new research project recently has started with the aim to further explore and integrate the possibilities of CBR-related techniques into *reuse* applications and product catalogs. The READEE project[2] (*R*euse *A*ssistant for *D*esigns in *E*lectrical *E*ngineering) employs CBR-techniques to re-use designs of electronic circuits. Reusing such designs in general can require solving very complex configuration tasks. The reuse assistant is meant to help electrical engineers in selecting and configuring third party intellectual property to fit their personal requirements. Making such a reuse tool available over the Internet or on intra-nets of larger electronics companies can speed up the time-to-market of new electronic circuits significantly. Both the vendor and the consumer of the intellectual property benefit from this tool. READEE is funded by the Foundation for Innovation of Rhineland Palatinate and will last for a period of three years.

When interactive electronic sales support will become more common in future, interfaces for interactive adaptation and configuration will come into use. CBR-related techniques for indexing and clustering the case base can be used to help the customers refine their queries by a step-by-step analysis of their needs. By analyzing the structure of the case base, a CBR system can suggest which undefined parameters the user should define next in order to find a good solution as quickly as possible. It is also possible to *explain* to the customer the reasons why the retrieved results are suitable for her query. For instance, the previously mentioned example online catalog of holiday accommodations might tell the user who requested a 3-star hotel on Pariku: “I suggest this 3-star hotel on Pung-Chitu because Pung-Chitu is close to Pariku and has a similar landscape and infra structure.”

7 Conclusion

CBR technology introduces intelligent sales support to electronic commerce applications. The deficits of standard database techniques are overcome by adding expert knowledge to the retrieval system. CBR gets rid of the problem of near misses, the need of expert knowledge on the customer's side and resulting frustration. New solutions can be created by adapting old solutions, as we saw in our sliding roof example

[2] http://wwwagr.informatik.uni-kl.de/~readee/

for the catalog of used cars – and even complex configuration tasks can be simplified by Case-Based Reasoning. Finally the vendor benefits from the ability to offer nearly perfect solutions and from satisfied customers that do not need to look for other vendors any more just because the database search engine could not find a product matching the customer's requirements. There are already many positive experiences with currently available applications. New technologies can be expected to offer sophisticated adaptation and interactive configuration abilities for complex products. This leads to negotiation between the virtual sales agent and the customer on the Internet [9]. Interactive refinement of the customer's requirements and explanation of results are also possible. Some future improvements to intelligent sales support systems are current research topics.

References

1. Doorenbos R B, Etzioni O, Weld D S 1996 A Scalable Comparison-Shopping Agent for the World-Wide-Web. *Technical Report UW-CSE-96-01-03.* Department of Computer Science and Engineering. University of Washington
2. Leake D 1996 CBR in Context: The Present and Future. In: Leake D (ed) 1996 *Case-Based Reasoning, Experiences, Lessons, & Future Directions.* MIT Press, Menlo Park, Calif, pp 3-30
3. Althoff K-D, Auriol E, Barletta R, Manago M 1995 A Review of Industrial Case-Based Reasoning Tools. *AI Intelligence*
4. Wess S, Althoff K-D, Derwand G 1994 Using K-D Trees to Improve the Retrieval Step in Case-Based Reasoning. In: Wess S, Althoff K-D, Richter M M (eds) 1994 *Topics in Case-Based Reasoning.* Springer Press
5. Lenz M, Burkhard H-D 1996 Case Retrieval Nets: Basic Ideas and Extensions. In: Görz G, Hölldobler S (eds) 1996 *KI-96: Advances in Artificial Intelligence.* Springer Press
6. Wilke W, Bergmann R 1998 Techniques and Knowledge used for Adaptation during Cased-Based Problem Solving. On the *11th International Conference On Industrial and Engineering Applications of Artificial Intelligence and Expertsystems (IEA-1998).* Springer-Press
7. Wilke W, Cunningham P, Smith B 1998 Using Configuration Techniques for Adaptation. In: Burkhard H-D, Bartsch-Spörl B, Lenz M, Wess S (eds). *From Foundations to Applications in Case-Based Reasoning.* Springer Press
8. The Board of Analog Devices 1995 *Analog Devices Annual Report 1995.* Analog Devices
9. Wilke W, Bergmann R, Wess S 1998 Negotiation During Sales Support with Case-Based Reasoning. In: *Procceedings of the German Workshop on Case-Based Reasoning (GWCBR)*

A Genetic Algorithm for Generalized Rule Induction

Alex A. Freitas

CEFET-PR, DAINF
Av. Sete de Setembro, 3165
Curitiba - PR, Brazil, 80230-901
Tel: ++55 +41 322-4544 ext. 648
alex@dainf.cefetpr.br

Keywords: genetic algorithms, data mining, knowledge discovery, generalized rule induction.

Abstract.

Data mining consists of the efficient discovery of knowledge from databases. This paper presents a new genetic algorithm designed for discovering a few interesting, high-level prediction rules from databases, rather than discovering classification knowledge (often a large rule set) as usual in the literature. Three important data mining issues addressed by our algorithm are the interestingness of the discovered knowledge, the computational efficiency of the algorithm, and the trade-off between representation expressiveness and efficiency.

1 Introduction.

In essence, data mining consists of the efficient discovery of knowledge from large databases. The knowledge discovery process consists of several phases [1], but the core of this process is usually the application of machine learning-based algorithms to databases. However, in "conventional" machine learning, the emphasis is usually on finding accurate knowledge, as measured e.g. by the classification error rate, in the case of classification tasks. In contrast, in data mining, knowledge accuracy is often a secondary concern, and the emphasis is on finding novel, interesting, comprehensible knowledge, which will be ultimately used by a human being (e.g. a senior business analyst) for high-level decision making. In general it is not very desirable to return to the user all the discovered rules. They would be simply too many in the case of large databases, and this would hinder the user's decision-making process. In data mining it is usually better to return to the user only a small set of high-quality rules, the "knowledge nuggets", rather than all rules.

In this paper we present a genetic algorithm (GA) designed to discover a few high-quality rules from a given database. The data mining task addressed by our GA consists of predicting the values of a small number of goal attributes given the values of other attributes, called predicting attributes. Note that this is an extension of the conventional data mining task of classification, where there is only one goal attribute to be predicted. The data mining task addressed in this paper is also often called generalized rule induction.

It should be noted that there are post-processing, rule set-refinement algorithms that extract a small set of high-quality rules from a large set of discovered rules, so that conventional machine-learning algorithms can be used, together with these post-processing algorithms, for extracting knowledge nuggets [2], [3]. However, this approach has two drawbacks. First, the discovery of a large set of rules for a given problem is usually a very time-consuming activity, and probably a not very cost-effective one, since many of the discovered rules will eventually be discarded by the post-processing algorithm anyway. Second, the post-processing algorithms can be quite complex and computationally expensive themselves. Hence, the approach followed in this paper (discovering only a small set of high-quality rules) tends to be significantly more efficient.

In addition, it should be emphasized that most data mining-related GAs proposed in the literature – e.g. [4], [5], [6], [7], [8] – address the task of classification only, while our GA addresses the task of generalized rule induction, as mentioned above.

This paper is organized as follows. Section 2 discusses the design of our GA. Section 3 presents the results of computational experiments. Finally, Section 4 presents the conclusions.

2 Design of the Genetic Algorithm (GA).

This section presents our genetic algorithm (GA) designed for the task of generalized rule induction. Since this algorithm was designed to discover a few high-quality rules ("knowledge nuggets"), we will call it GA-Nuggets. The next subsections discuss several aspects of the design of GA-Nuggets.

2.1 Knowledge Representation.

There is a trade-off between a knowledge representation's expressiveness power and its associated computational cost. A propositional logic ("0-th order") representation allows efficient execution of a data mining algorithm, but it is rather limited in expressiveness. For instance, a propositional miner cannot discover a rule with the condition "Income > Expenditure", since this condition involves a relation (">") between two distinct attributes. On the other hand, a first-order logic representation has a much greater expressive power - allowing the discovery of conditions such as "Income > Expenditure" - at the expense of a much larger rule space to be searched, which significantly increases the processing time of the data mining algorithm.

Our solution for the above trade-off is to use data dictionary information (commonplace in real-world databases systems) as a kind of background knowledge that guides the algorithm to consider only a small number of first-order relationships between attributes, rather than all possible relationships. The basic idea is to use an attribute-typing system significantly stronger than the one normally used in data mining algorithms. In general, data mining algorithms handle attributes by classifying them into one out of very few attribute types, often just two types: categorical and continuous attributes. However, database systems offer much richer information about attribute types, which can be used to filter out non-sensible first-order relationships. For instance, suppose an employee database contains the attributes Salary and Age. Both attributes would be simply considered as "continuous" by most data mining algorithms. However, in principle it does not make sense to let the algorithm consider first-order relationships such as "Salary > Age". These two attributes have very different types and domains, they are not "compatible", and the data mining algorithm can be aware of that by consulting the data dictionary. On the other hand, from the data dictionary the algorithm can find out which attributes are "compatible" to, say, attribute Salary. In this example, Salary and its compatible attributes would belong to the same data type, namely "money".

	A_0	A_1	A_2	A_3	A_4	A_5	A_6	A_7
A_0	yes	yes	yes	no	no	no	no	no
A_1	yes	yes	yes	no	no	no	no	no
A_2	yes	yes	yes	no	no	no	no	no
A_3	no	no	no	yes	yes	no	no	no
A_4	no	no	no	yes	yes	no	no	no
A_5	no	no	no	no	no	yes	no	no
A_6	no	no	no	no	no	no	yes	no
A_7	no	no	no	no	no	no	no	yes

Figure 1: Example of an Attribute-Compatibility Table for 8 attributes.

Consulting information in the data dictionary of the database system, GA-Nuggets computes an Attribute-Compatibility Table (ACT), as shown in Figure 1. The number of rows and columns of this table equals the number of attributes being mined. In the example shown in Figure 1, there are 8 attributes, named $A_0...A_7$. If attributes A_i and A_j - i,j in [0..7] - are compatible, the corresponding entry i,j of the table has the value "yes", otherwise that entry has the value "no". An ACT is constantly used by GA-Nuggets, not only to generate individuals of the initial population but also to apply genetic operators during evolution (see below). We believe that, by using an ACT, GA-Nuggets achieves a good trade-off between knowledge representation expressiveness and computational cost for searching the rule space.

2.2 An Individual's Encoding.

In GA-Nuggets an individual corresponds to a single rule of the form "if A then C", where A is the rule antecedent, consisting of a conjunction of conditions on the values of the predicting attributes, and C is the rule consequent, specifying the value predicted for a given goal attribute. The rule antecedent and the rule consequent are encoded separately in an individual, as follows.

An individual's rule antecedent is encoded as a variable-length list of rule conditions. The rule antecedent can contain two types of rule conditions: (a) propositional logic (hereafter propositional, for short) rule conditions; (b) first-order logic (hereafter first-order, for short) rule conditions, in which the pair of operands is compatible, as defined in the Attribute-Compatibility Table (ACT) for the data being mined (see above).

More precisely, propositional conditions have the form A_i *op* V_{ij}, where A_i is the i-th predicting attribute, *op* is a relational operator (=, ≥, ≤) and V_{ij} is the j-th value of the i-th predicting attribute's domain. First-order conditions have the form A_i *op* A_j, where A_i and A_j (i≠j) are two compatible attributes, as defined in the corresponding ACT, and *op* is a relational operator (=, ≥, ≤). For instance, the first-order condition $A_3 \geq A_4$ is satisfied by a database tuple iff the value of attribute A_3 in the tuple is greater than or equal to the value of attribute A_4 in the tuple.

To avoid the production of individuals with many conditions in the rule antecedent (which would reduce the degree of comprehensibility of the corresponding rule, going against the philosophy of data mining), the user specifies a maximum number of antecedent rule conditions for an individual.

An individual's rule consequent is encoded as a single condition of the form $G_i = V_{ij}$, where G_i is the i-th goal attribute and V_{ij} is the j-th value of the i-th goal attribute's domain. Note that although each rule predicts the value of a single goal attribute, in a given population different rules can predict the value of different goal attributes.

The genetic material of an individual's rule antecedent is randomly generated when the initial population is created and is thereafter subject to the action of crossover and mutation, as will be seen below. In contrast, the production of the genetic material of an individual's rule consequent is treated in a special way, due to its strategic importance in determining the quality of the rule represented by the individual. The basic idea is to delay the generation of the rule consequent until fitness-computation time, when the rule consequent is generated in such a way that the rule's predictive accuracy is maximized. A more elaborated description of this idea is as follows.

First of all, crossover and mutation operators are applied only to an individual's rule antecedent. Once these genetic operators have been applied, the just-produced rule antecedent of an offspring individual is matched against all the tuples of the database, to compute statistics to be used for calculating the individual's fitness. These statistics contain the frequency of tuples satisfying the rule antecedent, distributed per goal attribute and per goal attribute value, as illustrated by the table shown in Figure 2. In this Figure there are two binary goal attributes, denoted G_1 and G_2, each of them with two attribute values, denoted V_1 and V_2 . Each entry of the table shown in Figure 2, denoted F_{ij} (i=1,2; j=1,2,), contains the frequency of database tuples that satisfy the rule antecedent and have value V_j for goal attribute G_i.

	goal attribute values	
	V_1	V_2
goal attribute G_1	F_{11}	F_{12}
goal attribute G_2	F_{21}	F_{22}

Figure 2: Distribution of the frequency of tuples satisfying a rule antecedent, per goal attribute and per goal attribute value.

Once a table like the one shown in Figure 2 is computed, GA-Nuggets will select, to make up the rule consequent, the pair of <goal attribute, goal attribute value> whose frequency F_{ij} maximizes the predictive accuracy of the rule. As an example, suppose that predictive accuracy maximization is achieved by selecting the maximum value F_{ij} in the table. Suppose also that the maximum frequency value is F_{12}. Then the GA will choose the goal attribute G_1 and its value V_2 to make up the rule consequent "$G_1 = V_2$".

It is interesting to note that this delayed generation of an individual's rule consequent has two advantages. First, it maximizes the individual's predictive accuracy for a given rule antecedent, by design. Second, it leads to a gain in computational efficiency, since by matching a single rule antecedent against the database just once we are computing, in a *single database pass*, the predictive accuracy associated with several possible rules. This is significantly more efficient than matching the entire rule (antecedent and consequent) against the database for each possible rule. Note also that this is important because predictive accuracy computation usually by far dominates the processing time of a GA mining a large database.

A simple example of a rule encoded by an individual of GA-Nuggets is illustrated in Figure 3. In this Figure, A_i denotes the value of the i-th predicting attribute, while G_i denotes the value of the i-th goal attribute. In this example the rule antecedent has two conditions. The first one is a first-order condition, while the second one is a propositional condition. The rule consequent predicts a given value ("0") for the goal attribute G_1.

(rule antecedent) IF $(A_0 = A_1)$ AND $(A_2 \geq 0)$
(rule consequent) THEN $(G_1 = 0)$

Figure 3: A simple example of a rule encoded by an individual.

2.3 Fitness Function.

The fitness function of GA-Nuggets consists of two terms. One term measures the predictive accuracy of the rule associated with the individual, while the other term measures the degree of interestingness of the rule consequent. These terms are elaborated in the following.

One of the simplest, most intuitive estimates of a rule's predictive accuracy is its "confidence factor" on the training data, calculated as the ratio X / Y, where X is the number of tuples that satisfy both the rule antecedent and the rule consequent and Y is the total number of tuples that satisfy the rule antecedent. However, in the early experiments with GA-Nuggets we observed that this term had a strong bias towards the discovery of "small disjuncts", i.e. rules that cover a very small number (say about 5) of tuples. It is relatively easy to find small disjuncts that have a confidence factor of 100% on the training data. However, these rules are likely to reflect idiosyncrasies of the data which may not constitute the knowledge nuggets we are looking for, and are likely to have a significantly smaller confidence factor on the test set. (For a more comprehensive discussion about the problem of small disjuncts, see e.g. [9], [10].)

Therefore, we decided to modify this term of the fitness function in such a way that its bias towards small disjuncts were reduced, but still keeping as much as possible of the simplicity of the confidence factor definition. Hence, we have borrowed the idea of subtracting 1/2 from the numerator X in the above formula for confidence factor, i.e. we use the term

$$(X - 1/2) / Y. \qquad (1)$$

This idea was proposed by Quinlan [11] in the context of decision tree pruning. To see why this approach reduces the bias towards small disjuncts, note that the smaller the size of the disjunct associated with a rule, the greater will be the reduction of the numerator X in comparison with the denominator Y in the above formula, and the smaller will be the value of the above expression.

The second term of GA-Nuggets' fitness function is a measure of the degree of interestingness of the consequent ("THEN part") of the rule associated with the individual. The rationale behind the introduction of this term in the fitness function is the following. Let G_i denote a given goal attribute and g_i denote a given value in the domain of this attribute. The higher the "prior probability" (relative frequency) of the value g_i in the training set, the easier is to discover a rule predicting the value g_i (i.e. having consequent $G_i = g_i$), and, in some sense, the less interesting the discovered rule is. The challenge is to discover highly accurate and general rules predicting rarer goal attribute values. These rules tend to be more interesting for the user.

Based on the above argument, the second term of GA-Nuggets' fitness function is defined as

$$(1 - \mathrm{Prob}(g_i))^{1/k}, \qquad (2)$$

where $\mathrm{Prob}(g_i)$ is the prior probability (relative frequency) of the goal attribute value g_i, and k is a user-defined parameter, empirically set to 2 in our experiments.

The exponent 1/k in the above formula can be regarded as a way of reducing the influence of the rule consequent interestingness in the value of the fitness function.

The higher the value of k, the closer the value of the rule consequent interestingness term gets to one, and so the smaller the influence of this term in the fitness function (formula 3 below). The value of k was empirically set to 2 in the experiments reported in this paper.

Finally, the fitness function of an individual i is defined as the product of the above two terms, that is:

$$Fitness_i = ((X - 1/2) / Y) * (1 - Prob(g_i))^{1/k} \quad (3).$$

Note that each of the terms takes on values between 0 and 1, so that the fitness of an individual also takes on a value between 0 and 1.

2.4 Genetic Operators.

We have designed a crossover operator that takes into account our problem-dependent individual encoding. In essence, we use a one-point crossover. However, a straightforward application of the conventional crossover operator would lead to the production of invalid genetic material. In particular, a predicting attribute might occur in two or more conditions of the rule antecedent, with inconsistent values between the conditions. To consider a concrete example, suppose that the first rule condition of one of the mating individuals is "$A_1 = 0$", while the last rule condition of the other mating individual is "$A_1 = 1$". A crossover operation would produce an offspring individual whose conjunctive rule antecedent would include the two rule conditions "$A_1 = 0$" and "$A_1 = 1$" - which obviously characterizes an invalid genetic material, since attribute A_1 cannot take on values 0 and 1 at the same time.

To avoid this kind of problem, our crossover operator includes a method to repair invalid genetic material. After crossover, GA-Nuggets checks all the rule conditions of each of the two just-produced offspring individuals. Invalid conditions are replaced with a new rule condition. For simplicity and uniformity, this replacement is performed by calling the mutation operator (see below), which produces only valid rule conditions.

The mutation operator produces only rule conditions that are valid genetic material, avoiding inconsistency with other rule conditions of the same individual. The mutation operator is applied to an entire rule condition, rather than to the rule condition's components only (an operand or operator), and it has access to all the rule conditions of the corresponding individual. When a rule condition is to be mutated, its operands and operator are discarded and a triple of operand-operator-operand is randomly generated, subject to the restriction that the newly-generated condition is not inconsistent with other rule conditions.

The mutation operator can produce either propositional or first-order conditions, so that it also has access to the Attribute-Compatibility Table for the data being mined. Note that the new condition produced by the mutation operator can have either one or two attributes, depending on the new condition having a propositional or first-order representation (see section 2.2). Hence, the mutation operator replaces a rule condition with a new condition where the operands and operator are randomly generated, subject to the restriction that the attribute(s) occurring in the new condition do not occur already in other rule conditions of the same individual. We have also devised an adaptive mutation rate mechanism, to be explained in section 2.5.

It is important to clarify that in GA-Nuggets the mutation operator is used in two situations: (a) After crossover, in order to introduce new genetic material in an individual. In this case mutation is applied only with a mutation probability, which is dynamically adapted during the search (again, see section 2.5). (b) *Within* crossover, in order to repair invalid genetic material produced by crossover. In this case mutation is applied whenever necessary, regardless of any mutation probability.

Finally, both crossover and mutation act on the rule antecedent only. Recall that the rule consequent is automatically re-generated to maximize the rule's predictive accuracy, whenever the individual undergoes any modification in the genetic material of the rule antecedent.

2.5 Adaptive Mutation Probability and Result Designation.

In many GA applications, only the best individual of the last generation is designated as the GA's result. In our case, however, each individual corresponds to a single rule, and it is desirable to designate more than one rule (but still a *small* number of rules) as the result. Hence, it is desirable to avoid that the population be dominated by a single highly fit individual.

Our approach to cope with this problem is to use an adaptive mutation probability, where the value of the mutation probability is increased as the population becomes dominated by an individual. More precisely, the higher the number of copies of the fittest individual in the current population, the higher the mutation probability used in that generation. In the current version of GA-Nuggets, the value of the mutation probability varies from 5% (when the fittest individual has a single copy in the current population) to 90% (when all individuals of the current the population are copies of the fittest individual).

Although these mutation probability values seem to be high at first glance, we should bear in mind two points. First, we use an elitist strategy (see below), so that the high mutation probabilities cannot cause the loss of the fittest individual of a generation. Second, even though mutation introduces a seemingly unfavorable (with respect to predictive accuracy) change in the rule antecedent, recall that after mutation the rule will have its consequent re-generated in such a way to maximize the rule's predictive accuracy. Hence, in GA-Nuggets mutation promotes diversity with a reduced risk of significantly degrading the quality of an individual. Note that GA-Nugget's adaptive mutation approach is significantly different from the one used by some "conventional" adaptive mutation probability GAs, where the mutation probability is often reduced in the final generations.

The above method allows that the last generation contains several different, high-quality individuals, as will be shown in section 3. In this sense, the above adaptive mutation probability method achieves an effect similar to niching, with one advantage: the former tends to be more computationally efficient than the latter. In any case, in future versions of GA-Nuggets, we might implement some niching mechanism and compare its effectiveness against the one of the above adaptive mutation probability method.

GA-Nuggets returns to the user not only the best individual of the last generation but also other individuals of the last generation that have fitness greater than an user-specified value.

3 Preliminary Computational Results and Discussion.

We have done a preliminary, "proof-of-concept" experiment to evaluate the GA proposed in this paper. (We intend to do more extensive experiments in the near future.) In this experiment the parameters of GA-Nuggets were set as follows. The population size was 100, and each run consisted of 200 generations. The maximum number of conditions in the rule antecedent was set to 4 (half the number of predicting attributes). For reproduction we used tournament selection, where the number of individuals in the tournament was 2, a common default in the literature. An elitist reproduction strategy was also used, where the best individual of a generation was copied unaltered to the next generation. We used problem-specific crossover and mutation operators, as discussed in section 2.4. The crossover probability was 0.8 and the mutation probability was varied from 0.05 to 0.9, as explained in section 2.5. The next two subsections discuss the data set and the computational results of our experiment.

3.1 The Data Set.

We have used a synthetic database, consisting of 1,000 tuples and 10 attributes. The first 8 attributes, denoted $A_0 ... A_7$, are predicting attributes, while the last 2 attributes, denoted G_1 and G_2, are goal attributes. The data mining task consisted of predicting the values of the goal attributes, given the values of the predicting attributes. The values of the predicting attributes were randomly generated according to a uniform probability distribution. More precisely, predicting attributes A_0, A_1, A_2, A_3, A_4, A_5 were defined as ordinal categorical attributes, each of them with a domain size of 5. Since their values are ordinal, the operators applied to them in rule conditions can be not only "=" but also "≥" and "≤". Predicting attributes A_6 and A_7 are categorical, the former with a domain size of 2 and the latter with a domain size of 4. The only operator that can be applied to these attributes in rule conditions is "=".

To implement the idea of using strong attribute typing to focus the search for first-order attribute relationships (see section 2), the above 8 predicting attributes were classified into 5 different attribute sets, each associated with a distinct attribute type. Within each of these attribute sets, the attributes are compatible with each other, and first-order rule conditions can be formed only when the condition's attributes belong to the same attribute set. The attribute sets were defined according to the Attribute-Compatibility Table shown in Figure 1. As can be deduced from that table, the first set of compatible attributes is composed by attributes A_0, A_1 and

A_2; the second set by attributes A_3 and A_4, and the last three sets are unary sets, composed respectively by attributes A_5, A_6 and A_7 (i.e. each of these 3 attributes can be used only in propositional rule conditions, not in first-order conditions, since they are not compatible to any other attribute).

The two goal attributes G_1 and G_2 had a binary domain, taking on the value 0 or 1. For each tuple of the database, once its predicting attribute values were generated, two classification functions were applied to those values, to determine the values of the goal attributes G_1 and G_2. To introduce some noise (commonplace in the real-world), the wrong value of a goal attribute was assigned to a tuple with a probability of 5%. The two classification functions used are shown in Figure 4, where A_i denotes the value of the i-th predicting attribute and G_i denotes the value of the i-th goal attribute.

IF ((($A_0 < 3$) AND ($A_2 > 1$)) OR
(($A_0 = A_1$) AND ($A_3 < A_4$)) OR
(($A_2 > A_0$) AND ($A_1 > 2$) AND ($A_6 = 1$)))
THEN ($G_1 = 1$) ELSE ($G_1 = 0$)

(a)Function used to compute value of goal attribute G_1

IF ((($A_4 = 3$) AND ($A_6 = 0$) AND ($A_7 > 1$)) OR
(($A_0 < 2$) AND ($A_2 > 2$) AND ($A_3 = A_4$)) OR
(($A_0 < A_1$) AND ($A_5 > 1$) AND ($A_6 = 1$) AND ($A_7 > 0$)))
THEN ($G_2 = 1$) ELSE ($G_2 = 0$)

(b)Function used to compute value of goal attribute G_2

Figure 4: Classification functions used to compute the values of the goal attributes in the synthetic database.

3.2 Computational Results and Discussion.

The discussion presented in this section focuses mainly in three points. First, we analyze whether or not the adaptive mutation probability mechanism was able to avoid convergence of the population to a single highly fit individual and allow the evolution of other good individuals. Second, we analyze the predictive accuracy of the discovered rules on a separate test set (unseen by the algorithm during the search for rules). Third, we analyze the generality of the discovered rules (recall that we have introduced in the fitness function a certain bias against small disjuncts).

Let us start with the adaptive mutation probability mechanism proposed in the paper. This mechanism was able to avoid the convergence of the population to a single highly fit individual. The best individual of the run was created in generation 21. A few generations later this individual started to dominate the population. For instance, in generation 28 there were 27 copies of this individual. However, as the number of copies of this individual increased, the adaptive mutation probability was correspondingly increased. As a result, the number of copies of the best-of-the-run individual in any generation was almost always less than 50 (half the population size), and the number of copies of this individual in the last generation was 38.

This allowed other good individuals to evolve and to be present (sometimes with a few copies of themselves) in the last generation too. The result returned to the user consisted of all individuals (rules) of the last generation with fitness greater than or equal to 0.7. In our case, this result consisted of seven individuals.

Let us turn now to fitness and predictive accuracy issues. On the average, the seven discovered rules had a fitness value of 0.7595, resulting from an estimated predictive accuracy on the training set - computed by formula (1) - of 0.9573 and a rule consequent interestingness measure of 0.7935.

It should be noted that, although the value of the rule consequent interestingness measure is approximately the same in both the training and the test sets (with the conventional assumption that these two data sets follow the same probability distribution), the value of the predictive accuracy in the test set (unseen data) can be quite different from the one in the training set. Hence, we have measured the actual prediction accuracy rate of a rule on unseen data (the test set), a new database with 1,000 synthetically-generated tuples. The test set data has the same attribute definitions and probability distributions as the training set, but the two data sets have of course different data (this is easily achieved by changing the random seed used in the synthetic data generator). The prediction accuracy rate was calculated, on the test set, as the ratio X / Y, where X is the number of tuples that satisfy the rule antecedent and have the goal attribute value predicted by the rule consequent and Y is the total number of tuples that satisfies the rule antecedent. In our experiment, the seven rules returned to the user had an average predictive accuracy on the test set of 0.8534. At first glance, this value is significantly smaller than the above-mentioned estimated predictive accuracy (on the training set) of 0.9573. However, we should bear in mind two points.

First, predictive accuracy is just one term of the fitness function, and GA-Nuggets is maximizing fitness, i.e. the product of predictive accuracy and rule consequent interestingness. Actually, rule consequents with higher degrees of interestingness are more difficult to find, and so their predictive accuracy tends to be somewhat lower. Second, the value of 0.8534 was averaged over the seven discovered rules. It turns out that two of these rules had a relatively small predictive accuracy (0.667 and 0.571) on the test set. If we ignore these two "outliers", the remaining five rules had an average predictive accuracy of 0.947 on the test set. This is a very high value, considering that, in theory, the best predictive accuracy on the test set would be 0.95, due to the existence of 5% noise in the data. (For a general discussion on how noise limits prediction accuracy, see e.g. [12]).

To summarize, one can say that GA-Nuggets discovered five high-quality rules (concerning both predictive accuracy on the test set and rule consequent interestingness) and two not so good rules (with relatively small predictive accuracy on the test set).

It is also interesting to mention that the five high-quality discovered rules had a predictive accuracy on the test set higher than the best-of-the-run individual, which confirms that it is important to avoid the convergence of the population to a single highly fit individual.

Turning to the generality of the discovered rules, on the average the seven rules returned to the user covered 19.6 tuples. Two of the discovered rules could be considered as small disjuncts, covering 7 and 9 tuples. (One of these small disjuncts had a good predictive accuracy of 83,3% on the test set, but the other one had a relatively bad predictive accuracy of 57,1% on the test set.) Out of the seven discovered rules, the largest disjunct was a rule covering 56 tuples, which had a very good predictive accuracy of 97,9% on the test set. Overall, these results show that GA-Nuggets is still discovering relatively small disjuncts. Hence, although the use of formula (1) has the effect of somewhat biasing the search towards larger disjuncts, this formula alone seems not to be enough, and it seems desirable to include another term in the fitness function with an explicit bias towards rules with higher generality (larger number of tuples covered). This will be the object of future research.

4 Conclusion.

We have presented a new genetic algorithm, called GA-Nuggets, designed for the efficient discovery of a few interesting, high-level prediction rules ("knowledge nuggets") from databases. We have focused on the task of generalized rule induction, which can be regarded as an extension of the well-known task of classification.

GA-Nuggets searches for rules that have both a high predictive accuracy and a high degree of rule-consequent interestingness (defined in the algorithm as a small prior probability of the value predicted by the rule consequent).

GA-Nuggets uses a problem-dependent individual representation which represents a good trade-off between the expressiveness power and the computational efficiency of knowledge representations. On one hand it allows the discovery of first-order logic rules, with significantly greater expressiveness power than the propositional representation used by the vast majority of data mining systems. On the other hand, to reduce the computational complexity associated with the larger rule space of first-order systems, GA-Nuggets uses strong attribute typing information found in data dictionaries (commonplace in real-world database systems). This kind of information allows the algorithm to limit the search for first-order relationships to small groups of compatible attributes, rather than to all attributes.

Two other efficiency-related aspects of GA-Nuggets are as follows. First, as discussed in Section 2.2, in a *single* database pass we compute the information necessary to calculate the predictive accuracy of several potential individuals, whose corresponding rules predict different values of different goal attributes. This is clearly much more efficient than performing one database pass for computing the fitness of each of those potential individuals.

Second, since GA-Nuggets addresses the task of generalized rule induction, a single run of GA-Nuggets can discover rules predicting the value of different goal attributes. If we were to use classification algorithms in this task, we would need two run a classification algorithm n times, once for each of the n goal attributes to be predicted. While the required processing time can still be reasonable for two goal attributes, the situation quickly gets worse as the number of goal attributes is increased. GA-Nuggets avoids the excessively long time associated with running one classification algorithm for each goal attribute, since it searches for rules for predicting any of the goal attributes at the same time.

Of course, the advantages of GA-Nuggets do not come for free. Recall that classification algorithms usually discover a "complete classification knowledge", in the sense that every unseen tuple can be classified by the discovered knowledge. In contrast, GA-Nuggets discovers only a few interesting, high-level rules, without bothering for discovering a complete classification knowledge. In some sense, GA-Nuggets trades knowledge completeness for knowledge comprehensibility and computational efficiency. This can be undesirable in some applications, but we believe that overall, in the context of data mining and knowledge discovery applications, knowledge comprehensibility and computational efficiency are more important than knowledge completeness.

Finally, recall that the experiments reported in this paper are somewhat preliminary, since they have used only a synthetic database and that GA-Nuggets is still evolving. Data mining is mainly an empirical area, and for a validation of the promising results reported in this paper it is necessary to apply GA-Nuggets to real-world databases. We also intend to extend our algorithm in several ways, including a more comprehensible evaluation of different fitness functions and their effects on the kind of knowledge nuggets that are discovered by GA-Nuggets, as well as investigating some niching scheme to improve the control of the variety of rules produced by the algorithm.

References.

[1] U.M. Fayyad, G. Piatetsky-Shapiro and P. Smyth. From data mining to knowledge discovery: an overview. In: U.M. Fayyad, et al. (Eds.) *Advances in Knowledge Discovery and Data Mining*, 1-34. AAAI/MIT Press. 1996.

[2] J.A. Major and J.J. Mangano. Selecting among rules induced from a hurricane database. *J. Intelligent Information Systems* 4(1), Jan./95, 39-52.

[3] F. Gebhardt. Choosing among competing generalizations. *Knowledge Acquisition*, 3, 1991, 361-380.

[4] S. Augier, G. Venturini and Y. Kodratoff. Learning first order logic rules with a genetic algorithm. *Proc. 1st Int. Conf. Knowledge Discovery & Data Mining*, 21-26. AAAI Press, 1995.

[5] F. Neri and A. Giordana. A parallel genetic algorithm for concept learning. *Proc. ICGA-95*, 436-443.

[6] C. Anglano, A. Giordana, G. Lo Bello and L. Saitta. A network genetic algorithm for concept learning. *Proc. 7th Int. Conf. on Genetic Algorithms*, 434-441. Morgan Kaufmann, 1997.

[7] C.Z. Janikow. A knowledge-intensive genetic algorithm for supervised learning. *Machine Learning*, 13, 1993, 189-228.

[8] B. de la Iglesia, J.C.W. Debuse and V.J. Rayward-Smith. Discovering knowledge in commercial databases using modern heuristic techniques. *Proc. 2nd Int. Conf. Knowledge Discovery & Data Mining*. AAAI, 1996.

[9] R.C. Holte, L.E. Acker and B.W. Porter. Concept learning and the problem of small disjuncts. *Proc. Int. Joint Conf. AI (IJCAI-89)*, 813-818.
[10] G.M. Weiss. Learning with rare cases and small disjuncts. *12th Int. Conf. Mach. Learning*, 558-565. 1995.
[11] J.R. Quinlan. Generating production rules from decision trees. *Proc. IJCAI-87*, 304-307.
[12] C. Cortes, L.D. Jackel and W.-P. Chiang. Limits on learning machine accuracy imposed by data quality. *Proc. 1st Int. Conf. Knowledge Discovery & Data Mining*, 57-62. 1995.

Part 6: Optimisation

Papers:

Multiobjective Optimization by Nessy Algorithm

Mario Köppen[1], Stephan Rudlof[1]

[1]*Fraunhofer-Institute for Production Systems and Design Technology*
Pascalstr. 8-9, 10587 Berlin, Germany
email: {mario.koeppen|stephan.rudlof}@ipk.fhg.de

Abstract

This paper presents the extension of the Neural Evolutional Strategy System (Nessy) to the multiobjective optimization case. The neural architecture of the Nessy algorithm is extended by using more than one output neuron, one neuron for each objective. The learning law of Nessy is modified according to the presence of multiple measures of performance. Each hidden neuron of the generation layer randomly selects an objective for one cycle of the network. From this, the multiobjective ranking of the population (or neurons of the solutions layer) is stochastically approximated. The modified Nessy algorithm (Monessy) is able to search for the Pareto set of a multiobjective optimization problem. A test function from literature with well-known Pareto and trade-off set is examined. The newly proposed algorithm effectively searches for the Pareto set by switching between explorational and exploitational search phases. This was compared with random search, which did not hit the Pareto set as nearly as often as the Monessy algorithm. Also, the replacement of a weighted-sum matching measure with multiple matching measures in a framework for texture filter design is considered as a second example.

1 Introduction

In real world problems, one is often faced with the problem of multiple, possibly competing, measures of performance, which should be optimized simultaneously. Conventional optimization techniques, such as steepest descent or simulated annealing, are difficult to extend to the true multiobjective case. However, evolutionary algorithms had been pointed out to be possibly well-suited to multiobjective optimization since early in their development [1]. This is based on the ability of evolutionary algorithms to search for multiple solutions in parallel and to handle complicated tasks such as discontinuities, multimodality and noisy function evaluations.

Recently, the Nessy algorithm has been proposed [2,3]. Nessy (Neural Evolutional Strategy System) is a neat integration of an evolutionary algorithm into a neural network architecture. The neural architecture is based on a multilayer backpropagation neural network. The state of every neuron equals a chromosome. One iteration cycle of the Nessy network equals one generation of the underlying evolutionary algorithm. The three layers of Nessy are: input layer, hidden layer and output layer. Input and hidden layer and hidden and output layer are fully connected, weights are assigned to these connections. What Nessy makes different from Neuro-GA approaches are its redefined genetic operators. To ensure the neuron based evaluation mode of a neural network, genetic operators are modified in a suitable manner. The essential steps of the Nessy algorithm are: selection, computation of error, weights modification, transduction and mutation.

There is evidence for the Nessy algorithm to maintain memory in the evolutionary algorithm in a unique manner [3]. Besides of keeping search space "knowledge" in the population's schemata, the weights of Nessy can be employed as a memory, changing at a lower time scale than the population's schemata.

However, the current implementation of the Nessy algorithm uses only one neuron in the output layer. This is due to the fact, that there is only one value for the fitness function, from which the error of the selection operation is calculated. By using more than one output neuron, the Nessy algorithm is well-suited to multi-objective optimization in the context of multiple, possibly competing, fitness functions.

There is one essential difference between single objective optimization and multiobjective optimization. This difference is based on the fact, that there is no natural sort order of points of the n-dimensional Euclidian space, if $n \geq 2$. Hence, a solution of an optimization problem can not be directly compared with some other. There is a special subspace structure of optimal solutions in multiobjective optimization, referred to as the Pareto set. A solution is Pareto-optimal, if this solution is not dominated by some other solution, i.e. if no change in the optimization problems' domain variables gains increase in all fitness values at once. The set of all Pareto-optimal solutions is the Pareto set (or Pareto-front). The task of multiobjective optimization is generally considered as the search for the Pareto-front.

Few evolutionary algorithms for Pareto-optimization have been designed and proposed so far, starting with the proposal of the VEGA system [4] (see [1] for a comprehensive overview and [5] for a new approach). The essential problem here is that of multiobjectively ranking the individuals of a population. In [6], a study is given on different approaches to multiobjective ranking. The main problem is the sorting of individuals, which are on the same multiobjective ranking level. Nessy offers a surprisingly simple approach to this complicated task, bypassing the ranking problem, because it can be characterized as an individual evolutionary algorithm [7] (fitness values of individuals in the input layer are never directly competed against each other based on their fitness values - hence, there is no need for ranking). For doing so, Pareto-optimal is considered as a probabilistic feature. Instead of saying, that the modification of one domain variable gains no increase for all fitness function simultaneously, it is considered, that there will be no fitness increase for a randomly chosen fitness function value out of the set of all fitness function values. By this means, the weights update and transduction decision are restricted to only one of the error measures of the output layer neurons. Which one, is randomly assigned to every hidden neuron for every network cycle.

This paper is organized as follows. In section 2, the Nessy algorithm is recalled. The definition of Pareto set and the extension of the Nessy algorithm to the multiobjective case (Monessy) are given in section 3. Section 4 studies a well-known multiobjective optimization, using the Monessy algorithm, section 5 demonstrates, how the proposed Monessy algorithm can be employed for minimizing multiple performance measures simultaneously within the Lucifer (Lookup-Compositional Inference) framework for the automated design of 2D-Lookup texture filters. The paper ends with the conclusions.

2 The Nessy Algorithm

The Nessy algorithm [2,3] is a neat integration of an evolutionary algorithm into a multilayer backpropagation neural network. The Nessy architecture is shown in figure 1. In this section, the Nessy algorithm will be recalled for the single objective case. Its extension to the multiobjective case will be given in the next section.

Nessy is composed of three layers: a solution (or input) layer, a generation (or hidden) layer and an output layer. The state of every neuron of the solution layer is a chromosome. Altogether, the neurons of the solution layer are the counterparts of the individuals of a population in conventional genetic algorithms (GA). The generation layer is of the size of the number of genes of a chromosome, as will be explained below. Every neuron of the generation layer points to a neuron of the solution layer, i.e. it refers to its state. The functionality of the generation layer is the counterpart to the selection operation in conventional GA. The output layer has only one neuron with state 0 (representing a lower bound of the fitness values of a single objective optimization problem).

Solution layer and generation layer are fully connected, every connection has a weight assigned to it. These weights are considered as relative selection probabilities. The symbol p_{ij} stands for the weight of the connection from the generation layer neuron i to the solution layer neuron j. Generation layer neurons and the only output neuron are also connected, but no special weights are assigned.

Once initialized with random chromosomes and weights, Nessy works autonomously. The outline of the Nessy algorithm is as follows:

1. Each generation layer neuron *i* randomly chooses exactly one solution layer neuron *j* with the probability:

$$\frac{p_{ij}}{\sum_k p_{ik}}.$$

2. The backpropagated error of every generation layer neuron is simply the normalized fitness value of its state, i.e.

$$h_i = \frac{f_i}{\sum_k f_k}.$$

The training value Zero is the assumed limit of the optimization.

3. The weights of the solution-generation layer connections are updated according to the learning law:

$$p_{ij}^{new} = p_{ij}^{old} - \alpha \frac{f_j - g_i}{O},$$

where α is the learning rate, g_i is the fitness value of the state of generation layer neuron i and O the state of the output neuron. The new weights are restricted to the range [0,1].

4. The neuron states of the solution layer are modified by the transduction operator. For transduction, the modification of every solution is performed fitness-proportionately. Each solution compares its fitness with the fitness of generation layer neuron state i. If the second one is lower than the solutions' fitness, the solution takes gene i of the generation layer neuron state. This operator was introduced in [9] as implantation operation, later [10] it was referred to as transduction operator and deduced from bacterial genetics. Transduction is a directed and asymmetric genetic operator.

5. The solutions are mutated. Mutation is necessary if the transduction operator is applied. The whole population would tend to premature convergence, because parts of successful chromosomes are copied over unsuccessful ones. The mutation is performed by adding a Gaussian distributed random number to the actual gene. Two parameters control the mutation: mutation probability p_μ and the standard deviation of the Gaussian distribution σ_μ.

Altogether, Nessy uses one structural parameter (size of solution layer) and three learning parameters (learning rate α, mutation probability p_μ and mutation standard deviation σ_μ).

Essential insights into the backgrounds of this algorithm were given in [3]. On its base, there is the interplay of weights and the so-called diagonal prototype, which is the chromosome constructed from the diagonal of the chromosomes of the hidden layer. Please note, that the number of hidden neurons is equal to the number of genes of a chromosome. The weights are used as a memory for the genetic search, which stochastically approximates the fitness ranking of the population, while the diagonal prototype acts like an independent "explorer" of the search space. Hence, the Nessy algorithm maintains the exploration/exploitation phases of a GA in a unique manner, which results into faster optimization and better avoidance of local extrema.

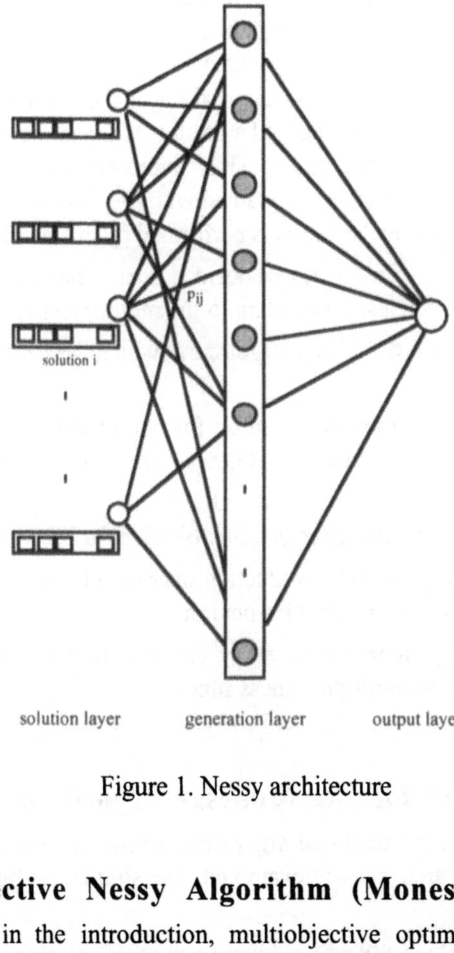

Figure 1. Nessy architecture

3 Multiobjective Nessy Algorithm (Monessy)

As pointed out in the introduction, multiobjective optimization is generally considered as search for the Pareto set. If a multiobjective optimization problem is given in terms of n fitness functions of m independent domain variables x_1, x_2, ..., x_m, or a vector $\mathbf{x}$ in the m-dimensional Euclidian space R^n, then a solution $\mathbf{x}_u$ is said to be Pareto-optimal, if it is not dominated by any other $\mathbf{x}_v$ from R^m, i.e. there is no $\mathbf{x}_v$ from R^m such that

$$\underset{i\in\{1,\dots,n\}}{\forall}(f_i(\vec{x}_v)\leq f_i(\vec{x}_u))\wedge\underset{i\in\{1,\dots,n\}}{\exists}(f_i(\vec{x}_v)<f_i(\vec{x}_u)).$$

Often, there is more than one Pareto-optima. The set of all such Pareto-optima comprises the Pareto-front or Pareto set. All function values of the Pareto set are referred to as trade-off set. In practical applications, this notion is only a first step towards the solution of a multiobjective optimization problem. Additional selection criteria have to be defined in order to select one solution out of all solutions of the Pareto set.

The use of multiple fitness functions in an evolutionary algorithm has been intensively discussed in [1]. The basic concept is fitness sharing. The population is divided into *n* subsets, each of which uses exactly one out of all *n* fitness functions. This is basically the famous VEGA approach, as proposed in [4]. However, this approach depends on the scaling of the objectives. Fourman [11] further examined the case and proposed a version, wherein the objectives for each comparison are chosen randomly. This approach was reported to work surprisingly well. In [1], this was explained with the stochastic approximation of the ranking of the individuals, which is performed by this modification.

Starting from these experiences, it was decided to use the ranking approximation for the extension of the Nessy algorithm to the multiobjective case, too.

Three modifications of the Nessy algorithm, which was presented in the last section, were introduced:

1. There are *n* output neurons, each neuron for one objective. Now, in step 2 of the algorithm, each output neuron determines the error value for its objective independently.
2. Each hidden neuron randomly selects an objective for every network cycle.
3. Weights update (step 3) and transduction operation (step 4) are performed due to the objective chosen by each hidden neuron.

From this, the Nessy algorithm is easily extended to the Monessy algorithm, which is able to process multiple fitness functions.

4 A Case Study for the Monessy Algorithm

In order to learn about the modified algorithm, a test function from literature [1] with a well-known Pareto set was examined. The simple test function is

$$f_1(x,y) = 1 - \mathbf{exp}[-(x-1)^2 - (y+1)^2]$$
$$f_2(x,y) = 1 - \mathbf{exp}[-(x+1)^2 - (y-1)^2]$$

Both functions, if considered independently, have a minimum function value of 0 at (1,-1) and (-1,1) for (x,y), respectively. For the Monessy run, both functions are scaled with an exponential γ, i.e. $\mathbf{f}_1(x,y)^\gamma$ and $\mathbf{f}_2(x,y)^\gamma$ were considered for simultaneous minimization. Figure 2 shows the trade-off sets for different values of γ.

Four experiments were performed, two using the Monessy algorithm and two using "blind" random search. Reals from -4 to 4 were encoded into a bitstring of size 44 by rescaling the binary numbers represented by the bitstrings onto this intervall. The Monessy network was composed of 30 solution neurons, 44 generation neurons and two output neurons. The learning rate was 3.0, the mutation used a (0, 0.5)-normal distribution. The network run 1000 cycles. The value for γ was chosen as 9, for which the trade-off set is nearly circular.

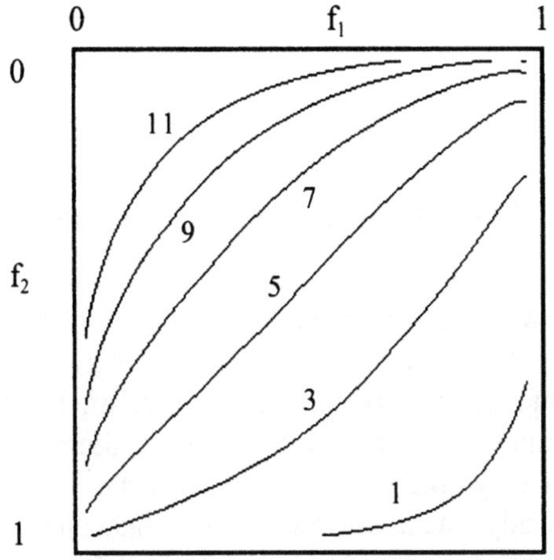

Figure 2. Trade-off sets for different γ-values

In the first experiment, the (x,y)-positions examined by the network were plotted to see how the Monessy algorithm approximates the trade-off set. In the second experiment, the same was done for 30000 random numbers (i.e. the same number as test points of the Monessy network during 1000 cycles). In the third experiment, it was counted, how often the Pareto-front was hit during the run. As reference set, the interior part of the Pareto-set for the γ value 9 was used. Finally, in the fourth experiment, the same was done for random numbers.

The results for the first experiment can be seen in figure 3. The Monessy algorithm approximates the trade-off set. Moreover, the region below the trade-off set is nearly empty. The Monessy algorithmus allocates most of its trials nearby the two single objective optima (either f_1 is 0 or f_2 is 0), or nearby the Pareto set. This should be compared with the result of the second experiment, as shown in figure 4. For x and y, random values from [-4,4] were chosen. The figure shows the distribution of the resulting function values in the f_1,f_2-plane.

What can not be seen from the figures 3 and 4 is, how often the Pareto front was reached by the Monessy algorithm and how this could be compared to randomly "guessing." Experiments 3 and 4 were carried out in order to reveal this. Figure 5 shows the number of solution layer neurons for cycle 1 to 1000, which were nearby the Pareto set. For reference, the plot of figure 2 was used with a γ value of 9. From figure 5, it can be seen, that the Monessy algorithms switches between explorational and exploitational phases. During exploration, the evolving population covers the fitness space in a manner similar to a standard genetic algorithm (SGA). Once a good initial region is found, the algorithm switches to the exploitational phase.

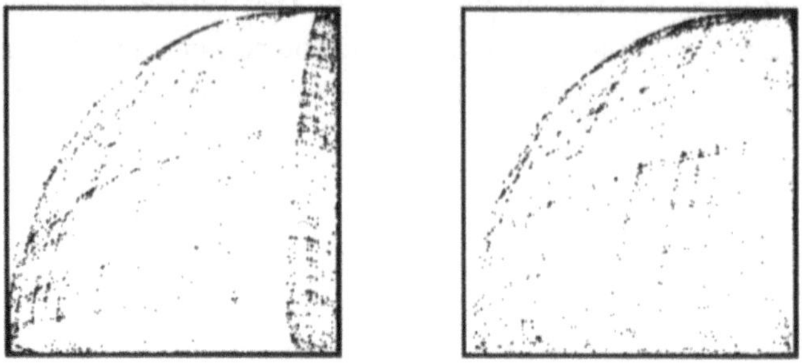

Figure 3. Trade-off set approximated by Monessy algorithm

For a SGA, this switch is performed due to the schemata theorem. If a SGA started exploitation, this process can't be stopped, because the number of trials allocated to exploitation exponentially grows in relation to the number of trials allocated to exploration (the number of schemata decreases to one). This also holds for the (Mo)nessy algorithm, but the algorithm may not converge at this point. Monessy quickly reaches the Pareto-front, but, after a while, the stability of the population breaks down and the individuals "fall back" to the simple single objective optimization stage. Then, the exploration phase restarts, and, after a while, the Pareto-front is revisited. During 1000 cycles, the Pareto set was hit 1657 times by the 30 neurons of the Monessy algorithm. The number of hits on the Pareto-front for experiment 4, the random search, was 80 in 30000 in average! Hence, the Monessy algorithm is able to effectively search for the Pareto-front of this multiobjective optimization problem.

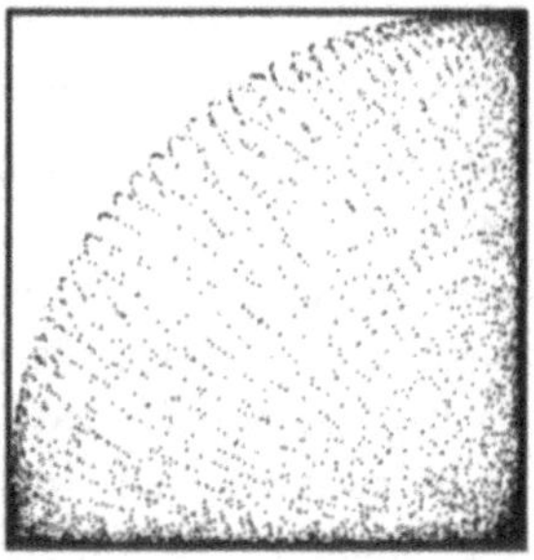

Figure 4. Trade-off set randomly approximated

The result of the experiments are surprisingly well. The Monessy algorithm is able to approach the Pareto-front. From this, it can be used as a tool for the needed refinement of the optimization goal in multiobjective optimization problems. After the Pareto-front is revealed, its subregions can be qualified according to its relevance for the (true) real optimization goal. Also, stopping criterias for a second run of the Monessy algorithm can be designed.

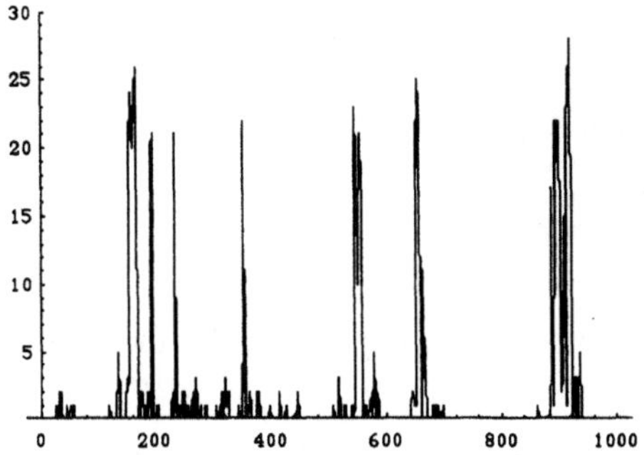

Figure 5. Number of solution layer neurons within the Pareto-set vs. cycle number

5 Texture filtering example

The Monessy algorithm was applied to the Lucifer framework [8]. This framework generates texture filters by employing the 2D-Lookup algorithm. For 2D-Lookup, a twodimensional matrix is given as lookup table for pairs of grayvalues. Two grayvalued images are scanned at every position for obtaining such a pair of grayvalues. The entry of the matrix at that position is a new grayvalue, which is the entry at that position in the result image. In the Lucifer framework, there are only entries for black (grayvalue 0) and white (grayvalue 255) in the matrix. Hence, the result of the algorithm is a binary image. The two input images of the algorithm are the results of the application of two image processing operations onto the original image. The optimization task here is a twofold one: the image processing operations have to be chosen out of a set of predefined image processing operators (e.g. convolution operators, morphological operators, rescaling operations, differences-of-gaussians, texturenumbers, statistical operators, together 256 operators), and the 2D-Lookup matrix has to be specified.

For using evolutionary search for this, a fitness function has to be formulated. In the Lucifer framework, this is done by means of a so-called goal image, which is given by the user. Consider figure 6, where the original image and the goal image are shown. The recognition task is to design a filter for extracting the surface faults. The binary image, which is the result of the application of the 2D-Lookup algorithm for some setting of the operators and the matrix, has to be compared with the binary goal image. This is the point where multiobjective optimization comes into the play. From the users point of view, goal and result image are similar, if most of the reference points of the goal image are also black in the result image. To obtain a measure for this, three quantities are considered: the amount of white pixels of the result image, which are also white in the reference image (**whiteok**), the amount of black pixels of the result image, which are also black in the goal image (**blackok**) and the amount of black reference pixels, which are also black in the result image (**blackrefok**). Please note, that **blackrefok** and **blackok** are not identical.

original image with low-contrats texture fault

goal image

Figure 6. Images given by the user for the Lucifer framework

For the Lucifer system, the fitness function was a weighted sum of these three quantities. For optimization, the initial Nessy algorithm was used. For multiobjective optimization, these three measures (**whiteok, blackok, blackrefok**) can not be used directly. If so, the completely black (fitness (0,1,1)) and completely white images (fitness (1,0,0)) would belong to the Pareto-set! Instead of this, the measures were combined in two ways. In the first version, the fitness vector was given as (2-(**blackrefok+whiteok**), |**blackrefok-whiteok**|), in the second version in a similar manner as (2-(**blackok+whiteok**), |**blackok-whiteok**|). The Pareto-fronts were scanned for 300 network cycles. The results are given in figure 7. As can be seen there, the two fitness functions behaves different. In the first version, both measures can simultaneously reach the value (0,0). In the second version, the first quantity can not go beyond a value of about 0.5. For the first version, it has to be considered, that all result images with at least one black pixel in the reference and all other pixels white has a fitness vector of (0,0)! Hence, the Pareto set contains all framework settings leading to such a situation, and nothing more. In the second version, the demands for zero measures can not be fulfilled. It is impossible to generate a result image with all black pixels correctly set. Here, the trade-off set is the vertical line with (blackok+whiteok) about 0.5. This can be used to design a stopping criterion for further runs of the Monessy algorithm.

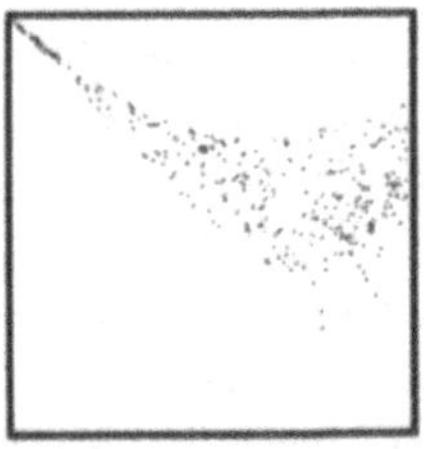

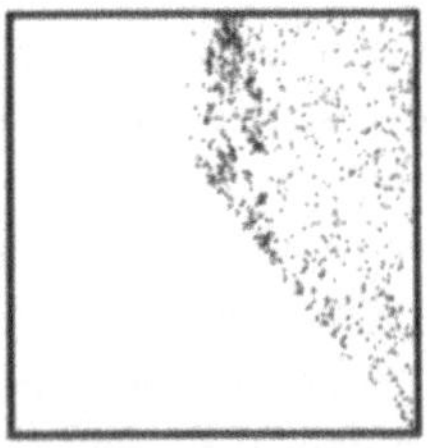

Figure 7. Trade-off sets approximated by Monessy algorithm for the problem given with figure 6, first and second version (see text for details)

Figure 8 shows one result of the run and the two operation images. Altogether, the Monessy algorithm could be used for framework adaptation as well as the

Nessy algorithm. However, the Monessy algorithm gains more insight into the way, the best possible solution can be yielded within this framework, if the plots of the trade-off sets are considered.

Also, this example has demonstrated, that care has to be taken into account, if multiple quality functions are designed for multiobjective optimization. Often, the Pareto set contains solutions which do not comprise solutions of the real problem. This also shows, that much more research is necessary in order to become "familiar" with the complicated task of Pareto-optimization.

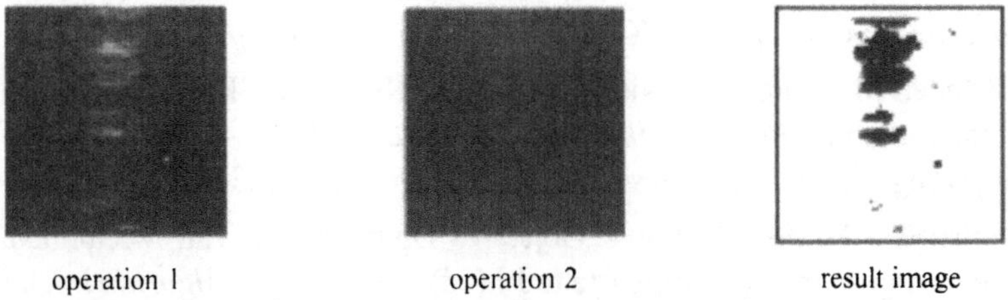

Figure 8. Operation images and result image for a run of Monessy, using version 2 for the fitness vector

6 Conclusions

In this paper, the Monessy algorithm for multiobjective optimization was proposed. It is an extension of the Neural Evolutional Strategy System (Nessy) by using more than one output neuron, one neuron for each objective. The multiobjective ranking of the individuals of the population (or the neurons of the solution layer) is stochastically approximated, if each hidden neuron of the generation layer chooses one objective in every cycle at random. All internal decisions of the algorithm (transduction decision, weights modification), which were based on one fitness measure, are now based on the randomly selected fitness measure. By this means, the Monessy algorithm is able to search for the Pareto set of a multiobjective optimization problem. The search can be divided into explorational and exploitational phases. Exploitational phases end up at the Pareto-front, but the stability of the population breaks down after a while and the explorational phase restarts. Therefrom, the Monessy algorithm often revisits the Pareto set during a long run. If the trade-off sets (for the biobjective case) are plotted, insight can be gained for the refined formulation of the optimization problem. Also, stopping conditions can be derived. The Monessy algorithm was examined on a test function from literature. The effective search for the Pareto-front of the Monessy algorithm was demonstrated and compared with the low performance of random search. Also, the replacement of weighted-sum approaches with multiple objectives was considered, employing a real-world application from the field of texture filtering. It comes out, that such replacements should be performed with caution. Often, the Pareto set contains unwanted solutions. The objectives have to be designed in a manner, which lowers the performance of such unwanted solutions.

Reference

1. Forensica C M, Fleming P J 1995 An Overview of Evolutionary Algorithms in Multiobjective Optimization. *Evolutionary Computation*, **3**(1):1-16

2. Köppen M, Teunis M, Nickolay B 1997 A Neural Network that uses Evolutionary Learning. *Proc. of 1997 IEEE International Conference on Evolutionary Computation, ICEC'97, Indianapolis, IN, April*, pp 635-639

3. Köppen M, Teunis M, Nickolay B 1997 Nessy - an Evolutionary Learning Neural Network. *Proc. of the 2nd International ICSC Symposium on Soft Computing, SOCO'97, Nîmes, France, September*, pp 243-248

4. Schaffer J D 1985 Multiple Objective Optimization with Vector Evaluated Genetic Algorithms. In: Grefenstette J J (ed) 1985 *Genetic Algorithms and Their Applications: Proc. of the First International Conference on Genetic Algorithms*, Lawrence Erlbaum, pp 93-100

5. Clergue M, Collard P, Gaspar A 1997 DGA and Pareto elitism: Improving Pareto optimization. *Proc. of the 2nd International ICSC Symposium on Soft Computing, SOCO'97, Nîmes, France, September*, pp 315-321

6. Forensica C M, Fleming P J 1993 Genetic Algorithms for Multiobjective Optimization: Formulation, Discussion and Generalization. *Proc. of the 5th International Conference on Genetic Algorithms, Illinois at Urbana Champaign, NM, July*, pp 416-423

7. Zhao Q 1996 Efficient Learning of NN-MLP based on Individual Evolutionary Algorithm. *Neurocomputing* 13 2-4:201-215

8. Köppen M, Teunis M, Nickolay B 1997 A Framework for the Evolutionary Design of 2D-Lookup Filters. Technical Report ME-MK-97-03, Department for Pattern Recognition, FhG IPK Berlin, Berlin, Germany

9. Furuhashi T, Nakaoka K, Uchikawa Y 1994 A New Approach to Genetic Based Machine Learning and an Efficient Finding of Fuzzy Rules. *Proc. WWW'94, Nagoya, Japan*, pp 114-122

10. Hashiyama T, Furuhashi T, Uchikawa Y 1995 Design of Fuzzy Controllers for Semi-Active Suspension generated through the Genetic Algorithm. *Proc. ANNES'95, Dunedin, New Zealand*, pp 166-169

11. Fourman M P 1985 Compaction of Symbolic Layout Using Genetic Algorithms. In: Grefenstette J J (ed) 1985 *Genetic Algorithms and Their Applications: Proc. of the First International Conference on Genetic Algorithms*, Lawrence Erlbaum, pp 141-153

The Scout Algorithm applied to the Maximum Clique Problem

Fabio Abbattista, Francesco Bellifemine, Donato Dalbis
Dipartimento di Informatica - Universita' di Bari
Via E. Orabona, 4 - 70126 Bari - Italy
Tel: +39 (0)80 5443298, Fax: +39 (0) 80 5443196
E-mail: fabio@di.uniba.it, donato@di.uniba.it

Keywords: Maximum Clique Problem, Natural Algorithms, Optimization.

Abstract

In this paper a new optimization algorithm is presented. It is inspired by the natural behavior of human or animal scout in exploring unknown geographical regions, and in their ability to exploit information coming from past experience. The algorithm has been applied to the maximum clique problem, and favorably compared with other more specialized technique.

1. Introduction

In this paper, we present a new natural algorithm, called *Scout*. The Scout algorithm has been inspired by the human (and animal) scouts. They are able to find, in unknown areas, useful places such as rivers, food, refuges, enemies and preys. They exploit the trails leaded by other scouts or by the prey, and they remember the safe path visited in previous explorations.

As stated in [1], a *natural algorithms* should possess the following five distinctive characteristics:

1- inspired from a natural process;
2- easy to implement on a parallel computer (intrinsically parallel);
3- not deterministic;
4- based on the Darwinian concept of evolution and selection;
5- it should use the positive feedback.

Scout is a natural algorithm in which near all the five characteristics are present; as it will be clear in the following, Scout uses the selection concept but, in its current implementation, a weak form of evolution occurs.

The paper is organized as follows: section 2 describes the details of Scout; in section 3, we analyze some properties of the algorithms, using some simple instances of the Maximum Clique Problem (MCP) as a benchmark. In section 4, we present the results of the application of Scout to more complex instances of the MCP; finally, conclusions and further work are presented in section 5.

2. The Scout algorithm

The Scout algorithm, in its current implementation, explores binary search spaces, represented by all the possible binary strings **S** = $<S_i>$ i = 1,..., l (l is a feature of the problem). The basic element of the algorithm is represented by the probability vector **P** = $<P_i>$, in which each element P_i gives the probability that the i-th element S_i will be 1.

The main goal of Scout is to identify a vector $\mathbf{P^{opt}}$ = $<P^{opt}_i>$ in such a way that, applying the P^{opt}_i it is possible to generate an optimal solution, $\mathbf{S^{opt}}$ = $<S^{opt}_i>$, for the problem at hand.

The vector $\mathbf{P^{opt}}$ is produced by iterating the Scout algorithm for a prefixed number of cycles, says Max_Cycles. For each cycle, *c*, the vector **P** is left unchanged (it will be denoted with P(c)); at the beginning of a run, $P_i(0)$'s are all set equal to 0.5.

The algorithm is iterative and, in each cycle, it performs the following operations:

1 - Generation. Generation of *K* solutions **S** using P(c). The value of *K* is constant in all cycles.

2 - Selection. Evaluation of each **S** by using a function f(S). Definition of the set $\{S^P(c)\}$ (its cardinality is denoted by Num_S) composed by all the positive solutions, that is all the solutions for which $f(S^P(c)) > f^{max}(c-1)$ (where $f^{max}(c-1)$ represents the maximum value of f(S) in the preceding cycle).

3 - Updating of P(c). To compute all the new $P_i(c+1)$ the subsets $\{S_i^P(c)\}^j$ has to be considered, whose cardinality are denoted by $N_{i,j}(c)$. They are composed of all the solutions in $\{S^P(c)\}$ in which the i-th element is equal to j (j = 0 or 1). Next, it is necessary to compute $\bar{f}_{\{S^P(c)\}}$ and $\bar{f}_{\{S_i^P(c)\}^j}$ representing the average function value on the sets $\{S^P(c)\}$ and $\{S_i^P(c)\}^j$, respectively.

Let $\lambda_i^j = \dfrac{\bar{f}_{\{S_i^P(c)\}^j}}{\bar{f}_{\{S^P(c)\}}}$, then it follows:

$$P_i(c+1) = \frac{P_i(c) + \lambda_i^1}{1 + \lambda_i^0 + \lambda_i^1}$$

These 3 steps are iterated until the stop condition occurs; in our experiments we used a fixed number of cycles (Max_Cycles).

A last few words are needed about the choosing of each S_i in step 1. Following the above-mentioned step 1, Scout generates solutions corresponding to the binary configuration with higher bit probability value. On the basis of the performance of these solutions, the vector **P** is updated. The updating, in this situation, will increase P_i's having high values and it will decrease P_i's with low values. After a very few cycles, P_i's will have only values 1 or 0 and the solutions will tend to be equal each other (each solution will correspond to the bit configuration with the highest probability value), leading to the stagnation of the algorithm. To overcome this situation, the step 1 is also driven by a factor E measuring the tendency to explore alternative paths, similar to the mutation operator in evolutionary algorithms. When the probabilistic choosing has been performed, for each S_i, the algorithm generates a random number *r*. If this number *r* is less than the parameter value E (that is the event "To explore alternative path" occurs) the chosen S_i value is inverted. In such a way, the algorithm has the possibility to take unlikely decisions, and to explore new solutions.

The following algorithm synthesizes the preceding description of Scout. In this algorithm we assume a maximization problem. Only slight changes are needed in the minimization case.

- c:=0;
- For each i
    - $P_i(c)$:=0.5;
- $f^{max}(c-1)$:=-∞;
- Repeat
    - $f^{max}(c)$:=- ∞;
    - f^{tot}:=0;
    - $\bar{f}_{\{S^p(c)\}}$:=0;
    - Num_S:=0;
    - For each i and j
        - $\Delta P_{i,j}(c)$:=0;
        - $N_{i,j}(c)$:=0;
    - For each of the K solutions S
        - For each i
            - $S_i \approx P_i(c)$; where $Prob(S_i=1) := P_i(c)$;
            - *r*:=Random();
            - IF (*r*<E)
                - $S_i := 1 - S_i$;
        - f:=f(S);
        - IF ($f > f^{max}(c)$)
            - $f^{max}(c)$:=f;
        - IF ($f > f^{max}(c-1)$)
            - Num_S:=Num_S+1;
            - $f^{tot} := f^{tot} + f$;
            - For each i
                - $j = S_i(c)$;

• $\Delta P_{i,j}(c) := \Delta P_{i,j}(c) + f$;
• $N_{i,j}(c) := N_{i,j}(c) + 1$;

• IF (Num_S>0)

• $\bar{f}_{\{S^p(c)\}} := f^{tot}/\text{Num\_S}$;

• For each i

• For each j

• IF ($N_{i,j}(c)>0$)

$$\bullet\ \lambda_i^j \frac{\Delta P_{i,j}(c)}{N_{i,j}(c) * \bar{f}_{\{S^p(c)\}}};$$

else

• $\lambda_i^j = 0$;

• For each i

$$\bullet\ P_i(c+1) = \frac{P_i(c) + \lambda_i^{\ 1}}{1 + \lambda_i^{\ 0} + \lambda_i^{\ 1}};$$

• c:=c+1;

until (c>=Max_Cycles)

where:

f^{tot} = the sum of the function values of all the positive solutions in a cycle;

$\Delta P_{i,j}(c)$ = is the increment of each P_i, not yet normalized, evaluated on the basis of all the positive solutions in a cycle.

A last few words about the complexity of the Scout algorithm. The temporal complexity is strongly dependent from the number of scouts as well as the problem to be optimized. It should be easily calculated that the temporal complexity is (Max_Cycles*K*Problem), where problem represents the maximum complexity between the size of the problem (l) and temporal complexity of the function to be optimized (f(S)). The spatial complexity, instead, is linearly dependent on the size of the problem at hand (l)

To conclude, it could be stated that despite its simple conceptual structure, Scout merges the advantages and the important features of natural groups or algorithms: the competition, the cooperation and the evolutionary character.

The competition performed at each cycle among the different solutions is very similar to the selection operator of evolutionary algorithms. In these, selection operates on the basis of the fitness function, that is, the objective function evaluated on individuals (or a derived value). The cooperation is represented by the "parallel updating" of vector **P** performed by the positive solutions. This step is conceptually similar to the pheromone trace in the Ant System [1]. The evolutionary character is present, in a reduced form, by means of the parameter E, a form of mutation of the probabilistic behavior of the agents.

3. Basic Properties of the Scout Algorithm

To analyze the behavior of Scout we use the Maximum Clique Problem as benchmark, in order to compare it with other approaches.
The Maximum Clique Problem is a well-known NP-hard optimization problem [2], and it is computationally intractable even with respect to good approximation [3]. It can be formulated as follows: let G=(V,E) be an undirected graph, where V={1, ..., n} is the set of vertices, and $E \subseteq V x V$ is the set of edges. A clique of G is a subset of V in which every pair of vertices is connected by an edge. The problem of finding the maximum clique consists of finding the clique of maximum size, that is the clique having largest cardinality. An example of graph, and some of its cliques, is shown in figure 1.

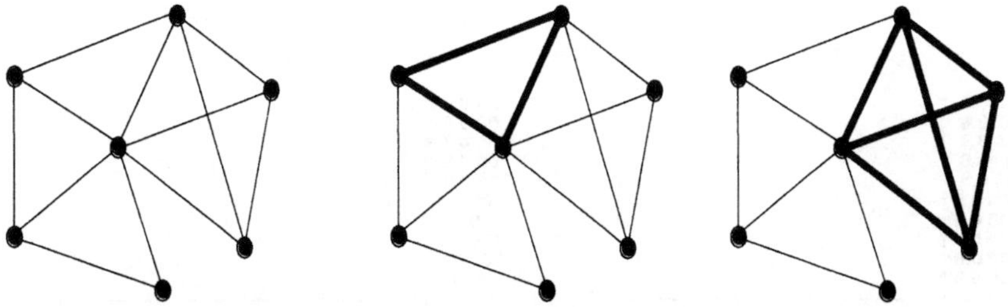

(a) (b) (c)

Figure 1. (a) An example of graph. (b) An example of clique (in bold) of graph (a). (c) The maximum clique (in bold) for graph (a).

Many algorithms have been proposed to approach MCP, either specifically designed for the problem or more generals; Scout belongs to the second category.
We have used as objective function f, the same as in [4], which is a derivation of the theorem due to Motzkin/Straus; in this way MCP is formulated as an indefinite global quadratic optimization problem:

$$\text{global min } f(x)=x^T Ax$$

with $x \in \{0,1\}^L$, A=Ag-I, where Ag is the adjacency matrix of the graph G, I is the identity matrix and x^T the transposed solution vector x; if x* solves this equation, then C, defined by the set of edge i where $x^*_i=1$ and $i \in V$, is a maximum clique of G with |C|=f(x*).
The main goal of the experimentation is to identify the optimal parameter configuration for the MCP. The parameters directly influencing the behavior of the Scout algorithm are S, the number of scouts and E, the tendency of the Scout algorithm to explore alternative solutions.

In all the experiments, the number of evaluations performed, that is the number of scouts times the number of cycles, has been kept constant (5,000). The values used for the two analyzed parameters were s∈{2, 5, 10, 20} and 0.000≤E≤0.200. The experimentation is based on Sanchis graphs [], with 50 and 100 nodes and with 5 different density values (0.5, 0.6, 0.7, 0.8, 0.9). Each of the ten graphs has been randomly generated five times, leading to 50 different instances of the MCP. All the tests were averaged over 10 trials.

The analysis of the results was based on the two following measures:

- the number of evaluations needed to reach the global maximum;
- the number of trials (out of 10) in which the global maximum is found.

In the following graphs, we highlight the value of the two parameters, s and E, corresponding to optimal test, i.e. the test with the minimum number of evaluations (graphs 1 to 4) and the maximum number of successful trials (graphs 5 to 8).

Each graph corresponds to a different value of s (2, 5, 10 and 20) and synthesizes the results over the ten different graphs (50 and 100 nodes with five different density values).

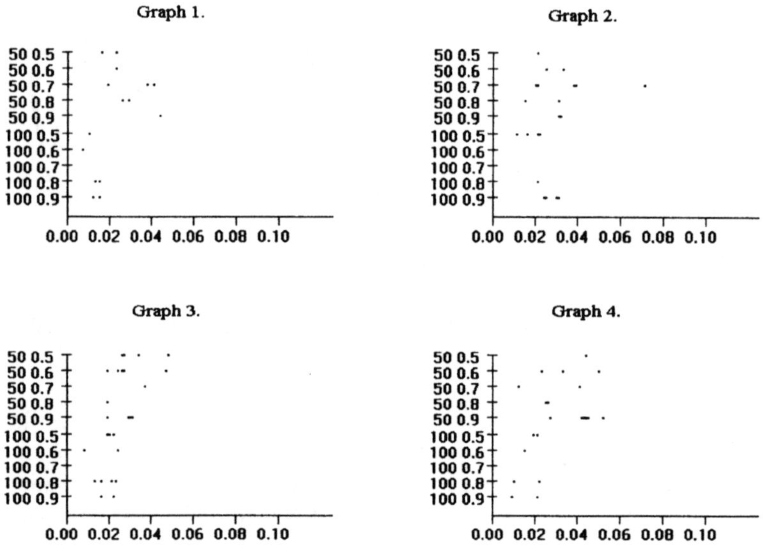

Figure 2. Values of parameter E (x axes) corresponding to the minimum number of evaluations needed to reach the optimum for the ten graphs (2 different sizes (50 and 100 nodes) times 5 different densities (0.5, 0.6, 0.7, 0.8a and 0.9)) with s=2 (graph 1), s=5 (graph 2), s=10 (graph 3) and s=20 (graph 4).

Graph 5.

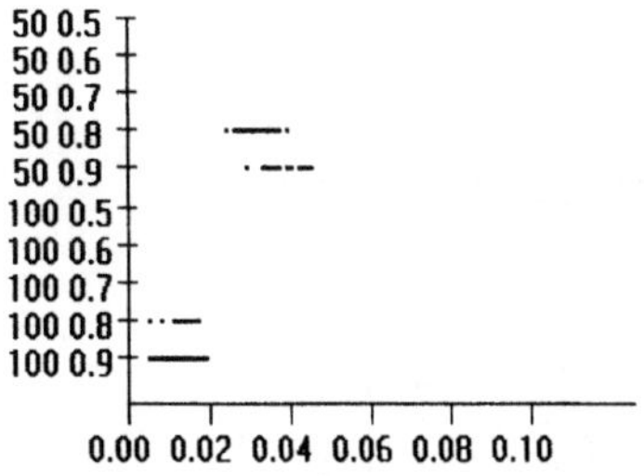

Graph 6.

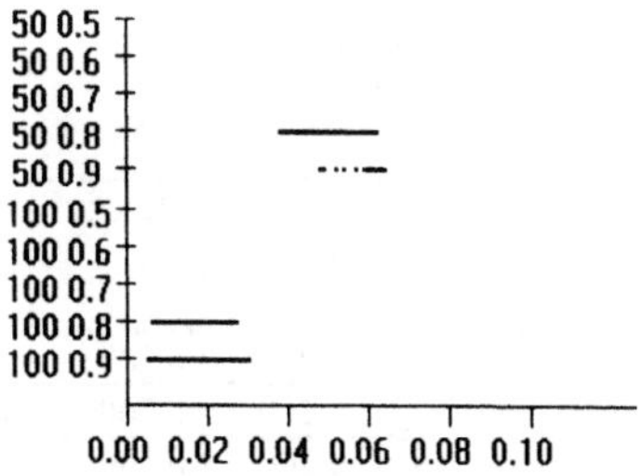

Graph 7.

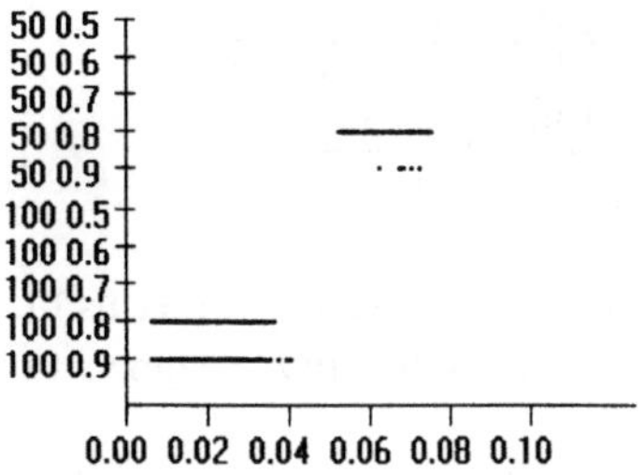

Graph 8.

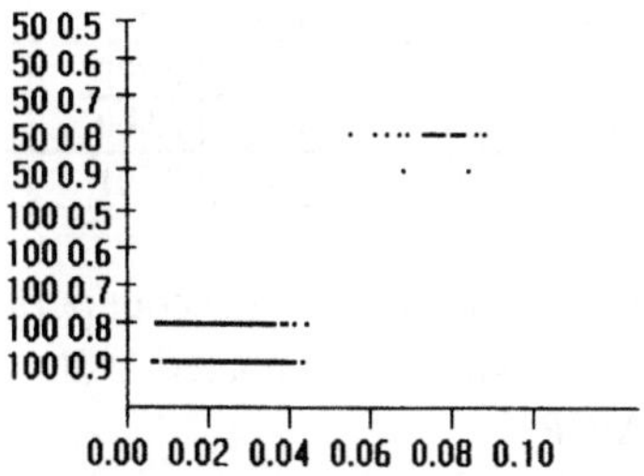

Figure 3. Values of parameter E (x axes) corresponding to the maximum number of trials reaching the optimum for the ten graphs (2 different sizes (50 and 100 nodes) times 5 different densities (0.5, 0.6, 0.7, 0.8a and 0.9)) with s=2 (graph 1), s=5 (graph 2), s=10 (graph 3) and s=20 (graph 4).

4. Comparison with other techniques

In this section we compare the Scout algorithm with three other different algorithms, namely M/S [6], CBH [7], RLS [8] over the DIMACS benchmark graphs.

The first experimentation was performed on the 11 DIMACS Sanchis graphs, with different sizes (200, 400 and 1000 nodes) and different densities (0,5, 0,7 and 0,9). In all the experiments we set the number of evaluations to 20,000, 40,000 and 100,000 for the three graph sizes 200, 400 and 1000 respectively. On the basis of the results shown in the graphs, increasing the size of graphs, it is necessary to reduce the E parameter values. Following this indication, the parameters used in all the experiments were: K=10 and E ranging from 0.000 to 0.015 step 0.001, leading to 16 different experiments for each graph; each experiment was run with 10 different, randomly chosen, initial conditions. Table 1 reports, for the 11 DIMACS graphs, the best result, in terms of clique size, for all the four algorithms.

Table 1: Comparison of the performance of Scout vs. other algorithms on the 11 Sanchis DIMACS graphs.

				Algorithms			
Graph file	Nodes	Density	Global Max	M/S	CBH	RLS	Scout
San200_0.7_1	200	0.700	30	15	15	30	20
San200_0.7_2	200	0.700	18	12	12	18	13
San200_0.9_1	200	0.900	70	45	46	70	48
San200_0.9_2	200	0.900	60	35	36	60	46
San200_0.9_3	200	0.900	44	33	30	44	35
San400_0.5_1	400	0.500	13	7	8	13	8
san400_0.7_1	400	0.700	40	20	20	40	22
san400_0.7_2	400	0.700	30	15	15	30	17
san400_0.7_3	400	0.700	22	12	14	22	14
san400_0.9_1	400	0.900	100	55	50	100	81
San1000	1000	0.501	15	8	8	15	8

As it can be seen from table 1, Scout performance are quite satisfactory in terms of clique size found, also compared with algorithms more specifically designed to solve the MCP.

In the next table, the same comparison was made among Scout and the three other techniques, on graphs with characteristics different from the Sanchis graphs. The hyphen (-) in some rows indicates that the corresponding algorithm has not been applied to the particular graph.

Table 2: Comparison of the performance of Scout vs. other algorithms on some DIMACS graphs.

				Algorithms			
Graph file	Nodes	Density	Max known	M/S	CBH	RLS	Scout
Brock200_1	200	0.745	21	18	20	21	20
Brock200_2	200	0.496	12	8	12	12	10
Brock200_3	200	0.605	15	10	14	15	13
Brock200_4	200	0.658	17	13	16	17	15
Brock400_1	400	0.748	27	21	23	25	21
Brock400_2	400	0.749	29	22	24	25	21
Brock400_3	400	0.748	31	20	23	31	21
Brock400_4	400	0.749	33	21	24	33	21
Mann_a9	45	0.927	16	12	16	-	16
Mann_a27	378	0.990	126	117	121	126	120
SanR200_7	200	0.697	18	16	18	18	16
SanR200_9	200	0.898	42	40	41	42	38
SanR400_5	400	0.501	13	11	12	13	11
SanR400_7	400	0.700	≥21	18	20	21	19

The table clearly shows that Scout performances are not so satisfactory as in the table 1. This is clearly due to the parameters, previously tuned on Sanchis graph, not able to support the Scout algorithm in the exploration of different graphs.

5. Conclusions

A new optimization algorithm, inspired by a natural process has been presented. The preliminary results, when applied to the MCP, are quite encouraging.
As the Scout algorithm does not need a priori knowledge about the problem to be solved, it could be favorably compared with other well-known optimization techniques such as evolutionary algorithms that, when applied to MCP performed quite poorly [9].
Moreover, Scout is a very general optimization algorithm, in the same way as evolutionary algorithms. It needs only a way to measure the goodness of found solutions (similarly to the fitness function of evolutionary algorithms). In order to verify the generality of the Scout approach to optimization, we are experimenting Scout on very different problems, such as: real functions optimization, edge detection, TSP, vector quantization, forecasting, routing, etc.
Further work is in progress to investigate the possibility to introduce a stronger evolutionary factor in Scout; on the basis of this enhancement, each scout will be able to adapt its own E parameter value (that is its tendency to explore new paths).

References

[1] M. Dorigo, V. Maniezzo and A. Colorni, 1994, Introduzione agli algoritmi naturali, *AICA Rivista di Informatica.* **XXIV, 3,** 179-197.

[2] M. R. Garey and D. S. Johnson, 1979, *Computers and Intractability: A Guide to the Theory of NP-Completeness*, W.H. Freeman.

[3] U. Feige, S. Goldwasser, L. Lovasz, S. Safra and M. Szegedy, 1991, Approximating Clique is Almost NP-Complete, *Proc. 32nd Ann. Symp. Found. Comput. Sci.*, San Juan, Puerto Rico, 2-12.

[4] P. M. Pardalos and J.Xue, 1994, The Maximum Clique Problem, *J. Global Optim.*, **4**, 301-328.

[5] L. Sanchis and A. Jagota, Some Experimental and Theoretical Results on Test Case Construction for the Maximum Clique Problem, DIMACS Tech. Rep. 93-69, October 1993, DIMACS Series in Discrete Mathematics and Theoretical Computer Science.

[6] I. M. Bomze, M. Pelillo and R. Giacomini, 1997, Evolutionary Approach to the Maximum Clique Problem: Empirical Evidence on a Larger Scale, in *Developments in Global Optimization*, I. M. Bomze et al. (eds.), Kluwer Academic Publishers, 95-108.

[7] L. E. Gibbons, D. W. Hearn and P. M. Pardalos, 1995, A Continuos Based Heuristic for the Maximum Clique Problem, in *Second DIMACS Implementation Challenge,* D.S. Johnson and M. Trick (eds.). DIMACS Series in Discrete Mathematics and Theoretical Computer Science.

[8] R. Battiti and M. Protasi. Reactive Local Search for the Maximum Clique Problem, Tech. Rep. TR-95-052, International Computer Science Institute, 1995.

[9] B. Carter, K. Park. How Good are Genetic Algorithms at Finding Large Cliques: an Experimental Study, Tech. Rep. BU-CS-93-015, Boston University, Computer Science Department, 1993.

Unconstrained Optimization Using Genetic Box Search

S. Lozano[1], J.J. Domínguez[2], F. Guerrero[1], J. Larrañeta[1] and L. Onieva[1]

[1]*Escuela Superior de Ingenieros, Universidad de Sevilla,*
[2]*Escuela Superior de Ingeniería , Universidad de Cádiz*

Abstract

This paper presents an iterative method that performs a succession of box searches. For multimodal functions, provided that the size of the box is larger than that of the valleys of the objective function, the method is able to jump across those valleys without getting trapped in local optima. To carry out each box search we propose using a Genetic Algorithm, which are known to be robust and efficient search methods. Our implementation of the Genetic Box Search concept uses floating point-coded individuals, convex linear combination crossover and gaussian mutation. Computational experiences show that this approach is very robust with respect to the starting point, is very effective at converging to the global optimum and has low computational requirements.

Keywords: Multimodal Optimization, Box Search, Genetic Algorithms

1. Introduction

Unconstrained optimization deals with the problem of minimizing (or maximizing) a function in the absence of any restrictions, i.e. $Min\ f(\bar{x})\quad \bar{x}\in\mathbb{R}^n$ We do not include as constraints the possible existence of lower and upper bounds on the different variables $l_i \le x_i \le u_i$ because they can be easily imposed within the optimization process.

Most solution methods involve line searches, i.e. minimizing the objective function along a properly chosen direction (coordinate axis, gradient, conjugate gradient, etc). There are many methods for performing such one-dimensional search ([1], [2]) but in the case of multimodal functions, since all of them require that the function be unimodal along the search interval these methods usually result in small step sizes and get trapped in local optima. In [3] a Genetic Line Search (GLS) approach was used to overcome such problems. Instead of restricting the line search to a small interval in the neighbourhood of the current solution, a look-further-ahead approach was proposed. The search interval was extended and typically contained multiple valleys. A Genetic Algorithm (GA) was used to perform the search, using niche sharing [4] to maintain diversity. The method allowed for jumping across valleys, thus accelerating the optimization process.

In this paper, instead of performing a line search, we perform a box search, i.e. we search a hypervolume centered on the current solution. The proper size of the box is problem-dependent and it can even be made adaptive. The larger the box, the smaller the number of box searches that may be required but the more computationally demanding will be each box search. On the contrary, small box sizes can be more easily searched but lead to small step sizes and local minima.

In section 2 a GA is presented to perform box searches. In section 3 the overall iterative process is described. In section 4 computational experiences regarding three multimodal functions are presented. Finally, in section 5 conclusions are drawn and further research outlined.

2. The Genetic Algorithm

In order to find the best solution within the box to be searched, a robust and efficient procedure must be used. GA are well-suited for this task. GA are probabilistic search techniques based on concepts from evolution theory and natural genetics ([4], [5]). Among the strong points of GA are their flexibility, simplicity and hybridization capabilities. Among their weak points are their heuristic nature and their inability at handling constraints.

There is not a single type of GA but whole families of algorithms that differ mainly in the coding scheme used and the way they renew the population in each generation. The specific GA that we have implemented for this problem codes individuals as vectors of floating point real numbers normalized between 0 and 1. Let y_{ijk} the i-th component of the j-th vector in the population after k generations, let a_i and b_i the lower and upper limits of the box along the i-th coordinate axis and let x_i^0 the i-th component of the center point of the box. Then, the j-th individual $\bar{y}_{jk}$ in the k-th population can be mapped to a solution $\bar{x}_{jk}$ with components

$$x_{ijk} = \begin{cases} x_i^0 + 2\cdot(y_{ijk} - 0.5)(b_i - x_i^0) & \text{if } y_{ijk} \geq 0.5 \\ x_i^0 - 2\cdot(0.5 - y_{ijk})(x_i^0 - a_i) & \text{if } y_{ijk} \leq 0.5 \end{cases}$$

which is guaranteed to lie within the box $[a_1,b_1]$x $[a_2,b_2]$x...x $[a_n,b_n]$.

Note that usually $x_i^0=(a_i+b_i)/2$ and $b_i-a_i=S^{1/n}$ where S is the size of the box and n is the problem dimension. However, whenever the box is close a boundary of the feasibility region the box is truncated appropiately. Mathematically,

$$a_i = \max\{l_i, x_i^0 - 0.5 * S^{1/n}\} \qquad b_i = \min\{u_i, x_i^0 + 0.5 * S^{1/n}\}$$

Only two operators are used, one for crossover and another for mutation. The crossover operator is a convex linear combination ([6]), also called arithmetical crossover ([7]): given two parents $\bar{y}_{jk}^1$ and $\bar{y}_{jk}^2$ and a random number $\alpha\in(0,1)$, two offsprings are computed as

$$\bar{y}_{jk}^3 = \alpha\cdot\bar{y}_{jk}^1 + (1-\alpha)\cdot\bar{y}_{jk}^2 \qquad \bar{y}_{jk}^4 = (1-\alpha)\cdot\bar{y}_{jk}^1 + \alpha\cdot\bar{y}_{jk}^2$$

The mutation operator consists in generating a random gaussian perturbation around the selected individual. The magnitude of the perturbation is controlled through the standard deviation σ, which is kept constant at a value proportional to one minus the probability of mutation. The rationale is that the more frequent mutations are performed, the smaller their magnitude and vice versa. Whenever a mutation causes an individual to step out of the box, it is repaired by setting its value equal to the closest feasible value. This reparation mechanism was not necessary after crossover since the convex linear crossover operator guarantees that if the two parents fall inside the box, so do their offsprings.

For a minimization problem, fitness is defined as an upper bound of the global optimum minus the objective function. For a maximization problem, fitness would be equal to the objective function. In that case, in order to keep fitness values positive adding a constant may be required.

Since the box size S controls the volume of the region to be searched, the population size is made proportional to that parameter. Thus, the larger the box, the larger the population size and, consequently, the more generations required for the population to converge. The initial population is chosen randomly within the box except for the individual (0.5, 0.5, ..., 0.5) which represents the center of the box and which is always included in the initial population.

A generational GA has been implemented. Given the previous population, the population for the next generation is formed in two steps. First, a percentage of individuals of the previous population are selected and copied to the new population. These individuals are selected with a probability proportional to their fitness. Elitism is also present since a copy of the best individual of the previous population is always included. The second step consists in completing the new population with individuals generated through crossover and mutation applied to individuals of the previous population. Thus, for crossover, two parents are selected (proportional to their fitness) from the previous population and the two offsprings are included in the new population. Analogously, for mutation, an individual is selected from the previous population and the mutated individual is included in the new population.

Summarizing, the basic parameters of the GA are: population size (N) , number of individuals from previous population (p_s), crossover probability (p_c), mutation probability (p_m=1-p_c) and number of generations (T). The effect of varying these parameters is the one expected from any GA. Thus, increasing p_s or p_c or decreasing N leads to faster convergence while the opposite helps maintain diversity longer. Many different combinations of these parameters can be used provided a proper balance between exploration and explotation is kept.

A pseudocode description of the GA follows:

```
procedure GA
begin
    t←0
    initialize Population(0)
    while t<T do
        t←t+1
        select best individual in Population(t-1) and copy to Population(t)
        select p_s-1 additional indiv. from Population(t-1) and copy to Population(t)
        while card(Population (t))< N do
            select operator (crossover or mutation)
            if crossover then
                select two parents from Population(t-1)
                generate two offsprings and insert in Population(t)
            else
                select one individual from Population(t-1)
                generate mutated individual and insert in Population(t)
            endif
        endwhile
    endwhile
end
```

3. Genetic Box Search

The Genetic Box Search (GBS) method has two inputs: the starting point $\bar{x}^0$ and the default box size S. In the first iteration, a box of size S centered in $\bar{x}^0$ is searched using the GA described on the previous section. Assuming a better solution is found in that box search it will be the center of the next box $\bar{x}^1$. If no better solution is found, the box is augmented and searched again. This process is repeted until either a better solution is found, in which case it will be used as the center of a new box, or the method stops. A pseudocode description of the method is:

```
procedure GBS
begin
    input starting point x̄^0
    input default box size S
    input maximum number of box size doubling steps allowed D
    input (if any) bounds l_i and u_i on decision variables
    k←0
    call GA to search box of size S centered in x̄^0
    while not-global-optimum do
        if improvement then
            k←k+1
```

```
            set box size to default value S
            call GA to search box centered in best solution of previous search x̄^k
      else
            double box size
            call GA to search augmented box centered in x̄^k
      endif
   endwhile
end
```

Some remarks on the method are due:

1. Since the initial population of each box search is seeded with the box center and since the GA uses elitism, the best solution in the succesive generations of a box search are not worse than the box center, which is the best solution found in the previous box search. This leads to monotonous improvements.

2. Assuming the box size is large enough to contain multiple valleys, GBS is able to jump across valleys and, hopefully, reach the global optimum.

3. The method would get stuck if for a box search no better solution than the box center can be found. That happens when the global optimum has been found but it can also happen if the box size is not large enough. To discriminate between both cases, the box size is doubled and the search resumed. If after a number of box size doubling steps the method is still stuck, convergence to the global optimum is presumed. On the contrary, if the box center was a local optimum and could be escaped from, the method continues on the new box center and with the box size reset to its default value.

4. The maximum box size S_{max} used is a function of the default box size and the maximum number of times the box size can be doubled ($S_{max}=S*2^D$)

4. Illustration

In this section, the results obtained with GBS for two of the problems tested are presented.The first test problem is a trigonometric function (TRIG) which for the bidimensional case is shown in Figure 1 and which in the interval under consideration $[0,100]^n$ has many local minima and a global minimum in the vicinity of the point (100,100,...,100). That global minimum has a value of approximately 280.

Figure 2 shows the evolution of the best solution found in each box search GBS for five runs starting at point (2,5). Note that all five runs converge to the global optimum. CPU times correspond to an Intel Pentium 166 PC. The parameters used are shown in Table 1

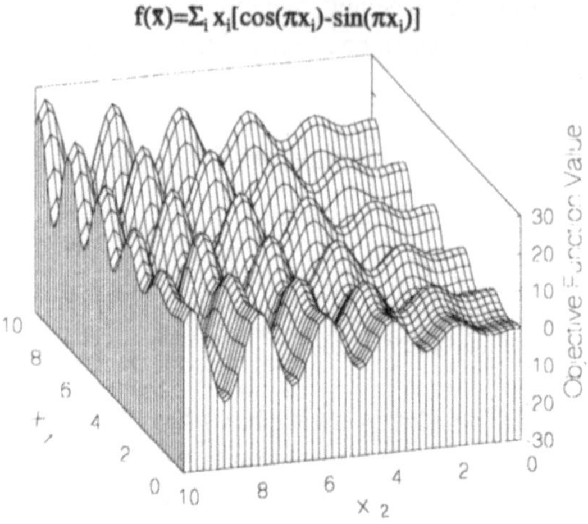

Figure 1. Bidimensional TRIG function

Parameter	Value
Default box size (S)	49(=7x7)
Population size (N)	0.5 S
No. of individuals from previous population (p_s)	0.3 N
Crossover probability (p_c)	7
Mutation probability (p_m=1-p_c)	3
Number of generations (T)	300
Maximum number of box size doubling steps (D)	2

Table 1. GBS parameters for problem TRIG

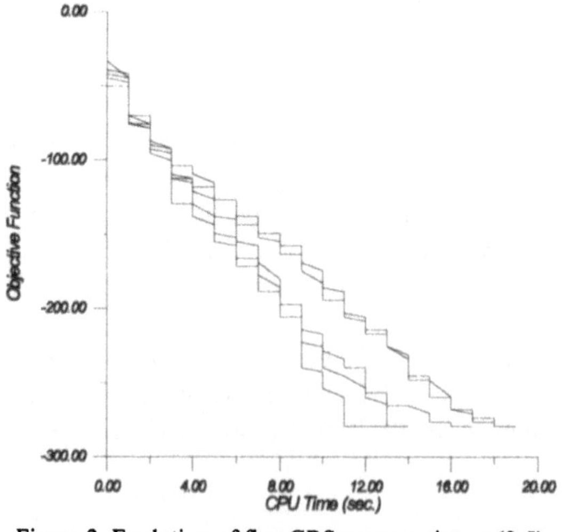

Figure 2. Evolution of five GBS runs starting at (2,5).

Figure 3 shows for the same five runs starting from (2,5), the evolution of the objective function versus for the successive box searches but in order to assess the sensitivity of the method to the default box size, five additional runs with S=25(=5x5) and another five runs with S=81(=9x9) are shown. The larger box size runs converge to the global optimum and in a smaller number of box searches. On the contrary, the small box size runs require a bigger number of box searches and can get trapped in local optima.

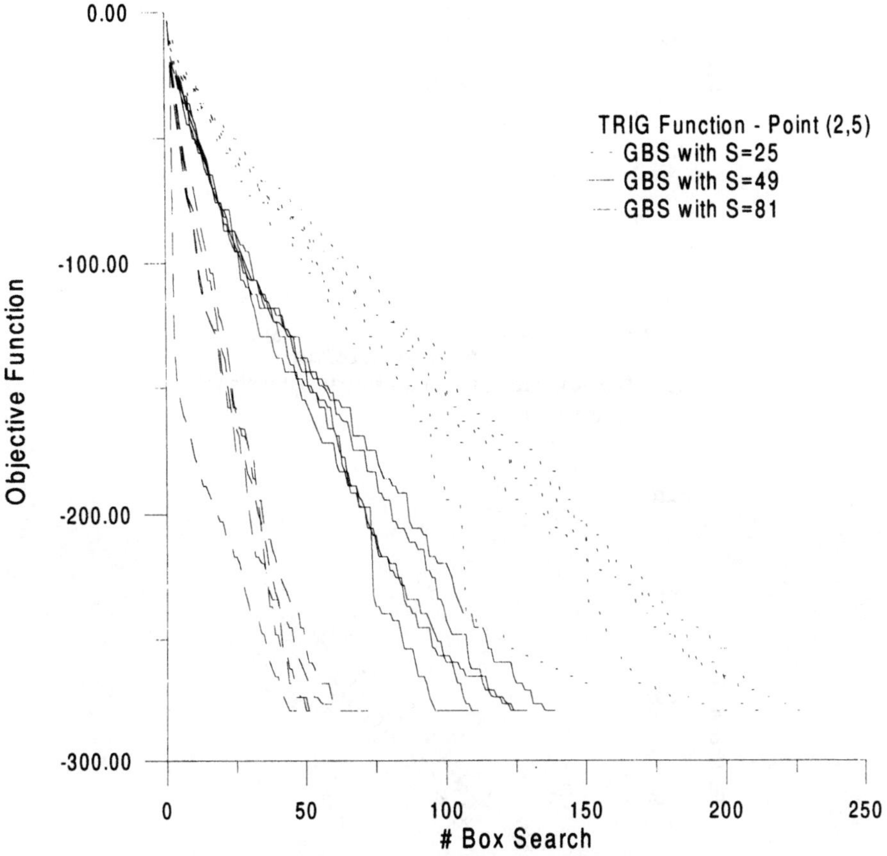

Figure 3. Evolution of GBS for different default box sizes

Problem TRIG was also solved starting from 100 different random points using the parameters shown in Table 1. In all cases GBS converged to the global optimum. Figures 4 and 5 show, respectively, the number of box searches and the CPU times required as a function of the initial distance to the optimum. Although the global optimum was reached even if the method started far from it, the number of box searches required increases with the initial distance to the optimum.

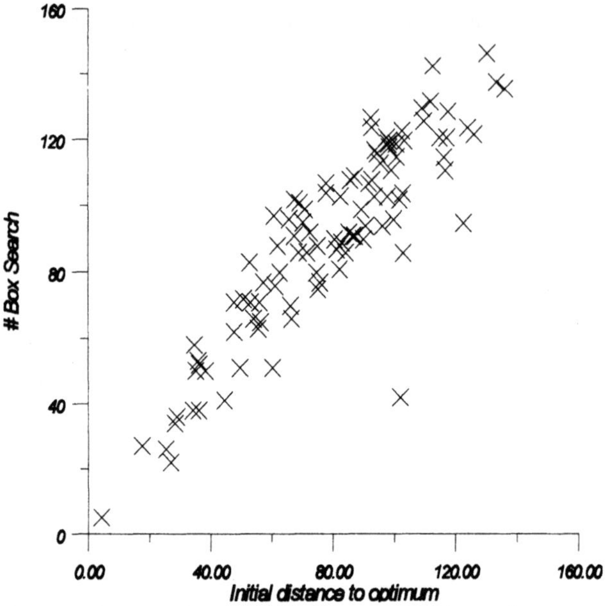

Figure 4. Box searches vs initial distance to optimum for problem TRIG

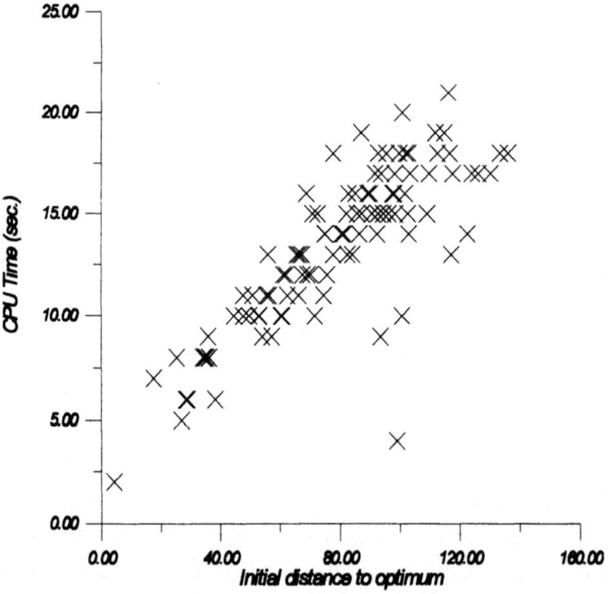

Figure 5. CPU time vs initial distance to optimum for problem TRIG

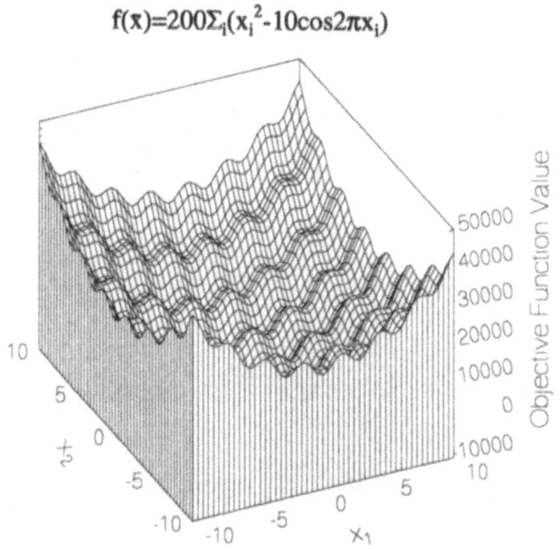

Figure 6. Bidimensional RAST function

The second test problem is the Rastrigin function (RAST), which for the bidimensional case is shown in Figure 6 and which in the interval under consideration $[-50,50]^n$ has many local minima and a global minimum at (0,0,...,0). That global minimum has a value of exactly -4,000.Problem RAST was solved starting from 100 different random points using the parameters shown in Table 2. Results were similar to those observed for the first test problem. In all cases GBS converged to the global optimum in a small amount of box searches and CPU time. Figures 7 and 8 show, respectively, the number of box searches and the CPU times required as a function of the initial distance to the optimum.

Parameter	Value
Default box size (S)	36(=6x6)
Population size (N)	0.5 S
No. of individuals from previous population (p_s)	0.3 N
Crossover probability (p_c)	5
Mutation probability (p_m=1-p_c)	5
Number of generations (T)	300
Maximum number of box size doubling steps (D)	2

Table 2. GBS parameters for problem RAST

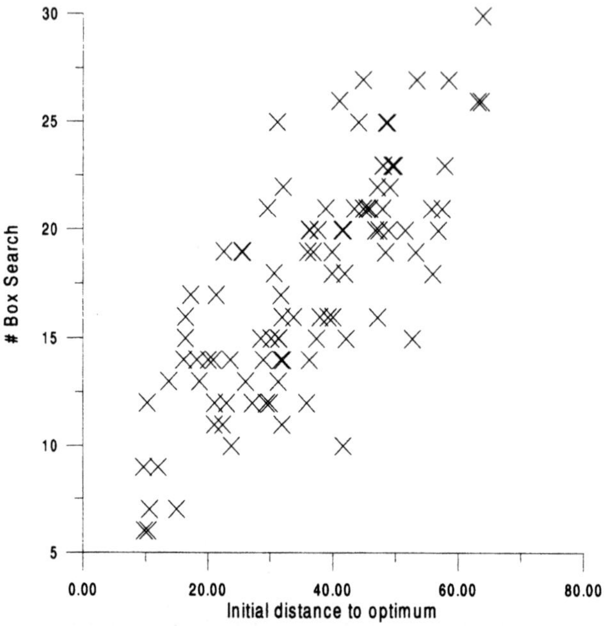

Figure 7. Box searches vs initial distance to optimum for problem RAST

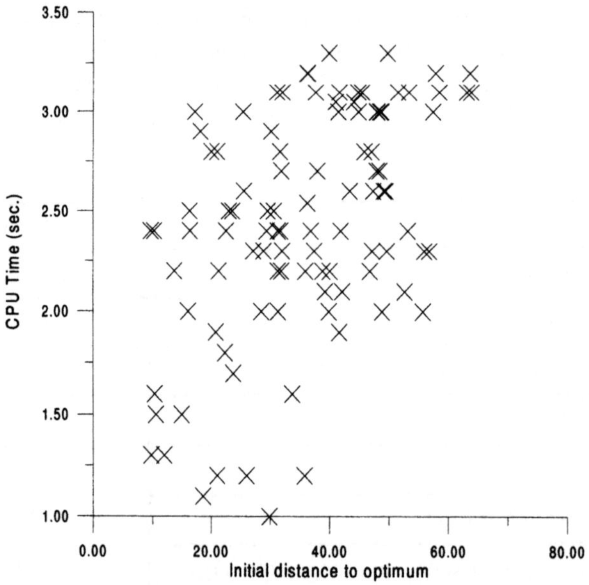

Figure 8. CPU time vs initial distance to optimum for problem RAST

5. Conclusions

In this paper a new optimization strategy GBS has been presented. It consists in solving a succession of box-constrained subproblems. The center of each box is the best solution of the previous subproblem. Using a box of enough size, the method is able to escape from local optima. For solving each subproblem a GA is used. In order to properly sample the box, the population size has been made proportional to the box size. A floating-point codification with convex crossover and gaussian mutation has been used. The method has been tested on two multimodal functions and, for both problems, it has been able to find the global optimum independently of the starting point and requiring small CPU times.

References

1. Bazaraa M S, Sherali H D and Shetty C M 1993 *Nonlinear Programming. Theory and Algorithms*. Wiley, Chichester
2. Gill P E, Murray W and Wright M 1981 *Practical Optimization*. Academic Press, London
3. Lozano S, Dominguez J J, Guerrero F, Onieva L and Larrañeta J 1997 Training Feedforward Neural Networks Using a Genetic Line Search. In: Dagli CH et al (eds) 1997 *Intelligent Engineering Systems Through Artificial Neural Networks Vol 7*. ASME Press, New York, pp 119-124
4. Goldberg D E 1989 *Genetic Algorithms in Search, Optimization and Machine Learning*. Addison-Wesley, Reading
5. Holland J H 1992 *Adaptation in Natural and Artificial Systems*. MIT Press, Cambridge
6. Wright A H 1991 Genetic Algorithms for Real Parameter Optimization. In: Rawlins G J E (ed) 1991 *Foundations of Genetic Algorithms*. Morgan Kaufmann, San Mateo, pp 205-218
7. Michalewicz Z 1992 *Genetic Algorithms + Data Structures = Evolution Programs*. Springer-Verlag, New York

Improvement of Simple Genetic Algorithm for Solving the Uncapacitated Warehouse Location Problem

Jozef Kratica

Faculty of Mechanical Engineering, Department of Mathematics,
27. marta 80., 11 000 Belgrade
e-mail: ejozef@ubbg.etf.bg.ac.yu

Keywords: Genetic Algorithms, Uncapacitated Warehouse Location Problem

Abstract

This paper investigates the applicability of the improvement of simple genetic algorithm (SGA) method for solving the uncapacitated warehouse location problem (UWLP). Function for computing the item's objective value is improved depending upon the number of established warehouses. It is efficiently implemented, giving excellent results in specified environment. Mutation rate is also changed and now depends on test problem size. Duplicate item strings in population are discarded, which makes the population more diversified.

Overall performance of implementation is finally tuned by caching SGA. Through caching technique relatively smaller profit in performance is obtained when compared to previous techniques, but it is a general technique, and can be directly applied to other problems, not only to UWLP.

Computational experience with given problem examples indicates that, when compared to code in [1], the implementation of the modified algorithm is faster several times. For large size test problems the increase of computational speed may even exceed factor 10, and quality of obtained solutions is also significantly better.

1. Introduction

1.1 Genetic Algorithms

Genetic algorithms (Gas), the development of which is inspired by the analogy of natural evolution model, are robust and adaptive methods for solving search and optimization problems. GAs traditionally works with a population of items, as candidate solutions of search space. Population most commonly contains 10-200 items, that are usually fixed-length strings over some given, finite alphabet. Fitness function is defined to evaluate the quality of the items in population. The goal of the GA is to produce better items by using the stochastic genetic operators. Most commonly used genetic operators are selection, crossover and mutation.

The selection operator that produces a new population, favors the reproduction of the items of better fitness value more than the one of the worse ones. As children inherit pieces of their gene patterns from parents, GA implicitly exploits the correlation between inner structure and quality of items. Crossover applies to one or more parents and exchange genetic material between them, with the possibility that good parents can generate better new candidates (childrens). Mutation involves the modification of the value of each solution gene with some small probability p_{mut}. The role of mutation is to prevent the premature convergence of the GA to suboptimal solutions, by restoring lost or unexplored genetic material into the population. Initial population is in most cases randomly initialized, but some GA implementations allow some heuristic initialization of the population. The basic principles of GAs were first laid down rigorously by Holland [2], and are well described in many texts. Some of interesting survey articles are [3], [4], [5] and [6].

Extensive studies of solving the combinatorial optimization problems, with main focus to NP-hard problems, through GAs, are given in [7], [8], [9], [10] and [11]. For detailed overview of this scientific area, reader may consult some comprehensive bibliographies, for example [12] and [13].

1.2 Uncapacitated Warehouse Location Problem (UWLP)

The uncapacitated warehouse location problem (UWLP) deals with the selection of warehouse locations from a set of possible locations with the intention of serving a given set of customers from the chosen locations. The objective is to minimize the sum of the fixed charges for establishing the warehouses and the transportation costs corresponding to the supply of the demands.

Consider a set $I = \{1, ..., m\}$ of candidate places to locate warehouses, and a set $J = \{1, ..., n\}$ of customers. Each warehouse place $i \in I$ has a fixed cost f_i. Each customer $j \in J$ has a demand b_j and c_{ij} is the unit transportation cost from warehouse i to customer j. Without loosing a generality we can normalize customer demand to $b_j = 1$ (see [14]).

We should find which warehouses are established, as well as the amount supplied from warehouse i to customer j so that the total cost, including fixed and variable cost, is minimized.

The problem may be formulated mathematically as:

$$\min\left(\sum_{i=1}^{m}\sum_{j=1}^{n} c_{ij}x_{ij} + \sum_{i=1}^{m} f_i y_i\right) \qquad (1)$$

subject to:

$$\sum_{i=1}^{m} x_{ij} = 1, \qquad \text{for every } j \in J; \qquad (2)$$

$$0 \le x_{ij} \le y_i \text{ and } y_i \in \{0,1\}, \qquad \text{for every } i \in I \text{ and every } j \in J; \qquad (3)$$

where x_{ij} is the amount supplied from warehouse i to customer j. Array y_i indicates whether warehouse i is established ($y_i = 1$) or not ($y_i = 0$) providing that a set of established warehouses is $E = \{i \mid y_i = 1\}$, with cardinality $e = |E|$.

1.3 Methods for solving the UWLP

The location problems have been dealt with in a great number of research problems during the past three decades. The review of all the important contributions in that respect is beyond the scope of this paper. Some of recent survey articles are [15], [16], and [17].

The UWLP is a well-known NP-hard combinatorial optimization problem ([18]), also known in literature as uncapacitated facility location problem or simple plant location problem. This problem has been studied previously by many researchers (see survey articles [18], [19] and [20]). Some efficient implementations for solving UWLP are given by Erlenkotter [21], Guignard [22], Simao and Thizy [23], Koerkel [24] and Holmberg [25]. In these implementations are mainly used dual-based algorithms or Lagrangean relaxation, and obtained solutions are very good. If an optimal solution is not found through a particular method, Branch-and-Bound procedure completes the solution process.

Although UWLP is NP-hard, in some special cases it is solvable in polynomial time. For detailed information see [26], [27] and [28].

One method for solving given problem by simple genetic algorithm (SGA) is presented in [1]. In this paper the method for improving that implementation is described.

2. Basic implementation of SGA for solving the UWLP

In this section is concisely presented SGA implementation for solving UWLP, that was described in [1]. Presented results in [1] are good for test problems with size up to 100×1000, with reasonably good running time in those cases (see Table 4.3).

2.1 GA representation and objective value function

For given problem binary coding of possible warehouse locations is used. Every bit in the item binary string, which value is 1 denotes that particular warehouse is established, while 0 denotes it is not established. For faster execution the item string is allocated in 32-bit words.

Example 1:

For m=5, item string **01011 | 000000010000000010001110101**

1. - 5.   6. - 32.

denotes that:

- Warehouses **2, 4** and **5** are established;
- Warehouses **1** and **3** are not established;
- Bits 6-32 are ignored.

Established warehouse locations can be obtained from item binary string, by objective value function. That function finds minimal cost warehouse for every customer. Since the capacity of the warehouses has no limit, if every customer assigns most suitable warehouse (transportation cost from that warehouse to him is minimal), the total cost is minimal. Note that, in the assignment procedure for a specific customer, participate only established warehouses ($y_i = 1$).

2.2 Parameters of the SGA

Population size is 20 items, and generational replacement of population is implemented. In every generation 19 items of population are replaced, and only the best item is passed directly to next generation without selection, crossover and mutation. Maximal number of iterations originally in [1] has been 2 000, but that number is not enough for solving test problems of larger size ($n \geq 100$), and we increase it to 20 000. If optimal solution is obtained before 20 000 iterations, execution is stopped, and results are immediately printing. By GA it is not possible to prove optimality of given solution, but for test purposes only optimal solution can be used if it is known in advance by some other methods..

As it is a minimization problem, the fitness of every item is reversed to its objective value. Formula for that scaling into the interval (0, 1) is given in (4).

$$f_i = \frac{\max - p_i}{\max - \min} \tag{4}$$

In equation (4) numbers *max* and *min* are maximal and minimal item objective values in current generation, and p_i is an objective value of item *i*.

Simple GA uses simple selection, one-point crossover, by crossover rate $p_{cross} = 0.85$, and simple mutation by mutation rate of $p_{mut} = 0.01$ for small size test problems ($n \leq 50$), and $p_{mut} = 0.005$ for test problems of larger size ($n \geq 100$). Initial population is randomly initialized, because in that case we maintain maximal diversity into the population. For more details about SGA parameters for solving UWLP see [1].

3. Improvements of the SGA for solving UWLP

3.1 Objective value function

For notational and computational convenience, in the initialization part of program, we reindex the transportation cost matrix for each customer in non-decreasing order.

Important part of fast implementation the objective value function is the choice of the applied data structure. Our implementation beside transportation cost matrix, also used the indexed lists of warehouse locations. For every customer, particular list contains warehouse locations arranged in nondecreasing order of the transportation costs.

In the initialization part of program, from transportation matrix are produced indexed lists by *qsort()* built-in function. Indexed lists have about 50% additional memory requirements, but conduce to several times faster computation of the objective value function.

Dependent of number of established warehouses ***e***, is proposed two different methods for computing objective value. Threshold is $e_0 = c \cdot \sqrt{m}$, where ***c*** is a constant which value of 0.4 - 0.5 is obtained experimentally to achieve best performance.

If number ***e*** is large ($e > e_0$), we search indexed list of a given customer, for the first warehouse with $y_i = 1$. That warehouse is chosen, because it is established warehouse ($y_i = 1$) with minimal transportation cost to given customer. Choosing procedure is fast because the array ***y*** is dense, and only few steps are needed. Number of steps for every customer is approximate ***m/e***, and overall run-time complexity is $O\left(m + n \cdot \frac{m}{e}\right)$.

Previous procedure in the case of small number ***e***, gives poor results, and it is slow, because the array ***y*** is sparse, and many steps on searching indexed list are needed to find first warehouse with $y_i = 1$. In that case we perform another strategy, and do not use the indexed lists in choosing procedure, but form array of ordinals of established warehouses ***o***, with condition $o_i = j$, if $y_j = 1$ is *i*-th occurrence of 1 in the array $\boldsymbol{y}$. Length of array $\boldsymbol{o}$ is small because array $\boldsymbol{y}$ is sparse.

For every customer we find warehouse with minimal cost, searching only array of ordinals. Array $\boldsymbol{o}$ has small size, and guarantee fast computation. Construction phase for array $\boldsymbol{o}$ has a large O(m) time requirement. But for every item, program constructs array of ordinals only one, and use it n times (for every customer). For every customer are necessary $\boldsymbol{e}$ steps, which give $O(m + n \cdot e)$ run-time complexity.

It is easy to prove that total run-time complexity of given objective value function in worst case is $O\left(m+n\cdot\sqrt{m}\right)$. Although theoretical possibilities for $O\left(m+n\cdot\log_2 m\right)$ run-time complexity exists, in cases that m is less than few thousand, such methods produce worse results in practice.

3.2 GA parameters

Initial population is randomly initialized, also as it is mentioned in previous section. We experiment with initialization by some heuristics, and fitnesses in first generation are better, but by using that method is produced worse gradient in the fitness function of subsequent generations, and worse overall results are obtained.

Improvements of the GA parameters are performed by making two major changes: mutation rate that depends on test problem size and discarding duplicate item strings from population.

Mutation rate applied in previous section, (also in [1]), are sufficient for solving test problems of the smaller size (up to 100 potential warehouse locations). Although, results obtained by testing larger size test problems (100 - 500 potential warehouse locations), are not sufficient. In those cases, applied GA slowly converges, and final results are not enough. Mutation operator is too disruptive, and searching abilities of GA are worse.

From that reasons are proposed mutation with variable rate given in (5).

$$p_{mut} = \begin{cases} 0.01, & n \le 50 \\ \dfrac{1}{2n}, & n \ge 100 \end{cases} \tag{5}$$

Example 2: Test problems MP with size 200×200 require p_{mut} = 0.0025, test problems MQ with size 300×300 require p_{mut}=0.00166, and test problems MR with size 500×500 require p_{mut} = 0.001 mutation rate.

As we see from experimental results presented in Table 4.4, this mutation rate is good compromise between exploration and exploitation components of GA search, applied to our problem.

3.3 Discarding duplicate strings from population

Duplicate item strings are discarded from population, and we practically prohibited the multiplied of same string. We discard duplicate item strings in every generation, by setting its fitnesses to zero. That method does not remove duplicate item string physically, but discard its occurrence in next generation.

This technique will be effective against the lack of variety in the population, and maintaining the diversity of genetic material. Discarding of duplicate strings decreases the appearance of dominating items in population, and effectively decrease possibility of premature convergence in local optimum. This is a very important factor of successful working GA, particularly in cases of large size test problems.

Example 3: If population contains 5 items *0101, 1010, 0101, 0001, 1010* with objective values p_1, p_2, p_1, p_4, p_2 respectively and $p = 2p_1 + 2p_2 + p_4$, then appropriate fitnesses are respectively equal to $p_1/p,\ p_2/p,\ 0,\ p_4/p,\ 0$.

3.4 Caching the GA

We also optimize run-time performance of the genetic algorithm by caching. The idea of caching is used so as to avoid the attempt to compute the same objective value repeatedly many times. Instead of that, objective values are remembered and reused. If program has computed objective value for a particular item string, and the same string appears again, the cached values are used to avoid computing it twice.

First impression for caching possibilities is negative, because duplicate strings are discarded from population, and same item strings practically do not exist in population in the same generation. Although, steady-state generational replacement provides appearance of some items in the several subsequent generations, and improve caching possibilities.

We used Least Recently Used (LRU) caching strategy, that is simple but effective. In practice, for all our UWLP test problems, improvement of run-time is smaller than previous improving methods, but it is also significant. For detailed informations about caching GA see [29].

4. Computational results

The described SGA and its improvement are written in the ANSI C programming language. Kernel of improved GA implementation consists of approximately 5 000 lines of source code (including comments). Specific functions in GA only for solving UWLP, have up to 1 000 lines. That part contains I/O operations, initialization and objective value function.

4.1 Test problems

Test problems 41-134 and A to C which will be used in this section are taken from the ORLIB [30]. The random problems MO, MP, MQ and MR of larger size, are generated by author. Input parameters for this problem generator are shown in the Table 4.1 and Table 4.2.

Table 4.1 Type of parameters for problem generator

Description of parameter	Minimal	Maximal	Type
Transportation cost	c_{min}	c_{max}	real
Fixed cost of warehouse	f_{min}	f_{max}	real
Customer demand	b_{min}	b_{max}	integer

Table 4.2 Input parameters for problem generator

Name of test problems	Problem size	f	c	b	Size of files
MO1-MO5	100×100	50 - 300	2 - 10	1 - 5	100 KB
MP1-MP5	200×200	100 - 600	2 - 10	1 - 5	400 KB
MQ1-MQ5	300×300	150 - 900	2 - 10	1 - 5	900 KB
MR1-MR5	500×500	100 - 600	0.5 - 5	1 - 5	2.4 MB

First we generate randomly customer demands b_i from interval $[b_{min}, b_{max}]$, then members of transportation cost matrix from interval $[c_{min}, c_{max}]$ multiplied by b_i (normalizing customer demands). After that, for every warehouse i, we compute sum $S_i = \sum_{j=1}^{n} c_{ij}$, which denote cumulative cost of completion all customer demands only from given warehouse. Then we perform inverse scaling into the interval $[f_{min}, f_{max}]$ by formula (6).

$$f_i = f_{\max} - \frac{(S_i - S_{\min}) \cdot (f_{\max} - f_{\min})}{S_{\max} - S_{\min}} \tag{6}$$

That is similar to some real situations, where smaller transportation costs c_{ij} produce higher warehouse establishing cost f_i, and vice versa.

4.2 Experimental results

The testing process is performed by PC compatible computer AMD 80486DX5 at 133MHz with 32 MB of memory. Results of running our algorithm are summarized in Table 4.4, comparing to results obtained by base implementation described in Table 4.3.

The columns in Tables 4.3 and 4.4 contain:

- Names of test problems;
- Number of problems in particular test problem set;
- Number of runs for each test problem;
- Average number of generations for finishing GA execution;
- Average execution time in seconds;
- Quality of solutions, i.e. how many solutions are optimal, how many solutions have relative error to optimal solution less than 0.2%, less than 1%, less than 5%, and over 5%.

Since genetic operators of selection, crossover and mutation are undeterministic, every test problem is running 10 times and we are computed average value.

Table 4.3 Results of running SGA

Test problem names	Num. of problems	Num. of runs	Avg. num. of gen.	Avg. time	Opt.	<0.2%	<1%	<5%	>5%
41-44, 51, 61-64, 71-74	13	10	44.6	0.355	118	9	3	-	-
81-84, 91-94, 101-104	12	10	416.8	4.13	113	7	-	-	-
111-114,121-124, 131-134	12	10	1 820	25.6	116	4	-	-	-
A - C	3	10	8 749	3358.2	19	7	4	-	-
MO1-MO5	5	10	6 476	247.3	35	8	7	-	-
MP1-MP5	5	10	6 003	774.5	44	-	6	-	-
MQ1-MQ5	5	10	12 424	3486.9	30	7	13	-	-
MR1-MR5	5	10	18 439	13693.0	9	2	10	29	-

Table 4.4 Results of running improved GA

Test problem names	Num. of problems	Num. of runs	Avg. num. of gener.	Avg. time	Opt.	<0.2%	<1%	<5%	>5%
41 - 44, 51, 61 - 64, 71-74	13	10	51.5	0.228	124	4	1	1	-
81 - 84, 91 - 94, 101 - 104	12	10	179.0	0.803	120	-	-	-	-
111-114,121-124,131-134	12	10	1 056	5.36	120	-	-	-	-
A - C	3	10	10 608	267.5	16	9	5	-	-
MO1-MO5	5	10	8 263	64.6	33	10	7	-	-
MP1-MP5	5	10	3 120	46.5	47	0	3	-	-
MQ1-MQ5	5	10	5 078	116.2	47	1	2	-	-
MR1-MR5	5	10	8 728	330.3	40	1	8	1	-

Improved implementation solved all test problem sets significantly faster than the base implementation. The speedups, which increased with the problem size, are: 1.56 (for test problems 41-74), 5.1 (81-104), 4.8 (111-134), 12.6 (A - C), 3.8 (MO), 16.7 (MP), 30.0 (MQ) and 41.5 (MR).

Note that the quality of solutions for problems sets MQ and MR is significantly better. For example, testing MR problem set, during 5×10 runs, are obtained 40 optimal solutions, compared to 9 optimal solutions, in case of testing base implementation. This behaviour corresponded to a smaller average number of iterations.

5. Conclusions

In this paper, improvement of SGA implementation for solving UWLP is explored. Base implementation produce good results, and optimal solution is at least once given for every particular test problem, but with high running time for large size test problems.

Implementation is improved by:

- New, more efficient function for computing objective value;
- Variable mutation rate;
- Removing duplicate item strings in population;
- Caching GA.

According to UWLP benchmark, the use of described techniques brings significant improvements. In general, run-time produced by improved implementation is smaller than the one produced by base implementation. For test problems of larger size, speedup is more than one order of magnitude (speedup is over the 40 times for MR test problem set).

Also, the quality of solutions is better when good convergence is maintained, especially with test problems with size 300×300 and more (problem sets MQ and MR). During testing process for every particular test problem at least one-half optimal solutions are obtained.

This implementation, with described improvements, is now able to solve large-scale uncapacitated warehouse location problems, with good running time.

The research presented in this paper can be improved, generalized and extended in several directions:

- Testing presented approach to test problems of very large size (over the 1000 warehouses and customers);
- Parallelization of the genetic algorithm and implementation to distributed and multiprocessor systems;
- Apply approach presented in this paper to the other related location problems: p-median, capacitated warehouse location, dynamic warehouse location, multi-product uncapacitated warehouse location and data file location/allocation problems.

Acknowledgements

This research was partially supported by the Serbian Ministery of Science and Technology under project 04M01/F.

References

1. Kratica J., Filipovic V., Sesum V., Tosic D., 1996, Solving of the Uncapacitated Warehouse Location Problem using a Simple Genetic Algorithm, In: *Proceeding of the XIV International Conference on Material Handling and Warehousing,* Faculty of Mechanical Engineering, Belgrade, pp 3.33 - 3.37.
2. Holland J.H., 1975, *Adaptation in Natural and Artificial Systems,* The University of Michigan Press, Ann Arbor.
3. Beasley D., Bull D.R. and Martin R.R., 1993, An Overview of Genetic Algorithms, Part 1: Fundamentals, *University Computing,* **15**:58-69.
4. Beasley D., Bull D.R. and Martin R.R., 1993, An Overview of Genetic Algorithms, Part 2: Research Topics, *University Computing,* **15**:170-181.
5. Goldberg D.E., 1989, *Genetic Algorithms in Search, Optimization and Machine Learning*, Addison-Wesley Publ. Comp., Reading, Mass.
6. Srinivas M., and Patnaik L.M., June 1994, Genetic Algorithms: A Survey, *IEEE Computer,* pp 17-26.
7. De Jong K.E., and Spears W.M., 1989, Using Genetic Algorithms to Solve NP-Complete Problems, In: *Proceedings of the Third International Conference on Genetic Algorithms*, Morgan Kaufmann, San Mateo, Calif., pp 124-132.

8. Khuri S., Back T., and Heitkotter J., 1994, An Evolutionary Approach to Combinatorial Optimization Problems, In: *Proceedings of CSC'94*, Phoenix, Arizona.
9. Michalewicz Z., 1995, *Genetic Algorithms + Data Structures = Evolution Programs*, Third Edition, Springer-Verlag, Berlin - Heidelberg.
10. Davis L., 1991, *Handbook of Genetic Algorithms*, Van Nostrand Reinhold, New York.
11. Reeves C.R., 1995, Genetic Algorithms, In: Reeves C.R (ed.), *Modern Heuristic Techniques for Combinatorial Problems*, McGraw-Hill Book Company, London, Chapter 4, pp 151-196.
12. Osman I.H., Laporte G., 1996, Metaheuristic: A bibliography, *Annals of Operations Research*, **63**:513-623.
13. Alander J.T., 1995, Indexed bibliography of genetic algorithms in operations research, Report 94-1-OR, Department of Information Technology and Production Economics, University of Vassa, Finland. *ftp://ftp.uwasa.fi/cs/report94-1/gaORbib.ps.Z*
14. Beasley J.E., 1993, Lagrangean heuristic for location problems, *European Journal of Operational Research*, **65**:383-399.
15. Aikens C.H., 1985, Facility location models for distribution planning, *European Journal of Operational Research*, **22**:263-279.
16. Dearing P.M., 1985, Location problems, *Operations Research Letters*, **4**:95-98.
17. Francis R.L., McGinnis L.F., and White J.A., 1983, Locational analysis, *European Journal of Operational Research*, **12**:220-252.
18. Krarup J., and Pruzan P.M., 1983, The simple plant location problem: Syrvey and synthesis, *European Journal of Operational Research*, **12**:36-81.
19. Cornuejols G., Nemhauser G.L. and Wolsey L.A., 1990, The uncapacitated facility location problem, In: *Mirchandani P.B. and Francis R.L. (eds), Discrete Location Theory*, John Wiley & Sons, chap. **3,** pp 120-171.
20. Gao L.L., Robinson E., and Powell Jr., 1994, Uncapacitated facility location: General solution procedure and computational experience, *European Journal of Operational Research*, **76**(3):410-427.
21. Erlenkotter D., 1978, A dual-based procedure for uncapacitated facility location, *Operations Research*, **26**:992-1009.
22. Guignard M., 1988, A Lagrangean dual ascent algorithm for simple plant location problems, *European Journal of Operational Research*, **35**:193-200.
23. Simao H.P., and Thizy J.M., 1989, A dual simplex algorithm for the canonical representation of the uncapacitated facility location problem, *Operations Research Letters*, **8**(5):279-286.
24. Koerkel M., 1989, On the exact solution of large-scale simple plant location problems, *European Journal of Operational Research*, **39**:157-173.
25. Holmberg K., 1995, *Experiments with primal-dual decomposition and subgradient methods for the uncapacitated facility location problem*, Research Report LiTH-MAT/OPT-WP-1995-08, Optimization, Department of Mathematics, Linkoping Institute of Technology, Sweden.
26. Aqeev A.A., and Beresnev V.S., 1990, Polynomially Solvable Cases of the Simple Plant Location Problem, In: *Proceeding of the First Integer Programming and Combinatorial Optimization Conference, University of Waterloo Press, ON, Canada*, pp 1-6.

27. De Simone C., and Mannino C., 1996, *Easy Instances of the Plant Location Problem*, Technical Report R-427, Gennaio 1996, University of Roma, Italy.
28. Grishukhin V.P., 1994, On polynomial solvability conditions for the simplest plant location problem, In: Kelmans A.K. and Ivanov S. (eds), *Selected topics in discrete mathematics*, American Mathematical Society, Providence, RI, pp 37-46.
29. Kratica J., 1998, Improving Performances of the Genetic Algorithm by Caching, *Computers and Artificial Intelligence* (accepted for publication in volume 17).
30. Beasley J.E., 1996, Obtaining test problems via internet, *Journal of Global Optimization*, **8**:429-433, *http://mscmga.ms.ic.ac.uk/info.html*

Optimizing Neural Networks for Time Series Prediction

I. De Falco, A. Della Cioppa, A. Iazzetta,
P. Natale and E. Tarantino

Institute for Research on Parallel Information Systems
National Research Council of Italy (CNR)
Via P. Castellino 111, Naples, ITALY.

Keywords: Time Series Prediction, Artificial Neural Networks, Evolutionary Algorithms, Breeder Genetic Algorithms, Mackey–Glass delay differential equation.

Abstract

In this paper we investigate the effective design of an appropriate neural network model for time series prediction based on an evolutionary approach. In particular, the Breeder Genetic Algorithms are considered to face contemporaneously the optimization of (i) the design of a neural network architecture and (ii) the choice of the best learning method. The effectiveness of the approach proposed is evaluated on a standard benchmark for prediction models, the Mackey–Glass series.

1. Introduction

The main motivation for time series research is to provide a prediction when a mathematical model of a phenomenon is either unknown or incomplete. A time series consists of measurements or observations of the previous outcomes of a phenomenon that are made sequentially over time. If these consecutive observations are dependent on each other then it is possible to attempt a prediction. Clearly it is supposed that the process is somehow predictable.

The time series prediction problems are usually faced by stochastic methods which are fast but of limited applicability since they commonly employ linear models [1]. In practice many processes are nonlinear and hence linear models can be inadequate. Among nonlinear methods Artificial Neural Networks (ANNs) have been widely used for time series prediction problems [2, 3, 4]. Though ANNs are more powerful than stochastic methods, their drawback is that the design of an efficient architecture and the choice of the parameters involved require high processing time. In fact, learning neural network weights can be considered a hard optimization problem for which the learning time scales exponentially becoming prohibitive as the problem size grows [5]. Besides in facing any new problem a lot of time can be wasted to find an appropriate network architecture. It seems then natural to devote attention to automatic procedures which exploit more powerful techniques for efficiently searching the space of network architectures. Moreover even if a suitable set of parameters is found under the guidance from training inputs, the ability of the resulting network to generalize on unknown data may be far from optimal.

In most cases heuristic techniques have been used for the training process [6, 7, 8]. Whitley has faced the above tasks separately [9]. A different approach is followed in [10]. In it Genetic Algorithms (GAs) [11, 12] have been utilized both for architecture optimization and choice of the best variant of the Back Propagation (BP) method among four proposed in literature. In [13] the GAs and the Simulated Annealing have been employed with the aim to optimize the neural network architecture and to train the network with a particular model of BP. In [14] Breeder Genetic Algorithms (BGAs) [15, 16, 17] have been demonstrated to be superior to GAs in designing neural networks for non linear system identification.

The objective of this paper is to investigate the possibility of employing an evolutionary approach for designing appropriate ANNs for nonlinear stationary time series prediction problems. The most popular neural network, the Multi–Layer Perceptron (MLP) [18], has been considered for the time series analysis. The best known and successful training method for MLPs is the BP [5]. Evolutionary Algorithms, like the BGAs, have been taken into account to efficiently design the neural model.

Namely, BGAs have to deal with the network topology optimization and the choice of the best technique to update the weights in the BP at the same time. Furthermore, BGAs have to optimize the related parameters for time series prediction problem. Of course this procedure does not guarantee to find the optimal architecture, anyway it allows to attain rapidly an architecture close to optimal performance. A standard benchmark for prediction models, the Mackey–Glass (MG) time series [19], is considered to evaluate the effectiveness of the proposed approach.

The paper is organized as follows. In section 2 the state of the art in time series prediction is briefly described. In section 3 a description of the aforementioned benchmark problem is reported. In section 4 some implementation details are outlined. In section 5 the experimental results are shown and discussed. Section 6 contains final comments and prospects of future work.

2. A Brief Review of the State of the Art in Time Series Prediction

Different approaches to time series prediction problems have been investigated over the years [1]. The methods can be divided into *global* and *local models*. In the global model approach only one model is used to characterize the phenomenon under examination. The local models are based on dividing the data set into smaller sets, each being modeled with a simple model. The global models give best results with stationary time series, i.e. series with statistics which do not change with time. The first approaches were devoted to linear models for which the theory is known and consequently many algorithms for model building are available. The most used linear regression methods have been the autoregressive (AR) and the autoregressive moving average (ARMA) models [20, 21]. An example of more complex regression method is the multivariate adaptive regression splines [22].

However, since almost all measured phenomena are nonlinear, the linear modeling methods are often inappropriate to describe their behavior. This is the reason why the nonlinear methods have become widely utilized. Among nonlinear methods ANNs represent an attractive approach for time series prediction problems [23, 24, 25, 26, 2, 3, 4]. In fact, in addition to their ability of representing complex nonlinear functions, many different types of ANNs, such as MLP and Radial Basis Function networks (RBF), can effectively construct approximations for unknown functions by learning from examples [27]. In time series prediction with ANNs the main problems are the design of the structure of the network and the effectiveness of the learning phase. Usually the topology of the network is fixed and the learning phase is performed using various methods [23, 4]. Among neural models MLP is used frequently [4]. Its limitation is that it can only learn an input–output mapping which is static. Hence this network can be used to perform a prediction of a stationary time series. The novelty of our work is the use of BGAs to provide simultaneously an efficient architecture and an optimization of the learning process.

3. The Mackey–Glass Series

In many situations in science and technology the future evolution of a system is predicted by past measurements on it. In systems with a small number of degrees of freedom, a mathematical model is made by writing equations of motion which, integrated forward in time, allow to predict the future state. This procedure fails in systems with many degrees of freedom. In this case we cannot keep track of the motion in such a high-dimensional space. However studies of the dynamics of apparently chaotic systems with many degrees of freedom reveal that dissipation can reduce the number of effectively relevant degrees of freedom to a small number.

In fact, let us take into account the set H of all "possible states". Given the evolution function:

$$\{\phi_t : H \to H\}_{t \in \mathbb{R}^+}$$

such that if $h \in H$ represents the state at time $t = 0$, then $\phi_{t=1}(h)$ represents the state at time $t = 1$, we assume that there is an "attractor" $\Lambda \subset H$ to which (almost) all evolution curves $\phi_t(h)$ tend as $t \to \infty$.

There are several ways to define the dimension for the attractor. Takens [28] has used the so-called limit–capacity. Let (S, ρ) be a compact metric space. Given $\epsilon > 0$ we define the maximal ϵ–separated set as the maximal cardinality $s(S, \epsilon)$ of a subset of S such that no two points have distance less than ϵ. Now we can define the limiting capacity $D(S)$ of S as

$$D(S) = \liminf_{\epsilon \to 0} \frac{\ln(s(S,\epsilon))}{-\ln \epsilon}$$

This attractor is often an object with a dimensionality D that is not even an integer [29]. The identification of the coordinates which characterize the attractor Λ makes the problem like one in simple physics, with only a few degrees of freedom. The choice of the combinations of the few relevant variables seems at first difficult or impossible. However, a sequence of n past measurements of a chaotic system can be used to form the state–space representation of the systems' dynamics provided that n is sufficiently large compared to the attractor dimensionality D [30]. This sequence contains enough information to predict the motion on the attractor. This is also proved by Takens' theorem which says that there exists a smooth function defined on a subset of H that correctly predicts the future value of the variable in question [28].

As shown in [24], we can attempt to predict the behaviour of a time series generated by a chaotic dynamical system as it follows: the time series is transformed into a reconstructed state space using a delay space embedding [30]. In this latter space, each point in the state space is a vector $\mathbf{x}$ composed of time series values corresponding to a sequence of n delay lags:

$$x(t), x(t-\delta), x(t-2\delta), \dots, x(t-(n-1)\delta) \tag{1}$$

The next step is to assume a functional relationship between the current state $\mathbf{x}(t)$ and the future state $\mathbf{x}(t+P)$,

$$\mathbf{x}(t+P) = f_P(\mathbf{x}(t)). \tag{2}$$

The aim is to find a predictor $\hat{f}_P$ which approximates f_P, so that we can predict $x(t+P)$ based on n previous values. If the data are chaotic, the f_P is necessarily nonlinear. P represents the number of time steps ahead we wish to perform our prediction.

Based on Takens' theorem, an estimate of the dimension D of the manifold from which the time series originated can be used to construct an ANN model using at least $2D+1$ external inputs [31]. For a noise–free system of dimension D, it is sufficient to choose $n = 2D+1$. It is obvious that for a D–dimensional attractor, n must be at least as large as D [24].

The time series used in our experiments is generated by the chaotic Mackey–Glass differential delay equation [19] defined below:

$$\frac{dx}{dt} = a\frac{x(t-\tau)}{1+x^c(t-\tau)} - bx(t) \tag{3}$$

where a, b, c and τ are costants. The value of τ is very important, because it determines the system behavior. In fact, for initial values of x in the time interval $[0,\tau]$, the system approaches a stable equilibrium point for $\tau < 4.53$, a limit cycle for $4.53 < \tau < 13.3$, and after a series of period doublings for $13.3 \leq \tau \leq 16.8$ the system becomes chaotic. From the value 16.8 on, τ controls how chaotic the series is. The larger τ, the larger the dimensionality of the attractor. For example, for $\tau = 17$ D is 2.1, while for $\tau = 23$ D is 2.4, for $\tau = 30$ D is 3.5, and for $\tau = 100$ D is 7.5.

For the Mackey–Glass series, the higher the value of P the more difficult the problem, especially for values of P greater than its characteristic period of approximately 50.

4. Multi–Layer Perceptron Optimization by Evolutionary Algorithms

The basic idea is to provide an automatic procedure to find the most appropriate neural network structure for time series prediction. Generally speaking, the optimization of an MLP which must be trained to solve a given problem is characterized by the need to determine:

1. **The architecture**
   (a) *The number of hidden layers* NHL,
   (b) *The number of nodes* NHl *for each layer l.*
2. **The activation function f_i^l to be used for each layer l.**
3. **Presence or absence of the bias b.**
4. **The most appropriate technique for weight change.**

We wish to point out that in our approach the training phase is effected with the BP algorithm. There are a number of heuristic modifications to the basic approach which may speed up the training time or enable completion of the learning process. These approaches propose different ways to proceed in the change of the weights [18, 32]. Since the values of the parameters involved depend on the problem under examination, it is not possible to take an *a priori* decision by a qualitative analysis about which of the variants is the best. Rather, for the solution of any given problem it is firstly necessary to fix a particular configuration on the basis of empirical considerations, then to train the network with the patterns and to evaluate its quality. The algorithmical analysis demonstrates that this search presents an exponential complexity if performed with an exaustive technique. Furthermore, in addition to the time complexity, there is the possibility that the training fails because the network configuration runs into a local minimum.

with momentum, so that the learning mechanism choice is not a parameter for our search. On the basis of this decision the following parameters are to be fixed:

1. *the training coefficient* η,
2. *the momentum coefficient* α.

We have utilized heuristic methods to reduce the search time and the probability of getting stuck in local minima. Namely, BGAs have been used to determine the appropriate set of parameters listed above. The aim is to provide the most general possible technique to determine the network structure. It is of high importance to point out that when we face time series prediction problems we have to decide how many previous outcomes of the phenomenon must be considered. This value is called *window size.* Moreover we also have to establish how to sample these past outcomes. This means we must find a constant lag sampling, called *sampling lag.* We have decided to insert also these features among those to be found thanks to evolution. It is important to note that in most cases these two parameters are evaluated experimentally, although for the former some guidance is given in [31]. In our approach, instead, we aim to determine the best values for both the parameters, thanks to the evolutionary search power.

The literature on the Mackey–Glass problem helps us in imposing limitations on some of the above parameters. For instance, it has been reported in [23] that two hidden layers are adequate for the task to be faced. This allows us to reduce the number of parameters to be taken into account.

4.1. Encoding

The neural network is defined by a "genetic encoding" in which the genotype codes for the different characteristics of the MLP and the phenotype is the MLP itself. Therefore, the genotype contains the parameters related to the network architecture, i.e. the number of hidden layers (NHL) and the numbers of neurons in each layer l (NHl), and other genes representing the activation function type for the output layer f_i^{out}, the parameters related to BP (η and α), the number of neurons in the input layer N^{in}, presence or absence of bias b, sampling lag δ.

For our Evolutionary Algorithm, the chromosome structure $\mathbf{y} = (y_1, \dots, y_\ell)$, constituted by $\ell = 9$ loci, is reported in Fig. 1.

Locus	y_1	y_2	y_3	y_4	y_5	y_6	y_7	y_8	y_9
Gene	NHL	η	α	b	N^{in}	NH1	NH2	δ	f_i^{out}
Set	A_1	A_2	A_2	A_3	A_4	A_5	A_5	A_6	A_7

Table 1. The genetic encoding.

Each allele is defined in the subset A_i, $i \in \{1, \dots, 7\}$ reported in the third row of the Table 1. The value of NHL has been allowed to vary from 1 to 2 while the maximum value for NHl has been fixed equal to 30. Namely, the alleles are defined within the following subsets:

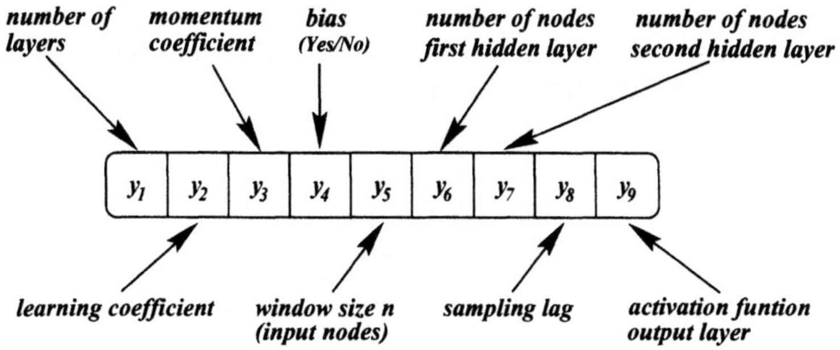

Fig. 1. The graphical representation of the genetic encoding.

$$\begin{cases} A_1 = \{1,2\} \\ A_2 = [0,1] \subset \mathbb{R} \\ A_3 = \{0,1\} \\ A_4 = [1,80] \\ A_5 = [1,30] \\ A_6 = [1,15] \\ A_7 = \{1,2,3\} \text{ with } \{1 \equiv f_1,\, 2 \equiv f_2,\, 3 \equiv f_3\} \end{cases}$$

where A_7 means that the activation function used for the output layer can assume the value f_1 for the *sigmoid*, f_2 for *tanh* and f_3 for the *semi–linear*.

In the chromosome there are no loci for the choice of activation functions of the hidden layers. This is because we performed a preliminary set of runs in which also these features had to be found by the Evolutionary Algorithm. In those runs the best solutions always contained the *tanh* for both the hidden layers. The presence of those loci caused a high execution time due to the higher search space, the loci under examination containing always the *tanh* . So we decided to cancel those fields out from the final chromosome, depicted in Fig. 1, and to use *tanh* for the hidden layers.

As a counter–experiment for this, we also performed some more runs by modifying the final solutions yielded by the system. We changed the proposed network by using the *sigmoids* instead of the *tanh*. The results were always worse. In fact, the predicted values always kept close to the average value of the series.

4.2. The Fitness Function

Whenever an ANN is used, the total pattern set is divided into two subsets: the *training set* and the *verifying set*. The former is used to train the network, while the latter is used to evaluate the generalization ability of the trained network.

To evaluate the goodness of an individual, the parameter which seems to describe better its goodness is the normalized mean square error (NMSE) on the training set, E_t. We define NMSE as:

$$\mathrm{NMSE} = \frac{\sqrt{\frac{1}{N-1}\sum_{i=1}^{N}(x_i - o_i)^2}}{\sqrt{\frac{1}{N-1}\sum_{i=1}^{N}(x_i - \bar{x})^2}}$$

Since the ANN generalization ability is computed with reference to the verifying set, it seems natural to take into account also the normalized mean square error on this latter set, E_v. Specifically the fitness function we use is the following:

$$\mathcal{F}(\mathbf{y}) = E_t + E_v \tag{4}$$

4.3. The Optimization Algorithm for the Neural Network Structure

A population of μ networks which represent potential candidate solutions is let free to evolve. These solutions are individually evaluated to determine how well they solve the problem. A selection process allows to establish the survivors. They mate by means of genetic operators to create the individuals of the next generation. The process is repeated until an adequate network is achieved or a stopping criterion is fulfilled.

Let MLP($\mathbf{y}_i$) with $i \in \{1, \ldots, \mu\}$ be the algorithm for training the MLP related to the individual $\mathbf{y}_i$ representing a neural network configuration. The general scheme for the BGA is the following:

Given a training set T_s;
Procedure *Breeder Genetic Algorithm*
begin
  randomly initialize a population of μ neural network structures;
  while (termination criterion not fulfilled) **do**
    train $\mathbf{y}_i$ by means of MLP($\mathbf{y}_i$) on T_s;
    evaluate the trained network $\mathbf{y}_i$;
    save the best trained network in the new population;
    select the best λ neural network configurations;
    for $i = 1$ **to** $\mu - 1$ **do**
      randomly select two structures among the λ;
      recombine them so as to obtain one offspring;
      perform mutation on the offspring;
    od
    update variables for termination;
  od
end

The procedure is constituted by an evolutionary algorithm (a BGA), and by a procedure for training the MLP encapsulated in the main program. This latter procedure returns the values for the fitness evaluation. Its action is completely transparent to the main program which can only evaluate the goodness of the result.

5. Experimental Results

It is well known from literature that we cannot speak of Mackey–Glass as of one single problem. Rather, it there exists a whole family of Mackey–Glass problems. Any instance of this family is determined by setting the values τ, a, b, c, in the equation (3), and deciding the value of P. This latter describes how far in the future we predict values, starting from the ones of the current window. For our experiments we have decided to set the above values as it follows: $\tau = 17$, $a = 0.2$, $b = -0.1$, $c = 10$. As regards P, we have performed two sets of experiments, one with $P = 6$ and another with $P = 85$. By doing so, we can evaluate the effectiveness of our system for both short–term and long–term predictions. We have used the above reported values because they are very often used in literature and we wish to compare our results against those achieved by other researchers.

After this preliminary phase, by using a Runge–Kutta method, we have been able to obtain the related time series, consisting of 1500 samples. The available data have been divided into two sets: the training data set T_s used to train the prediction system, and the verifying set V_s to evaluate the performance of the trained system. The former consists of 1200 samples, while the latter is made of 300 values.

As we have already said in the previous section, we wish that our system be able to find on its own the best window size and the best sampling lag which are needed to effectively perform the prediction.

In the following we describe both the sets of experiments for $P = 6$ and for $P = 85$.

5.1. Experiments with $P = 6$

In this case we have set the BGA as it follows: population size $\mu = 10$, truncation rate $\lambda = 40\%$, maximum number of generations $g_{\max} = 30$, fuzzy recombination [33] as recombination operator, discrete mutation with mutation probability $p_m = 0.06$. A one–elitist strategy has been allowed in order to save in the next generation the best individual of the current generation. The maximum number of training epochs is 200, the minimum NMSE for learning is 0.02, the stopping criterion for the genetic evolution is a fitness value equal to or lower than 0.02. We have performed ten runs by using different seed numbers for the random number generator.

The value of μ has been kept very small because the training and the evaluation of one ANN take at least 30 minutes by means of a 166-Mhz Pentium even for a small ANN, and this time hugely increases with the number of neurons in the network. The same reason has led us to limit $g_{\max}$. The number of training epochs is just 200 because of the same reason. By using the above values we have been able to limit the execution time to some days.

The results of all the runs are quite similar. In the following we report plots for the best run. Namely, Fig. 2 shows the results achieved on the training set (a window of size 300 is taken into account for clarity's sake) and the prediction obtained on the verifying set. In both cases also the real MG series is shown.

The MLP corresponding to this figure has 12 nodes in the input layer, 11 in the first hidden layer, no second hidden layer present, and, of course, one node in the output layer. The learning coefficient is 0.323285, and the momentum coefficient is 0.273095, the sampling lag is 7, bias is absent, the activation function for the output layer is *tanh*.

The most direct way to evaluate the quality of a solution is the NMSE. For the best run the NMSE value is 0.387703 on the training set and 0.237364 on the verifying set.

One more way to evaluate the quality of the solution found is by means of a parameter taking into account the shape of the predicted series with reference to the shape of the real Mackey–Glass series. Roughly speaking, such a parameter should be such that the closer the shape of the predicted curve with respect to the MG curve, the higher the value of the parameter. This can be done as follows: let us take into account two consecutive points in the real MG series at time steps t and $t+1$. If $x(t) < x(t+1)$ we set a value $\texttt{slope}_{\mathrm{MG}}$ to 1: $\texttt{slope}_{\mathrm{MG}}(t) = 1$. If, instead, $x(t) \geq x(t+1)$ we set $\texttt{slope}_{\mathrm{MG}}(t) = 0$. By acting this way we can write a new series consisting of 0s and 1s. This series will have a number of points equal to that of the MG series minus one. We can do the same for the predicted series, creating a new series $\texttt{slope}_{\mathrm{pr}}(t)$. Then we compare these two newly created series. If $\texttt{slope}_{\mathrm{MG}}(t) = \texttt{slope}_{\mathrm{pr}}(t)$, we consider the prediction correct, otherwise it is incorrect. Finally, we can consider the percentage of correct predictions with reference to the total number of points, C_p. We can evaluate C_p on the whole set of points of the MG series, or on the training set only (C_p^t), or on the verifying set only (C_p^v). The results for the best run are $C_p = 84.666667\%$, $C_p^t = 85.083333\%$, and $C_p^v = 83.000000\%$.

5.2. Experiments with $P = 85$

The parameter setting is the same as in the previous set of experiments. Also in this case we have performed ten runs and we have found solutions similar one another. In the following we report for the best run in Fig. 3 the results achieved on the training set and the training set itself (actually a window of size 300 for graphical reasons). Moreover Fig. 3 also shows the prediction compared against the actual series. The values for window size and sampling lag in this best run are 63 and 11 respectively. The network has one hidden layer with 16 neurons. The learning coefficient is 0.200747 and the momentum

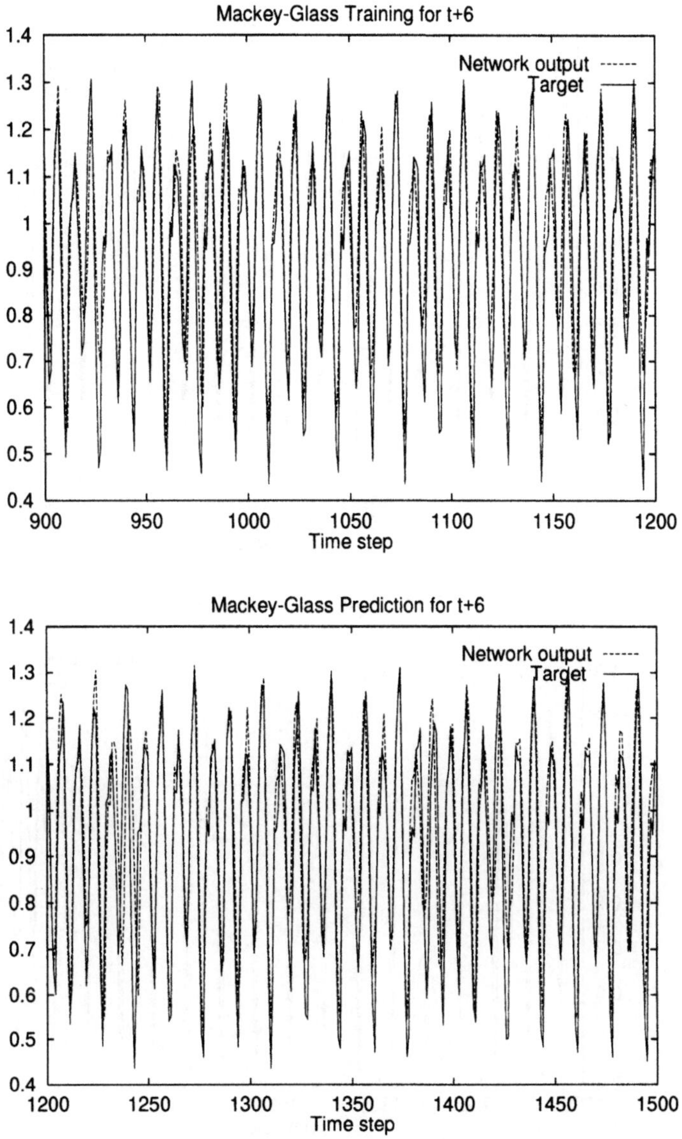

Fig. 2. The results on the training set (up) and on the verifying set (down) for $P = 6$.

coefficient is 0.786918. Bias is absent, the activation function for the output layer is the *semilinear*.

For the best run the NMSE value is 0.381985 on the training set and 0.266626 on the verifying set. These values can be compared against those reported in [34]. It is reported there that the relative errors obtained by using an MLP with 400 learning epochs are 0.54 on the training set and 0.55 on the verifying set in the best case. Thus, our approach allows to achieve better values by using a lower number of learning epochs.

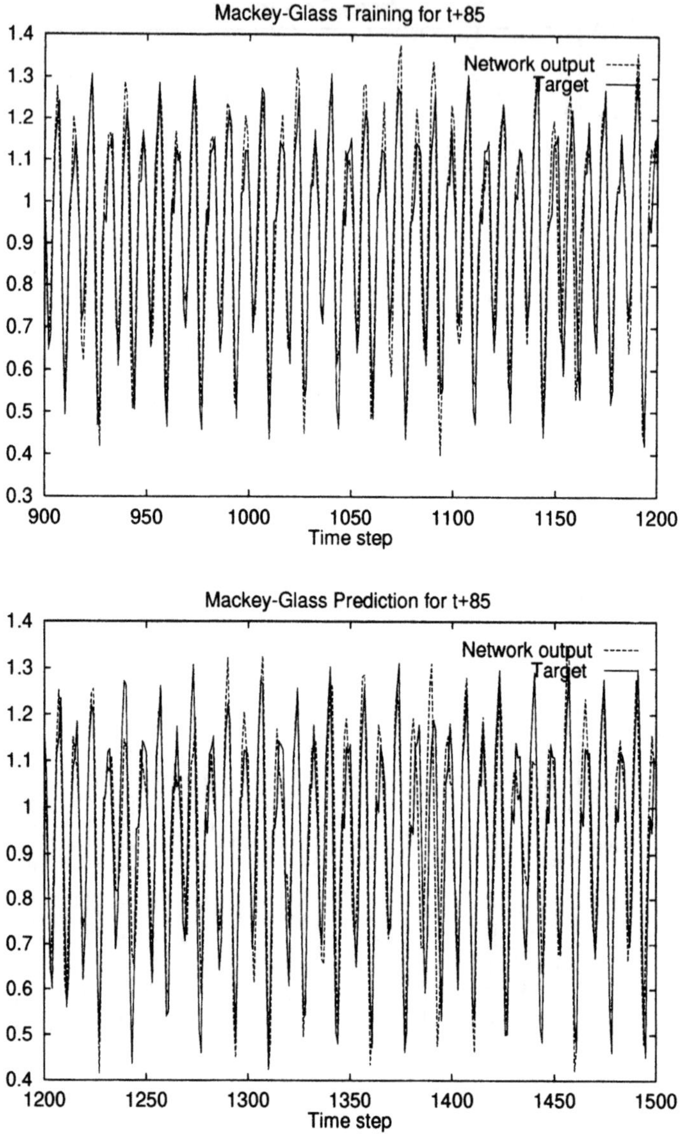

Fig. 3. The results on the training set (up) and on the verifying set (down) for $P = 85$.

As regards the percentages of correct predictions, the results for the best run are $C_p = 90.466667\%$, $C_p^t = 93.333333\%$, and $C_p^v = 79.000000\%$.

As regards the window sizes in both experiments, it should be remarked that Takens' Theorem states that $2D + 1$ past measurements are sufficient to perform prediction. Nonetheless, a larger window will do. It is to note that in our Algorithm we haven't inserted any pressure in the fitness function which could lead to windows as small as possible. Because of this, we have achieved values for the window size which are higher than $2D + 1$ (which yields 5 and 8, for $P = 6$ and $P = 85$, respectively).

6. Conclusions

In this paper an evolutionary approach to effective design of a neural network model for time series prediction has been investigated. The BGAs have been used in facing the simultaneous optimization of the architecture design and the best learning method choice. In particular, the fuzzy recombination operator, which often improves the performance of the BGA [14], has been employed. The approach has been tested on a standard benchmark for prediction techniques, the Mackey–Glass series. The experimental results have proved the effectiveness of the BGA proposed for both the prediction and the optimization of the neural model for the problem at hand. It is to point out that the automatic procedure provided has allowed to achieve a network architecture for time series prediction close to the optimal performance. Of course, in order to ascertain its generality, this procedure has to be analyzed on different time series prediction problems.

In this paper we have evaluated the predictions for $t+85$ directly. We know that in literature some papers report of performance improvements obtained by iterating the $t + 6$ prediction for 14 times, so as to achieve the $t + 85$ prediction (actually, the $t + 84$). We wish in the future to perform the same with our system.

Further work will be focused on complex real life prediction problems. Furthermore, since the BGA method can also be easily distributed among several processors, we intend to perform the implementation of a parallel version of the evolutionary approach proposed with the aim at improving the performance and the quality of neural networks, while decreasing the search time.

References

1. Weigend A, Gershenfeld N (eds) 1993 *Time Series Prediction: Forecasting the Future and Understanding the Past*. Addison–Wesley, Reading, Massachussets.
2. Mozer N 1993 Neural net for temporal sequence processing *Time Series Prediction: Forecasting the Future and Understanding the Past*. Weigend A, Gershenfeld N (eds), Addison–Wesley, Reading, Massachussets
3. Tsoi A, Back A 1994 Locally recurrent globally feedforward networks: a critical review of architectures. *IEEE Trans. on Neural Networks* 5 (2):229–239
4. Koskela T, Lehtokangas M, Saarinen J, Kaski K 1996 Time series prediction with Multilayer Perceptron, FIR and Elman neural networks *Proceedings of the World Congress on Neural Networks*. INNS Press, pp 491–496
5. Rumelhart D E, McLelland J L 1986 *Parallel Distributed Processing*. MIT Press

6. Montana D J, Davis L 1989 Training feedforward neural networks using genetic algorithms *Proceedings of the Eleventh International Joint Conference on Artificial Intelligence* pp 762–767
7. Hiestermann J 1990 Learning in neural nets by genetic algorithms *Parallel Processing in Neural Systems and Computers.* North–Holland, pp 165–168
8. Battiti R, Tecchiolli G 1995 Training neural nets with reactive tabu search. *IEEE Trans. on Neural Networks* 6 (5): 1185–1200
9. Whitley D, Starkweather T, Bogart C 1990 Genetic algorithms and neural networks: optimizing connections and connectivity. *Parallel Computing* 14:347–361
10. Reeves C R, Steele N C 1992 Problem–solving by simulated genetic processes: a review and application to neural networks. *Proceedings of the Tenth IASTED Symposium on Applied Informatics* pp 269–272
11. Holland J H 1975 *Adaptation in Natural and Artificial Systems.* University of Michigan Press, Ann Arbor
12. Goldberg D E 1989 *Genetic Algorithms in Search, Optimization and Machine Learning.* Addison-Wesley, Reading, Massachussetts
13. Stepniewski S, Keane A J newblock 1996 Pruning back propagation neural networks using modern stochastic optimization techniques. *Neural Computing & Applications*
14. De Falco I, Della Cioppa A, Natale P, Tarantino E 1997 Artificial neural networks optimization by means of evolutionary algorithms *Soft Computing in Engineering Design and Manufacturing.* Springer–Verlag, London
15. Mühlenbein H, Schlierkamp–Voosen D, 1993 Analysis of selection, mutation and recombination in genetic algorithms. *Neural Network World* 3:907–933
16. Mühlenbein H, Schlierkamp–Voosen D, 1993 Predictive models for the breeder genetic algorithm I. Continuous parameter optimization. *Evolutionary Computation* 1 (1):25–49.
17. Mühlenbein H, Schlierkamp–Voosen D, 1994 The science of breeding and its application to the breeder genetic algorithm. *Evolutionary Computation* 1:335–360.
18. Rumelhart D E, Hinton G E, Williams R J 1986 Learning internal representations by error propagation *Parallel Distributed Processing: Explorations in the Microstructure of Cognition.* MIT Press
19. Mackey M, Glass L 1977 Oscillation and chaos in physiological control system *Science* 197–287
20. Tong H 1990 *Nonlinear Time Series: A Dynamical System Approach.* Oxford University Press
21. Box G, Jenkins G, Reinsel G 1994 *Time Series Analysis: Forecasting and Control.* Prentice Hall, Englewood Cliffs, New Jersey
22. Friedman J, 1991 Multivariate adaptive regression splines. *Annals of Statistics* 19:1–142
23. Lapedes A, Farber R 1987 Nonlinear Signal Processing using Neural Networks: Prediction and System Modeling. Los Alamos National Laboratory Technical Report LA-UR-87-2662
24. Farmer D, Sidorowich J, 1987 Predicting chaotic time series. *Physical Review Letter* 59:845–848
25. Farmer D, Sidorowich J 1988 Exploiting chaos to predict the future and reduce noise in *Evolution, Learning and Cognition.* World Scientific, Singapore

26. Casdagli M, Eubank S 1992 *Nonlinear Modeling and Forecasting*. Reading MA, Addison–Wesley
27. Cybenko G 1989 Approximations by superpositions of a sigmoidal function. *Mathematics of Control, Signals and Systems* 2:303–314
28. Takens F 1981 Detecting strange attractors in turbulence in *Lecture Notes in Mathematics*. Springer–Verlag, Berlin
29. Mandelbrot B B 1982 *The Fractal Geometry of Nature*. Van Nostrand Reinhold, New York
30. Packard N H, Crutchfield J D, Farmer J D, Shaw R S 1980 Geometry from a time series. *Physical Review Letters* 45:712–716
31. Casdagli M, Eubank S, Farmer J D, Gibson J 1991 State space Reconstruction in the presence of noise. *Physica D* 51:52–98
32. Wassermann P D 1989 *Neural Computer Theory and Practice*. San Francisco, Freeman
33. Voigt H M, Mühlenbein H, Cvetković D 1995 Fuzzy recombination for the continuous breeder genetic algorithm. *Proceedings of the Sixth International Conference on Genetic Algorithms*. Morgan Kauffmann
34. Hartman E, Keeler J D 1991 Predicting the future: advantages of semilocal units. *Neural Computation* 3 (4):566–578

Part 7: Optimisation for Industrial Applications

Papers:

Maximum Entropy Image Restoration by Evolutionary Algorithm

Vitaly. G. Promislov

Astro Space Center, Lebedev Institute of Physics, Moscow, Russia,
Profsouznia st 84/32 117810
Email: vitaly@diogen.asc.rssi.ru

Keywords: Evolution Strategies , Maximum Entropy , Image Restoration.

Abstract

The Evolution Strategies algorithm for maximum entropy image restoration is proposed. Evolutionary algorithms are widely used to search for solutions of "ill-posed" optimization problems. These are based on simulating the natural evolution process within a population of individuals. The statistical properties of the mutation operator are discussed. The approximate formulas to choose the reasonable mutation parameters for a population in the case of a maximum entropy objective function are presented. The algorithm is tested on a real and simulated astrophysical data. For the purpose of comparison, restoration results, based on Evolution Strategies MEM *and gradient* MEM *for synthetic images are shown.*

1. Introduction

Image restoration problem refers to a finding of an image which has been corrupted by noise and probably has undergone blurring. This is disastrous task since finite support for initial data leads to existing of "invisible" part of a true image distribution and no linear method will recover it. Consequently, the image reconstruction always involves procedure for optimizing some non-linear objective function and applying extra information to find a true image estimation. The maximum entropy method for image deconvolution (MEM) was introduced in [1], and utilizes a kind of objective function which allows to choose between all admissible image distributions the one which has maximum entropy in a resulting solution . Narayan et al. [2] argue that entropy assumption forces producing an image with sharpen peaks and flatten background.

The feature to get in a maximum entropy image the details were missed in the initial corrupted image together with clear physical interpretation and the ability to take a priori information into account makes MEM attractive for image synthesis but encounters difficulties in optimizing highly nonlinear objective function. The problem of MEM restoration else is challenging since the matrices involved in optimization often are singular and ill-conditioned, and inverse problem would be "ill posed". The various optimization techniques were invented to solve this task, and a review of some of these can be found in [2]. In spite of a big number of published algorithms they bring limited success.

The idea for this work was to use as an optimization method the evolutionary type algorithm and links together the entropy driven nature of MEM with evolution process.

2. Maximum Entropy Formulation

We consider the problem of estimating a true image distribution from a blurred and probably corrupted by noise image given only that it is positive on the spectral support, zero outside and compatible with known data. A priori information about image distribution is assumed to be given , i.e. exact knowledge of a point spread function (PSF) corresponding to a shift invariant blur, the initial "dirty" image unknown on a full support but the one or several regions are known, probably with some measurement errors. The image observation model in a lexicographical order is $g = f \otimes h + e$, and then the origins of maximum entropy image restoration are based on the following assumptions in the 1-D index notation [3]:

$$\sum_{i=1}^{N} f_i = \gamma \tag{2.1}$$

$$g_k + e_k = \sum_{i=1}^{N} h_{ik} * f_i \tag{2.2}$$

$$\sum_{k \in A} \frac{e_k^2}{\sigma_k^2} = \Omega, \tag{2.3}$$

where f is restored maximum entropy image, g the observed image, h the PSF matrix , e is the estimated observation error, normally distributed with zero mean and σ_k variance, γ the total intensity, A is a region where the data are defined, $\sum_{l \in A} 1 = L$, and Ω is confidence level.

The (2.1–2.3) consider a priori information about brightness distribution in an initial "dirty" image. The equation (2.1) restricts a total amount of energy in the restored image, (2.2) fits the initial data within some error and (2.3) assumes that RMS is plausible with some confidence level $\Omega = L \pm \sqrt{2L}$. The entropy to be maximized is the Shannon's entropy

$$E_{\text{ent}}(f) = -\sum_{i=1}^{N} f_i \log f_i \tag{2.4}$$

The Lagrangian for (2.1–2.4) is

$$E_{\text{unv}} = E_{\text{ent}} + E_{\text{fit}}\,, \tag{2.5}$$

where

$$E_{\text{fit}}(\lambda, e, f) = E'_{\text{fit}} + E_{\text{pow}}\,, \tag{2.6}$$

$$E'_{\text{fit}} = -\lambda_0\{\sum_{i=1}^{N} f_i - \gamma\} + \sum_{k\in A} \lambda_k\{g_k + e_k - \sum_{i=1}^{N} h_{ik} * f_i\}\,, \tag{2.7}$$

$$E_{\text{pow}} = \mu \sum_{k\in A} \frac{(g_k - f_k)^2}{\sigma_k^2} - \Omega \tag{2.8}$$

E_{fit} corresponds to the "fitness" to the measured data, E_{ent} to the entropy in the resulting solution and λ , μ are Lagrange multipliers. The maximizing (2.4) in respect to (2.1–2.3) using the Lagrange multipliers and setting the partial derivatives dE_{unv}/df_i , dE_{univ}/de_i in (2.5) to zero leads to the well-known expression for f

$$f_i(\lambda) = \gamma \frac{\exp(-\sum_{i\in A} \lambda_k * h_{ik})}{\sum_{l=1}^{N} \exp(-\sum_{k\in A} \lambda_k * h_{lk})} \tag{2.9}$$

3. The Evolution Strategies Algorithm and Main Definitions.

The evolutionary algorithm (EA) is an iterative procedure that applies Darwin's evolutionary model to solve an optimization problem. Currently there are three main modifications of the EA: Evolutionary Programming (EP) [4], Genetic Algorithms (GAs) [5] and Evolution Strategies (ES) [6]. The ES algorithm was accepted for maximum entropy image restoration. The Evolution Strategies works with a real value objects that makes simple of the data representation for MEM, and demonstrates capability to get good approximation for solution even in case of complicated objective function (for overviews of existed applications see [7]).
Let us introduce the main definitions to be used.

– **Definition 3.1.** *The analog elements of* $\{\lambda_i\}$ *are chromosomes.*

- **Definition 3.2.** *An individual is a set of parameters* $\lambda_i\, dim\{\lambda_i\} = N$, *that are defined for a given objective function* E_{λ_i}. *An individual is fully described by his chromosomes.*
- **Definition 3.3.** *A population* $\{\Lambda_j\}\, dim\{\Lambda_j\} = P$ *is a set of individuals with the same objective function* E_{λ_i} .
- **Definition 3.4.** *The selection operator* S *selects* K , $K \leq P$ *individuals at evolution step* t *to be reproduced at the step* $t+1$ *in accordance with a given criterion* $\Psi = min\{E_{\lambda_i}\}$.
- **Definition 3.5.** *The crossover operator* C *is applied to two individuals at step* t *to produce two new individuals at step* $t+1$.
- **Definition 3.6.** *The mutation operator* M *randomizes small number of chromosomes in the individual.*
- **Definition 3.7.** *The energy distribution in the population at step* t *is defined as* $\rho_t(E_\lambda)$.
- **Definition 3.8.** *The moments of population* $m^n =< E_\lambda >^n$, *where* $< ... >$ *denotes the averaging over population,* $n = 1, 2..$.
- **Definition 3.9.** *The normalized second moment of distribution in the population at step* $t, t+1$ *is defined as* $V = \frac{m_{t+1}(E_\lambda)^2}{m_t(E_\lambda)^2}$.

The full evolution in the chain can be represented as the composition of operators:

$$\rho_t(E_\lambda) \longrightarrow^S \rho_t(E_\lambda)^S \longrightarrow^C \rho_t(E_\lambda)^C \longrightarrow^M \rho_t(E_\lambda)^M = \rho_{t+1}(E_\lambda) \qquad (3.1)$$

The objective function for ES MEM is constructed accompanying the usual procedure to minimize only the power component of E_{fit} neglecting the exact amplitude fitness. So we seek a solution for f in the form (2.9) with the constraints (2.3), (2.1). Then, the ES can be represented as an iteration process aimed at minimizing the objective function $E_\lambda = E_{\text{pow}}$ with individual $\Lambda_i = f(\lambda)$. Thus way the image that is the result of (3.1) will be the maximum entropy solution for g. The global convergence property for proposed ES with an objective function (2.8) follows from the Borel-Cantelli lemma [8]. Further, we must distinguish the entropy of two different objects:

- the first entropy (2.4) defines the entropy in a single individual . This entropy is subject to maximize as a matter of the maximum entropy postulate.
- the second entropy is a property of the population distribution itself. This entropy is proportional to the second moment (m^2) of population.

Interestingly enough, Shapiro et al. [9] show that between individuals with equal values for the objective function, the individual that has the larger entropy E_{ent} should be chosen for further reproduction. It reflects the fact that physical systems evolve from less probable state to more probable state agree to the second thermodynamic law [10] and it tends to increase entropy. The constructed evolution algorithm with the objective function (2.8) will implicitly choose individuals with "better" entropy. We expect that this will lead to a good convergence rate for the algorithm. This is simply intuitive observation and would be interesting to have some analytical justification.

4. Algorithm Discussion

4.1 Initialization

The initialization step must define the maximum number of iterations N_{iter}, the accuracy ϵ, and the confidence level Ω. The initialization creates a population $\{\Lambda_j\}$ with $dim\{\Lambda_j\} = P, dim\{\lambda_i\} = N$. A good value for P may be $P = 0.2 - 0.8 \sum_{n \in A} 1$ for small data set and $P \leq 10^2$ is suitable for many larger cases. Our experience shows that the initial values for chromosomes are not important, and only a high initial value for m^2 in population must be checked.

4.2 Selection

The selection increases the mean fitness m^1 in the population. For each element in Λ_i the $E_{\lambda_i} = E^i_{\text{pow}}$ is computed taking (2.1)– (2.9) into account. Individuals with better values are considered for further reproduction, and a "roulette-wheel" selection method (see [11]) with Boltzmann weighting is used to generate the intermediate population $\{\Lambda'_j\}, dim\{\Lambda'_j\} = 2P$, $\{\Lambda'_j\} \in \{\Lambda_i\}$. The parameter p^a is the probability to choose the individual (a) from $\{\Lambda_i\}$

$$p^a = \frac{e^{-\beta * E_{\Lambda_i}}}{Z} \tag{4.1}$$

where $Z = \sum_{i=1}^{P} \exp^{-\beta * E_{\Lambda_i}}$, β is the selection parameter. The lost of diversity can be estimated through V^s [11]

$$V^s(\beta) \approx ln(1.2 + \frac{2.8414}{2.225\beta - ln(\beta)}), \tag{4.2}$$

where s index vs. *selection*. The selection operator is irrelevant to the problem formulation so we can reuse the result by Shapiro et al. [9] for an optimal selection parameter. Thus the parameter β is chosen assuming that the selection must improve the mean fitness in the population as much as possible $(\beta \to max)$, but the loss of entropy $\frac{m_2^{t+1}}{m_2^t}$ (diversity) in the population must be small after selection. For β_s, where $\beta_s = \beta\sqrt{\frac{m_2}{2\log P}}$ is the normalized selection parameter the optimal value is about 0.1-0.3. Then, the balance between a fast convergence rate and small losses in an entropy in the population is achieved.

4.3 Crossover

The crossover constitutes the main difference between random search methods and evolutionary algorithms. This operator focuses the evolution on the most promising paths. The crossover operates on a wide range of genetic material, and it makes possible to generate new candidates with properties that were not found in the parents. The two new individuals are produced by randomly taking chromosomes from the two parents (*discrete crossover*). The parents are individuals (s, t) that were chosen for reproduction at the selection step. The new child gets chromosomes: $\{\lambda_i\} = \{\lambda_i^t \, or \, \lambda_i^s \, , i \in A\}$
In terms of the optimization problem, the selection operator will change the search direction in the tangent plane, making it coherent but different from the parents' directions, and "hopefully" better.

4.4 Mutation

Mutation is fully described by following parameters:

- p_t – probability of changing a chromosome at step t, $p_t << 1$.
- $A^{'}$ – mutation space. The $A^{'}$ is a support region to be defined as a subspace of A , constructed from mutated elements of A.
- ξ – sparse mutation vector $dim\{\xi_l^{'} \, l \in A^{'} \, or \, 0\} = L$, $\xi_l^{'}$ is a random value with probability density $N(0, \delta)$.
- $D(f_i^t)$ – global step size meta control in a f domain, $D(f_i^t)^2 =< f_i(\xi) >^2$.

The new value for the mutated chromosome is:

$$\lambda_l^{t+1} = \lambda_l^t + \xi_l \, , \tag{4.3}$$

where l is the index of the chromosome in the set, t and $t+1$ are steps before and after mutation and $\xi_l = \frac{\xi_l^{'}}{f_l h_0}$ The element of the image f_i^t during mutation of $M \approx p_t * N$ chromosomes obtains the new value:

$$f_i^{t+1} = f_i^t \sum_{l \in A'} \exp(-h_{li} * \xi_l) \tag{4.4}$$

if we restrict analysis to the case of $||\xi|| << \frac{1}{||h||}$, $M << N$, so the ξ is sparse "enough" ,then the second image moment:

$$D(f_i^{t+1})^2 = \begin{cases} D(f_i^t)^2 + \delta^2 \, , i \in A^{'} \\ 0 \, \text{otherwise} \end{cases} \tag{4.5}$$

The worse to see that after mutation $\rho(E_{\text{pow}}) \sim \delta^2 \chi_M^2(k)$,where $\chi_M^2(k)$ is the χ^2 noncentral distribution with M degrees of freedom and $k = \frac{||E_{\text{pow}}^t||^2}{\delta^2}$
We take interest only in the second moment of E_{pow}, since the mutation is the only operator that increases the second moment, and thereafter increases

the entropy in the population. The maintaining diversity in a population is important to avoid premature convergence and the main task for mutation is compensate losses of diversity $\Delta m^s(\beta)^2 \sim V^s(\beta)$ after selection. From (4.4)–(4.5) the second moment after mutation increases:

$$m_{t+1}^2 = m_t^2 + 2M\delta^2 \tag{4.6}$$

Consequently the reasonable estimation for p_t and δ are taken from equation

$$\delta^2 \approx \frac{m_t^2}{2Np_t}(1 - V^s(\beta)) \tag{4.7}$$

The (4.7) together with 1/5 *successive rule* by Rechemberg [5] allow to choose values for δ and p_t, at the each iteration step.
The "simulated annealing" method was used to adjust value of δ^2 from (4.7) for a population during the convergence process. The algorithm starts with a large initial value $m_{t_0}^2 = D_0$ and gradually makes it smaller as the population grows. The δ^2 is adjusted with coefficient $\alpha = 1 + \log(\varsigma t + 1)$,where t is the population age and ς is the normalization coefficient $\sim \frac{1}{P}$. This strategy allows to obtain a high convergence rate at the beginning of the iterations with a solution that is close to optimal at the end.

4.5 Stopping criteria

The objective function (2.3) is computed at the end of every iteration step for the full population and compared with the confidence level Ω. If $|\Omega_i - \Omega| < \epsilon$ or the maximum number of allowed iteration is exceeded, the algorithm finishes; otherwise, the algorithm goes to step (4.2).

5. Simulation

An algorithm was developed for restoration single dish radio astronomy images. In this case the brightness distribution over the sky has to be estimated on the basis of a small data set ($L \leq 10^2$) of blurred and corrupted image. The initial array usually is incomplete and has many "holes" in the data. In this section we present a result of applying algorithm for degraded images consisting of four "astronomy like" sources (Fig. 5.1), where the sources are Gaussian with the parameters are listed in the Table 5.1.
For comparison the same images were processed by MEM algorithm of [3] based on a gradient optimization.
The dependence of quality of restoration on degradation factors (additive noise, PSF, percent of holes in the data) is investigated (Table 5.2). The quality of reconstruction for MEMs images for both algorithms were estimated proceeding from the following criteria:

Table 5.1. Source parameters

Source	Peak Position , $F_d(\alpha, \delta)$ [arcsec]		Peak flux, F_p [mJy]
	α	δ	
1	0	0	2.6
2	1	0.9	1.3
3	-0.5	0.8	1.7
4	0.5	0.8	0.6

- t_c - time to convergence,
- the fitness to true distribution by parameters $Q_f = \sum_{i=1}^{N}(f_i - f_i^{\sim})^2$, where $^{\sim}$ stands for true (original) image, and the peak position mismatching $Q_p = \sum_{i=1...4}(F_{d_i} - F_{d_i}^{\sim})^2$

All images have been processed on a lattice with dimension 64x64 cells. For evolutionary algorithm the set of parameters were used $\beta = 0.2\,, P = 65\,,$ $N_{iter} = 200\,, \epsilon = 1.e^{-3}$. The results are discussed below and shown in figures 5.2–5.4. The superscript (^{e}) stands for an evolution MEM and (^{g}) for the standard gradient based MEM algorithm.

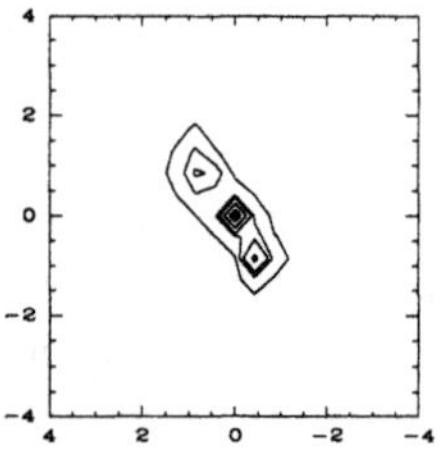

Fig. 5.1. The original (clear) image

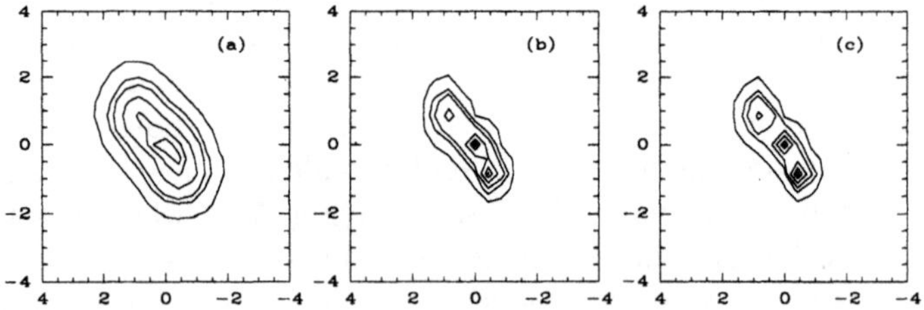

Fig. 5.2. (a) dirty, (b) MEM^{e}, (c) MEM^{g} image, simulation 1

Based on the simulations (Table 5.2), the following observations have been made:

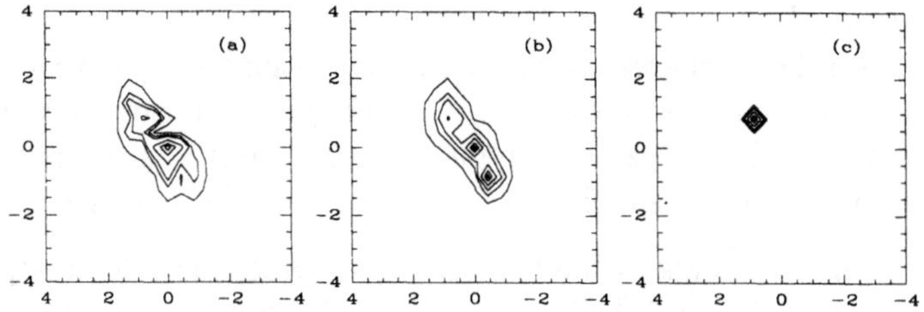

Fig. 5.3. (a) dirty, (b) MEM^e, (c) MEM^g image, simulation 2

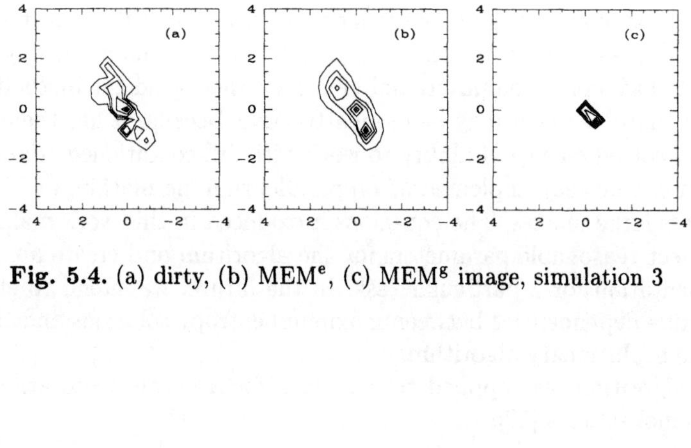

Fig. 5.4. (a) dirty, (b) MEM^e, (c) MEM^g image, simulation 3

Table 5.2. Simulation summary

Simulation parameters				*Simulation results*					
				ES MEM			Gradient MEM		
N	HPBW* [arcsec]	F_c^{**} [%]	SNR [dB]	t_c [hrs]	Q_f	Q_p	t_c [hrs]	Q_f	Q_p
1	0.7	100	12	≤ 18	4.3	0.3	≤ 0.5	4.21	0.4
2	0.6	60	20	≤ 12	10.2	0.9	–	–	–
3	0.6	40	20	≤ 4	12.3	1.3	–	–	–

* *half power beamwidth*

** *percent of coverage for the observed data*

I For synthetic Fig. 5.2(a), simulation (1) both MEMs algorithms show good restoration result with significantly faster convergence for the gradient MEM,

II For synthetic figures 5.3(a), 5.4(a) simulations (2),(3) with a sparse images and large errors the MEM^g convergence was not reached while evolutionary MEM^e was still applicable even with less computational time comparing to the simulation (1) due to a smaller data set.

Comparing the ES algorithm with a simple random search strategies the simulated images 5.2(a), 5.4(a) were additionally processed by the algorithm is constructed from graph (3.1) excluding the crossover operator and setting $P = 1$. The random search method shows no reasonable result in rational time.

6. Conclusion

This maximum entropy evolutionary strategies algorithm can be used to restore images in cases when more traditional maximum entropy algorithms fail. It has worse computational efficiency then gradient methods, but in a some "hard" cases it may be more attractive, because it has the advantage of global convergence and ability to work with "ill-conditioned" data. The algorithm can be easy implemented on parallel running machines which allows to process large images. The equations introduced in this work make it possible to select reasonable parameters for the algorithm and create an efficient implementation for a particular task. In the future, we intend to elucidate the intuitive dependencies between maximum entropy solutions and the behavior of the evolutionary algorithm.
The algorithm was applied to real data for accurate restoration images of methanol sources [12].

Acknowledgement. The author is grateful for helpful conversations on this subject with Sergey Likhachev of Astro Space Center.

References

1. Burg Joan 1975 PhD thesis, Stanford University.
2. Narayan R, Nityanada R, 1986 Maximum entropy image restoration in astronomy. *Ann. Rev. Astron. Astrophys* 24:597
3. Wilczek R, Drapatz S 1985 A high accuracy algorithm for maximum entropy image restoration in the case of small data sets. *A&A*. 142:9-12
4. Fogel L J 1966 Artificial Intelegence through simulated evolution. John Wiley, New-York.

5. Holland J H 1975 Adaptation in natural and artificial systems. The University of Michigan Press. Ann Arbor.
6. Rechemberg I 1973 Evolutionsstrategie, Optimierung technisher Systeme nach prinzipen der biologischen Evolution. Frommann Holzboog Verlag, Stuttgart.
7. Back T, Schwefel H 1993 An Overview of Evolutionary Algorithms for Parameter Optimization. *Evolutionary computations.* V.1 1:1-23.
8. 1981 Convergence of random optimization methods for constrained optimization methods. *Journal of optimization theory and Applications* 33:451-461.
9. Shapiro L J, Prugel-Bennet A 1994 A Statistical Mechanical Formulation of the Dynamics of Genetic Algorithm. *Physical review letters*, V. 72, N. 9, pp. 1305-1308
10. Thomson, C.J, 1988. Classical equilibrium statistical mechanics. Oxford Univ. Press.
11. Bickle Tobias , Thiele Lothar 1995 A comparison of Selection Schemes used in Genetic Algorithms, Computer engineering and communication networks lab., Swiss Federal Institute of Technology, TIC-Report Nr 11, December
12. Dzura A, Promislov V.G 1997 The methanol sources restoration by the maximum entropy method. Proc. XXVII Radio astronomy conference, St'Petersburg

The Finite Element Method and Soft Computing

Larry Manevitz[1] and Dan Givoli[2]

[1]*Department of Computer Science*
University of Haifa
[2]*Faculty of Aerospace Engineering*
Technion - Israel Institute of Technology
Haifa, Israel
email: manevitz@mathcs2.haifa.ac.il givolid@givoli.technion.ac.il

Keywords: Finite Element Method, Soft Computing, FEM, Mesh, Neural Networks, Expert Systems, Mesh Placement, Dynamic Meshes, Mesh Numbering

Abstract

A programme to automate the finite element method is discussed, and the implementations of three specific sub-problems (node numbering, mesh placement, and adaptive meshing) are described. It is also argued that the overall architecture of an "intelligent finite element package" can serve as a "test-bed" for many soft-computing techniques.

1 Scientific Background

The finite element method (FEM) is a computationally intensive method for the numerical solution of partial differential equations. It is a widely used tool and in many cases is the method of choice. However, in practice one can not use the FEM as a "black-box" solver; i.e. it is not sufficient to know the governing equations, the geometry, and the boundary and initial conditions in order to obtain high-quality numerical results. It is well appreciated among finite element users in industrial and scientific communities that the successful application of this technique requires substantial amount of experience and expertise in order to make the computation feasible and the results accurate at the same time.

Partially supported by joint U.Haifa-Technion Research Grant

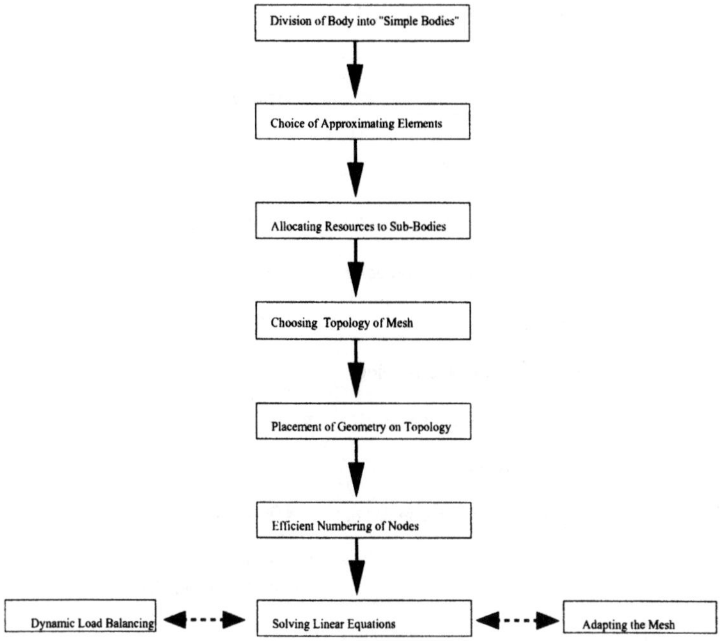

Figure 1: **Some general tasks in the Finite Element Method.**

This is because there is a very large number of parameters that need to be chosen; computational limitations make the choice of these parameters crucial for successful use. These are knowledge based requirements; and current usage requires much human expertise. There are no known effective algorithms that can replace a human user in all cases; and many of the problems are known to be computationally intractable in the general case [11]. Figure 1 1 lists some of these tasks.

On the other hand, the complexity of some of the problems are beyond the realm of efficient manual control (e.g. in more than two dimensions or in quite complicated situations like parallel implementations which are not directly amenable to human intuitions).

In recent years there has been a large development of tools for artificial intelligence, neural networks, fuzzy logic and related disciplines, (called by some "Soft Computing" [34], [33] [32]), which our work and analysis indicate are applicable to the problems under discussion here. Each problem can in fact be attacked by a number of methods, but *usually* a certain approach suggests itself as the most promising one in each case. Thus expert system technology [6] seems most appropriate to replace human numberers, self-organizing neural networks [19] and fuzzy logic "critics" [32] seem appropriate for mesh placement, neural network non-linear predictors are appropriate for dynamic mesh placement for solutions to time dependent PDEs [31] , distributed artificial intelligence [27] and genetic algorithms [14] can be appropriate for load balancing in parallel computation, and so on . The specific problems and tools we have

PROBLEMS AND METHODS

Problem	Approach
Partitioning given "body"	Computational Geometry
Choice of approximating Elements	Expert System
Allocating resources to sub-bodies	Expert System; NNs; GAs (?)
Choosing Topology of Mesh	Expert System
Assigning Geometry to Topology	NN(Kohonen-style); fuzzy critic
Efficient Numbering of Nodes	Expert System
Adaptive Solving	
Prediction of Error Indicator	NN (learning network)
Load Balancing	Distributed AI; Automatic Negotiation; GA

Figure 2: **Table of Problems and Soft Computing Solution Approaches**

in mind are described in Figure 2.

So far, together with our students Miha Margi and Malik Yousef, we have implemented two of these solutions: (automated mesh numbering via an expert system and mesh placement via a self-organizing neural network). In addition a third sub-problem (mesh adaptivity) is under current development using feed-forward neural network run in a temporal predictive mode. In the following sections, we sketch the results of these sub-problems (full details appear or will appear elsewhere [13], [35]).

In addition, we have come to realize [35] [13] that the FEM serves as a rich test-bed appropriate for the serious use of these techniques in real world (i.e. non-"toy") problem settings. Evaluating these techniques in realistic settings is in itself an important research goal. In fact, our initial work has already resulted in advancing somewhat [21] one of the most classical neural network algorithms as a result of evaluating the needs in the realistic setting.

We point out that since one can compare results with analytic ones in certain settings, it allows for quantitative evaluation of the effectivity of the techniques. In principle, this allows to compare and evaluate competing techniques and directions both between themselves and with current commercial packages.

By a test-bed we have in mind the a system which allows one to implement a variety of techniques on various problems, run it easily on "real-world" applications, and be able to do some sort of evaluation/comparison studies with other techniques.

Intelligent automation of the FEM can be appropriate for this because

- There are a wide variety of optimization/satisfaction sub-problems to be solved. Various techniques are applicable to the different problems.
- The architecture of the automation is modular. This means that each of the sub-problems can be solved more or less independently. ("More or less" means that when all the solution steps are in place, one has to consider another optimization problem concerning trade-offs. However, this affects only the potential optimality of the FEM solution, and not the effectivity of the individual steps.) Each of the steps can be evaluated individually or how it affects the global quality of the numerical solutions.
- The nature of the FEM, a numerical solver for partial differential equations, makes it easy to generate test problems. In fact any partial differential equation with boundary conditions defines a test case. This means it is almost as easy to work on real world problems as on artifical simple or "toy" ones. One can also provide test cases with analytic solutions to provide a natural "gold standard" for evaluations.
- The importance of the FEM means that there are commercial codes available for comparison.

2 Heuristically Numbering the Nodes

When applied to linear boundary value problems, the finite element discretization of the governing partial differential equations leads finally to a linear system of algebraic equations (see [16]),

$$\mathbf{Kd} = \mathbf{F}$$

Here $\mathbf{K}$ is the so-called *stiffness matrix*, $\mathbf{F}$ is the load vector, and $\mathbf{d}$ is the vector of unknowns. The dimension of the system (1) in typical industrial applications is of order 100 or 1000 or even 10000. This dimension arises from the chosen discretization of the spatial domain; i.e. the spatial domain is divided into *elements* which are connected to each other by nodes. These nodes are numbered and $\mathbf{K_{ij}} \neq \mathbf{0}$ only if nodes i and j belong to a common element.

Since the mesh is typically imposed on a physical domain and thus can be thought of as a planar or spatial graph, most $\mathbf{K_{ij}}$ are zero, i.e. $\mathbf{K}$ is a *sparse* matrix. This in principle reduces the computational resources needed to solve such a system; however this is sensitive to how "near-diagonal" all the non-zero entries are.

The discussion above implies that the order in which the nodes are numbered is crucial, since this numbering determines how near-diagonal $\mathbf{K}$ is. Note first that optimal numbering of a given mesh is an NP complete problem ([11]). Thus for a mesh containing a large number of nodes and elements it is of course impractical to find an exact solution for the optimization problem by performing a full enumeration. However, there are so-called "algorithmic" solutions which have been incorporated in commercial software. Anecdotal evidence about this software reports that: (i) a human expert can often outperform it. (ii) The best such algorithms typically result in rather complicated numberings involving substantial "windings" and for some applications there is an advantage in the simplicity of the numbering. (These algorithms include the methods described by Cuthill and McKee [4], King [18], Collins [3], Akin and Pardue [1] Gibbs [12] Razzaque [26] [20], Pina [24] Sloan and Randolph [28], Fenves and Law [8], and Sloan [29].)

According, we decided to test the feasibility of developing an expert system to try and mimic the heuristic mechanisms of a proficient human numberer. Actually only some of the possible heuristics were implemented, but the system proved to compare favorably with the human user on a variety of test examples. (Full details appear in [13] .)

In order to state the optimization problem mathematically we introduce some notation. Let the dimension of the matrix $\mathbf{K}$ be $N \times N$. Let b_i be defined as $b_i = i + 1 - \max_{K_{ji} \neq 0} j$. In other words, b_i is the height of column i starting from the diagonal and up to the skyline. Now, let the *average half bandwidth* AHB and the *root-mean-square bandwidth* RMSB be defined as AHB $= 1/N \sum_{i=1}^{N} b_i$, RMSB $= \left(1/N \sum_{i=1}^{N} {b_i}^2\right)^{1/2}$. It is apparent that the bandwidth depends on the nodal numbering system. This dependence becomes strong for large meshes. The optimization problem under consideration can thus be stated as follows: Find a numbering system for the nodes (from 1 to N) such that AHB (or RMSB) will be *minimal.*

2.1 A brief description of the expert system characteristics

One of the hardest steps in producing an expert system is to construct the heuristics according to which the expert system will perform, and which will mimic the considerations of a human expert. To this end, we note the following functions typically performed when numbering the nodes of a finite element mesh manually and briefly comment on our system's approach to these tasks.

0. Preparation of numbering strategies for "paradigm simple blocks" with different geometries and topologies. Varying parameters of freedom are associated with each strategy. These strategies are achieved by experience, by trial and error, and sometimes by full analysis. This step is not "performed" for each mesh separately, but is rather a data base of knowledge an expert has accumulated. Examples of simple blocks are rectangles, annuli, discs.

1. Subdivision of the mesh into a disjoint union of simple blocks. Our system does not perform this, but receives it as given.

2. Choosing the order in which these blocks are picked for numbering.

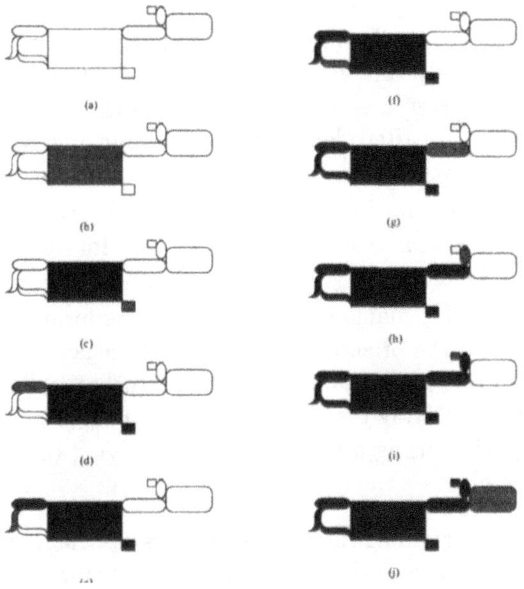

Figure 3: Ordering the simple blocks

In this step the order of the simple blocks is determined. To optimize the numbering a user tries to keep the node numbers as continuous as possible. However when passing from block to block this is often impossible. Thus the goal is to keep the "jumps" across block interfaces as small as possible. In the discussion that follows, a *component* is a topological component, i.e. a maximal connected sub-body.

In the implemented system the procedure for block ordering is as follows:

• Choose the first block to be numbered as the largest simple block, based on the number of nodes.

• Remove this simple block from the mesh. (This may cause the remaining mesh to have several disconnected components.)

• Place all resulting separate components in increasing order in a stack.

• Until the stack is empty do the following: • Remove the smallest component from the stack.

• From the chosen component, choose the next simple block to be numbered in the following way:

• Consider all simple blocks in the component with an interface with the previously numbered block (there are always such).

• Choose the one with the largest interface.

• Remove the chosen block from the component. (This may divide the remaining blocks into several connected components.)

• Place the new components on the stack in increasing order.

See figure 3 for an example of how the blocks are numbered.

3. For each block, choosing a numbering strategy, depending on the topology and the geometry of the block. That is, choosing the best match to a paradigm simple block. Essentially this has two parts: (1) solution of a pattern recognition problem; i.e. given a simple block find which of the paradigm simple blocks it is closest to; (2) after identificat ion choosing a numbering method for the block form amongst the possible methods mentioned in step 0.

In the implemented version, all the blocks are assumed to be pseudo-rectangles and there is only one numbering strategy for numbering rectangles, so this step is to a large extent vacuous.

Notwithstanding, it remains for the system to perform the following analysis: (a) to decide on the orientation of the block, regarded as having a rectangular shape; (b) to be able to ignore small perturbations. The first is done using an algorithm involving the convex hull of the block and regression ; the second is accomplished using a heuristic (also involving the convex hull) that defines a "grain size".

4. For each block, determining the free parameters associated with the strategy chosen for that block, such as the node from which to start the numbering. For the rectangular strategy which is handled in the implemented version, the only free parameter is the node from which to start the numbering, and there are four possibilities, corresponding to the four corners of the pseudo-rectangle. This is selected by a heuristic that is based on both topological and geometrical considerations.

It is interesting and crucial to note that while the mesh is a purely topological object as far as node numbering is concerned, the user decides on the division based not only on topological criteria, but also on *geometric* criteria. This is an obvious logical fallacy and it is easy to devise artificial mesh examples where the results are extremely bad. Nonetheless for real meshes the results seem acceptable.

The reason why human users can be so successful in numbering meshes although basing their decisions on geometric considerations is that they intuitively rely on the fact that *actual meshes are designed with implicit geometric constraints.* For example, a good practice in mesh design is to use elements which have aspect-ratios close to unity, namely quadrilaterals and triangles which are nearly equilateral. Another good practice is to pass from a crude region (large elements) to a refined region (small elements) in a gradual manner. (See e.g. Irons and Ahmad [17].)

Our expert system tries to mimic a human user; therefore it uses geometric considerations as well as topological ones. This is in contrast to the "algorithmic approach", where only the topological properties of the mesh play a part in the node reordering procedure.

The system was tested on a number of meshes taken from the literature; here we mention the largest mesh tested to date similar to one that appears in [36] which has 359 nodes and 559 triangular elements. Here the average half band width of the system was 17.6; that of the human expert was 15.9 while for comparison a random numbering resulted in a value of 210.2.

Some other meshes were taken from applications in large deformation continuum mechanics discussed in [15] and solved by the finite element code NIKE2D. The results were are quite comparable to that of the human expert. (In one case the expert system actually outperformed the human expert.)

The results show that although the human expert performs better than the expert system, the differences between the two AHBs are reasonably small. Other examples have been tested as well and the results compared favorably with those of a human expert.

3 Mesh Placement via Self-Organizing Neural Networks

Once a topological mesh has been defined (a problem we have suggested is appropriate for an expert system), the mesh has to be given its geometry; i.e. it has to be placed appropriately on the body. Moreover, there are points and regions which one wants to cover with a finer mesh than other areas. These are regions or points of "interest" where the approximation used in the finite element method is intrinsically worse. Essentially it is an optimization problem; given a fixed amount of computation that one wants to expend, which translates into a fixed amount of elements; how should one best distribute these elements so as to obtain the best approximation using the finite element method. A better mesh results in a better approximation.

There are several requirements for the quality of a mesh. For example, one wants the proportions of the elements to be as close as possible to those with good aspect ratios (the aspect ratio is the ratio of the radii of circumscribed circle to that of the inscribed circle; "good" means here close to one). For a quadrilateral then, for the best results from the finite element method one wants quadrilaterals close to squares and triangles close to equilaterals. In addition, one wants the change in size between elements determined by the mesh to be gradual. Quantitatively, one wants to keep the ratio of radii of circumscribed circles of adjacent elements to be close to one; globally that means that the maximum and minimum of such ratios of all pairs of adjacent elements should be as close to one as possible.

Moreover, in a typical finite element mesh, the density of the nodes is taken to vary from being very high near certain critical regions or points to more sparse where simpler approximations will suffice.

This can be handled using a modification of Kohonen's self organizing the neural network approach ([35], [22]) by setting the probability density function of the input to correspond to the desired density of the network. The "self-organizing" feature of the Kohonen map will then place the highest density of nodes according to the sampling of data according to this density function. In this implementation, one identifies the weight of the neurons with the geometric coodinates of the body. The Kohonen map results in an equiprobably response map of the neural net (constrained by its given topology) subject to the sample data. Actually, the algorithm we use is a substantial adaptation of

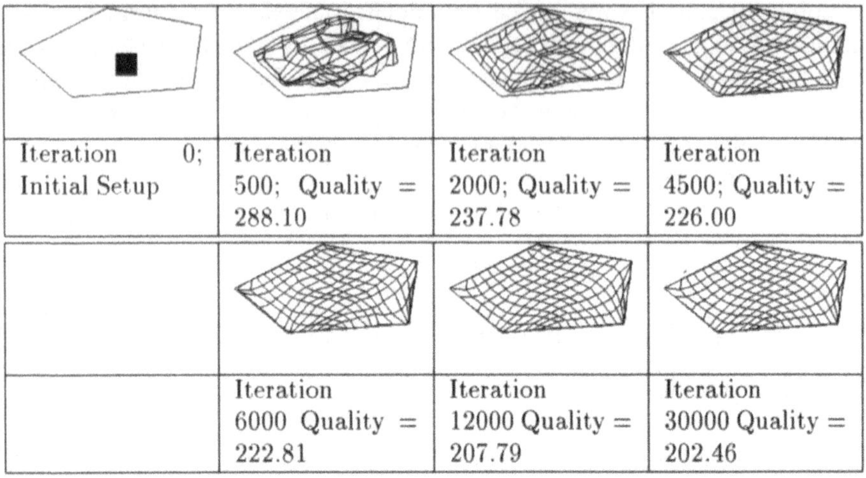

Figure 4: **A sequence of snapshots of a mesh being placed via a NN algorithm**

the basic Kohonen map since we require that the network map boundaries to boundaries. This is accomplished by using a weaving of one dimensional and two dimensional Kohonen maps. This is discussed further in [21]. Figure 4 gives a view of the algorithm in action.

In a series of experiments this algorithm was compared with one of the most popular finite element packages over a series of problems, and was found to be generally superior. Figure 5 gives a sample comparison with the PLTMG [2] package. Full details of this experiment can be found in [35].

4 Dynamic Mesh Generation for Temporal PDEs

Time is often treated distinctly from spatial dimensions in the solution phase of PDEs . That is, the typical method of choice for solution of such equations is *not* to treat time as simply another dimension, but to "simulate time"; i.e. to repeatedly solve the equations for different times; using the previous solution as the starting conditions for the next one.

However, in a dynamic system, this implies that one should not use the same mesh at different times since the "areas of interest" are, of course, changing with time.

For example, when the solution of a hyperbolic problem involves a shock wave which propagates through the mesh, the location of the "area of interest," namely the shock vicinity, keeps changing in time. Another example is a problem of fluid flow in a cavity, where flow cells are generated and undergo continuous changes in their shapes and size as time proceeds [9] .

Thus the mesh choice should be dynamic; varying with time. In current usage, the method is to use the solution at time t_n to indicate where the mesh

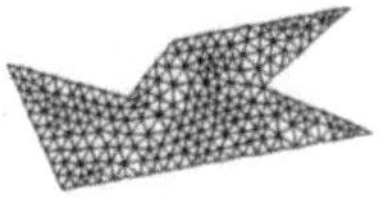	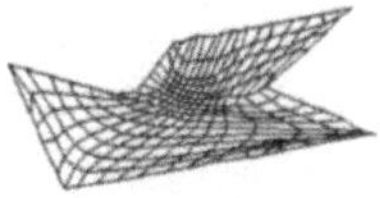
PLTMG 249 nodes	NN 225 nodes; Quality = 278.713715; ''Hot-spot" (2,2) near center; 46650 Iterations

7-Sided Non-Convex Domain PLTMG (249 nodes, 437 elements) N N (225 nodes, 196 elements)					
		Error/Node		Error/Value	
$u(x,y)$	$f(x,y)$	PLTMG	NN	PLTMG	NN
$e^{-(x-2)^2}e^{-(y-2)^2}$	$-u_{xx}-u_{yy}$	2.412143E-02	7.530449E-03	4.515054E-02	9.097765E-03

Figure 5: **A typical comparison between the NN and a commercial package**

should be modified (where it should be refined and where it can be made coarser) at time t_{n+1}. The work [9] used this mechanism.

However, this suffers from the obvious defect that one is always one step behind. If the area of interest is propagating (a common phenomenon) then one may be always refining directly behind the most interesting phenomenon. This is also assuming that one does not miss the "action" altogether. One can look at this as a special instance of a control problem.

It would be preferable to "predict" the area of interest at time t_{n+1} based on the solution at t_k for $k \leq n$. There is what we propose to try using one of several neural network approaches. Since a NN is a universal approximator [5, 10, 25, 20] the intention is to use as input the areas of "interest" at recent times and predict the areas of interest at the next time stage.

The simplest such mechanism is to use a feed-forward neural network trained with a delay to do the prediction. This is what we propose to try using a neural network plant predictor, which has been developed in the context of control theory and signal analysis [31, 30] . The basic idea can be seen in Figure 6. Various parameters; (e.g. function values at the node at several prior timesteps, values at neighboring nodes, distance from "anomalies" in the body) will be entered to the network; the output will be an encoding on the need for refinement following a delay.

The problem in such an approach, is (1) to decide on the appropriate information to train the network with, and (2) to decide on the appropriate breakdown into sub-problems. That is, although it is known that simple NNs are in principle universal approximators; in practice, one should not attempt to make one network do too many things, and the best results in the literature are obtained when the NNs are applied to specific sub-problems.

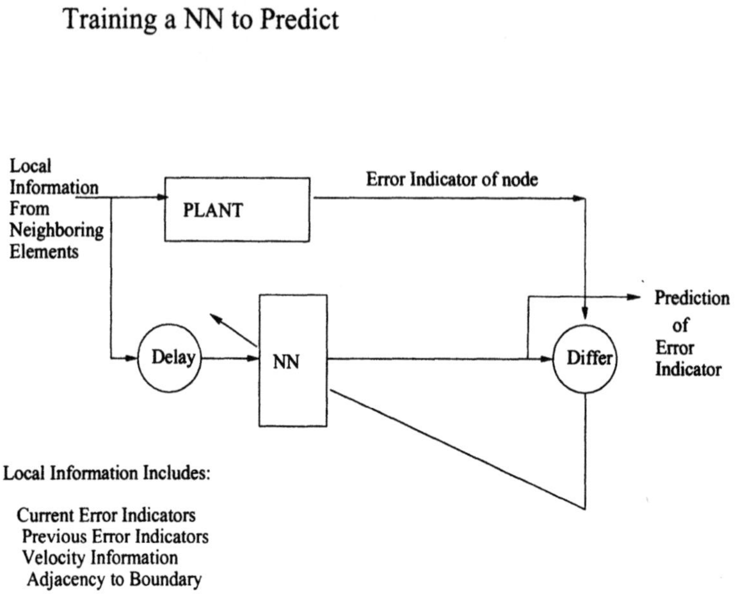

Figure 6: **A neural network designed to predict**

Such a mechanism was used to good affect in the recent work "GloveTalk" [7] where a neural network was taught to translate hand-language to phonemes by using varied information; including rate of change of hand position. We mention this work, because we think its use of velocity; computed by including the values of a function over several time steps (and acceleration; by including the velocity over several time steps) as input to a NN is something that will be useful for our implementation. We anticipate including the solution over several previous time steps in order to predict where the solution will have a large change in the *next* time step. Substantial experimentation will be needed to decide on the correct encoding of information for this prediction.

Another breakdown we anticipate is the use of different such neural networks for different parts of the mesh. That is, we expect that training such a network will be more effective by dividing nodes in the mesh into different classes and using separate nets for the different classes. The underlying philosophy here is that the more similar the elements of a class are; the simpler the representation of the predictor function should be; and the easier and faster the training. For example, one would use a different network for nodes near boundaries than ones in the interior of a body. On the other hand, one does not want to divide the body into too many subclasses. (In the limit, this would lead to having a different network for *every* node.) A predictor should work quickly, and thus use the "experience" gained at some nodes to predict density needs at others. A priori, it is our estimate that such a method will work well on propagation solutions; such as occur in models measuring, e.g. the heat equation, or shock [23].

References

[1] J.E. Akin and R.M. Pardue, *Element resequencing for frontal solutions*, Mathematics of Finite Elements and Applications (New York) (J.R. Whiteman, ed.), Academic Press, New York, 1975.

[2] R.E. Bank, *Pltmg: A software package for solving elliptic partial differential equations*, SIAM publications, Philadelphia, 1994.

[3] R.J. Collins, *Bandwidth reduction by automatic renumbering*, Int. J. Numer. Meth. Eng. **6** (1973), 345–356.

[4] E. Cuthill and J. McKee, *Reducing the bandwidth of sparse symmetric matrices*, Proceedings of the ACM National Conference, ACM, 1969.

[5] G. Cybenko, *Approximation by superpositions of a sigmoidal function*, Mathematics of Control Signals and Systems **2** (1989), 303 - 314.

[6] D.A. Waterman F. Hayes-Roth and D. Lenat (eds.), *Building expert systems*, Addison-Wesley, 1983.

[7] Sidney Fels and Geoffrey Hinton, *GlovetalkII: An adaptive gesture-to-formant interface*, Proceedings of ACM CHI'95 Conference on Human Factors in Computing Systems, 1995, pp. 456–463.

[8] S.J. Fenves and K.H. Law, *A two-step approach to finite element ordering*, Int. J. Numer. Meth. Eng. **19** (1983), 891–911.

[9] D. Givoli J.E. Flaherty and M.S. Shephard, *Simulation of czochralski melt flows using parallel adaptive finite element procedures*, Modelling Simul.Mater. Sci. Eng. (to appear).

[10] A. R. Gallant and H. White, *On learning the derivatives on an unknown mapping with multilayer feedforward networks*, Neural Networks **5** (1992), 129–138.

[11] M.R. Garey and D. S. Johnson, *Computers and intractability: a guide to the theory of np-completeness*, Freeman, 1979.

[12] N.E. Gibbs, *A hybrid profile reduction algorithm*, ACM Trans. Math. Software **2** (1976), 378–387.

[13] L. Manevitz D. Givoli and M. Margi, *Heuristic finite element node numbering*, Computing Systems in Engineering **4** (1993), 159–168.

[14] D. Goldberg, *Genetic algorithms*, Addison-Wesley, 1989.

[15] J.O. Hallquist, *Nike2d — a vectorized, implicit, finite deformation, finite element code for analyzing the static and dynamic response of 2-d solids*, Tech. Report UCID-19677, Rev. 1, University of California, 1967.

[16] T.J.R. Hughes, *The finite element method*, Prentice-Hall, New York, 1987.

[17] B. Irons and S. Ahmad, *Techniques of finite elements*, Ellis Horwood and John Wiley, Chichester, England, 1980.

[18] I.P. King, *An automatic reordering scheme for simultaneous equations derived from network systems*, Int. J. Numer. Meth. Eng. **2** (1970), 523–533.

[19] T. Kohonen, *Self-organization and associative memory, second edition*, Springer-Verlag, Berlin, 1988.

[20] M. Shoham M. Meltser and L. Manevitz, *Approximating functions by neural networks: A constructive solution in the uniform norm*, Neural Networks **9** (1996), no. 6, 965–978.

[21] L. M. Manevitz, *Interweaving kohonen maps of different dimensions to handle measure zero constraints on topological mappings*, Neuroprocessing Letters **5** (1997), 153–159.

[22] Y. Stephan O. Sarzeaud and C. Touzet, *Finite element meshing using kohonen's self-organizing maps*, Artificial Neural Networks (T. Kohonen et al, ed.), North Holland, 1991, pp. 1313–1317.

[23] J.T. Oden, *Adaptive methods in computational fluid dynamics*, Finite Elements in Fluids, Vol. 8 (Washington) (T.J. Chung, ed.), Hemisphere Publishers, Washington, 1992, pp. 3 – 30.

[24] H. L. Pina, *An algorithm for frontwidth reduction*, Int. J. Numer. Meth. Eng. **17** (1981), 1539–1546.

[25] M. Leshno V. Lin A. Pinkus and S. Schocken, *Multilayer feedforward networks with non-polynomial activation functions can approximate any continuous function*, Neural Networks **6** (1993), 861–867.

[26] A. Razzaque, *Automatic reduction of frontwidth for finite element analysis*, Int. J. Numer. Meth. Eng. **15** (1980), 1315–1324.

[27] J. Rosenschein, *Rules of encounter*, MIT Press, 1994.

[28] Sloan and Randolph, *Automatic element reordering for finite element analysis with frontal schemes*, Int. J. Numer. Meth. Eng. **19** (1983), 1153–1181.

[29] S.W. Sloan, *An algorithm for profile and wavefront reduction of sparse matrices*, Int. J. Numer. Meth. Eng. **23** (1986), 239–251.

[30] B. Widrow and M. Lehr, *30 years of adaptive neural networks: perceptron, madaline, and backpropagation*, Proceedings of the IEEE **78** (1990), no. 9, 1415–1442.

[31] B. Widrow and S. Stearns, *Adaptive signal processing*, Prentice Hall, 1985.

[32] R. R. Yager, *Essentials of fuzzy modeling and control*, J. Wiley, 1994.

[33] R. R. Yager and B. Bouchon-Meunier, *Fuzzy logic and soft computing*, World Scientific, 1995.

[34] R. R. Yager and Lofti A. Zadeh (eds.), *Fuzzy sets, neural networks, and soft computing*, World Scientific, 1995.

[35] L. Manevitz Malik Yousef and D. Givoli, *Finite element mesh generation using self-organizing neural networks*, Microcomputers in Civil Engineering **12** (1997), 233–250.

[36] O.C. Zienkiewicz, *The finite element method*, 3 ed., McGraw-Hill, New York, 1977.

A Tabu Search Approach for the Tool Assignment and Machine Loading Problem in Flexible Manufacturing Systems

F. Guerrero, S. Lozano, J. Racero, L. Onieva and J. Larrañeta

Escuela Superior de Ingenieros, University of Seville
C/ Camino de los Descubrimientos, s.n. 41092. Seville. Spain
Phone: 34-95-4487202, Fax: 34-95-4463153

Abstract

This paper presents a new tabu search approach to the solution of a mixed integer linear programming formulation of the tool assignment and machine loading problem in flexible manufacturing systems. Our heuristic search method uses candidate list strategy, diversification, intensification and strategic oscillation elements. To test our heuristic approach, a set of test problems has been randomly generated and pseudoptimally solved by a mathematical programming software package. Computational results disclose that our tabu search approach is most effective, in terms of both solution quality and efficiency, for larger problem instances.

Keywords: flexible manufacturing systems, loading problem, mixed-integer linear programming, tabu search.

1. Introduction

Since the late seventies Flexible Manufacturing Systems (FMS) have become increasingly popular in manufacturing. The possibility of manufacturing a variety of different part types on a versatile automated system seems to be especially attractive for an environment with low to medium-sized yearly production requirements. The key to the success of FMS is the degree of flexibility incorporated into it. One of the most important flexibilities in FMS is the routing flexibility, i.e. the ability to process a given part mix on alternative machines. A good introduction to FMS is [1]. Extensive reviews and methodology surveys of FMS can be found in [2], and [3]. In this paper we will address the tool assignment and machine loading problem in FMS, which arises as the most important production planning problem within an off-line planning scheme. Other approaches proposed for this problem can be found, for example, in [4]. This production planning problem could be defined as follows:

Given the production requirements which will be produced concurrently in a given FMS, allocate the operations and required tools among the machine groups subject to the technological and capacity constraints of the FMS, where the objective function is workload balancing.

The problem will be modelled using the notion of production route. A part may be manufactured in a number of different ways each of which constitutes a production route. Each production route may be described as a set of operations each of which requires a machine type, a tool type and a processing time. Given a part type, its alternative production routes arise from alternative process plans and where there are two or more machines capable of performing the same operation. Thus, in our formulation the allocation of individual operations to machines consists then in determining a routing mix for each part type. From these routing decisions, machine workloads and tooling requirements can be derived. The tool assignment and machine loading problem is modelled as a mixed-integer linear program (MILP) formulation which is given below. First, the notation of all the variables and parameters used in the development of the model are defined. Second, the model is presented in the form of an objetive and its associated constraints.

Subscripts and input data:	
i	part type
r	route
j	machine group
k	tool type
D_i	production requirements (demand) for part type i
m_j	number of machines in machine group j
C_j	tool magazine capacity of each machine in machine group j
A_k	number of available tools of type k
L_k	life of tool type k
S_{jk}	number of tool slots occupied by one tool of type k in tool magazine of each machine in machine group j
p_{irjk}	processing time of part type i along route r on machine group j and using tool type k
$J(k)=\left\{j: \sum_i \sum_r p_{irjk} > 0\right\}$	set of machine groups that use tool type k
$K(j)=\left\{k: \sum_i \sum_r p_{irjk} > 0\right\}$	set of tool types used in machine group j
Decision variables:	
θ_{ir}	proportion of units of part type i to be assigned to route r
u_j	workload of every machine in machine group j
u_{min}, u_{max}	minimum and maximum workloads
ξ_{jk}	number of tools of type k assigned to each machine in machine group j

Table 1. Tool assignment and machine loading problem notation

Model

$$\text{minimize } u_{max} - u_{min} \tag{1}$$

subject to:

$$\sum_r \theta_{ir} = 1 \quad \forall i \tag{2}$$

$$\sum_i \sum_r D_i \, p_{irjk} \, \theta_{ir} \le m_j \, L_k \, \xi_{jk} \quad \forall k, J(k) \tag{3}$$

$$\sum_{j \in J(k)} m_j \, \xi_{jk} \le A_k \quad \forall k \tag{4}$$

$$\sum_{k \in K(j)} S_{jk} \, \xi_{jk} \le C_j \quad \forall j \tag{5}$$

$$\sum_i \sum_r \sum_k D_i \, p_{irjk} \, \theta_{ir} = m_j \, u_j \quad \forall j \tag{6}$$

$$u_{min} \le u_j \le u_{max} \quad \forall j \tag{7}$$

$$\xi_{jk} \ge 0 \text{ integer} \quad \forall k, J(k) \tag{8}$$

$$\theta_{ir} \ge 0 \, \forall i, r; \, u_j \ge 0 \, \forall j; \, u_{max} \ge 0; \, u_{min} \ge 0$$

Constraints (2) impose that for each part type, the sum of proportions assigned to each alternative route must be one. Constraints (3) bound from below the number of tools of each tool type to be assigned to each machine group. Constraints (4) impose the limitations on the number of tools of each tool type that are available. Constraints (5) correspond to the limitations on tool magazine capacities. Constraints (6) compute the workload of each machine in each machine group. Constraints (7) define u_{max} and u_{min} as maximum and minumum, respectively, of the workloads of the machines. The integrality of variables ξ_{jk} is imposed in constraints (8).

Since constraints (4) and (5) are of a multiknapsack nature, it follows that the problem may be not easily solved in the worst case. Furthermore, the solution space is highly constrained due to the large number of constraints. Consequently, it will be also difficult to find out fesible solutions for the stated MILP. The aim of this paper is to develop local search mechanisms, in which the existing constraints will be dropped. However, in order to drive the search process to feasible regions, the evaluation of solutions will modified by introducing strategic oscillation mechanisms

The plan of this paper is as follows: Section 2 defines a modified solution evaluation function based on strategic oscillation concepts. In Section 3, we present a new Tabu Search (TS) algorithm for solving the presented MILP formulation. Section 4 presents computational results which are followed by a summary in Section 5.

2. Strategic Oscillation by Altering the Solution Evaluation Function

In this section, we develop a modified solution evaluation function that uses strategic oscillation concepts. The general concept is one of allowing feasible as well as infeasible solutions to occur in the search process. The idea is to drive the search through feasible boundaries in an oscillating manner. This will be done by altering the solution evaluation function.

There are two basic reasons for considering the use of strategic oscillation when solving problems like our optimization problem. One, if the feasible solution space of the problem is disjoint, the strategic oscillation provides a mechanism for crossing regions of infeasibility in the course of searching for the optimal solution.

The second reason is that the proposed MILP formulation presents NP-complete feasibility problem in addition to their optimality requirement. In other words, it is of the same complexity to find out a feasible solution that the optimum. Strategic oscillation is a flexible technique that provides diversity by simultaneously managing feasible and infeasible solutions and that can be used to locate high quality, feasible solutions.

The strategic oscillataion designed here is based upon a hiearchical division of the solution space. Let ξ be a vector of integers which represents a tool assignment to machines, we define three possible solution types: (a) feasible in the original MILP formulation, (b) feasible in the multiknapsack coinstraints (4)-(5) but infeasible in the original MILP, and (c) infeasible in the multiknapsack constraints (4)-(5). It is clear that the best solution type are the feasible in the original problem, i.e. type (a). Within this type, solutions will be ranked according the original objective function (1). After this solution type, we consider those in the type (b). Given a solution ξ in the type (b), and in order to evaluate it, we calculate the total infeasibility in the original problem by solving the following Linear Problem (LP) formulation:

Problem B

$$\text{minimize} \sum_{k} \sum_{j \in J(k)} \omega_{jk}^{-} \tag{9}$$

subject to:

$$\sum_{i} \sum_{r} D_i \, p_{irjk} \, \theta_{ir} + \omega_{jk}^{+} - \omega_{jk}^{-} = m_j \, L_k \, \xi_{jk} \quad \forall k, J(k) \tag{10}$$

$$\sum_{r} \theta_{ir} = 1 \quad \forall i$$

$$\theta_{ir} \geq 0 \; \forall i, r; \omega_{jk}^{+}, \omega_{jk}^{-} \geq 0 \quad \forall k, J(k)$$

where each constraint (10) has been modified by introducing two non-simultaneously basic variables, ω_{jk}^{+} and ω_{jk}^{-}, which measure the amount of feasibility or infeasibility, respectively, in each constraint (10). Those solutions of this type are ranked according to (9), i.e. the total amount of infeasibility. Note that if the objective function (9) is equal to zero, then the considered solution is feasible for the original problem. Finally, solutions in the type (c) will be evaluated according to the violation of the multiknapsack constraints (4)-(5). For each one of the multiknapsack constraints, we calculate a measure of its violation as follows:

$$ic_k = \max \left(\frac{\sum_{j \in J(k)} m_j \, \xi_{jk} - A_k}{\sum_{k} A_k}, 0 \right) \quad \forall k \tag{11}$$

$$ic_j = \max \left(\frac{\sum_{k \in K(j)} S_{jk} \, \xi_{jk} - C_j}{\sum_{j} C_j}, 0 \right) \quad \forall j \tag{12}$$

Note that when an individual multiknapsack constraint is not violated, then the corresponding individual infeasibility is equal to zero. The evaluation of these solution will be the sum of individual infeasibility corresponding to each multiknapsack constraint (4)-(5). In the same way, those solutions of this type are ranked according to them evaluations.

Following this evaluation scheme, we have an specific value associated with not only feasible solutions, but also infeasible solutions. The next step will be to re-scale these values in order to ensure the previously commented ranking between solution types. Figure 1 shows an scheme with the modified objective function for each individual solution taken into account its solution type.

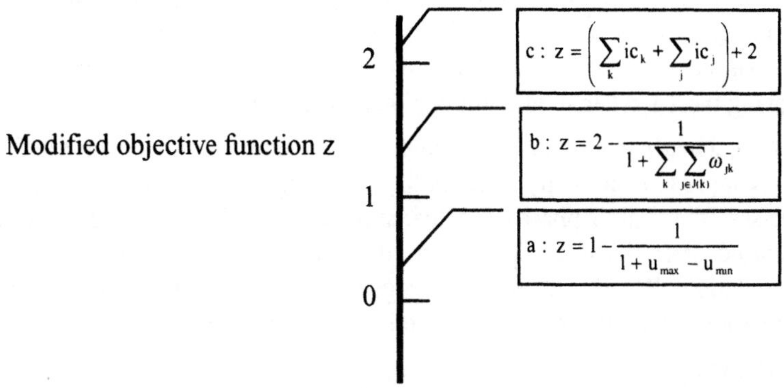

Figure 1. Modified objective function for each solution type

Thus, as we are dealing with a minimization problem, we assign the best modified objective function values to those in the solution type (a). These solutions are ranked between them according the original objective function (1). On the contrary, the worst solutions (placed on the top) are those of the solution type (c). These solutions are ranked according to the total infeasibility in the multiknapsack constraints (4)-(5). In the middle, we place the solution type (b) which are ranked taken into account the objective function of problem (B), i.e. the total infeasibility of constraints (10). Given a solution ξ, we first have to check the multiknapsack constraints (4)-(5) according (11)-(12). If the solution is multiknapsack feasible, then we solve the linear problem (B). Otherwise, the modified objective function is computed as in the scheme. When we solve the problem (B), we determine whether the solution belongs to type (b) or not (the solution belongs to type (a)). In the first case we use the objective function of (B) to compute the modified objective function. On the other hand, i.e. when the solution belongs to type (a), we have to compute the imbalance measured as the difference between the maximum and minimum workloads. To that end we solve the following LP formulation:

Problem A

$$\text{minimize } u_{max} - u_{min}$$

subject to:

$$\sum_i \sum_r D_i \, p_{irjk} \, \theta_{ir} \leq m_j \, L_k \, \xi_{jk} \qquad \forall k, J(k) \tag{13}$$

$$(2), (6), (7)$$

$$\theta_{ir} \geq 0 \, \forall i, r; \; u_j \geq 0 \, \forall j; \; u_{max} \geq 0; \; u_{min} \geq 0$$

in which ξ_{jk} is one of the coordenates corresponding to the specified solution represented by the vector ξ. By using the objective function of (B), the modified objective function is calculated following the formulae of the Figure 1. Note that using the previous scheme, we can visit in a search process both feasible and infeasible solutions, but in a controlled way, i.e. the search process is driven to feasible solutions but with the possibility to cross through infeasible regions.

3. Tabu Search Approach

TS is a metaheuristic that was introduced in [5] and has been used for various classes of integer programs as in scheduling, planning and design (for surveys, see [6, 7]. TS has also been applied to solving general zero-one MILP problems by superimposing the TS scheme on traditional heuristics [8], [9].

In this section, a heuristic procedure based on the TS methodology is presented. The designed TS approach externally controls the tool assignment variables and consists of two phases: the search phase, and the improvement phase. The improvement phase is started when the search phase has found a fesible solution for the original MILP formulation. We first present the algorithm of the search phase and then explain its various components. After this, the improvement phase will be presented.

SEARCH PHASE

Step 1: Using the relaxed LP solution to the original MILP, generate a starting solution by rouding to the nearest integer those non-integer tool assignements. If the solution is feasible for the MILP formultion, then STOP. Otherwise, GO TO step 2.

Step 2: Evaluate all moves on the *candidate list* and choose the move that improves the modified objective function (described in the previous section) the most (or degrades it the least) and is not on the tabu list. Perform the move. If the solution has improved, with respect to the modified objective function, in the last *n_iter_without_improvemnt* iterations, then repeat step 2. Otherwise, GO TO step 3.

Step 3: Perform a move in order to diversify the search using a *frecuency matrix*, and GO TO step 2.

The search phase is started by obtaining the relaxed LP solution (dropping the integrality constraints (8) in the MILP formulation) to the original MILP. The initial solution is obtained by rouding to the nearest integer those non-integer tool assignments. In most cases, this rounding solution leads to infeasible solutions for the original problem. Consequently, additional manipulation of the solution is required to achieve feasibility. Step 2 is a standard implementation of TS with both short-term and long-term memory components. The moves are defined as follows:

$$\text{a)}\quad \xi_{jk} \leftarrow \xi_{jk} + 1; \qquad \text{b)}\quad \xi_{jk} \leftarrow \xi_{jk} - 1$$

Move (a) adds a copy of tool type k to each one of the machines of machine group j. Similarly, move (b) drops a copy of tool type k in machines of machine groups j. The candidate list is a dynamic list that specifies all of the tool assignments that are available for such moves. This list varies according to the type (defined in the previous section) of the current solution, in a way such that priority is given to those moves that reduce the modified objective function (defined in the previous section). Thus, if the current solution belongs to type (c), then only those drop moves that are involved in violated multiknapsack constraints (4)-(5) are considered. Otherwise, if the current solution belongs to type (b), then are first explored those add moves involved in the violated constraints (3). If there are no improvements moves, the heuristic explores the other moves.

The tabu list is a short-term memory structure that prohibits not only entries that have recently been visited, but also those in which are involved the tool types corresponding to the previously visited solutions. Thus, for instance, if an add move is performed, then are prohibitted those drop moves that are involved with the same tool type. It is done in this way in other to reduce the number of moves to explore in each iteration as well as to diversify the search process. A move remains tabu (forbidden) only during a certain number of iterations determined by the parameter *tabu_list_size*. The tabu list is updated circularly, i.e. before the next moves are stored, the oldest ones must be removed. A tabu status is overridden when a move results in an modified objective function better than that of the best solution found so far. Furthermore, if the current solution is not improved in a fixed number of iterations, a diversification move is considered in order to explore those less explored regions. These moves are performing using a matrix that contains the frecuencies of occurrences of moves in the search process.

Once the first feasible solution for the MILP formulation has been obtained, the search process continues according to the improvenment phase. In this phase, the previously defined elements are similar except two of them: the tabu list and the frecuency matrix. Thus, in this phase no diversification moves are performing since we are interested in a more intensificated search in feasible regions. For the same reason, in the tabu list are prohibits only those moves corresponding to previously visited solutions. As stopping criterium for this phase, a maximum number of iterations is fixed, which obviously will depend on the problem size (number of integer variables).

4. Computational Results

The designed algorithm was coded in Visual Basic-4 and run on a PC-100 Mhz under Windows 95 operating system. The LP formulations associated with the designed TS approach were solved using the XA Callable Library [1]. To test our heuristic, we have defined six problem sizes: 40, 60, 80, 100 and 120 integer variables (IV). For each problem size, seven data sets were randomly generated (a total of 35 problem instances were tested). Preliminary investigation reveled the following parameters for the TS approach:

- *tabu_list_size:* all tests were done with a dynamic tabu list length size starting at $\left|\sqrt{\mathrm{IV}}\right|$. The list size is then varied randomly each 10 iterations between this number and its double.
- *n_iter_without_improvemnt:* this parameter has been fixed to 10.

Furthermore, as stopping criterium for our TS approach, a maximum number of iterations is fixed, which depends on the problem size.

The main issue to research is that, although standard branch and bound techniques based upon continuous lower bounds would perform extremely well on relatively small problem instances, our conjeture is that as the problems increase in size, the TS method would become more attractive in terms of the tradeoff between solution quality and computational effort. Since finding and confirming optimal solutions to the test problems requires an unreasonable amount of computer time, we compare the solutions offered by our TS apprach, with those offered by an standard branch and bound method with the $\in$-optimality option [11], where $\in$ has been fixed to 1%. This means that the gap from these solutions with repect to the optimal solutions are less than 1%.

Table 2 summarizes the results for each one of the problem instances. Thus, both objective function and CPU time are shown for both methods. In 8, out of the 35 problem instances, the TS solutions are better than those $\in$-optimal. In 6 problems, the $\in$-optimal solutions are better than those offered by the TS approach. In these problems, the maximum gap between the TS solution with respect the $\in$-optimal solution is equal to 1.31%. For 21 test problems, both methods offer the same solution quality.

The average CPU times (in seconds) for both methods were computed for each problem size and the results are shown in Figure 2. Note that the vertical axis is a logarithmic scale. Hence, it is clear from this figure that for problem size larger than 80, the TS procedure consistently finds optimal or near-optimal solutions in less CPU time than the standard branch and bound method with the $\in$-optimality feature. Furthermore, it can be seen from the figure that the difference between CPU times becomes more extreme as the problem size increases.

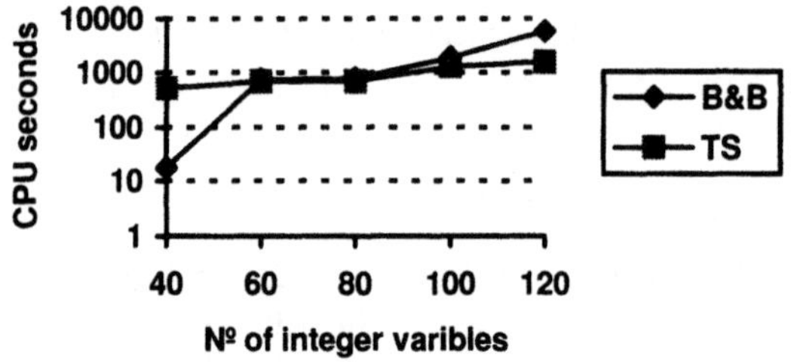

Figure 2. Comparison of average time to $\in$-optimal ($\in$=1%)

Figure 3 shows the evolution of objective function values with the iteration of the TS algorithm for the problem number 9. "O.F.(9)" represents the objective fuction (9) corresponding to infeasible solutions of type (b). In the same way, "O.F.(1)" denotes the original objective function (1) for feasible solutions. The dis-continuous line represents the boundary of the feasible region. As can be seen from the figure, the oscillation through feasibility boundary allows to reach the best solution in the search process.

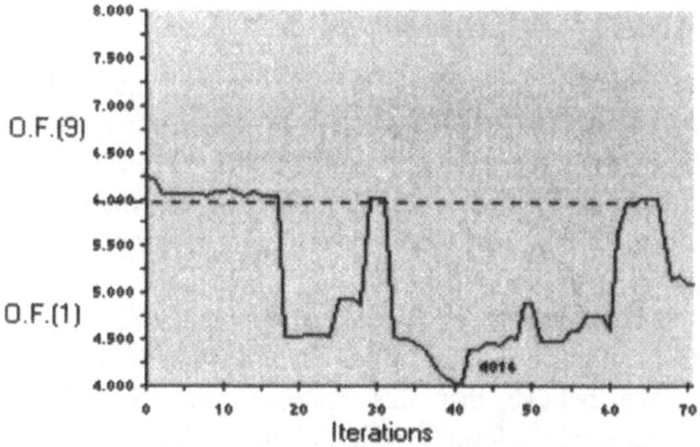

Figure 3. Typical Evolution of our TS approach

Problem	Integer variables	TS Approach CPU time	TS Approach Objective function	∈-optimal solutions ∈=1% CPU time	∈-optimal solutions ∈=1% Objective function
1	40	298	0.00	11	0.00
2	40	374	0.00	6	0.00
3	40	509	525.47	49	525.47
4	40	519	0.00	6	0.00
5	40	594	0.00	6	0.00
6	40	561	1960.85	29	1973.95
7	40	657	3219.76	15	3219.76
8	60	501	2135.80	44	2004.49
9	60	1171	4016.12	2830	4037.67
10	60	345	0.00	10	0.00
11	60	844	2293.18	1009	2236.48
12	60	627	115.85	79	82.96
13	60	932	13050.14	1180	12881.03
14	60	326	0.00	21	0.00
15	80	395	1152.57	476	1152.57
16	80	590	0.00	616	0.00
17	80	1301	0.00	1525	0.00
18	80	819	0.00	900	0.00
19	80	495	4060.52	563	4032.41
20	80	842	3675.34	940	3687.67
21	80	397	8737.80	412	8737.80
22	100	382	0.00	490	0.00
23	100	599	0.00	727	0.00
24	100	2241	1069.83	3618	1069.83
25	100	1194	4201.74	1493	4176.68
26	100	859	0.00	1045	0.00
27	100	1224	3530.027	2004	3563.96
28	100	2228	0.00	3287	0.00
29	120	1015	0.00	3757	0.00
30	120	1526	2254.84	2634	2260.20
31	120	1752	4436.05	3565	4465.88
32	120	814	3866.85	926	3897.77
33	120	965	9180.22	827	9253.51
34	120	4147	0.00	18908	0.00
35	120	698	0.00	11135	0.00

Table 2. Summary of results for the test problems

5. Conclusions

In this paper we present a TS heuristic to solve a mixed-integer formulation of the tool assignment and machine loading problem in FMS. Strategic Oscilation strategic elements by altering the move evaluation function have been introduced to drive the search toward feasible regions. Our TS approach externally controls the tool allocation variables of the MILP and uses a linear programming optimizer to evaluate the quality of the individual moves. Furthermore, both intensification and diversification elements have been introduced. Computational results disclose that our TS approach is effective in terms of both solution quality and efficiency in large size problems. We believe that this approach may be applicable to other optimization problems where feasibility is difficult to obtain.

References

1. Maleki R A 1991 *Flexible Manufacturing Systems: The Technology and Management.* Prentice Hall, Englewood Cliffs
2. Buzacott J A, Yao D D 1986 Flexible manufacturing systems: a review of models. *Management Science* 32: 890-905
3. Kouvelis P 1992 Design and planning problems in flexible manufacturing systems: a critical review. *Journal of Intelligent Manufacturing* 3: 75-99
4. De Werra D, Widmer M 1990 Loading Problems with Tool Management in Flexible Manufacturing Systems: A Few Integer Programming Models. *The International Journal of Flexible Manufacturing Systems* 3: 71-82
5. Glover F 1977 Heuristics for Integer Programming using Surrogate Constraints. *Decision Sciences* 8: 156-166
6. Glover F 1990 Artificial Intelligence, Heuristic Frameworks and TS. *Managerial and Decision Economics* 11: 365-375
7. Hertz A, De Werra D 1991 The Tabu Search Metaheuristic: How we used it. *Annals of Mathematics and Artificial Intelligence* 1: 111-121
8. Aboudi R, Jörnsten K 1994 Tabu Search for General Zero-One Integer Programs Using the Pivot and Complement Heuristic. *ORSA Journal on Computing* 6: 82-93
9. Lokketangen A, Glover F 1995 Tabu search for zero-one mixed integer programming with advanced level strategies and learning. *International Journal of Operations and Quantitative Management* 1: 89-109
10. XA Callable Library 1994 *Reference Manual.* Sunset Software Technology, CA, USA
11. Garfinkel R S 1979 Branch and Bound Methods for Integer Programming. In: Christofides, Mingozzi, Toth, Dand (eds) 1979 *Combinatorial Optimization.* John Wiley & Sons, pp 1-20

Investigating Evolutionary Optimisation of Constrained Functions to Capture Shape Descriptions from Range Data

C.Robertson[1], **R.B.Fisher**[1], **D. Corne**[2], **N. Werghi**[1], and **A. Ashbrook**[1]

[1] *Research Vision Group, Department of Artificial Intelligence, University of Edinburgh, 5 Forrest Hill, Edinburgh, EH1 2QL.*
[2] *Parallel, Emergent and Distributed Architectures Laboratory, Department of Computer Science, University of Reading, Reading, Berkshire, RG6 6AY.*

Keywords: Genetic algorithms, shape fitting, constrained fitting

Abstract

*This paper examines the application of an evolutionary algorithm (*GENOCOP III*) to the problem of fitting surfaces and lines to both 2D synthetic and real 3D range data. The fitting is performed with both non-linear (domain) constraints and with non-linear (geometric and relational) constraints. Example fittings are given as well as an explanation of experimantal caveats that should be addressed when using this kind of approach. We have discovered, for example, that the time to generate starting reference points on the solution manifold is a significant part of the experiment time.*

1. Introduction

Shape analysis of objects from range data (captured three dimensional coordinates of surface points) is a key problem in computer vision with several important applications in manufacturing, such as assembly, quality control and reverse-engineering. The problem is generally formulated as a nonlinear programming problem (NLP), which tries to optimally fit the data to candidate shape descriptions. The NLP optimises a function subject to several constraining equations and inequalities. Especially with nonlinear constraints, it is notoriously difficult to optimise and there is no known method to guarantee a satisfactory solution. Only if the optimisation function and the constraints are satisfied are the optimal point (or points) likely to be found. Traditional techniques, such as gradient descent, are unsatisfactory for the solution of NLPs, due to the local nature of their search methods and the reliance on smooth derivatives in the search-space. The formulation of constraints and the application of constraint-based correction and optimisation of surface fitting has been achieved previously [1] with notable success using the optim1 and optim2 constraint application strategies. There are, however, some associated problems with this approach, such as the complex formulation of the objective function and reliance on the convexity of the solution space. The first problem is addressed here by using only the least-squares metric as an objective and using the constraints to define the manifold of allowable solutions. In this work we examine the applicability of evolutionary strategies to the problem of fitting lines and surfaces to both synthetic and acquired object range data. This is a traditional NLP since a priori constraints exist between the fitted surfaces and lines. The GENOCOP III algorithm [2, Ch.7] was used in this paper. It is an evolutionary algorithm system which is specialised to handle constrained function optimisation. GENOCOP III uses real-valued genes, and included methods to deal with linear, non-linear, domain and inequality constraints. We have examined the use of GENOCOP III, with a specialised fitness function (described later), applied to the problems of fitting both 2- and 3- dimensional surface equation constructs to range data while simultaneously applying several necessary geometric and domain constraints. The constraints cover the two typical types : domain, the restriction on the parameter size; relational, relationships between surfaces that are known a priori.

2. Chromosome Representation and Fitness Functions

Fig. 2.1 is a typical example of the kind of data that is derived from a laser range-finder. The data is from a typical engineered part comprising a number of planes and degenerate quadric surfaces arranged in a fairly well defined way. There are, for example, several places where angles of 90° have been used as relationships. There is also a cylinder at one end whose axis lies orthogonal to the normal of some planes and parallel to the normals of others. It is these kinds of relationships we are seeking to preserve.

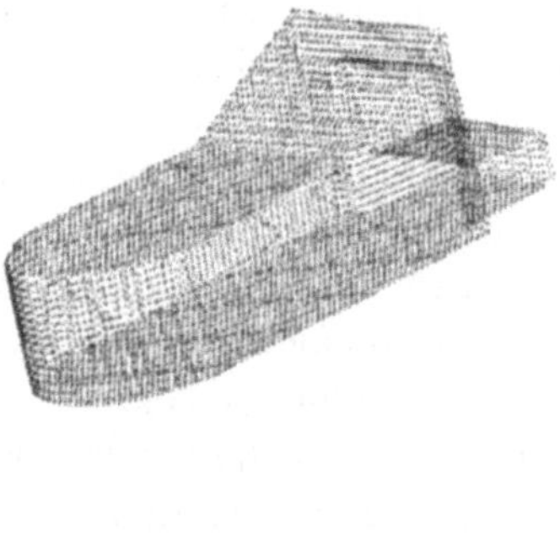

Fig. 2.1. Example 3d range data points

We use a real-valued chromsome representation in which each gene corresponds to a coefficient in a shape equation. In the cases we have investigated typical chromosomes are shown in Fig 2.2(a)-(c).

(a) | m^1 | c^1 | $y + m^1x + c^1 = 0$

(b) | v_1^1 | v_2^1 | v_3^1 | d^1 | $v_1^1x + v_2^1y + v_3^1z + d^1 = 0$

(c) | a_1^1 | a_2^1 | a_3^1 | a_4^1 | a_5^1 | a_6^1 | a_7^1 | a_8^1 | a_9^1 | a_{10}^1 |

$$a_1^1x^2 + a_2^1y^2 + a_3^1z^2 + a_4^1xy + a_5^1xz + a_6^1yz + a_7^1x + a_8^1y + a_9^1z + a_{10}^1 = 0$$

Fig. 2.2. Chromosomes representing (a) a line (b) a plane (c) a general quadric

The surfaces were fitted using a standard least-squares metric over a large number of surface points. The value at each point gives an indication of the probability that the point belongs to the parameterized surface; when summed this gives a good overall estimation of how well the parameterized surface matches the data.

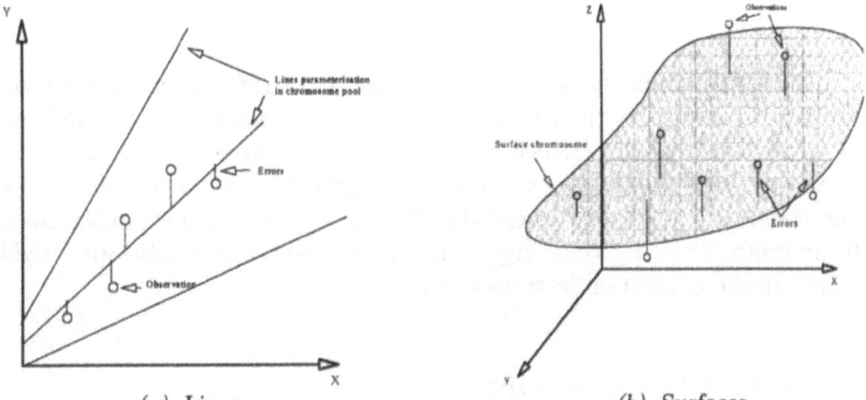

(a) Lines *(b)* Surfaces

Fig. 2.3. Line and surface parameterization and error measures

For lines, the best line is defined as the least squares fit through the observation data:

$$E = min \sum_{i=1}^{i=N} (y_i - mx_i - c)^2 \tag{2.1}$$

In this case the parameterized line is (m,c) and the observation data is (xi, yi). In the case of two separate lines, where some constraint exists, this error is expressed as the sum of the two errors. In the case of planes, the least-squares error is given by :

$$E = min \sum_{i=1}^{i=N} (\mathbf{v}.\mathbf{x}_i + d)^2 = minF(\mathbf{v}, d) \tag{2.2}$$

where $\mathbf{v}$ is the plane parameterization and x_i is the observation range data. Notice that F is homogeneous with respect to the parameters of $\mathbf{v}$ and d. This means that the solution must be constrained in order to avoid the trivial null solution. The most natural constraint is $\mid v \mid= 1$, the unit normal, which also has the advantage of being invariant with respect to translations and rotations. For quadrics the geometric distance from a point the surface is non-trivial to compute so the most straightforward distance measure to use as an error measure is the algebraic distance:

$$E = min \sum_{i=1}^{i=N} (\mathbf{x}_i^t A \mathbf{x}_i + \mathbf{x}_i.\mathbf{v} + d)^2 = F(A, \mathbf{v}, d) \tag{2.3}$$

where the quadric parameters are given as a symmetric matrix A, the vector $\mathbf{v}$ and the constant where :

$$A = \begin{bmatrix} a_1 & a_4/2 & a_5/2 \\ a_4/2 & a_2 & a_6/2 \\ a_5/2 & a_6/2 & a_3 \end{bmatrix}$$
$$\mathbf{v} = [a_7, a_8, a_9]$$
$$d = a_{10}$$

Once again, it can be seen in equation 2.3 that the function F is homogeneous with respect to the parameters $a_1..a_{10}$. This means that the problem of fitting must be constrained to avoid the trivial null solution. The constraint we have employed is the same as suggested in [3], where the first six parameters of the quadric essentially the shape parameters) are constrained to be unity: $Tr(AA^t) = 1$. Again, this is invariant with respect to a rigid transformation, rotation or translation.

3. Geometric Constraints

In the real world all optimization problems are constrained. We need to employ constraints in our fitting for the following reasons :

- *Avoiding null solutions.* To avoid convergence of parameter values to zero, virtually every surface parameterization has a unit normal constraint or equivalent.
- *Improved fitting.* They improve the overall fit of the data since boundary constraints can be used to ensure that solutions do not fall outside a required range. This means that virtually every parameterization requires a set of inequality constraints.
- *Inclusion of a priori information.* For example, if two planes are known to be orthogonal then the chromosome represents a complex form as shown in Fig. 3.1. This means that although the two lines of best fit may be at some angle F to each other, the actual value may be known and used in order to fix the value at the correct one. Usual examples are $\phi = 90^o$ or

$\phi = 0^o$. Such errors often result from mis-registration of multiple views of data, which is also known to be a difficult problem. In the case of reverse-engineering, these kinds of constraints are very often known, whereas the parameterization of the actual system components are not.

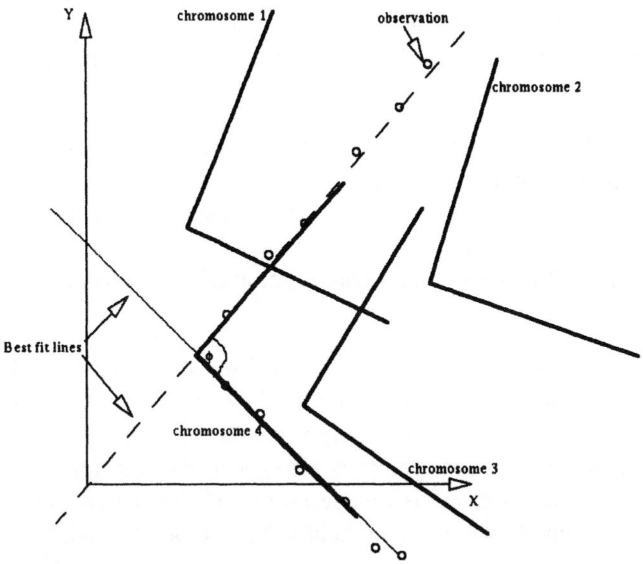

Fig. 3.1. Candidate chromosomes for an orthogonally-constrained line-fitting case

3.1 The GA Constraints Strategy

Overall flow of control in a GA normally consists of the following operations:

1. Random generation of initial population
2. Evaluation of fitness of chromosomes in the population
3. Application of genetic operators
4. Repeat 2-4 as required

The question of how constraints are implemented as a method of population shaping, (keeping solutions on the constraint manifold), is slightly more complex. If non-conforming chromosomes are simply culled, the populations can become severely reduced at each cycle, thus scuppering the diversity needed for effective search. A penalty function can be introduced into the fitness function which inhibits the growth of individuals straying from the constraints or dependent upon the number of constraints violated. This is the most common approach, but is difficult to implement effectively since the penalties for all constraints must be considered independently. The use of repair operators, to re-map infeasible chromosomes back to the feasible space has also become popular recently. It is sometimes (but not always) easy to repair chromosomes and those repaired may be used for evaluation or may be placed (with some probability) back into the population.

GENOCOP III uses a repair and replace heuristic as well as a specialized set of operators. In [2, Ch. 7.5], Michalewicz suggests an empirical figure of 15% replacement seems optimal for repaired individuals. GENOCOP III also uses operators that ensure that any mutation operation or crossover operation always produces a child that is from the constrained solution space. Example operations are:

- *Uniform mutation.* Where the value of a gene may mutate, but only into the range specified by its domain constraint.
- *Boundary mutation.* Where the mutation pushes the gene toward its domain boundary.
- *Arithmetical crossover.* This is the linear combination of two vectors such that closure is guaranteed.

There are several other such operators. As well as these, GENOCOP III maintains two sets of populations, a search population and a set of reference points. The feasibility of the search population is maintained by using appropriate operators. A set of search points is also generated and repaired as necessary to provide new potential reference points. If the search point generated is better than one of the currently held reference points then it is replaced.

4. Experiments

This section details some of our experiences performing fittings for both 2D and 3D data with constraints. Note that the 3D points were all scaled into a unit sphere.

4.1 2D- Line Fitting with Constraints

In the first instance, 2-dimensional linear features with orthogonality and parallel-ness were fitted to noisy synthetic data. With conventional techniques this is not a particularly hard problem when low numbers of lines are used. In our case we examined sets of two lines at various constrained angles to each other. Gaussian noise was applied to the data at varying values. In each case there were 400 (x,y) data points in two sets of 200. The initial population of 100 chromosomes can be seen to converge to a useful fit after about 35000 evaluations, taking around 12 seconds on an average load 200MHz Sun Ultrasparc 1. It should also be noted that convergence in the case of parallel lines was not made more difficult by the fact that local solutions where the two lines are the same were reached. Whilst constraints could have been applied

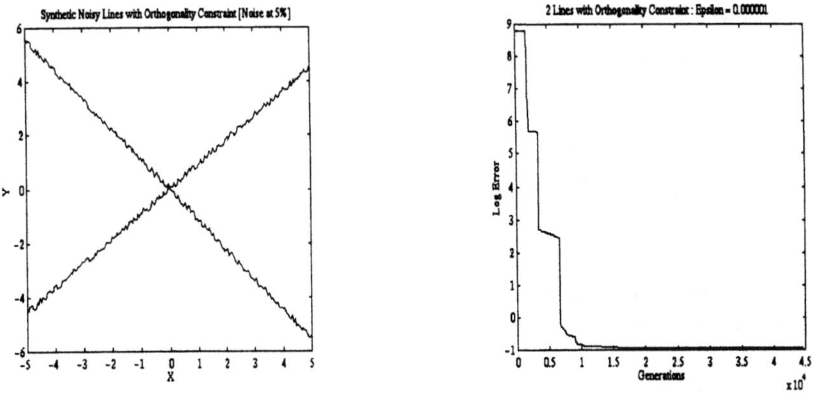

(a) Noisy constrained line *(b)* Orthogonally constrained lines

Fig. 4.1. Orthogonal Line Fitting

to avoid these solutions, it was not found necessary. In each case the degree to which the chromosome could stray from the constrained solution manifold was defined to be e = 0.000001 units. Example data and graphs are shown in figures 4.1 and 4.2.

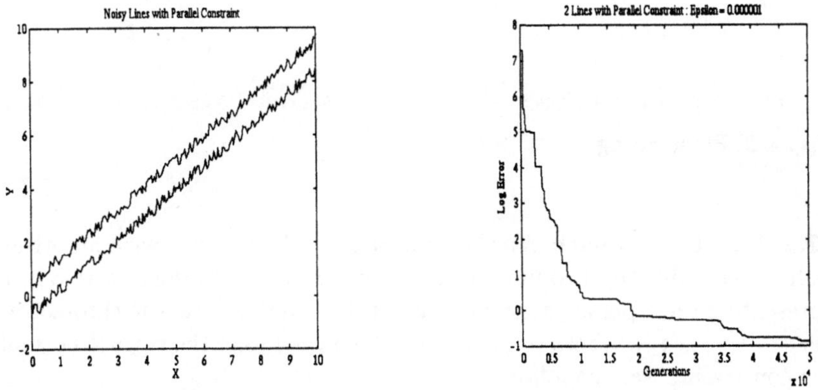

(a) Synthetic lines with noise *(b)* Constrained lines

Fig. 4.2. Parallel Line Fitting

4.2 3D- Plane and Quadric Fitting with Constraints

In the 3-dimensional case the following sections describe the fittings that have been performed.

4.2.1 Plane with unit normal constraint. A single plane with a unit normal constraint, chromosome with 4 genes representing the plane parameters was fitted. This plane data is from a real object and has a error of around 0.3mm. In this case the following chromosome was used: v = (v1, .., v4), where (v1, v2, v3) represented the surface normal. The constraints were as follows:

1. $(v_1^2 + v_2^2 + v_3^2) = 1$
2. $\mid v_1 \mid, ..., \mid v_4 \mid < 1$

The original plane is shown in Fig. 4.3(a) and the surface fitting error is shown in Fig, 4.3(b). Experience has shown that in the case of plane fitting often the first parameter to stabalize is the constant, followed by the rotation of the normal. This can be seen in the error graph in Fig. 4.3(b) where the initial steep descent is the fixing of this parameter.

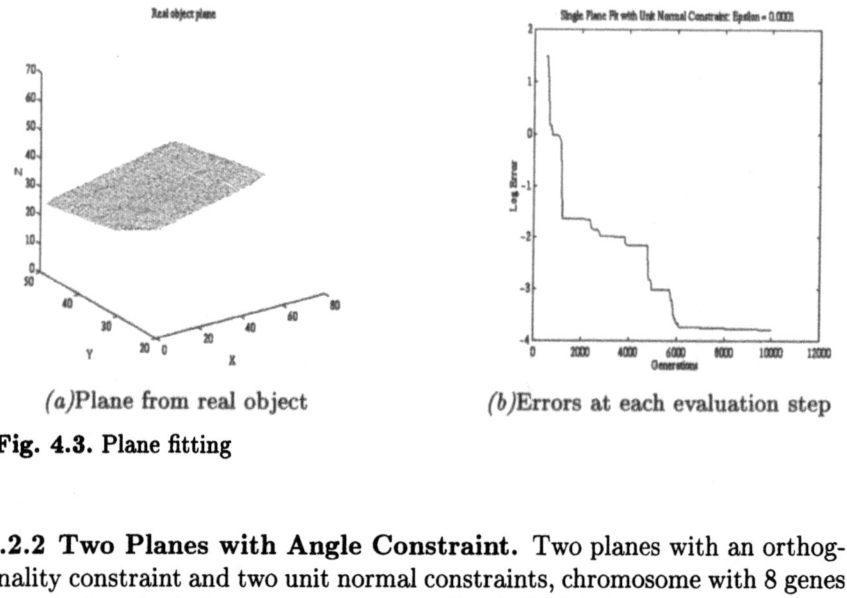

*(a)*Plane from real object *(b)*Errors at each evaluation step

Fig. 4.3. Plane fitting

4.2.2 Two Planes with Angle Constraint. Two planes with an orthogonality constraint and two unit normal constraints, chromosome with 8 genes representing the 2 plane parameters, was fitted. In this case, the chromosome was $v = (v_1, .., v_8)$, where (v_1,v_2,v_3) and (v_5,v_6,v_7) were the surface normals. The constraints were as follows:

1. $(v_1^2 + v_2^2 + v_3^2) = 1$
2. $(v_5^2 + v_6^2 + v_7^2) = 1$
3. $v_1 * v_5 + v_2 * v_6 + v_3 * v_7 = 0$ - two plane orthogonality constraint
4. $\mid v_1 \mid, ..., \mid v_8 \mid < 1$

It can be seen from the graph in Fig. 4.4(b) that the errors in the fitting decrease to a very good level after only 2000 evaluations.

4.2.3 Degenerate General Quadric with Unit Constraint. A general quadric fit of a sphere with a unit constraint on the first six parameters, chromosome of length 10 genes representing the quadric parameters, was fitted. In this case there was no explicit use of the fact that the data represented was from a sphere.

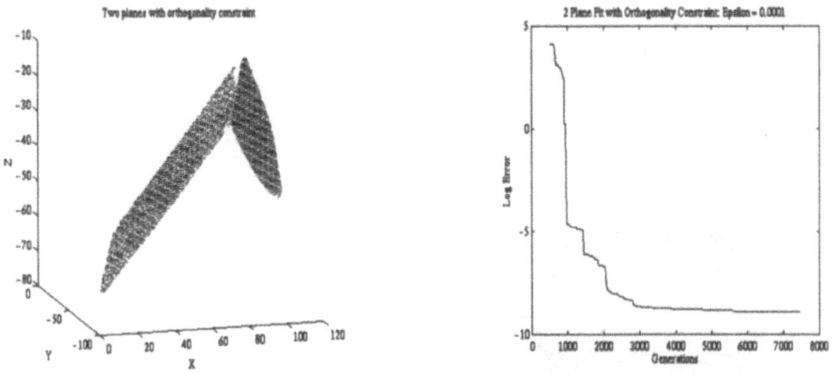

(a) Planes from real object *(b)* Errors

Fig. 4.4. Two constrained planes fitting with orthogonality constraint

In this case, the chromosome was $a = (a_1, .., a_{10})$ where each gene was one of the quadric parameters. The constraints used were as follows :

1. $(a_1^2 + a_2^2 + a_3^2 + a_4^2 + a_5^2 + a_6^2) = 1$ -unity constraint to avoid null solution
2. $\mid a_1 \mid, ..., \mid a_{10} \mid < 1$

After the constant terms found a good approximation to their best values each of the shape and pose genes $(a_1, ..., a_6)$ in the fittest solution chromosome converged gradually to their optimal value.

(a) sphere (+) and fitted sphere (.) *(b)* Fitting errors

Fig. 4.5. Noisy synthetic sphere fitting

4.2.4 General Quadric (Ellipsoid) with Unit Constraint. A general quadric fit of an ellipsoid with a unit constraint on the first six parameters, chromosome of length 10 genes representing the quadric parameters. The methodology was as in the last section. The data is shown in Fig. 4.6(a) and the resultant graph of fitting errors for the winning chromosome at each evaluation is shown in Fig. 4.6(b).

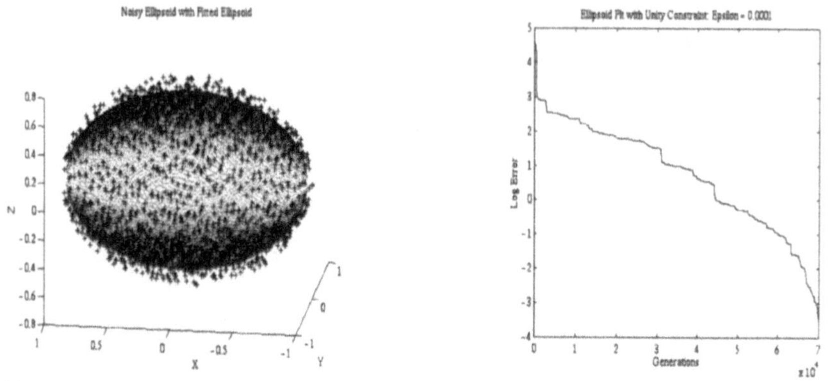

(a) ellipsoid (+) fitted sphere (.) *(b)* Fitting errors

Fig. 4.6. Noisy synthetic ellipsoid fitting

4.2.5 Compound - Degenerate Quadric and Plane with Geometric Constraints. In this experiment we fitted a cone with a planar bottom, where the cone is constrained to be both right and circular with axis along the normal of the plane. A chromosome of 14 genes represents the quadric parameters of the cone and the parameters of the plane, $G = (v_1, .., v_4, a_1, ..., a_{10})$, where (v_1, v_2, v_3) is the normal of the plane, v_4 is a constant and $a_1, ..., a_{10}$ are the quadric parameters.

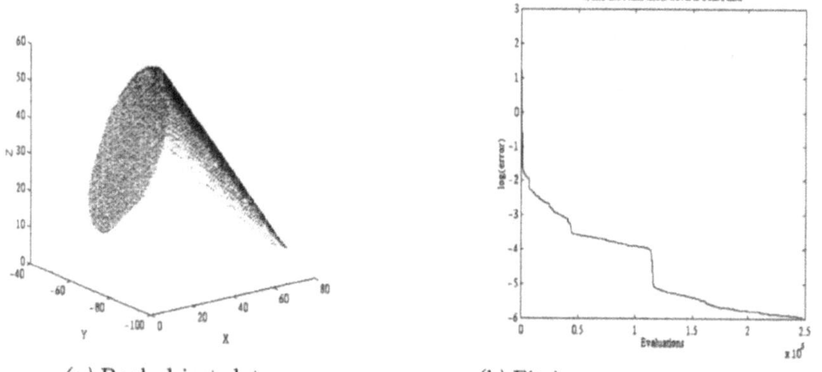

(a) Real object data *(b)* Fitting errors

Fig. 4.7. Cone and plane fitting

The constraints used in fitting the two surfaces in this object were::

1. $(a_1^2+a_2^2+a_3^2+a_4^2+a_5^2+a_6^2) = 1$ -unity constraints to avoid null solution
2. $(v_1^2 + v_2^2 + v_3^2) = 1$
3. $a_1 - a_2 = v_1^2 - v_2^2$ - circularity constraints
4. $a_1 - a_3 = v_1^2 - v_3^2$
5. $a_2 - a_3 = v_2^2 - v_3^2$
6. $a_4 = v_1.v_2$
7. $a_5 = v_1.v_3$
8. $a_6 = v_2.v_3$
9. $|v_1|, ..., |v_4|, |a_1|, ..., |a_{10}| < 1$

4.3 Experiment Summary and Caveats

The following table summarises the experiements performed, where 'DC' stands for 'Domain Constraints' and 'GC' stands for 'Geometric Constraints'.

Type	No. points	e value	DC	GC	Evaluations
Lines	400	0.000001	4	1	30,000
Plane	2217	0.0001	4	1	10,000
Two planes	4913	0.0001	8	3	5,000
Sphere	1941	0.0001	10	1	70,000
Ellipsoid	1776	0.0001	10	1	70,000
Cone + plane	8199	0.0001	14	8	250,000

One typical problem is that given a large chromosome with many parameters it is often particularly hard to find an initial start condition unless the original least-squares fitting for each of the surfaces is available. Even if so, a start-point may sometimes not be readily calculated. Several fixes for this were tried, for example using vector cross-products to ensure orthogonality, but in the future a more rigorous approach will be formulated. Another problem was that given an arbitrary quadric surface from a range image, the data origin may be some distance from the origin. This means that the traversal distance in the search-space of the last 4 parameters of the quadric is high relative to the first six (which are unit normalized). Often this makes a satisfactory solution unlikely. In order to achieve a fitting, all of the sampled range data were normalized to a unit sphere, centred at the origin, and a scale factor extracted. This proved a good remedy since the traversal distances for all parameters was at least the same order of magnitude.

5. Conclusions

First and foremost, this paper represents a proof of concept - an evolutionary algorithm can be used to fit surfaces where a priori domain and geometric constraints are important. All of the fittings in the last section satisfied both the constraints and the least-squares fit to a high degree. This initial success has indicated to us that evolutionary strategies are feasible in this domain, but the time complexity to convergence (each fitness evaluation involved calculations for each of several thousand coordinate pairs or triplets) has suggested possible poor scaling to larger and more complex surfaces. To deal with these we did further experiments which involved evaluating candidates in early stages against only a small proportion of the available data. This succeeded in significantly speeding up the process without unduly misleading the future direction of search. One other approach we are investigating involves the off-line computation of datasets and parameterisations which has been shown [1] to significantly speed up least-squares fitting. Work so far thus indicates the viability of evolutionary approaches to fitting range data from complex real-world components with engineered surfaces. Future work in this area is as follows:

- *Segmentation.* We are assessing whether a 3-dimensional segmentation of the data (that is the decision of which points belong to which surfaces) may also possible using a priori constraints on the range of possible shapes and their relationships.
- *Failure points.* We are assessing at what point the system breaks down under the constraint loading and surface complexity.
- *Feature identification.* We are examining the explicit coding of constraints for surface features, such as a slots or a drilled hole. This is a particularly difficult problem in 3D segmentation because often sufficient surface data does not exist to segment each of the individual elements of, say, a drilled hole. If constraints are taken into consideration then it may be possible to code the individual feature elements into a single chromosome and search for it in a scene.

Acknowledgments

This work was funded by UK EPSRC grant GR/H86986. GENOCOP III is available from Michalewicz's ftp site at ftp://ftp.uncc.edu/coe/evol . Updates are also available from http://www.coe.uncc.edu/ gnazhiya/gchome.html.

References

1. Werghi, N., Fisher, R.B., Ashbrook, A., Robertson, C. (1997): Improving model shape acquisition by incorporating geometric constraints. Proc. British Machine Vision Conference BMVC97, Essex, 520–529
2. Michalewicz, Z. (1996): Genetic Algorithms + Data Structures = Evolution Programs. Springer, 3rd edition
3. Faugeras, O., Hebert, M. (1986): The Representation, Recognition and Positioning of 3-D Shapes from Range Data, in Techniques for 3-D Machine Perception, ed. A. Rosenfeld, North-Holland, Amsterdam

Optimal Selection of Pressure Vessels

C.M.Kalker-Kalkman and M.Shahtaji*

Dept. of Mechanical Engg., Delft University of Technology,Delft,The Netherlands.
Tel:+31-15-2783983 Fax:+31-15-2781397,
E-mail:c.m.kalker-kalkman@wbmt.tudelft.nl
*Northeastern University, Boston, Mass., U.S.A.
E-mail: mash345@coe.neu.edu

Keywords : Pressure vessels, optimization, genetic algorithms.

Abstract

For storage of gases pressure vessels are used. The pressure, material, manufacturing method, sizes ,type and number of vessels are quantities that a designer has to decide on. Usually the cost is a function one wants to minimize. In this paper a general structure for the various variables is proposed. This structure can be regarded as an expert system for a certain class of pressure vessels. The problem of selecting the various variables and making the right choices can be formulated as an optimization problem with mixed discrete/continuous variables and with variables structured in tables. The optimization procedure is integrated in the whole design process. An existing optimization program based on genetic algorithms that is able to handle those kind of variables is used to arrive at a well-balanced choice. A case study is given to illustrate the method.

1. Introduction.

A pressure vessel is a capital equipment that has many variations and requires high safety. It is built by order and is not mass produced. Usually there is a long procedure with many partners, who have very different interests. In this paper we will confine ourselves to pressure vessels for the storage of gas. During the first phase of the design the different partners involved have several sessions where they negotiate and all try to have as many advantages as possible. Hopefully an agreement on the general design will be achieved. The second phase involves the last calculations and the drawing of the pressure vessel(s). The third phase is the manufacturing and construction, delivery is the fourth phase. Usually optimization takes place in the second phase, when many decisions have been taken.

In this paper we will show that the optimization procedure can be extended to the first and second phase. In this phase it is important to know more about the following points:

- What design is optimal with respect to all costs.
- What is a good estimation of the costs for the client.

If those things are known, it will save time during the first phase, as all partners know what the advantages for the other partners are. It is very difficult for a researcher to find out information about the different costs, as most private companies keep their data to themselves. In addition some information varies from one region to another region. Nevertheless we will try to formulate the problem in an independent way, but so flexible that it can be applied to different cases. It will be shown later that every partner in the process involved can add his own variables and ways of computation to the program that is developed. When a certain quantity of gas has to be stored, an expert has to make various decisions about the way the storage has to take place.

A number of questions have to be answered, like :
What pressure is the best ?
How many vessels do we take?
What is the best shape for the vessel ?
What properties do we need for the material ?
What are the costs for the ground, foundation, material and manufacturing?
What are the sizes and weight of the vessel ?
What are the temperatures allowed ?
What are the design stress and corrosion resistance of the material?
What are the desired safety factors and what are the demands of various (local) regulations, like ASME Pressure Vessel Code?
See [1], [2], [3] and [4].

It is clear that many variables play a role in the final decision. Usually the total cost will have to be minimized. To formulate the problem as an optimization problem,a number of variables and relations between them have to be stated. In section 2 it will be shown that many discrete and continuous variables can be structured in tables. A row to be selected from such a table (allele) can be regarded as discrete variable. The insertion of a new row in these tables is always possible. It is clear that an infinite number of choices is possible. In section 3 the objective function that is a measure for the total cost will be formulated. As optimization method a genetic algorithm is proposed. In section 4 this method is explained and applied to the problem at hand. It will be seen that the various tables, variables and relations can be considered as an expert system for a certain class of problems. In section 5 some results are shown and a conclusion is given in section 6.

2. Variables and relations.

2.1 Variables that are structured in tables.

The following five different tables can be distinguished:

- gas
- material
- norm
- shape
- manufacturing

The gas table gives information about various gases to be stored, it is clear that a row in this table is specified. In the gas table the gastype (integer) and the critical pressure are given, in the example used it consists of 13 different gases, but in principle any number of gases can be used. The material table gives information about various materials to be used and their properties, see table II, that has 16 rows. The norm table III [3] gives different norms to be satisfied, one can specify a norm, but it is also possible to select an optimal norm with the aid of the optimization program. The shape table IV gives various shapes that can be realized. The manufacturing table V gives various ways of manufacturing like welding, to be chosen from. The first six gases are always vapour, above the critical pressure they are liquid. In this case a hydrostatical pressure must be added to the formulation of the problem.

Table I. Type of gas.

gast [--] : Gasttype
pcr [bar] : Critical pressure at 30 °C

1. Oxygene O2	200
..	..
..	..
..	..
13. Propane C3H8	10.1

Table II. Material of vessel.

A larger number for weldability means more difficulty in welding.

mat [--] : Material number
wel [--] : Weldability
Sd [N/mm^2] : Design stress

X3 [$/kg] : Cost per kg

Fe37-2KP [NEN]	1	1	235	1.7
	..	..	..	..
	..	..	..	..
	..	..	..	..
A535%9Ni [ASTM]	16	1.3	380	2.65

Table III Norms

Norm [--] number norm

1. Stoomwezen (dutch norm for steam)
2. NEN-EN 286-1, Dutch norm for simple unfired pressure vessels, designed to contain air of nitrogen.
3. API-ASME VII Norms.
4. BS 1501 to 1515 and BS 2654
5. DIN 28011 to 28014 and DIN-EN 1442.

The fourth and the fifth row are not defined. The European Community has decided to introduce a new norm, the so-called CEN, that should be ready before 2000. This norm is supposed to uniform all European norms. It will be easy to add this norm to table III. The optimization described here can help to recognize the type of behaviour of each norm for a certain class (i.e. different capacities) by comparing them with each other.

Table IV Shapes.

sht [--] shape type

1. cylinder vertical flat front
2. cylinder vertical half-elliptic front
3. cylinder vertical hemisphere front
4. cylinder horizontal flat front
5. cylinder horizontal half-elliptic front
6. cylinder horizontal hemisphere front
7. cone
8. sphere

Although theoretically a sphere is ideal, the choices 2 to 6 score practically better than others. Finally one can think of different ways of manufacturing, leading to the following table:

Table V. Manufacturing.

jt [--]	joint type
mind [m]	minimum thickness
maxd [m]	maximum thickness
we [---]	efficiency factor welding
wv [--]	prize factor welding
wvv [--]	butt factor welding

el.arc without butt	1	0.004	0.02	0.7	2	1.1
MIG without butt manual	2	0.004	0.02	0.75	1.8	1.1
MIG without butt auto	3	0.004	0.02	0.75	1.2	1.1
MIG with butt manual	4	0.02	0.035	0.8	1.9	1.2
MIG with butt auto	5	0.02	0.035	0.8	1.3	1.2
TIG without butt manual	6	0.004	0.02	0.8	1.6	1.1
TIG without butt auto	7	0.004	0.02	0.85	1	1.1
TIG with butt manual	8	0.02	0.035	0.9	1.7	1.2
TIG with butt auto	9	0.02	0.035	0.9	1.1	1.2
submerge (under powder)	10	0.035	0.120	0.85	1.3	1.4
Nut and bolt (flanged)	11	0.004	0.120	--	--	--

2.2 Other variables and relations between them.

The following variables play a role in the computations:

Rb1	[m]	inner radius sphere
Rb2	[m]	outer radius sphere
dk	[m]	thickness
Rc1	[m]	inner radius cylinder
Rc2	[m]	outer radius cylinder
vol	[m3]	volume of vessel
P	[bar]	pressure
CPV	[m3bar]	capacity per vessel
h	[m]	height
K1	[$]	cost ground
N	[--]	number of vessels
x1	[$/m2]	prize per m2 ground
G	[kg]	weight of empty vessel
Fw	[bar]	wind pressure
mb	[Nm]	bending moment
wf	[--]	weight factor bending moment
dfc	[m]	thickness front cylinder
K4	[$]	joint cost
x4	[$/kg]	joint cost per kg

K5	[$]	shaping cost
x5	[$/kg]	shaping cost per kg
K3	[$]	material cost
x3	[$/kg bar]	material cost / kg / pressure unity
K2	[$]	foundation cost
x2	[$/Nm]	cost per bending moment unity
Ktot	[$]	total cost

The number of pressure vessels chosen can play a role for the following reasons: The fabrication cost will decrease as the number of vessels decreases. However, the transportation cost can be very high for large vessels, manufacturing costs can be high and increasing the thickness means some extra cost (small oven for heat treatment, higher inspection cost like x-ray). In addition a huge number of workers at one vessel can be difficult with respect to planning and timetables. For safety reasons it can be better to have more small vessels than one large vessel as the maintenance and the disaster scenario will be easier. There are various relations between the variables in the above table and the variables occurring in the tables in section 2.1. We will not give all formulae in detail, but we give the dependencies in table VI.

Table VI Dependencies of various variables.

Capacity = pressure * volume
dk = Rb2-Rb1 in case of sphere
dk =Rc2-Rc2 in case of cylinder
vol=CPV/P
Rc1=function(vol,sht)
h=function(sht,vol,Rc1)
Rb1=function(vol,sht)
Rc2=function(sht,P,Rc1,h,Sd,norm,we)
Rb2=function(sht,P,Rb1,h,Sd,norm,we)
K1=function(sht,N,Rb2,Rc2,x1,h)
G=function(sht,we,Sd,Rc2,Rb2,Rb1,h,P)
Fw=function(sht,h,Rc2,Rb2)
mb=function(sht,G,Fw,h,Rb2,Rc2)
wf=function(mb,sht)
dfc=function(norm,sht,Rc1,Rc2,h,P,we,Sd,dk)
K4=function(mind,maxd,dk,wv,we,wvv,wel,N,G,x4)
K5=function(sht,wvv,N,sd,G,x5)
P < Pcr
P = (Sd * we * dk)/(Rc1+0.6*d) for thick-wall cylinder (LAME's formula)
P = 2 Sd we d/(Rc1-0.4 d) for thin-wall cylinder
P=(2 Sd we dk)/(Rb1+0.2 d) for sphere
d < 0.5 Rc1
P < 1.25 Sd we

K3=N*P*G*x3
K2=N*P*wf*mb*x2
wf=function(mb,sht)
Vm = function (thickness, shape, radius, inside radius)
K tot = K1 + K2 + K3 + K4 + K5

The theory of internal pressure vessels for thin wall vessels is limited for pressures up to 206.8 bar. Any pressure in excess of 206.8 bar would be called high pressure, and thick wall formulae should be used.

3. Costs

The following types of costs can be distinguished:

1.Cost of foundation.
2.Cost of ground.
3.Shaping cost.
4.Joint cost.
5.Material cost.

Foundation cost (K2).

The foundation of a pressure vessel is a civil engineering problem. The kind of construction used (column or skirt) will depend on the bending moment.

Cost of ground (K1).

The specification of the location and also the number of vessels and their hazard zones are the most important factors.

Shaping cost (K5).

The shaping cost involves the cost for sizing, the edging of the plate, the forming of the plate into a cylinder, the forming of the front and heat treatment.

Joint cost (K4).

The joint cost is mainly the welding cost and post welding treatment. The weldability of the material, the thickness and the labour cost are the most important factors.

Material cost (K3).

For the selection of the material one has to consider the following factors:

- corrosion resistance
- steel-making process
- mechanical behaviour dependent on temperature
- weldability
- local or state regulations [2].

Total cost.

It is clear that we want to minimize the total cost that is the sum of the ground cost, foundation cost material cost, joint cost and shaping cost.

Ktot=K1+K2+K3+K4+K5

This cost will be the objective function to be minimized. The formulation of the costs is very flexible and a designer is able to use his own specification, for example he/she is able to eliminate K1 and K2 or give a joint cost (x4) according to a certain region.

4. Optimization by a genetic algorithm

For optimization problems with mixed discrete/continuous variables a genetic or evolutionary algorithm is very appropriate, [5],[6] and [7]. Optimization algorithms that work with gradient methods use one initial solution and find better solutions with the aid of derivatives. For problems with discrete variables as we have here this method is not appropriate. A genetic algorithm is a stochastic optimization method that works with sets of solutions, called populations. It is based on evolution theory and is based on the principle of survival of the fittest. Many terms in the theory of genetic algorithms stem from evolution theory, like genes, chromosomes, generation, etc. From a population a new and better population is created with the aid of the operators: selection, cross-over and mutation. In [8] a general program is discussed that is used here. The set of solutions that form a population (also called generation) are sorted on increasing value of the objective function called the fitness function. The best solution of a current generation can be regarded as the latest optimum. The way of finding new generations is based on a number of parameters of the genetic algorithm.
Those parameters are for instance :

- population size
- scaling factors for fitness function
- mutation rate
- selection criteria

- method for cross-over

The program described in [8] is able to modify those parameters during runtime, depending on the intermediate results. Those results are shown on the screen with the aid of graphic functions. For all variables the optimum solutions are plotted on the screen, and it can be seen in what intervals of the variables the best solutions are found. In an interactive way the search intervals can be modified, and new populations can be started in those intervals. In this way a fast convergence to optimal solutions is found. To apply this program for the problem of finding a suitable way of storing gas, the variables and relations mentioned in section 2 and 3 have to be defined and implemented. In section 4.1 the chromosomes are shown.

4.1 Chromosomes and cross-over.

The designs that are possible for storing gases can be represented by the values of the independent variables mentioned in section 2, and the row numbers (alleles) of the tables mentioned in this section. These values are represented by chromosomes by the program [8] as follows:

For floating point variables the search intervals are divided into a number (a power of two) of intervals determined by a tolerance (or resolution) given by the user. For instance when we divide the search interval for the height h in 1024 steps, we can distinguish 1024 different values for h and it can be represented by a binary bit string of 10 bits. In general, if we use N bits, and the variable can take values between hmin and hmax, the resolution r is given by

$$r = (hmax - hmin)/ (2^N)$$

The integer variables and the row numbers (alleles) of the variables grouped in tables can be treated in the same way. For instance, if a table consists of 23 rows, the nearest power of two is searched (32) and the row can be represented by a bit string of length 5. If by cross-over or mutation a value over 23 is produced, this value is rejected by the program. All the variables represented this way are consecutively placed in a bit row, that represents a design for a vessel in an unique way. The user can determine the resolution, r, determining the length of the bit string. For production of new generations a single-point cross-over method is used. It is also possible to use floating-point strings and use heuristic cross-over, this can be done by setting

$$x3 = x2 + R\,(x2 - x1), \text{ where R is a random number in } (0,1),$$

x1 and x2 are the values of the parents (x2 is better than x1) and x3 is the value for the child variable. This way the values of the variables need not be coded and decoded, thus saving computing time. With the graphic interface used in [8] the cross-over methods, mutation rate, population size and scaling factors for the fitness function can be modified during runtime. Moreover the search intervals can

be modified and complete new populations can be created in the new intervals. The best member of a current population, being the optimum, is always kept. It is found that in this interactive way a satisfying optimum is sooner found.
Some results are given in section 5.

5. Results

To test the program with the data given in the preceding sections, we present the following problem:

Suppose we want to store CO_2 with a capacity of 1000.00 and we have the following values for the constants x1,x2,x4,x5:
x1= 100 (Cost ground per m^2)
x2=0.15 (Cost per bending moment)
x4=1.6 (Joint prize per kg)
x5=1.5 (shaping cost per kg)

In Table I we have to specify row no 9.
For the number of vessels N we take boundaries between 1 and 25, for the pressure P we take $2 < P < 65$, as the critical pressure is 65, see Table I. In the other tables the boundaries are set as given in Table VII.

Table VII Boundaries for tables

	Lower boundary row no	Upper boundary row no
Shape	1	6
Material	1	12
Norm	1	2
Manufacturing	1	11

All other variables are computed by the program from the relations in Table VI, for those variables certain boundaries can be given.
The results of the computations were that an optimum value for the total cost was found for P=65 and N=4.
For the tables the following rows were selected:
Shape : 5th row, cylinder horizontal, half-elliptic
Material : 6th row, Fe47-1KP(NEN)
Norm : second row, NEN(EN286)
Manufacturing : 7the row, TIGz.n.auto

The following values for the variables were found:
number of vessels: 4

design pressure : 65
volume of each vessel:3.846 m^3
height from welding line of fronts :1.3923 m
inside radius of cylinder : 0.8414 m
thickness of wall : 17 mm
thickness of front (netto) : 7.893 mm
weight of vessel (without saddle) : 2009.7 kg
ground cost : $ 10399.6
foundation cost : $ 8631.
material cost: $ 19695.
welding cost + inspection : $ 19974.
shaping cost : $ 15604.
Total cost: $ 74303.6
(cost of equipment like flange and painting cost are excluded)
An optimum was found after an average number of 300000 solutions, the computation speed is 1000 solutions in 1 sec. The population size was started with 10 members, and during runtime increased up to 80 members. Mutation rate was taken from 0.4 % to 20 %.

6. Conclusions

It is found that it is possible to find an optimal solution for the problem of selecting a suitable (number of) vessel(s) for the storage of gases. The computation time is a matter of several seconds. The structuring of the data needed for the program is very simple and is easy to extend to more gases, more materials, shapes, norms and manufacturing procedures. Various costs dependent on the situation at hand can easily be modified. The optimization program used and the input data and subroutines for computation of the various quantities can be regarded as an expert system for the storage of gases.

References

1. Nichols, R.W., 1971, *Pressure Vessel Engineering Technology*, Elsevier, Amsterdam.
2. *Rules for Pressure Vessels*, Government Publishing Office, The Hague,The Netherlands.
3. Norms (NEN,BSI,API,STOOMWEZEN,DIN)
4. *Papers of FME-NIL*, Metal Federation of the Netherlands, TNO (Centre of Steel, VM 84)

5. Davis, L., 1991, *Handbook of Genetic Algorithms*, Van Nostrand Reinhold,New York.

6. Goldberg,D.E., 1989, *Genetic Algorithms in Search, Optimization and Machine Learning*, Addison-Wesley, Reading.

7. Zbigniew,M., 1992, *Genetic Algorithms + Data Structures = Evolution Programs*, Springer-Verlag, New York.

8. Kalker-Kalkman,C.M., Offermans, M.F., 1995, A general design program based on genetic algorithms with applications, *Proceedings of the 21st ASME Design Automation Conference*, Boston, USA, September.

Part 8: New Topics in EA Basics

Papers:

Simulation of Baldwin effect and Dawkins memes by genetic algorithm

Vladimír Kvasnička[1] and Jiří Pospíchal[2]

Department of Mathematics, Slovak Technical University, 812 37 Bratislava, Slovakia
email: [1] kvasnic@cvt.stuba.sk, [2] pospich@cvt.stuba.sk

Key words: Genetic algorithm, learning, fitness landscape, Baldwin effect, Dawkins memes

Abstract

Three different levels of sophistication of genetic algorithms (GAs) are described. (1) The *first* level are GAs, where the chromosome fitness is determined exactly by the chromosome position on the fitness landscape. This first level of sophistication corresponds to standard GAs, where chromosomes are directly mapped on positive real numbers. (2) In the *second* level the chromosome fitness is determined by the nearest neighborhood of the chromosome position on the fitness landscape, i.e. chromosomes are capable of learning. The role of learning in evolution theory called the Baldwin effect was first studied by GAs in 1987 by Hinton and Nowlan. (3) In the *third* level the chromosome fitness is determined not only by the nearest neighborhood of the chromosome position on the fitness landscape but also by the so-called meme that determines an information that is able to increase the fitness of chromosomes. The idea of memes was introduced to evolutionary biology by Dawkins. The concept of chromosome is enlarged to a complex of chromosome and meme, both of them determine a fitness of chromosomes itself. All the above three different levels of GAs may be formally considered as evolutionary steps of Darwin's evolution. The purpose of this communication is to present a GA simulation of learning and Dawkins' memes.

1. Introduction

Genetic algorithms (GAs) [1,2] are computational models of biological evolution. GAs are useful both for solving optimization problems and for modeling evolutionary systems. The purpose of this communication is to present a GA simulation of learning [3-5] and Dawkins memes [6] in a population composed of chromosomes that are represented by linear strings of symbols.

In 1896 J.M. Baldwin [7,8,4] proposed a hypothesis, which pointed out that if learning helps survival, then the organisms best able to learn will have the most offspring, thus increasing the frequency of genes responsible for learning. If the environment remains relatively fixed, so that the most profitable information to learn remains constant, selection can lead to a genetic encoding of a trait that originally had to be learned. Baldwin called this mechanism "organic selection", but it was latter dubbed the "Baldwin effect". We will consider Dawkins theory of memes [6] as an extension of Baldwin effect. The memes are ideas - messages broadcasted throughout the whole population and are composed of useful information how to increase fitness of chromosomes. Dawkins assumed that complexes of chromosome with meme might be important for an acceleratation of evolution. Many social traits (e.g. altruism) are discussed in the framework of population genetic [9] only with some strong restrictions and assumptions that are not acceptable by all evolutionary biologists. An application of Dawkins' memes represents a plausible possibility how to explain and interpret many features of social behaviour of animals [10].

Darwin's theory of evolution [11,12] is based on a principle of natural selection, according to which only the best adapted individuals from a population survive. By reproduction of two strong individuals such offspring are produced, which will be with high probability well adjusted for a successful survival. A detailed analysis (mainly mathematical) shows, that sufficiently effective creation of well adjusted individuals with such new features, which would substantially increase a chance for survival, requires reproduction including both genetic information exchange as well as so called mutations. The mutations influence (positively or negatively) by a random manner genetic material of individuals, influencing in long term the whole population.

In the higher degree of abstraction [1] a notion of biological individual can be replaced by a notion of chromosome, which represents a linearly ordered information content of an individual (genotype). Then we can talk about a population of chromosomes, which reproduce with a probability directly proportional to their strength, while integral part of that reproduction are mutations. Mutations introduce into chromosomes a new information, which can increase fitness of chromosomes created by a crossover from parental chromosomes. New chromosomes push out from population chromosomes with small fitness. This basic reproduction cycle is continually repeated. After some time there is a high probability, that chromosomes with new properties substantially increasing fitness emerge in population, and push out older chromosomes without those properties.

Genetic algorithms (GAs) [1,2] are an abstraction of the above presented basic principles of Darwin's evolution theory [11,12]. According to the way of fitness calculation three different levels of GAs may be considered:

(1) *Fitness of chromosomes is determined exactly by their positions on the fitness landscape, i.e. it is determined entirely by chromosome compositions, no other effects are included in the fitness calculation. This way of fitness calculation corresponds to standard GAs, where chromosome are directly mapped on positive real numbers - fitness, without any intermediate or hidden considerations.*

(2) *Fitness of chromosomes is influenced by the nearest neighborhood of their positions on the fitness landscape. It means that chromosomes are capable of*

learning in which they search through nearest neighborhood on fitness landscape. As a result of this process chromosomes with the same fitness can be distinguished by a presence of higher fitness positions in their nearest neighborhood, see Fig. 1. These chromosomes are more effective for further evolution process than those ones that are not able to look for higher fitness chromosomes in the nearest neighborhood. The role of learning in evolution theory is called the Baldwin effect [7,8], its first study by GA was done by Hinton and Nowlan [3] (cf. also refs. [4,5]). However, better modeling of a learning process should also represent a quality of a learning ability by a size of a neighborhood searched around a chromosome. Really complex chromosomes then could theoretically code not only their position on a landscape, but to code their own learning algorithm as well.

(3) Fitness of chromosomes is determined not only by the nearest neighborhood of their positions on the fitness landscape but also by the so-called memes that determine an information that is able to increase the fitness of chromosomes. The idea of memes was introduced to evolutionary biology by Dawkins in his famous book Selfish Gene [6]. The concept of chromosome is enlarged to a complex of chromosome and meme, both of them determine a fitness of chromosomes itself. Moreover, there should be introduced a notion of the so-called memetic interaction, where a chromosome-meme complex called the donor creates a message (idea). Message is broadcasted throughout a subpopulation of the so-called recipients. These recipients incorporate the broadcasted message into their memes. Incorporation of memes into GAs was already studied by [12,13].

All these three different levels of sophistication of GAs may be formally considered as evolutionary steps of Darwin's evolution itself. This view of the evolution distinguishes a complexity of fitness evaluation process. On the first lowest level the fitness is fully determined by chromosome composition only. The second level corresponds to the previous first level that is enriched by a possibility of learning, i.e. chromosomes are able of local search in their nearest neighborhoods. Finally, the third level is nothing but the second level enlarged by memes, i.e. chromosomes are not only capable of learning but also capable of employing memes in the fitness evaluation process. We have to emphasize that in all these three levels the evolution is strictly Darwinian, there is not postulated that memes or traits acquired by learning are inherited by offspring. Offspring chromosomes are fully determined by "crossover" combination and random mutation of their parent chromosomes, no other information acquired by learning is included. The learning process is used only for evaluation of chromosomes by fitness but not for their modification. For memes is situation slightly different. In the framework of memetic interaction a special type of learning process is used for the creation of a donor meme broadcasted throughout a subpopulation of recipients, where the donor meme transforms in a random way the recipient memes. This fact is in accordance with Dawkins' concept, memes may be transformed not only in the course of a reproduction process but also in the course of their lifetime.

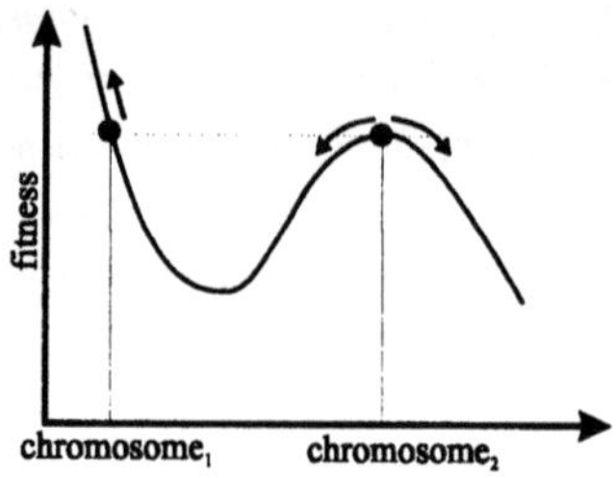

Figure 1. Diagrammatic illustration of a situation when two chromosomes are placed on a fitness landscape so that they have the same fitness. The second chromosome is situated at a local maximum and therefore it does not have a chance to increase its fitness in a nearest neighborhood. Though the first chromosome has the same fitness as the second one, it is situated on the fitness landscape in such a position that in its nearest neighborhood there exist positions with greater fitness. From the standpoint of classical Darwin's evolution theory, when learning processes are fully ignored, these chromosomes could not be distinguished and therefore would have the same chance to participate in a reproduction process. On the other hand, when the learning process is included into the evolution process, these chromosomes are distinguished, the first chromosome is evaluated by a greater fitness and natural selection prefers this chromosome.

2. Basic concepts of GAs

One of basic concepts of GAs is a notion of *chromosome*. A chromosome $\boldsymbol{x}$ is an abstract mathematical object, which in actual applications may be identified to a string of symbols, or to a matrix composed of real or integer entries, or to a graph with evaluated vertices and edges, etc. Let X be a finite set composed of all possible chromosomes that may be constructed so that chromosome specification rules are satisfied

$$X = \{\boldsymbol{x},\boldsymbol{x}',\boldsymbol{x}''...\} \tag{1}$$

An *objective function* $f: X \to R$ assigns to each chromosome $\boldsymbol{x} \in X$ a real number $f(\boldsymbol{x})$. We look for its global minimum over the set X

$$\boldsymbol{x}_{opt} = \arg\min_{\boldsymbol{x}\in X} F(\boldsymbol{x}) \tag{2}$$

In GAs the notion of objective function will serve as a representation of *environment* in which chromosomes live and are evolved by Darwinian evolution. The goal of the evolution is to "construct" such chromosomes that are closely related (or even equal) to an optimal chromosome $\boldsymbol{x}_{opt}$ specified by (2).

A *mutation* is an operator $O_{mut}: X \to X$ that assigns stochastically to each chromosome $\boldsymbol{x} \in X$ a new chromosome $\boldsymbol{x}' = O_{mut}(\boldsymbol{x}) \in X$. In a similar way, a *crossover* is an operator $O_{cross}: X^2 \to X^2$. This operator assigns stochastically to a pair $\boldsymbol{x}_1, \boldsymbol{x}_2 \in X$ a new pair $(\boldsymbol{x}_1', \boldsymbol{x}_2') = O_{cross}(\boldsymbol{x}_1, \boldsymbol{x}_2)$. An actual specification of both these stochastic operators depends on the form of chromosomes and on a problem to be solved by

GA. Mutation and crossover operators are unified in an operator called the *reproduction* $O_{repro}: X^2 \rightarrow X^2$ determined as follows:

$$(x_1', x_2') = O_{repro}(x_1, x_2)\ ,\ \ (\tilde{x}_1, \tilde{x}_2) = O_{cross}(x_1, x_2)$$

$$x_1' = O_{mut}(\tilde{x}_1),\ \ x_2' = O_{mut}(\tilde{x}_2) \qquad (3)$$

This means that in the first stage of reproduction the crossover is performed and then the resulting chromosomes are mutated.

A stochastic decision, whether the operators O_{mut} and O_{cross} will be applied or not, is based on two probabilities P_{mut} and P_{cross}, respectively. An application of the operator O_{mut} onto a chromosome x is realized so that

$$\lim_{P_{mut} \rightarrow 0} O_{mut}(x) = x \qquad (4)$$

Each symbol in the chromosome string will be changed with a probability P_{mut}. The requirement (4) means that after turning off the probability P_{mut}, the result of mutation is equal to the original (nonmutated) chromosome x. In the opposite case, if the probability P_{mut} is greater and greater, then the chromosome $O_{mut}(x)$ is more and more dissimilar from the original chromosome x. The crossover operator O_{cross} is stochastically applied in the framework of the reproduction operator O_{repro}. A stochasticity of this operator is determined by a probability P_{cross}, it determines whether in the framework of the above reproduction operator the crossover process will be used or not. If no crossover process is used, then in the second relation of (3) chromosomes are simply copied.

A *population* P is a multiset with fixed finite cardinality of chromosomes from the set X, $P = \{x_1, x_2, ..., x_p\} \subseteq X$. A *fitness* of chromosomes from the population P is a positive real number determined by a mapping $f: P \rightarrow R_+$ restricted by the following condition

$$F(x_1) \le F(x_2) \Rightarrow f(x_1) \ge f(x_2) \qquad (5)$$

Each chromosome from the population P is evaluated by a positive number called the *fitness*. It reflects a smallness of objective function values $F(x)$, a chromosome evaluated by a smaller function value $F(x)$ is evaluated by a greater fitness $f(x)$ so that the condition (5) is satisfied.

The fitness values are used for the calculation of probabilities of selection of chromosomes from the population P. Let $O_{select}(P)$ be a stochastic operator - *proportional selector* - which realizes a selection of chromosomes from the population P so that the probability of this selection is equal to a normalized fitness. A chromosome with a greater normalized fitness has a greater probability to be selected by the selector $O_{select}(P)$. Formally, the term $O_{select}(P)$ is understood as a chromosome $x \in P$ selected from the population P with the probability determined as follows

$$prob(x) = \frac{f(x)}{\sum_{y \in P} f(y)} \tag{6}$$

GA imitates the Darwinian evolution over a population of abstract chromosomes. A simple realization of GA is outlined in Algorithm 1.

```
procedure GA;
begin P:=a set composed of p randomly
          generated chromosomes;
      t:=0;
      while t<t_max do
      begin t:=t+1;
            evaluate all chromosomes by fitness;
            Q:=∅;
            while |Q|<p do
            begin x_1:=O_select(P); x_2:=O_select(P);
                  (x'_1,x'_2):=O_repro(x_1,x_2);
                  Q:=Q∪{x'_1,x'_2};
            end;
            P:=Q;
      end;
      x_opt:=best chromosome of P;
end;
```

Algorithm 1. An implementation of simple GA. Algorithm is started by a random generation of population *P*. Integer variable *t* is a counter of epochs in GA, outer while-do cycle is repeated a prescribed number of times determined by t_{max}. Inner while-do cycle forms a new population *Q*, which is initialized as an empty set, it is sequentially occupied by offspring created by the reproduction operator applied on a pair of quasirandomly selected chromosomes $\boldsymbol{x}_1$ and $\boldsymbol{x}_2$ from the population *P*. Input parameters are *p* (number of chromosomes in population), t_{max} (maximal number of generations), and probabilities P_{mut} and P_{cross}. As an output of GA is a chromosome $\boldsymbol{x}_{opt}$, determined as that one with smallest function value $F(\boldsymbol{x})$ over the population *P*.

3. GA with learning

Let a *chromosome* $\boldsymbol{x} = (x_1 x_2 \ldots x_n)$ be a linear string of the length *n* composed of symbols 0, 1, and #

$$\boldsymbol{x} \in \{0,1,\#\}^n \tag{7}$$

An *optimal chromosome* $\mathbf{x}_{opt}$ is randomly generated bit string of the length n

$$\mathbf{x}_{opt} = \left(x_1^{opt} x_2^{opt} \ldots x_n^{opt}\right) \in \{0,1\}^n \tag{8}$$

A *population P* is a multiset composed of p chromosomes

$$P = \{\mathbf{x}_1, \mathbf{x}_2, \ldots, \mathbf{x}_p\} \subseteq \{0,1,\#\}^n \tag{9}$$

Its *cardinality* $p=|P|$ is kept fixed throughout all generations of genetic algorithm.

Each chromosome $\mathbf{x} \in P$ is evaluated by a *fitness*. This evaluation process is connected with the *learning process* and it is determined as follows: Let us define a neighborhood of a given chromosome $\mathbf{x} \in P$

$$U(\mathbf{x}) = \{\mathbf{y}_1, \mathbf{y}_2, \ldots, \mathbf{y}_q\} \subseteq \{0,1\}^n \tag{10}$$

This neighborhood is composed of q binary vectors that are randomly created from the reference chromosome $\mathbf{x}$ so that its hash symbols are randomly substituted either by 0 or 1

$$\mathbf{y} = (y_1 y_2 \ldots y_n) \in \{0,1\}^n \tag{11a}$$

$$y_i = \begin{cases} x_i & (\text{if } x_i = 0 \text{ or } x_i = 1) \\ \text{random}(2) & (\text{if } x_i = \#) \end{cases} \tag{11b}$$

The cardinality $q=|U(\mathbf{x})|$ is called the *learning size*. For the studied case there was only one optimal chromosome $\mathbf{x}_{opt}$. A function $d(\mathbf{x})$, for $\mathbf{x} \in P$, corresponds to a minimal distance between the optimal chromosome $\mathbf{x}_{opt}$ and binary vectors from the neighborhood $U(\mathbf{x})$

$$d(\mathbf{x}) = \min_{\mathbf{y} \in U(\mathbf{x})} \left|\mathbf{y} - \mathbf{x}_{opt}\right| \tag{12}$$

A fitness of the chromosome $\mathbf{x}$ is determined as follows

$$f(\mathbf{x}) = f_{min} + (f_{max} - f_{min}) e^{-\xi d(\mathbf{x})} \tag{13}$$

where ξ is the so-called *slope coefficient* that determines a sharpness of the fitness function. The above specified evaluation of population chromosomes by fitness has a strong stochastic character. In particular, the neighborhood $U(\mathbf{x})$ is randomly constructed from the reference chromosome $\mathbf{x}$. This stochastic character of the fitness evaluation is suppressed if the learning size $q=|U(\mathbf{x})|$ is increased so that there exists a real chance to find the global minimum of (12).

The learning process (i.e. evaluation of chromosomes by fitness) is mainly determined by the cardinality q of the neighborhood $U(\mathbf{x})$ and the slope coefficient ξ. The cardinality q corresponds to a number of trial events in the learning process when we construct a binary vector y from a given chromosome $\mathbf{x}$ so that the binary vector y has a smallest Hamming distance from the optimal binary vector $\mathbf{x}_{opt}$. Increase of the cardinality q offers to the learning process a bigger chance to find the best binary vector with the smallest possible Hamming distance from $\mathbf{x}_{opt}$. The slope coefficient ξ is a positive real number that may be interpreted as a selector of acceptance of binary vectors that are created from the given chromosome $\mathbf{x}$ in the framework of the learning process. If this parameter tends to infinity, then the learning process evaluates by the maximal fitness the binary vectors identical to the optimal binary vector $\boldsymbol{x}_{opt}$, all other binary vectors are evaluated by the mimimal fitness. On the other hand, if the slope coefficients ξ approaches zero, the learning process evaluates all binary vectors by values of fitness close to the maximal value f_{max}. The above discussed properties of the fitness are formally expressed by the following limiting relations

$$\lim_{\xi\to\infty} f(\boldsymbol{x}) = \begin{cases} f_{max} & (\text{if } \boldsymbol{x} = \boldsymbol{x}_{opt}) \\ f_{min} & (\text{if } \boldsymbol{x} \neq \boldsymbol{x}_{opt}) \end{cases} \tag{14a}$$

$$\lim_{\xi\to 0} f(\boldsymbol{x}) = f_{max} \tag{14b}$$

The learning process can be interpreted as a searching in a nearest neighborhood of the given chromosome (cf. eqs. (10) to (12)), where the chromosome hash symbols are randomly substituted by 0s or 1s symbols. After Hinton and Nowlan [3] this is a "needle in a haystack" search problem, since there is only one correct solution in a space of 2^n solutions. If the learning is included, then GA can be loosely interpreted as a search method, where some one advises to the "chromosome" in which direction the needle (correct solution) is hidden. More realistic interpretation of GAs with learning is presented in Fig. 1, where the learning process is interpreted as a method that is able to give an information about the nearest neighborhood on fitness landscape, i.e. whether chromosome fitness can be potentially increased.

Finally, we have to specify GAs (outlined in the previous chapter, see Algorithm 1) applied to populations of chromosomes described in the first part of this chapter. An operation of *crossover* for two chromosomes $\boldsymbol{x}$ and $\boldsymbol{x}'$ is simply determined as a 1-point crossover. A crosspoint is randomly generated and then chromosome parts behind this point are mutually exchanged. An operation of *mutation* applied to a chromosome $\boldsymbol{x}$ is performed in such a way that going successively throughout all parts of chromosome we change with small probability P_{mut} chromosome entries to one of their complements, e.g. symbol 1 is changed randomly with the same probability either to 0 or to #.

Genetic algorithm with learning features specified in this chapter has the following set of parameters:

(1) n, chromosome length (n=20)
(2) p, population size (p=500)
(3) P_{cross}, crossover probability (P_{cross}=0.1)
(4) P_{mut}, mutation probability (P_{mut}=0.01)
(5) t_{max}, number of epochs (t_{max}=200)
(6) q, learning size (q=10)
(7) ξ, slope parameter (ξ=0.1 and 1.0)
(8) f_{max}, maximal value of fitness (f_{max}=2.0)
(9) f_{min}, minimal value of fitness (f_{min}=1.0)
(10) initial population P is randomly generated so that its chromosomes
contains 50% symbols #, 25% symbols 0, and 25% symbols 1

where in parentheses are given numerical values of parameters used in our illustrative calculations.

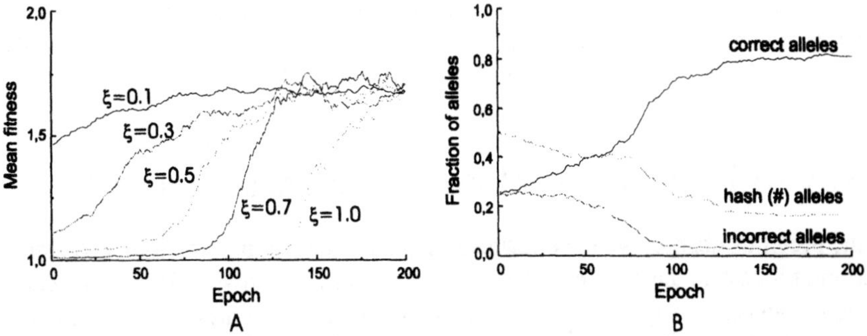

Figure 2. (A) Five plots of mean fitness vs. epochs for different values of slope parameters, from promoted learning (ξ=0.1) to suppressed learning (ξ=1.0). Other parameters of the GA are specified at the end of Chapter 3. We see that if the slope parameter ξ tends to greater values, then the learning process is more and more suppressed, the period when mean fitness is equal to the minimal value of fitness (f_{min}=1.0) is longer. If the slope parameter ξ is greater than a threshold value (say ξ=1.0), then the learning is entirely suppressed, the mean fitness in the course of whole evolution is equal to f_{min} (the optimization strength of a canonical GA is too small to show here). (B) Plot of fraction of correct alleles, incorrect alleles, and hash symbol (#) alleles for GA with learning (slope parameter is ξ=0.5). The fraction (frequency) of fixed correct alleles increased over epochs and the fraction of fixed incorrect alleles decreased almost to zero. On the other hand, the frequency of hash alleles (that can be substituted by 0s or 1s symbols in the learning process) stayed relatively very high. After Hinton and Nowlan [3] this may be caused by too insignificant selective pressure to fix all the hash alleles. This selective pressure may be substantially increased, e.g. by lowering the learning size q (it determines the cardinality of neighborhood $U(\boldsymbol{x})$, where the learning process is performed). According to our numerical experiences, when the learning size is set q=1, then fraction of hash alleles is monotonously decreasing in the course of evolution almost to zero value.

Different numerical results of our model calculations are outlined in Fig. 2. This figure is very instructive for interpretation of Baldwin effect in the framework of our GA simulation calculations. Five plots of mean fitness for different slope parameters ξ are presented in Fig. 2, diagram A . We see that the learning may be substantially suppressed by increasing the slope parameter ξ. The same effect may be also achieved by descreasing the learning size q, the learning is suppressed when the cardinality of the neighborhood is very small (i.e. $q\rightarrow 1$), i.e. the minimization problem (12) is restricted only to a few binary vectors. An evolution of fractions of alleles (assigned to 0, 1, and # symbols) is displayed in Fig. 2, diagram B. Very interesting observation is a decrease of the fraction of hash symbols in the course of evolution from the inital 50% value to the final 15% value. This means that the used GA with learning simulates properly classical Baldwin effect, it is able of a genetic encoding of a trait that originally had to be learned. This fact is manifested by monotonously decreasing value of fraction of hash symbols in the population in the course of its evolution. Hash symbols in chromosomes are introduced as learnable chromosome alleles, their descreasing fraction means that some of them were fixed by 0s or 1s alleles, i.e. learnable traits are fixed genetically.

4. GA with meme chromosomes (m-chromosomes)

In the previous chapter we have formulated simple GA with a possibility to simulate learning process in the course of the evolution process (Baldwin effect). The learning process consists in a replacing hash # symbols in chromosomes by 0s or 1s. These values are substituted randomly and in a subset of these binary vectors we look for a binary vector with minimal Hamming distance from the given optimal chromosome. The learning process assigns to each chromosome a binary vector that is the best one of all possible vectors that can be randomly constructed from the chromosome. If we compare this version of the genetic algorithm with a genetic algorithm where a learning process is considerably suppressed (e.g. slope parameter ξ is used with high numerical value or the learning size q is of very small value), then we see that an effectiveness of the genetic algorithm with learning is greater than those one with suppressed learning, see Fig. 2.

How to intensify the genetic algorithm with already included learning process? Further possibility is to include the so-called memetic approach initially introduced into evolutionary biology by Dawkins [6]. We use very simplified model of memes so that it can be easily introduced to genetic algorithms with learning. In our approach memes are linear strings (of the same length as chromosomes) and each population chromosome is automatically accompanied by a meme. The memes are carriers of information about best local solutions that have been already achieved (see Chapter 1). Employing of memes is very important in evaluation process as a further source of information about a best solution that should be found. At the beginning of evolution the memes are empty (i.e. they have zero information content), in the course of evolution process they may receive an information in the framework of the so-called memetic interaction. Here, a quasirandomly (with high fitness) selected chromosome–donor and its meme creates a message - meme broadcasted through whole population, and some chromosomes–recipients accept this meme and incorporate it in their memes. In the reproduction process, created offspring obtain memes from one of two parental chromosomes declared as a mother (we say that offspring are educated by mothers).

A *meme* is a string composed of n symbols 0, 1, and *

$$\boldsymbol{m} = (m_1 m_2 \ldots m_n) \in \{0,1,*\}^n \qquad (15)$$

A *width* of the meme $\boldsymbol{m}$, denoted by $|\boldsymbol{m}|$, is determined as a number of nonstar symbols, e.g. for a meme $\boldsymbol{m}$=(01**01***) its width is $|\boldsymbol{m}|$=4. In our forthcoming considerations memes are used as an auxiliary vehicle of information used in the learning process of chromosomes. If the width of a meme is zero, then we say that this meme does not contain any useful information and its application in the framework of a learning process is irrelevant.

A *meme chromosome* (*m-chromosome*) is determined as an ordered couple of a chromosome and a meme, both of the same length

$$\boldsymbol{\alpha} = (\boldsymbol{x},\boldsymbol{m}) = ((x_1x_2...x_n),(m_1m_2...m_n)) \in \{0,1,\#\}^n \times \{0,1,*\}^n \tag{16}$$

where $\boldsymbol{x}$ is a *chromosome part* of α and $\boldsymbol{m}$ is a *meme part* of α. A population P is a multiset composed of p m-chromosomes

$$P = \{\boldsymbol{\alpha}_1,\boldsymbol{\alpha}_2,...\boldsymbol{\alpha}_p\} \subseteq \{0,1,\#\}^n \times \{0,1,*\}^n \tag{17}$$

The learning process of fitness evaluation of simple chromosomes, specified in the previous chapter, will be now generalized for m-chromosomes. This learning process consists of two stages.

In the *first stage*, the meme part $\boldsymbol{m}$ of α is used for a repair of the chromosome part $\boldsymbol{x}$ of α , chromosome hash symbols are substituted by meme nonstar symbols. In particular, let $\hat{\boldsymbol{x}}$ be a chromosome created from the chromosome part $\boldsymbol{x}$ by a repair process performed with respect to the meme part $\boldsymbol{m}$

$$\hat{\boldsymbol{x}} = (\hat{x}_1\hat{x}_2...\hat{x}_n) = O_{mod}(\boldsymbol{x};\boldsymbol{m}) \tag{18a}$$

$$\hat{x}_i = \begin{cases} x_i & (\text{if } x_i \neq \# \text{ or } m_i = *) \\ m_i & (\text{if } x_i = \# \text{ and } m_i \neq *) \end{cases} \tag{18b}$$

For better understanding of this repair process let us consider an m-chromosome

$$\boldsymbol{\alpha} = \left(\underbrace{(00\#\#1\#0\#)}_{\boldsymbol{x}},\underbrace{(11***0**)}_{\boldsymbol{m}} \right) \tag{19}$$

The repaired chromosome $\hat{\boldsymbol{x}}$ is determined by (18a-b) as follows

$$\begin{array}{r} \boldsymbol{x} = (0\,0\,\#\,\#\,1\,\#\,0\,\#) \\ \boldsymbol{m} = (1\,1\,*\,*\,*\,0\,*\,*) \\ \hline \hat{\boldsymbol{x}} = (0\,0\,\#\,\#\,1\,0\,0\,\#) \end{array} \tag{20}$$

where in the sixth position the lower condition in (18b) is satisfied, i.e. hash symbol # is substituted by 0.

In the *second stage*, the repaired chromosome $\hat{\boldsymbol{x}}$ is used as a reference chromosome for construction of the neighborhood

$$U(\hat{\boldsymbol{x}}) = \{\hat{\boldsymbol{y}}_1,\hat{\boldsymbol{y}}_2,...,\hat{\boldsymbol{y}}_q\} \subseteq \{0,1\}^n \tag{21}$$

where binary vectors $\hat{\boldsymbol{y}}_i$ are created from the repaired chromosome $\hat{\boldsymbol{x}}$ so that its hash symbols are randomly substituted by 0 or 1 symbols, see eq. (11b). Applying the same procedure as in the standard GA with learning, we look for a minimal Hamming distance between binary vectors from the neighborhood and the optimal chromosome $\boldsymbol{x}_{opt}$ (cf. eq. (12))

$$\hat{d}_{min} = \min_{\hat{y} \in U(\hat{x})} \left| \hat{y} - x_{opt} \right| \tag{22}$$

Finally, a fitness assigned to an m-chromosome α is determined in the same way as in eq. (13)

$$f(\boldsymbol{\alpha}) = f_{min} + (f_{max} - f_{min}) e^{-\xi \hat{d}_{min}} \tag{23}$$

Summarizing, an evaluation of m-chromosomes by fitness is performed formally by the same procedure as simple chromosomes (see previous chapter) except for the fact that meme part is used for a repair process of the chromosome part. Memes are carriers of additional information that is used for a repair process that precedes the learning process. For a usefull meme the resulting value d_{min} may be smaller than in the case when the repair process is not used (or it is used but a width of the meme part is zero).

A *memetic interaction* is a term that will be used to denote a process in which an idea - meme produced by a donor m-chromosome (quasirandomly selected) is broadcasted to a subpopulation of recipients (randomly selected). These recipients accept the broadcasted meme so that their own meme parts are modified. Let α_{donor} be m-chromosome with high fitness quasirandomly selected from the population P

$$\alpha_{donor} = O_{select}(P) \tag{24}$$

where the operator O_{select} is a proportional selector described at the end of Chapter 2. A meme from this donor m-chromosome should be extracted so that it may be used as a message composed of a useful information for other recipient m-chromosomes $P_{recip} \subset P$, where $\boldsymbol{\alpha}_{donor} \notin P_{recip}$ and $r=|P_{recip}|<p=|P|$. This meme extraction is performed in a way very similar to the learning process within fitness evaluation of m-chromosomes (see initial part of this Chapter). Let $\hat{y}_{min}$ be a binary vector constructed by a minimization (22)

$$\hat{y}_{min} = \arg \min_{\hat{y} \in U(\hat{x})} \left| \hat{y} - x_{opt} \right| \tag{25}$$

where $\hat{x}$ is repaired chromosome part of the donor m-chromosome. From the binary vector $\hat{y}_{min}$ we construct a meme $\overline{m}_{donor}$ in such a way that only t randomly selected entries are taken into account, all other remaining meme entries are substituted by * symbol. We say that a width of the meme $\overline{m}_{donor}$ is prescribed by an integer t ($0<t<n$) and it will be kept fixed in the course of all epochs in GA. For instance, let $\hat{y}_{min} = (00110101)$ be a binary vector determined by (25), we select randomly t=3 in the first, second, and fifth positions, then we get $\overline{m}_{donor} = (00{*}{*}0{*}{*}{*})$. The meme $\overline{m}_{donor}$ will be used for a modification of meme parts of all m-chromosomes from the subset P_{recip} , so that nonstar symbols in $\overline{m}_{donor}$ are introduced to their meme parts. For illustration, let $m = (10{*}10{*}0{*})$ be a meme part, its modification by the message $\overline{m}_{donor} = (00{*}{*}0{*}{*}{*})$ is done by substituting the first, second, and fifth entry in m by the corresponding nonstar entries in $\overline{m}_{donor}$, we get

$$
\begin{array}{rl}
\boldsymbol{m} & =(1\,0*1\,0*0*) \\
\overline{\boldsymbol{m}}_{donor} & =(0\,0**0***) \\
\hline
\boldsymbol{m}_{mod} & =(0\,0*1\,0*0*)
\end{array}
\tag{26}
$$

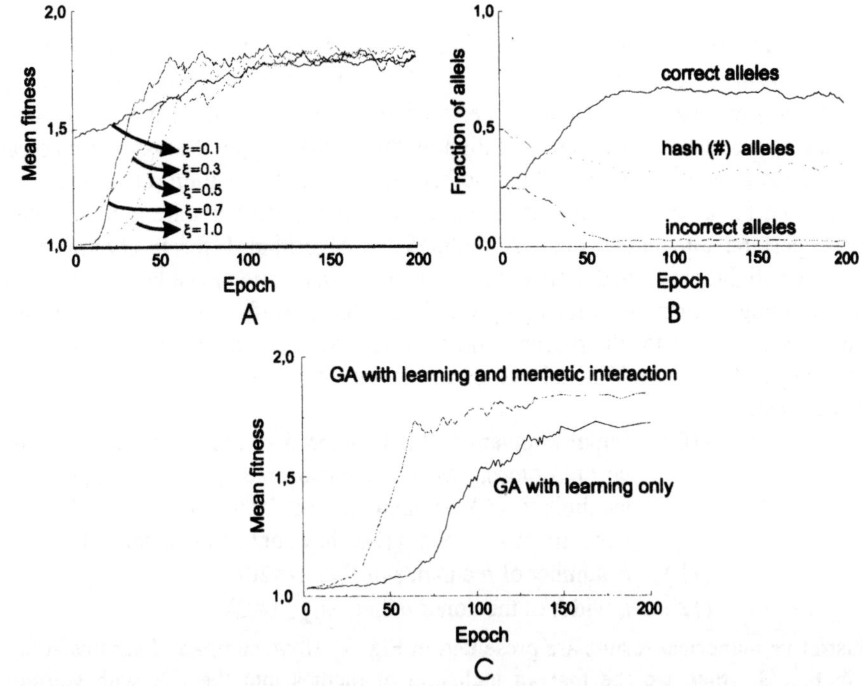

Figure 3. (A) Five plots of mean fitness vs. epochs obtained by GA with learning and memetic interaction, for different values of slope parameter, from promoted learning (ξ=0.1) to suppressed learning (ξ=1.0). Other parameters of the GA are specified at the end of Chapter 3. (B) Plot of fraction of correct alleles, incorrect alleles, and hash symbol (#) alleles for GA with learning and memetic interaction (slope parameter is ξ=0.5). (C) Plot of mean fitness for two different versions of GA with and without a memetic interaction when learning is included in both versions. Both calculations were done for slope parameter ξ=0.5. We see that GA with learning and memetic information outperforms GA with learning only.

Formally, this modification may be expressed as a noncommutative sum of memes, $\boldsymbol{m}_{mod} = \boldsymbol{m} + \overline{\boldsymbol{m}}_{donor}$, where single entries of $\boldsymbol{m}_{mod}$ are determined by

$$
m_i^{mod} = \begin{cases} m_i & \left(\text{if } \overline{m}_i^{donor} = *\right) \\ \overline{m}_i^{donor} & \left(\text{if } \overline{m}_i^{donor} \neq *\right) \end{cases}
\tag{27}
$$

In order to simplify the concept of this modification process we introduce the so-called *modification operator* that modifies all m-chromosomes from recipient subset P_{recip} with respect to a meme generated from the donor chromosome

$$P'_{recip} = O_{mod}\left(P_{recip};\boldsymbol{\alpha}_{donor}\right) \tag{28}$$

where the modified recipient subset is determined as follows

$$P'_{recip} = \left\{\left(\boldsymbol{x},\boldsymbol{m}+\overline{\boldsymbol{m}}_{donor}\right);\left(\boldsymbol{x},\boldsymbol{m}\right)\in P_{recip}\right\} \tag{29}$$

i.e. all m-chromosomes from P_{recip} are modified in such a way that their meme parts are modified by a meme $\overline{\boldsymbol{m}}_{donor}$, see example (26).

Finally, we have to specify two GA operators crossover and mutation for m-chromosomes. The crossover operator O_{cross} applied to a couple α_1 and α_2 of m-chromosomes (that are randomly declared as mother and father) is performed so that their chromosome parts are modified by the standard 1-point crossover whereas their meme parts are determined by mother meme part. A mutation operator O_{mut} is determined for m-chromosomes such that only chromosome parts are mutated while meme part remain untouched. GA outlined by Algorithm 1 for m-chromosomes should be slightly modified, in particular a memetic interaction must be introduced at the beginning of each evolutionary epoch. Parameters specified at the end of Chapter 3 are still valid also for the present modification. Moreover, a random generation of initial population and additional two parameters used in this version of GA should be specified:

(10') initial population P is composed of randomly generated m-chromosomes, where chromosome parts contains 50% symbols #, 25% symbols 0, and 25% symbols 1 and meme parts are of zero width (i.e. they contain only star symbols)

(11') r, number of recipients in P_{recip} (r=20)

(12') t, width of the donor meme $\overline{m}_{donor}$ (t=3)

Illustrative numerical results are presented in Fig. 3 . If we compare diagrams A and C in Fig. 3, then we see that an inclusion of memes into the GA with learning considerably increases its effectivity. In particular, transient phases where the mean fitness is kept to its minimal value f_{min}=1 is now substantially narrowed also for higher values of the slope parameter ξ. This fact is nicely illustrated by diagram C, where two plots of mean fitness are simultaneously displayed, one for GA with learning only and the other for GA with learning and memetic interaction. In general, we may say that an inclusion of memes to GAs means a substantial increase of the evolution effectivity, irrespective whether the simple learning (Baldwin effect) is costly or not. Plots of fractions of alleles for both versions of GA with learning only and GA with learning and memes are presented in Fig. 2, diagram B and in Fig. 3, diagram B, respectively. Comparing these plots we see that an introduction of memes decreases the selection pressure in GAs with learning and memes in comparison to GAs with learning only. In particular, a fraction of hash alleles (that are responsible for learning) for GAs with learning only asymptotically tends to 15% (see Fig. 2, diagram B) while this asymptotic value for GAs with learning and memes is substantially greater 35% (see Fig. 3, diagram B). This observation has the following simple interpretation: An inclusion of memes decreases a selection pressure onto learnable hash symbols, in other words, memes may substitute in some extent the learning process. If the learning process is costly, the memes might be very important for acceleration of evolution, they are able to overcome of learning barriers.

5. Future plans

The population of m-chromosomes will be "geographically" structured so that it will be decomposed onto disjoint subpopulations and each subpopulation has some other subpopulations as neighborhoods. We will examine the effect of erecting complete or semipermeable barriers between different subpopulations and the effect of migration. Until now, only one meme was attached to each chromosome, it will be interesting to study when this restriction is removed and a variable number of memes is attached to chromosomes. Many different strategies can be introduced which meme will be used in the process of evaluation of m-chromosomes by fitness, e.g. that one producing the best fitness or another one with greatest width, and so on. In our calculations we kept fixed the optimal chromosome $\boldsymbol{x}_{\mathrm{opt}}$, it means that the goal of evolution remained unchanged in the course of whole evolution. Interesting results might be obtained by slight variation of this optimal chromosome $\boldsymbol{x}_{\mathrm{opt}}$, in each generation it can be mutated with very low probability, e.g. $P_{mut}^{(opt)} = 0.001$. Then, we will compare an effectivity of GA with learning only and GA with learning and memes. According to our preliminary results it seems, that memes might be very important for permanent adapting of the best evolutionary solution when goal of evolution (or environment where populations exist) is slightly changing. In particular, if GA is performed only with learning, we have observed that the evolving system has a great inertia, it is unable to well adapt to a new evolution goal.

Acknowledgments

This work was supported by the grants # 1/4209/97 and # 1/5229/98 of the Scientific Grant Agency of Slovak Republic.

References

1. Holland J H 1975 *Adaptation in Natural and Artificial Systems.* MIT Press, Cambridge, MA

2. Goldberg D E 1989 *Genetic Algorithms in Search, Optimization, and Machine Learning.* Addison-Wesley, New York.

3. Hinton G. E, Nowlan S J 1987 How learning can guide evolution, *Complex System* 1:495-502

4. Belew R K, Mitchell M Eds. 1996, *Adaptive Individuals in Evolving Populations: Models and Algorithms.* Addison-Wesley, Redwood City, CA.

5. 1996 *Evolution, Learning, and Instict: 100 Years of the Baldwin Effect.* Special issue of *Evolutionary Computation* 4 (3)

6. Dawkins R. 1976 *The Selfish Gene.* Oxford University Press, Oxford
7. Baldwin J M 1896 A New Factor in Evolution. *American Naturalist* 30: 441-451.
8. Baldwin's article (Ref. 7) is available in electronic form on URL address
9. http://paradigm.soci. brocku.ca/~lward/SUP/Baldwin/BALD_002.html.
10. Hamilton W D 1964 The Genetical Evolution of Social Behaviour I and II. *Journal of Theoretical Biology* 7:1-16 and 17-52
11. Heylighen F Ed. 1998 *The Evolution of Complexity.* Kluwer Academic.
12. Darwin C 1859 *The Origin of Species.* Reprinted 1985, Penguin, London.
13. Darwin book (Ref. 11) is available in electronic form on URL address
14. http://www.tbi. univie.ac.at/Origin/index.html.
15. Gabora L M 1995 Memes and Variations: A Group Model of Cultural Evolution. In *Lectures in Complex Systems*, eds. Nadel L and Stein D L. Addison Wesley. Available at http://www.lycaeum.org/~sputnik/Memetics/ variations.txt
16. Hutchins E. Hazelhurst B 1992 Learning in the Cultural Process. In *Artificial Life II*, eds. Langton C. G. et al. Addison-Wesley, Redwood City, CA, pp. 689-706.

Approach to Structure Synthesis on the Base of Genetic Algorithms

Igor P. Norenkov, Oleg T. Kosachevsky, Georgie K. Pisarenko
Bauman Moscow State Technical University (Russia),
Phone: 7-095-2636290,
E-mail: norenkov@aicad.isrir.msk.su,
kot@src.samsung.ru,
pisarenko@mail.kbimpuls.msk.ru.

Keywords: CAD/CAM, Structure Synthesis, Genetic Algorithms.

Abstract

The paper is devoted to one possible approach to a generalization of formulation of structure synthesis problem by the genetic algorithms for some applications. The approach includes a substitution of design parameter representation in chromosome by representation of rules for choice of parameter values in accordance with Heuristics Combination Method. Number and sense of parameters are defined from information model of application similar IDEF1X model. A transition from one application to another requires replacement of rules (heuristics) set and algorithm of fitness function calculation. Also the paper includes the results of solving some problems such as OSSP, VRPTW and distribution of subnetworks in VLAN (application distribution problem).

1. Introduction

Structure synthesis is very important design procedure in many applications, but it is difficult for solving by formal way. Usually structure synthesis problems are solved by hand in interactive mode and therefore a solution is far from optimal value.

The main obstacle for formal solving is discrete and non-ordering character of many design parameters like physical principles, types of components, kinds of connections, etc. Sometimes a problem may be represented as a discrete mathematical programming problem. But usually it is found NP-hard and its solving by means of known exact methods is impossible.

For this reason different heuristic formulation and solving are applied. Unfortunately a realization of each new structure synthesis task grows into scientific problem. Furthermore an exactness of received solution often is not satisfactory one. The alternative approach based on expert system is not applied widely because of similar difficulties of creation and development of knowledge bases.

Genetic algorithms [1] are perspective approach to solve structure synthesis problems under these conditions. There are some examples of effective application of genetic algorithms (GA) including numerical optimization [2], partitioning problems [3], Steiner problem in a graph [4], Job-Shop Scheduling Problem [5], etc.

However there are some difficulties in using GA too. The main of them is mutual dependencies between design parameters. It leads up to incorrectness of arbitrary combinations of parameters. Since some special techniques are used to exclude unacceptable combinations. The known examples of the techniques are subsequent exchange crossover [6], harmonization algorithm [7] for schedule synthesis and PMX method [1] for traveling salesman problem and so on. Unfortunately, the reordering algorithms are special for each class of problems. Therefore generalization of approach and programming became more difficult.

This paper is devoted to the approach that aims to overcome these difficulties and to increase an effectiveness of GA application to some structure synthesis problems. The approach is based on an initial description of design synthesis problem in terms of information modeling method, for instance IDEF1X [8]. Another item of the approach is application of Heuristics Combination Method (HCM) [9,10], suggested early for schedule problems. Use of GA is local-genetic one [5,11]. We will use HCM abbreviation for future reference.

The basics of the approach are considered below in section 2. Sections 3...5 are devoted to the examples of formulation of some problems by the HCM. There are Vehicle Routing Problem with Time Windows (VRPTW), Application Distribution Problem (ADP), Open Shop Scheduling Problem (OSSP). The technique of heuristics formulation is discussed in the section 6. Section 7 includes the experimental results of application of HCM to the problem solving.

2. HCM basics

Implementation of HCM invokes creating of information model (IM) of given subject field, forming a set of heuristics for defining the vector A of design parameters and elaborating of structure-criterion model that is **Mod: A→ F**, where F is a fitness function.

IM is designed in accordance to methods like IDEF1X. A model is described as a set of entities. Attributes of entities are design parameters.

Forming a chromosome is by the following way.

Usually each i-th gene corresponds to one i-th design parameter A_i in chromosome **C1**. If relation between entities is 1:N (that is type "one-to-many") and entity-descendant has n attributes, then chromosome **C1** has the fragment for this descendant with $N*n$ genes. The peculiarity of HCM is that the chromosome consists of only N genes under these conditions. If relation in IM is non-specific that is M:N (type "many-to-many"), then corresponding fragment in chromosome C1 includes $M*N*$n genes. However length of the fragment in the HCM case is only $M*$N genes (here and below optional attributes and attributes that are not design parameters are not taking into account).

Lets consider a forming the fragment of chromosome for the example of one non-specific relation *M*N* between the entities **R1**(*A0*,**B0**) and **R2**(*A1*,**B1**) by HCM, where *A0* and *A1* are keys, **B0** and B1 are lists of non-key attributes. The substitution of relation *M:N* generates the entity-mediator **R3**(*A0,A1,A2...An*), where *A2...*An are the attributes that are not keys. The chromosome includes three fragments. One corresponds to the mediator and each gene in it corresponds to one instance of the mediator. Each gene in the second (third) fragment corresponds to one instance of the first (second) parent (if **B0** or B1 are the empty lists, then second or third fragments are absent in the chromosome).

After removal non-specific relation a relation $\mathbf{W}_k$ is introduced for each k-th fragment. These relations represent the set of heuristics and may be named a knowledge database. For example, the relation $\mathbf{W}_1(H,P_0,...P_n)$ is one for the first fragment. Here H is the key or, in the other words, the name (number) of heuristics. *Pi* is the name of the rule, intended for the choice of the attribute *Ai*.

More than one rule must be formed for each attribute. So, if there are m rules for each attribute and there are n attributes for an entity, then m^n heuristics may be created for this entity. However it is enough $L \ll m^n$ rational heuristics and usually L is in the range from 2 to 8.

Therefore the following components represents the structure synthesis problem in HCM:

The chromosome **C**(**E**,**H**), where **E** is the set of the names of the entity instances; the allele of i-th gene is the name of concrete heuristics (i.e. one value from the set **H**) for defining the attributes of the i-th entity instance; each chromosome in population represents one design solution;

The knowledge base, consisting of $\mathbf{W}_j(H,P_0,...P_n)$, j=1...K, where K is the number of entities with non-empty lists of non-key attributes;

The algorithm **Mod: A→ *F*** for transforming of the design parameter vector **A** to fitness function F.

If designed object includes one or more attributes, depended on mutual design solutions, then entity-categories appear in IM. In particular it is in IM, if an application model is represented by AND/OR tree. A hierarchy of chromosomes corresponds to a model in such case. The upper level has a chromosome with genes for general entity. The lower levels include chromosomes of categories. Genetic operators are fulfilled only in chromosome of upper level and in categories that activated by corresponding values of keys in the general entity.

The advantages of HCM are as follows:

Since there is general method for formulation of wide class of these problems, we have more abilities for structure synthesis problem solving automation.

High efficiency of solution search, because a chromosome length and a size of search space are decreased due to use only rational heuristics.

Clear separation of application models from genetic search procedures, so descriptions of applications and programming are simplified.

Preparation of synthesis task consists of following actions:
selection design parameters A_i;
working out the set of rules S_{ij} for defining design parameters;
forming the heuristics sets;
elaboration of structure-criterion model Mod.

3. Vehicle Routing Problem (VRP)

The tasks of this type and examples of GA application to their solving are described in some papers, in particular [12, 13]. One complex problem of this class is VRPTW (VRP with Time Windows).

The formulation of solved VRPTW is described below. The following data are given:

1) the number Q of product types;

2) the number E of nodes-sources of product and the quantities of every type product in each sources;

3) the number N of nodes-consumers of product and the quantities of every type product ordered by each sources;

4) the number M of transport servers with known characteristics:

velocity of server;
the maximum volume of transported product;
the cost of one unit of run by server;
the initial servers locations in the net of nodes.
the matrix **D**, where element dij is the distance between nodes i and j;

5) the time window $[t_{min}, t_{max}]$ for every order, t_{min}, and t_{max} are minimum and maximum time for the order satisfaction;

penalty assigned for every unit of product, received by consumers belong the time window.

It is required to compile the schedule with minimum product shipment expenses.

The following relations are formed in accordance to HCM:

R = (A_0 - the name of route, A_1 - the name of consumer, A_2 - the name of source, A_3 - the name of server, B- the time of run beginning on the route);

C = (A_0 - the name of route, H - the number of heuristics);

W = (H - the number of heuristics, P_1, P_2, P_3).

The variable P_1 is the name of the rule for defining the value of parameter A_1. The rules S_{1i} define the consumer, which has: S_{11} (maximum volume of ordered product; S_{12}) maximum lower boundary of time window; S_{13}) maximum order cost; S_{14}) minimum upper boundary of time window.

The variable P_2 is the name of the rule for defining the value of parameter A_2. The rules S_{2i} define the source with the following feature: S_{21}) minimum distance to chosen consumer; S_{22}) minimum summa of distances between the consumer and nearest server. Only sources with product quantity not less than ordered quantity are considered in these rules. If there is no such source, then the rule S_{23} is used: the source with minimum value of $G=(L_1+L_2)*Uo/U$ is chosen, where L_1 and L_2 are distances from the source to consumer and to server accordingly, Uo is the volume of ordered product, U is the volume of '

The variable P_3 is the name of rule for defining the value of parameter A_3. The rules S_{3i} define the server with following feature: S_{31}) minimum distance to chosen source; S_{32}) minimum release time after previous route; S_{13}) minimum route run time.

The fitness function is a sum of expenditures of server runs and penalties for break time limitations (the value of penalty depends on cost of undelivered product).

4. Application Distribution Problem (ADP)

Under Application Distribution Problem we assume a task of software distribution over subnetworks of virtual LAN.

Given a LAN or an Intranet that contains $m \leq M$ subnetworks. It is necessary to distribute N software application over this network. In information model we have "application" and "subnetworks" entities, and intermediate entity "assignment". The relations among them are as follows:

R1 = (A0, A1, A2), where A0 – assignment name, A1 – application name, A2 - subnetwork name.

The following data are also given:

Matrix **D** reflects interdependencies of applications. D_{ik} is a traffic between application i and application k, where i = 1.n, k = 1..n.

Internal traffic bounds Tj in subnetwork j, where j = 1.m.

It is necessary to determine:

Matrix **X** of application distribution over the network. Xpq = 1, if application q is installed in subnetwork (segment) p and Xpq = 0 otherwise, where p = 1..m, q = 1..n.

Intersegment traffic V.

An extreme problem statement is like layout (composition) one

$$\max_{\mathbf{X}} \sum_{j=1}^{m} Y_{jj}, \quad where\ \mathbf{Y} = \mathbf{X}^*\mathbf{D}^*\mathbf{X}^t.$$

The limits are as follows:

$$Y_{jj} \leq T_j; \quad \sum_{j=1}^{m} X_{ji} = 1, \quad i=1..n..$$

A chromosome is formed on the basis of the following relations.

R2 = (A0, H),

W = (H, P1, P2), where A0 – assignment operator, H – heuristics number, Pi – Ai calculation rule.

Fitness function is a sum of intersegment traffic V and penalty functions for bounds breaking. It is necessary to minimize fitness function.

In the solved problem P1 is the name of the rule for determining the name of the application (A1). The rule S1i selects an application A1 with the following features: S11) corresponding row in matrix **D** has maximum value over the matrix (only rows and columns which correspond to unassigned application count); S12) A1 is the same as A0 (if application with number (name) A0 has been already distributed, this rule is just skipped); S13) minimum sum of column elements of matrix **D** of not yet assigned applications.

The rules to determine the name of the segment A2 (variable P2) selects the following rule: S21) segment with minimum current workload; S22) segment with maximum current workload; S23) assume there is such a distributed application A in subnetwork SN that it has maximum traffic with distributing application; S24) segment which has maximum traffic with distributing application.

5. Open-shop scheduling problem (OSSP)

There are lots of publications on JSSP and OSSP. Let's consider a problem discussed in [14]. There are N jobs; S groups, each job belongs to a certain group; M servers; Q stages. Each server belongs to a certain stage. Every job must be served on a server of every stage in consequence. The following data are given: matrix of time expenditures of job service and servers adjustment, time limits, service and adjustment prices. The task is to create a schedule with minimum price in time limits.

Information model includes parent entities "job" and "servers" and intermediate entity "operation".

R1 = (A0, A1, A2, B1);

R2 = (A0, H);

W = (H, P1, P2), where A0 – operation number, A1 – job name, A2 – server name, B1 – operation beginning time, H – number of heuristics, P1 – job name determining rule, P2 – server name determining rule. The time B1 is max {server release time, job arrival at current stage}.

Number of operation g = N*Q. Chromosome consists of g genes. Gene value is heuristics number.

6. A Set of Heuristics

The set of heuristics is crucial to the solution quality. It should be said, however, that heuristics set developing in HCM is much less chargeable than in traditional heuristic methods. Specific rule can reflect just a certain characteristics of intermediate solution. (It is possible even to include in heuristics set a neutral L rules like $A_k := j(r)$, where j is one from L possible values of A_k, r=1...L). It is a GA that provides us with rational solution.

How close is a solution obtained to optimum one and how quick we can get it depends not only on heuristics set, but also on probability of their selection qi in initializing and mutation operators.

Assume that a number of design parameters is n and a number of rules per each parameter is m. Hence number of possible heuristics is m^n. Usage of a whole set of heuristics is beside the purpose. There is no formal procedure to get probability q_i. We can suggest the following procedure to work out q_i:

Make a trial run of GA. Initially every q_i has equal value. After K generation frequency at a particular allele of the best individual (under frequency we mean number of locuses with current allele to number of genes ratio) q_i in the main run is equal to frequency defined in trial run.

Self-adaptive q_i: q_i in population K is equal to frequency of i-th value appearance in genes of the best individual in the population K-1.

Heuristics test results usage: during a test a fitness function calculation is carried out for each heuristics separately. Assume we are to minimize fitness function F_i then

$$q_i = 1 / (F_i * \sum_{j=1}^{m} (1 / F_j)), \qquad (1)$$

where m is a number of heuristics. In most cases the best results are obtained when we use fitness function Zi; received when penalties are not count, instead of F_i.

7. Experimental results

The first objective of experiments were to show the advantages of HCM and applied hybrid algorithms in comparison with simple GA and traditional heuristic approaches in the field of discrete optimization. The second objective is to compare different techniques in order to provide recommendations on choosing heuristics probability q_i.

We used hybrid GA, which combines local (simple hillclimbing) and genetic search. We performed simple hillclimbing over each perspective child (here "perspective child" is an individual we got after crossover with best fitness). This hybrid local-genetic algorithm was described in [5] in detail.

One positive feature of HCM noted above is easy transition to new problem solving. Our experimental HCM program consisted of invariant component U1, which implemented GA, and application component U2 with the application model (**Mod: A→ *F***) and with the set of heuristics. All the problems considered were solved on the same U1. It is required only one-two days for transition to a new problem such as ADP or VRPTW from OSSP solving.

The initial data we used in described experiments are as follows:

VRPTW:

number of nodes-sources $E = 40$,
number of nodes-consumers $N = 25$,
number of servers M = 12,
number of product types Q = 2;
total number of nodes G for possible placement servers is 80.

ADP:

number of applications N = 39,
maximum number of subnetworks M = 10;
internal traffic for each subnetworks is not more than 540.

OSSP:

number of jobs N = 105,
number of job types S = 20,
number of servers M = 15,
number of stages Q = 4.

Heuristics set and test results for both VRPTW and ADP without HCM are shown in the table 1. Heuristics are subsets of rules pointed in the table and described above. The results represented in the i-th column were obtained when all values of genes in chromosome have the same heuristics number i. In the other words, the data in i-th column were received when only i-th heuristics were used. Here Fi is a fitness function with penalties; Zi is a fitness function when penalties is not taken into account (Z_i=0 in some experiments with ADP means that one subnetwork comprises all applications). Using a single heuristics means that data of the table 1 correspond to results of common (not GA) heuristic methods.

Table 1

	1	2	3	4	5
Rules (VRPTW)	S_{11},S_{22},S_{31}	S_{12},S_{21},S_{31}	S_{11},S_{21},S_{32}	S_{13},S_{22},S_{32}	S_{11},S_{21},S_{33}
F_i/Z_i (VRPTW)	705/505	1638/648	1407/1307	1516/1416	816/796
Rules (ADP)	S_{11},S_{21}	S_{12},S_{21}	S_{11},S_{23}	S_{13},S_{24}	S_{11},S_{22}
F_i/Z_i (ADP)	1857/1857	1870/1870	4040/ 0	4040/ 0	4040/ 0

The results of GA applications for VRPTW are given in the table 2. There were 5 runs for each variant of the task. Every value in the column 2 is the result of one run. These results were received after 30 generation for HCM and after 50 generations for simple GA (about 5400 evaluations of fitness function F). These data show:

1) significantly better results of HCM in comparison with simple GA,
2) that effectiveness of HCM implementation depends greatly on q_i.

Therefore we compared the techniques of qi calculation described above. The best results were received by the self-adaptive method.

Table 2

	Results (F) of runs	**Average value of F**
Simple GA	543; 578; 593; 598; 663	595
HCM with equal q_i	516; 529; 545; 548; 577	543
HCM with trial run for q_i	496; 503; 510; 522; 542	515
HCM with self-adaptive definition q_i	494; 495; 503; 504;504	500
HCM with definition q_i by (1)	495; 495; 496; 506; 569	512

Similar results received for ADP are represented in the table 3. They confirm the expressed appraisals. Note that using the self-adaptive algorithm for calculation of q_i more preferable despite of modest value of F in ADP case, because the trial run technique requires additional computer runtime, and the technique on the base (1) is not reliable in tasks with hard constrains.

Table 3

	Results (F) of runs	**Average value of F**
Simple GA	1378; 1395; 1488; 1518; 1575	1471
HCM with equal q_i	926; 1012; 1056; 1068; 1108	1034
HCM with trial run for q_i	913; 913; 919; 919; 929	919
HCM with self-adaptive definition q_i	913; 928; 981; 988; 1039	970

Experimental results for OSSP solving by HCM can be found in [5]. The experimental data with set of three perspective heuristics are following:

simple GA results in 22283 of average (in some runs) fitness function F_{av};

HCM with the same values of q_i gave F_{av}=22059, and

HCM with correction of qi gave Fav=21986 (after about 36000 fitness function calls).

Result's comparison (table 1, no HCM, best single heuristics; tables 2 and 3, HCM) show that fitness in HCM goes 40-50% better.

This proves HCM to be quite economical method. Calculation time is also acceptable, for example there are about 130 steps (evaluations of fitness function) per second for described VRPTW on PC Pentium 100 MHz.

Conclusion

A general approach to setting and solving structure synthesis and discrete optimization problem is presented. The approach combines ideas of known information database design methods with evolutionary genetic algorithms.

A formal procedure for describing the problem as a population of chromosome is given. Application models and procedures are separated from GA operators that make it possible to design open software. The approach proves to be effective in comparison with traditional GA implementations.

Bibliography

1. Goldberg D. Genetic Algorithms in Search, Optimization, and Machine Learning. - Adison-Wesley Publ., 1989.

2. Michalewicz Z. Genetic Algorithms, Numerical Optimization, and Constraints. - Proc. of 6th Int. Conf. on GA, Morgan Kaufmann Publ., San Francisco, 1995.

3. Jones D., Beltramo M. Solving Partitioning Problems with Genetic Algorithms. - Proc. of 4th Int. Conf. on GA, Morgan Kaufmann Publ., San Mateo, 1991.

4. Esbensen H. Finding (Near-)Optimal Steiner Trees in Large Graphs. - Proc. of 6th Int. Conf. on GA, Morgan Kaufmann Publ., San Francisco, 1995.

5. Norenkov I.P., Goodman E.D. Solving Scheduling Problems via Evolutionary Methods for Rule Sequence Optimization., Soft Computing in Engineering Design and Manufacture, P.K.Chawdry, R.Roy, R.K.Pant, eds., Springer Verlag, 1998.

6. Kobayashi S., Ono I., Yamamura M. An Efficient Genetic Algorithm for Job Shop Scheduling Problem. - Proc. of 6th Int. Conf. on GA, Morgan Kaufmann Publ., San Francisco, 1995.

7. Nakano R. Conventional Genetic Algorithm for Job Shop Problems. - Proc. of 4th Int. Conf. on GA, Morgan Kaufmann Publ., San Francisco, 1991.

8. Norenkov I.P. Scheduling and Allocation for Simulation and Synthesis of CAD System Hardware. / Proc. of EWITD'94, East-West Int. Conf. - Moscow: ICSTI, 1994.

9. Fang H., Ross P., Corne D. A Promising Hybrid GA/Heuristic Approzch for Open-Shop Scheduling Problems. / ECAI'94, 11th European Conf. on AI. - Jonh&Wiley&Sons Ltd., 1994.

10. Ishibuchi H., Murata D., Tomioka S. Effectiveness of Genetic Local Search Algorithms. - Proc. of 7-th Conf. on GA, 1997.

11. Integrated computer-aided manufacturing (ICAM): Information modeling manual, IDEF1 - Extended (IDEF1X). - Albany, New York: GEC, 1985.

12. Blanton J., Wainwright R. Multiple Vehicle Routing wuth Time and Capacity Constraints Using Genetic Algorithms. - Proc. of 5th Int. Conf. on GA, Morgan Kaufmann Publ., San Mateo, 1993.

13. Thanglash S., Vinayagamoorty R., Gabbi A. Vehicle Routing and Time Deadlines Using Genetic and Local Algorithms. - Proc. of 5th Int. Conf. on GA, Morgan Kaufmann Publ., San Mateo, 1993.

14. Tcheprasov V., Punch W., Goodman E., Ragatz G., Norenkov I. A Genetic Algorithm to Generate a Pro-Active Scheduler for Printed Circuit Board Assembly/ Proc. EvCA'96, Moscow, 1996

A Study of Altruism by Genetic Algorithm

Jiří Pospíchal[1], **Vladimír Kvasnička**[2]

Department of Mathematics, Slovak Technical University
812 37 Bratislava, Slovakia
[1]*email: pospich@cvt.stuba.sk*
[2]*email: kvasnic@cvt.stuba.sk*

Keywords: altruism, evolution, genetic algorithm, agents, sociobiology

Abstract

Altruism belongs to "social attributes" of individuals in a population. Model of "kinship" altruism is studied, where an altruistic individual is more likely to help its relative.

In our model simulations by a modified genetic algorithm the chromosomes were defined by a presence or an absence of an altruistic disposition, affiliation to a group (kinship relation), and a fitness, initially random. No objective function is used, new fitness is created in a reproduction process from an arithmetic mean of parental fitness modified by a random number with normal distribution.

Altruistic process increases fitness of a randomly chosen chromosome - recipient, and decreases fitness of a randomly chosen chromosome with an altruistic allele - donor. The increase/decrease is unproportional to an absolute value of the difference of their affiliated groups. Results suggest, that an altruistic behavior in the initial stage was created by a random genetic drift in a subgroup and initial division of a population into subgroups is crucial for its evolutionary (meta)stability.

The presented simulation can be useful for evolutionary biology as well as for a modification of genetic algorithms (e.g. in artificial agents) towards natural systems.

1. Introduction

The evolutionary computational models and algorithms insofar were built on fitness based selection. Such evolving replicators are argued to be intrinsically selfish [1]. Though the cooperation is generally advantageous, selfish local optimization is supposed to preclude the reaching of globally optimal cooperative arrangement, since evolution tries to optimize first at the subsystem level [2]. From the point of view of traditional evolutionary theory the altruistic chromosomes must die out during evolution because their fitness is systematically decreased. In this connection a question emerges, how is it possible, that in populations of some species altruism exists and is evolutionary stable, i.e. altruistic chromosomes do not become extinct. The most advanced study trying to explain how altruism can emerge during evolution were done in sociobiology [3] and genetics [4-8], but the problem is studied in economics [9-12] and game theory [13-15] as well. Although it is generally agreed that cooperation is advantageous for the group of cooperators as a whole, a satisfactory explanation of how such behavior can evolve is still missing.

There are several models of altruism. "Weak" altruism benefits more the other individual than the altruist, "strong" altruism benefits other individuals at altruist's own cost. Evolution of cooperation in nature is usually explained by a kin selection, group selection, reciprocal altruism ("tit for tat"), or moralism. Model of "kinship" altruism in the area of mathematical genetics was developed by Hamilton in 1964 [4], who deduced, that for so called "viscous" populations, where children are not far from their parents in space distribution, altruistic individuals are more likely to interact with their relatives, which would be probably also altruistic. It actually agrees with Dawkins idea of selfish gene [1], because by helping relatives one helps the survival and reproduction of genes at least partially identical with his own genes. Group selection [2,16] is based on comparison of two groups, where one group is more cooperative. The cooperative group gathers more resources, becomes more fit, and the non-cooperative group will be eliminated. Reciprocal altruism [17] means helping another individual, when one expects the other one to return the favor. In order to avoid cheaters, one would have to know the other individual, or have more that one encounter with him. Moralism [2] avoids necessity to know the other interacting individual by developing a general approach punishing a non-cooperative individual by individuals that have not been with him in direct interaction.

All the above mentioned mechanisms are currently by most biologists considered too unstable to resist the evolutionary pressure of "selfish gene". In evolutionary biology and sociobiology are therefore pursued more sophisticated explanations. However, arguments requiring complexity may not be so valid for the case of evolutionary models of processes in other media than a nature. In this study an altruistic individual in interaction with another individual increases his fitness at expense of his own fitness (strong altruism). The model is based on a mix of kinship/group selection, where kinship is defined only by an affiliation to a group and no deeper specification of relatedness is used. Individuals from different groups can mate, while their two offspring will belong to the same group as one (randomly chosen) of the two parents. Conditions, under which an altruistic behavior can be evolutionary stable, or at least metastable, are studied. The model is kept as simple as possible to highlight the causes behind different evolutionary directions and explain, when an altruistic behavior will be selectively favored. Although an actual implementation of a transfer of fitness would be complicated (if possible at all) for most cases of optimization, the transfer could be possible for special kind of artificial agents. We do not seek a model exactly copying nature, however, the results of the study should be also relevant for sociobiology, modeling at least partially forces behind creation and maintaining altruistic behavior.

In our model simulations by a modified genetic algorithms the chromosomes were defined by vectors composed of three entries. The first 1-bit entry determines presence (1) or absence (0) of altruism. The second nonnegative integer entry describes a similarity (belonging to a clan, kinship relation), two chromosomes are more similar, if an absolute value of the difference of their similarity descriptor is smaller, in the limit case of zero difference the chromosomes belong to the same clan. The third nonnegative real number entry determines a fitness of the chromosome. This is a very simplified approach to definition of chromosome, where in the canonical definition its fitness is determined as a consequence of the quality of solution coded in a linear sequence of symbols of chromosomes. The simplified approach is satisfactory for our study of fundamental principles of evolution of altruism. We do not have to seek a solution of some optimization problem, it is enough to presume, that there exist some hypothetical solution evaluated by a fitness.

In each epoch a new population is created, which totally replaces the old population. The selection of chromosomes into the reproduction is quasirandom, we use linear rank selection, where the probability of a selection is proportional to a rank in a sequence of chromosomes arranged according to increasing fitness. In a reproduction process two new chromosomes are generated from two parental chromosomes, altruistic gene is copied, similarity gene is determined by one of parental chromosomes, which is declared as a "mother", and new fitness is created from arithmetic mean of parental fitness. The first two entries of chromosomes can be modified with low probability by a mutation, the last entry is modified by a random number with normal distribution. The initial population is generated randomly, with prescribed initial frequency of an altruistic alleles in a population. Altruistic process is carried out between chromosome-donor (which has an altruistic allele) and a randomly chosen chromosome-recipient. A recipient's fitness is increased proportionally to its similarity to the donor, similarly to the decrease of fitness of the donor.

2. Specification of chromosomes

Chromosomes are vectors composed of three entries of different kinds (see Fig. 1)

$$\alpha = \left(\alpha_a, \alpha_s, \alpha_f\right) \in A \tag{1a}$$

$$A = \{0,1\} \times \{1,2,\ldots,m\} \times [0,\infty) \tag{1b}$$

where single entries are determined as follows:

(1) α_a is a 1-bit entry that specifies whether the chromosome α is an altruist (α_a=1) or not (α_a=0).

(2) α_s is an integer entry corresponding to a "family class" of the chromosome α, we say that the chromosome α belongs to the α_s-th class (where the number of these

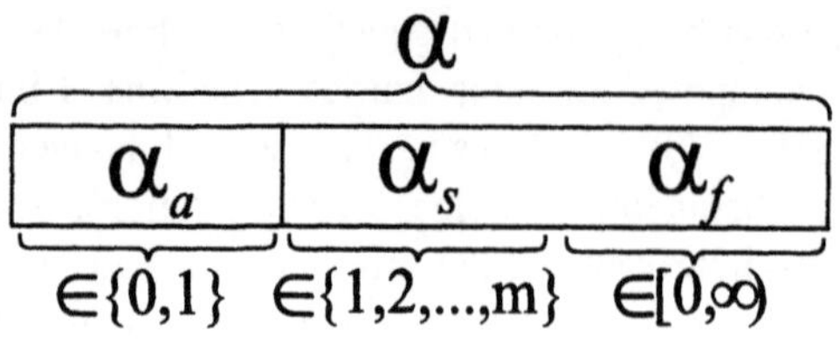

Figure 1. A schematic outline of the chromosome. It is composed of three parts. The first 1-bit part specifies altruistic (α_a=1) or notaltruistic (α_a=0) behavior, the second integer part specifies a similarity class, and the third nonnegative real part corresponds to the chromosome fitness.

classes is *m*). Moreover, for two chromosomes α and β a sum of differences $|\alpha_a - \beta_a| + |\alpha_s - \beta_s|$ represents a measure of similarity (or dissimilarity) between them. If this difference is smaller, then chromosomes are more similar, and in reverse, if this difference is greater, then chromosomes are more dissimilar.

(3) α_f is a nonnegative real entry assigned to a fitness of the chromosome α. The fitness will serve in our forthcoming considerations as a measure of success how the chromosome α takes a part in a reproduction process. Loosely speaking, a greater fitness implies that the chromosome has a greater probability to participate in a reproduction process.

For a given chromosome α its entries can be explicitly given by three functions

$$a : A \to \{0,1\}, \quad \alpha_a = a(\alpha) \tag{2a}$$

$$s : A \to \{1,2,\ldots,m\}, \quad \alpha_s = s(\alpha) \tag{2b}$$

$$f : A \to [0,\infty), \quad \alpha_f = f(\alpha) \tag{2c}$$

3. Population

A population is a multiset composed of a finite number of chromosomes

$$P = \{\alpha, \beta, \alpha', \beta', \ldots\} \tag{3}$$

Formally, we may say that the population is a subset of the set A, $P \subset A$. We postulate that the cardinality of the population *P* (i.e. the number of its elements - chromosomes) determined by an integer *p*, |*P*|=*p*, is always kept fixed.

4. Selection of chromosomes

A quasirandom selection of a chromosome $\alpha \in P$ is specified by a probability that is determined by a rescaled fitness

$$\text{prob}_{qr}(\alpha) = \frac{F(f(\alpha))}{\sum_{\beta \in P} F(f(\beta))} \tag{4a}$$

where F is a mapping that "rescales" chromosome fitness so that following condition is fulfilled (see Fig. 2)

$$f(\alpha) < f(\beta) \Rightarrow F(f(\alpha)) < F(f(\beta)) \tag{4b}$$

In general, this rescaling of chromosome fitness is worthwhile to use when there exist differences of many orders between chromosomes from the same populations. If this rescaling is introduced so that all rescaled fitness are not very different, e.g. if they are from the open interval (0,1), then we give a chance to be quasirandomly selected also to chromosomes that initially have fitness about few orders smaller than a best chromosome from the population. A quasirandom selection can be expressed as an application of stochastic operator onto the whole population P

$$\alpha = O_{qr,sel}(P) \tag{5}$$

This operator selects stochastically from the population P a chromosome α with probability specified by (4a). Although in further described calculations the difference of many orders in fitness is unlikely, we use rescaling function

$$F(f(\alpha)) = MIN + \frac{(MAX - MIN) * (rank(f(\alpha)) - 1)}{p - 1} \tag{6}$$

where MIN is a small number (in our calculations equal to $1/p$), $MAX=1$, p is a cardinality of a population P, and $rank(f(\alpha))$ is an index number of a place of $f(\alpha)$ in a sequence of fitness of all chromosomes arranged in increasing order.

Fully random selection of a chromosome $\alpha \in P$ is specified by a probability

$$\text{prob}_{fr}(\alpha) = \frac{1}{|P|} \tag{7}$$

This means that in the framework of this type of selection a chromosome α is selected with a probability specified as an inverse value of cardinality of the population P. In a similar way as above, we introduce a stochastic operator that selects a chromosome from the population P or from a subpopulation consisting of all altruistic chromosomes P_{altr}.

$$\alpha = O_{fr,sel}(P),\ \alpha = O_{fr,sel}(P_{altr}) \tag{8}$$

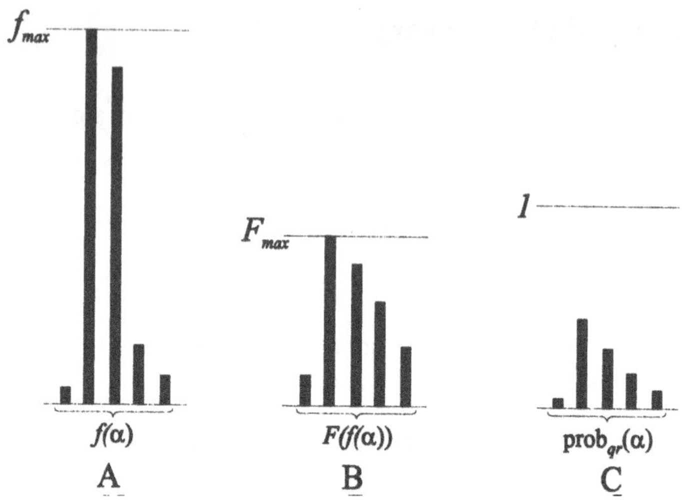

Figure 2. A rescaling of the fitness of chromosomes from the population *P*. Diagram A shows fitness of chromosomes so that it is displayed that there may exists a case when a minimal fitness is of many order smaller than a maximal fitness. The process of rescaling of fitness is outlined in diagram B, where it is possible to see that new rescaled fitness is gradually uniformly decreasing so that the condition (4b) is fulfilled. Last diagram C corresponds to probability of selections of chromosomes from the same population *P*, where these probabilities are normalized so that their summation is equal to one.

5. Reproduction process

In general, the term reproduction means that there exists a prescription how to assign to a pair of chromosomes a new pair of chromosomes

$$O_{repro} : A \times A \to A \times A \tag{9}$$

In order to specify this stochastic operator we have to introduce two different kinds of stochastic operators, in particular, mutation and crossover operator.

A crossover operator $O_{cross} : A \times A \to A \times A$ assigns stochastically to a pair of chromosomes $\alpha, \beta \in A$ another pair of chromosomes $\tilde{\alpha}, \tilde{\beta} \in A$, explicitly

$$\left(\tilde{\alpha}, \tilde{\beta}\right) = O_{cross}(\alpha, \beta) \tag{10}$$

Let us $\alpha = \left(\alpha_a, \alpha_s, \alpha_f\right)$, $\beta = \left(\beta_a, \beta_s, \beta_f\right)$ and $\tilde{\alpha} = \left(\tilde{\alpha}_a, \tilde{\alpha}_s, \tilde{\alpha}_f\right)$, $\tilde{\beta} = \left(\tilde{\beta}_a, \tilde{\beta}_s, \tilde{\beta}_f\right)$ be chromosomes from (10), they are called parents and offspring, respectively. Single entries of offspring chromosomes are determined as follows: (i) altruistic entries remain unchanged, (ii) similarity entries of offspring are determined by the parent - mother similarity entry of chromosome α, and (iii) fitness entries are determined as arithmetic means of parent values.

$$\tilde{\alpha}_a = \alpha_a, \quad \tilde{\beta}_a = \beta_a \tag{11a}$$

$$\tilde{\alpha}_s = \alpha_s, \quad \tilde{\beta}_s = \alpha_s \tag{11b}$$

$$\tilde{\alpha}_f = \frac{1}{2}\left(\alpha_f + \beta_f\right), \quad \tilde{\beta}_f = \frac{1}{2}\left(\alpha_f + \beta_f\right) \tag{11c}$$

The fitness entries have as a lower bound zero, but no upper bound is set, so that during evolution a medium fitness increases and its modifications by a random number has less and less effect. This change a very gradual one and should also correspond to evolution in nature, where in later stages of evolution of species usually no big advantageous mutations appear.

A mutation operator $O_{mut} : A \rightarrow A$ assigns stochastically to each chromosome $\tilde{\alpha} \in A$ another chromosome $\alpha' \in A$, explicitly

$$\alpha' = O_{mut}\left(\tilde{\alpha}\right) \tag{12}$$

It affects all three parts of the chromosome α. Let $\tilde{\alpha} = \left(\tilde{\alpha}_a, \tilde{\alpha}_s, \tilde{\alpha}_f\right)$ be one of two chromosomes resulting from (11) and $\alpha' = \left(\alpha'_a, \alpha'_s, \alpha'_f\right)$ be a resulting chromosome, then

$$\alpha'_a = \begin{cases} 1 - \tilde{\alpha}_a & (\textit{if}\ \text{random} < P_{mut}) \\ \tilde{\alpha}_a & (\textit{otherwise}) \end{cases} \tag{13a}$$

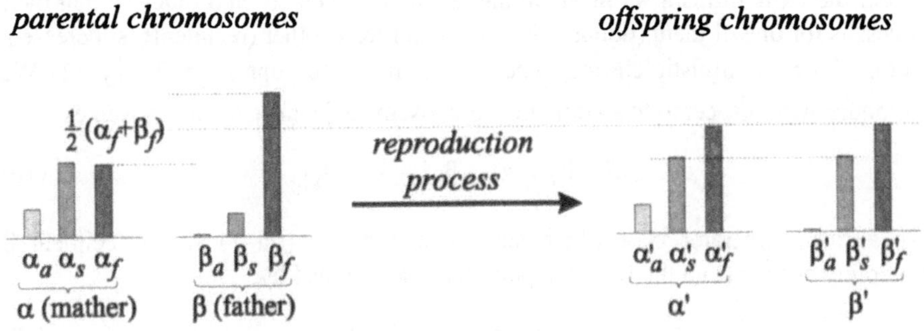

Figure 3. A reproduction process consists in a construction of offspring chromosomes from parental chromosomes, for simplicity mutations are here ignored. Altruistic parts of chromosomes are not changed in the process of reproduction, whereas similarity entries of offspring chromosomes are fully determined by the similarity group of the mother chromosome and fitness of offspring chromosomes are determined as an arithmetic mean of fitness of parental chromosomes.

$$\alpha'_s = \begin{cases} (1+\text{random}(m)) \neq \tilde{\alpha}_s & (\textit{if } \text{random} < P_{mut}) \\ \tilde{\alpha}_s & (\textit{otherwise}) \end{cases} \tag{13b}$$

$$\alpha'_f = \tilde{\alpha}_f + r(0, \sigma_{mut}) \tag{13c}$$

A parameter P_{mut} corresponds to a mutation probability (usually small positive number, e.g. $P_{mut}=10^{-2}$). A fitness entry is modified by random number $r(0,\sigma_{mut})$ (random number with normal distribution, zero mean, and standard deviation σ_{mut}).

Now we are ready to specify the reproduction operator O_{repro} by (14), (see Fig. 3)

$$(\alpha', \beta') = O_{repro}(\alpha, \beta) \tag{14}$$

Let us introduce intermediate chromosomes $\tilde{\alpha}$ and $\tilde{\beta}$ that are specified by

$$(\tilde{\alpha}, \tilde{\beta}) = \begin{cases} O_{cross}(\alpha, \beta) & (\textit{if } \text{random} < P_{cross}) \\ (\alpha, \beta) & (\textit{otherwise}) \end{cases} \tag{15a}$$

Then offspring are determined as results of application of mutation operators to these two intermediate chromosomes

$$\alpha' = O_{mut}(\tilde{\alpha}), \quad \beta' = O_{mut}(\tilde{\beta}) \tag{15b}$$

6. Altruism

Under the term altruism we mean an interaction between chromosomes so that their fitness is for one of them (donor) decreased and for another (recipient) is increased. Let $\alpha \in P$ be an altruistic chromosome - donor from the population P, $\alpha_a = 1$. We introduce with respect to this altruistic chromosome a distance of chromosomes

$$d(\alpha, \beta) = |\alpha_a - \beta_a| + |\alpha_s - \beta_s| \tag{16}$$

An altruistic operator Oalt which acts on a donor α (αa=1) and a recipient β randomly chosen from the whole population is determined by

$$(\alpha', \beta') = O_{alt}(\alpha, \beta) \tag{17}$$

where

$$\alpha'_a = \alpha_a, \quad \alpha'_s = \alpha_s, \quad \alpha'_f = \alpha_f - e^{-\xi \cdot d(\alpha,\beta)} |r(0, \sigma_{alt})| \tag{18a}$$

$$\beta'_a = \beta_a, \quad \beta'_s = \beta_s, \quad \beta'_f = \beta_f + e^{-\xi \cdot d(\alpha,\beta)} |r(0, \sigma_{alt})| \tag{18b}$$

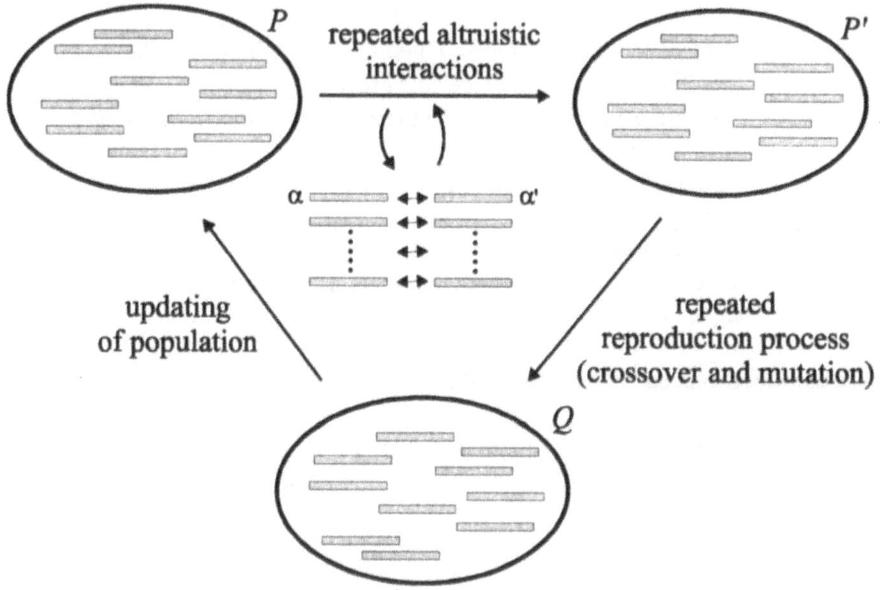

Figure 4. A diagrammatic visualization of genetic algorithm with altruism. Whole algorithm is composed of two steps that are repeated, reproduction with replacing a parental population by a new one and altruism.

where ξ is a real positive number and $r(0,\sigma_{alt})$ is a random number with a normal distribution and a standard deviation σ_{alt}. The same random number is used for both donor in equation (18a) and recipient in equation (18b). The parameter ξ controls the "strength" of relatedness relevant for an amount of fitness donation. If this parameter is small (e.g. ξ=0.01), than an altruistic process is applied strongly even for recipients, which are in relatedness very far from donor, in the other case, if this constant is sufficiently big (e.g. ξ=15.0), than an altruistic transfer of fitness is realized substantially only for closely related couple donor - recipient. We see that a fitness of the donor is randomly decreased while fitness of recipient chromosomes is increased by the same value, all other donor and recipient entries remain unchanged. The "canonical" genetic algorithm is modified in such a way, that an operation of altruism is included before a reproduction process. This operation is repeated for prescribed number of times for a randomly chosen donor with an altruistic chromosome and for totally randomly chosen recipient.

7. GA with altruism

A genetic algorithm with altruism was applied with the following parameters:

t_{max} ....maximal number of generations
P_{mut} .... probability of mutation, P_{mut}=0.01 (see eqs. (13a,b))

P_{cross} ....probability of crossover, P_{cross}=0.1 (see eq. (15a))
σ_{cross} ....standard deviation of random modifications of fitness in the reproduction, σ_{cross}=0.1 (see eq. (11c))
σ_{alt} ....standard deviation of random modifications of fitness in the altruism, σ_{alt}=0.1 (see eqs. (18a,b))
p ....cardinality of the population P, $|P|$=1000
m ....number of chromosome classes, m=10 (see eqs. (2b, 11b, 13b, 16))
u ....number of altruistic acts, u=100
ξ................. a measure of similarity between donor α and recipient β relevant for an amount of fitness donation (see eqs. (18a,b))

The pseudoPascal code of the genetic algorithm with altruism is:

```
P:=a multiset of p randomly generated chromosomes;
t:=0;
while t<t_max do begin
  t:=t+1;
  for i:=1 to u do
  begin α:=O_fr,select(P); β:=O_fr,select(P_altr);
        (α',β'):=O_alt(α,β); P:=P-({α}+{β})+({α'}+{β'});
  end;
  Q:=∅;
  while |Q|<p do begin x:=O_qr,select(P); y:=O_qr,select(P);
                        (x',y'):=O_repro(x,y); Q:=Q+{x',y'};
                 end;
  P:=Q;
end;
```

Typical results of the algorithm are displayed in figures 5, 6, and 7. Figure 5 shows, that for small values of the parameter ξ (ξ=0.001) the altruistic chromosomes are not asymptotically stable, they become extinct during the evolution. For a higher value of the parameter ξ (ξ=1.0) we get asymptotically evolutionary stable altruistic chromosomes, they do not die out. For intermediate levels of the parameter ξ (ξ=0.01-0.1) there is essential the initial frequency f_a of altruistic chromosomes. For small initial frequencies they die out, for high frequencies they occupy the whole population, and for medium frequencies the behavior is random at the beginning, before the altruism either occupies the whole space or becomes extinct. There does not seem to be any case of evolutionary stable intermediate level of altruism in the presented model.

Figures 6 and 7 present the evolution of numbers of chromosomes of 10 groups. Figure 6 shows, that the variation in population cease within 50 epochs, one group prevails, so that more than 90% of chromosomes is affiliated to it. Though the groups are not totally uniform for application of altruism (groups 1 and 10 have neighbors/closest relatives only on one "side"), it does not have a visible effect on the choice of the prevailing group. If altruistic chromosomes become extinct, the distribution of chromosomes in groups is not totally stable, figure 7 shows it for

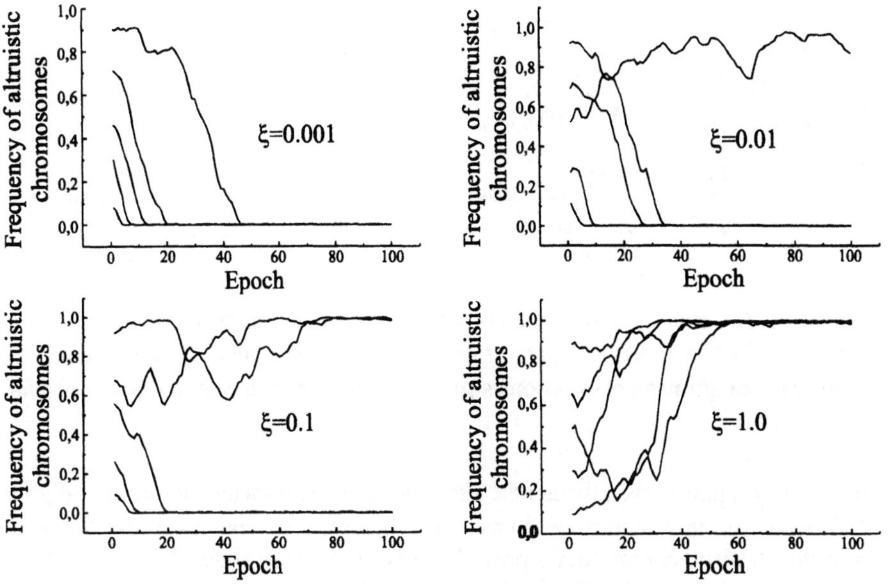

Figure 5. An evolution of altruism in a population for values of a parameter ξ=0.001, 0.01, 0.3, and 1.0, and for various initial frequencies of presence of altruistic chromosomes (α_a=1) with f_a=0.1, 0.3, 0.5, 0.7 a 0.9 (single curves are determined by their initial values).

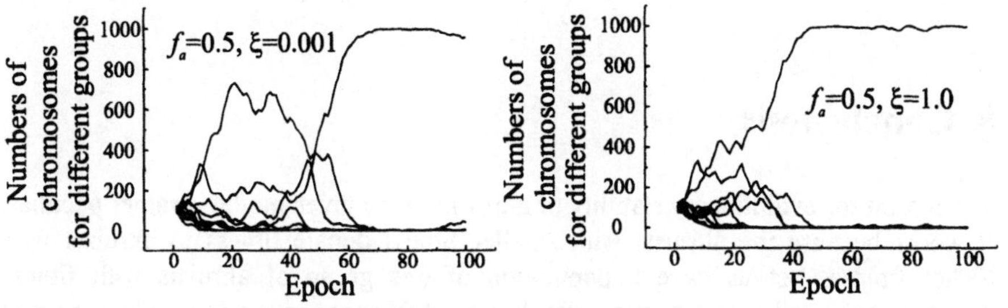

Figure 6. An evolution of numbers of chromosomes in different groups in a population for two values of a parameter ξ=0.001, and 1.0, and for initial frequency of presence of altruistic chromosomes (α_a=1) f_a=0.5, for 100 epochs.

ξ=0.001, where number of chromosomes of another group sometimes rises so high, that the previously prevailing group is replaced, even though most of the epochs only one of groups is prevailing. If altruistic chromosomes prevail, the distribution of chromosomes in groups is practically frozen, figure 7 shows it for ξ=1.0, where after one of the groups prevails, other groups do not rise more than 10 per cent to threat its "winning" position.

Altruistic chromosomes give greater amount of fitness to altruistic than to nonaltruistic chromosomes, which is caused by the term $|\alpha_a - \beta_a|$ in equation (16). Removing this term has an interesting results. The results for 100 epochs look

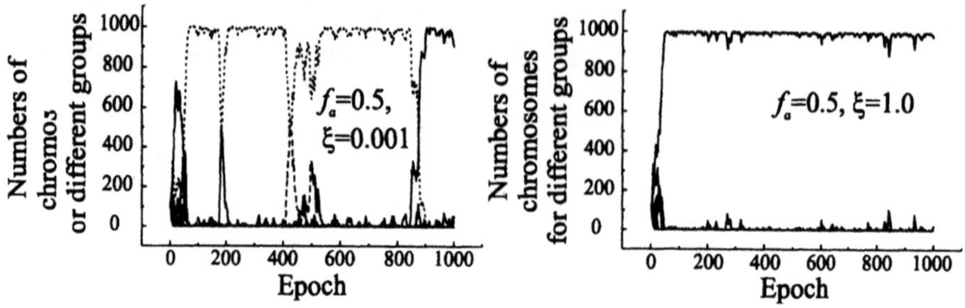

Figure 7. An evolution of numbers of chromosomes in different groups in a population for two values of a parameter ξ=0.001, and 1.0, and for initial frequency of presence of altruistic chromosomes (α_a=1) f_a=0.5, the same as in Fig. 6, only for 1000 epochs.

similar as in figure 5, even though the altruistic chromosomes tend to prevail only for $\xi \geq 10$. However, this dominance is unstable. After about several hundreds of epochs, when the altruistic chromosomes prevail nearly 100%, they suddenly die out. It can be related to the fact, that though in the beginning the altruistic chromosomes are preferable, because they help their own group to become more fit. When only one group prevails, this reason for existence of altruism terminates, so altruistic chromosomes die out. If altruistic chromosomes help to each other more, they actually divide the group into two, altruistic and a nonaltruistic one, in which competition the altruistic group wins.

8. Conclusions

The reason for evolutionary stability of altruism in the given model is rather peculiar. It works because the altruists with smaller fitness donate fitness to altruists with higher fitness. Let us have a population of one group of altruists with fitness distributed according to Gaussian distribution. After application of altruism the plot of the density of distribution has a lower peak with higher values at positions far from mean, this would correspond to a bigger standard deviation. However, the mean remains the same. In case of mixed population of altruists and nonaltruists the sum of probabilities of selection of altruistic chromosomes to reproduction after application of altruism is even smaller than before its application, because part of fitness was given to nonaltruists. However, if the sum of probabilities of selection is calculated to the third generation, selection of altruism is more probable, because probabilities for selection are squared before normalization. This means, that chromosomes with higher fitness gain more probability than those with lower fitness loose. The dependence of probability of selection plotted against rank changes from linear to quadratic (parabola) to cubic, for second, third, and fourth generations respectively. On the other hand, in case of a selection, where the dependence of probability of selection into the next generation plotted against rank would be highly nonlinear (concave) and increasing, the altruism could work only if the altruists with higher fitness would selectively donate mainly to altruists with lower fitness.

Evolution of altruism is sensitive to primarily to the parameter ξ and rather less to an initial frequency of presence of altruistic chromosomes. In a case of a small value of ξ chromosomes in a population, altruistic chromosomes become extinct during the evolution. Presence of altruistic chromosomes is asymptotically stable (i.e. they do not become extinct) only if a parameter ξ is big (altruism is realized practically only between chromosomes belonging to the same or closely related group). Only for medium levels of a parameter ξ the initial frequency of an altruism in the population controls its extinction or prevalence.

An existence of the altruism in a population also depends strongly on clustering of population into subgroups of similar chromosomes. If the population is not divided into subgroups, altruistic alleles practically become extinct during evolution. In the opposite case, when both a parameter ξ and an initial frequency of presence of altruistic chromosomes are high, altruistic behavior becomes evolutionary stable, or at least metastable.

On the bases of the above stated results one can speculate about emergence of an altruism in a population. The results suggest, that an altruistic behavior in the initial stage was created by a random genetic drift [18] in a subgroup. For sustaining its presence it is plausible to assume, that the population was small and the reproduction process was carried out among close relatives. With increase of the population and the number of subgroup members the altruism had a chance to remain in the population. By that way an altruism could establish itself even for larger populations, where it would not have a chance to survive because of its low initial frequency. These results are in accord with current conjectures about possibility of evolution of altruism only in highly structured populations clustered into subpopulations of related individuals.

The applicability of altruism can be further studied in models in economy, when the transfer of fitness can be easily translated into transfer of resources. Another application can be found for global optimization, appended by a local optimization. As an example could serve optimization of neural networks by genetic algorithms, followed by some kind of local optimization, like classical backpropagation, or something closer to hillclimbing. The transfer of fitness in such a cases could be replaced by a transfer of computational resources allocated to local optimization of an individual, like giving one individual's optimization more processor time at the expense of the other individual's time. For artificial agents gathering some resources this transfer could be even more straightforward. Optimization with altruism would probably intensify local optimization at the expense of global optimization. An altruism could also prove efficient, when a solution emerges from cooperation of more individuals, like in optimization by a kind of ant colony.

Acknowledgments

This work was supported by the grants # 1/4209/97 and # 1/5229/98 of the Scientific Grant Agency of Slovak Republic.

References

1. Dawkins R 1976 *The Selfish Gene*. Oxford University Press, New York

2. Heylighen F 1992 Evolution, selfishness and cooperation; selfish memes and the evolution of cooperation. *Journal of Ideas* 2 (4):70-84

3. Wilson E O 1975 *Sociobiology: The New Synthesis*. Harvard University Press, Cambridge

4. Hamilton W D 1964 The genetical evolution of social behavior I, II. *Journal of Theoretical Biology* 7:1-16, 17-52

5. Taylor P D, Frank S A 1996 How to make a kin selection model. *Journal of Theoretical Biology* 180 (1):27-37

6. Taylor P D 1996 Inclusive fitness arguments in genetic models of behaviour. *Journal of Mathematical Biology* 34:654-674

7. Maynard Smith J 1982 Optimization theory in evolution. *Annual Review of Ecology and Systematics* 9:31-56

8. Trivers R 1985 *Social Evolution*. Benjamin/Cummings, Menlo Park, CA

9. Becker G S 1976 Altruism, egoism, and genetic fitness: economics and sociobiology. *Journal of Economic Literature* 14 (3):817-826

10. Schelling T C 1987 Altruism, meanness, and other potentially strategic behaviors. *American Economic Review* 68 (2):229-230

11. Bergstrom T C, Oded S 1993 How altruism can prevail in an evolutionary environment. *American Economic Review* 83 (2):149-155

12. McConnell K E 1997 Does altruism undermine existence value? *Journal of Environmental Economics and Management* 32 (1):22-37

13. Axelrod R 1984 *The Evolution of Cooperation*. Basic Books, New York

14. Batali J, Kitcher P 1995 Evolution of altruism in optional and compulsory games. *Journal of Theoretical Biology* 175 (2):161-171

15. Bester H, Gueth W 1994 *Is Altruism Evolutionarily Stable*? Tilburg University, Center for Economic Research, Discussion Paper, http://greywww.kub.nl:2080/greyfiles/center/1994/doc/103.ps

16. Goodnight C J, Schwartz J. M, Stewens L 1992 Contextual analysis of models of group selection, soft selection, hard selection, and the evolution of altruism. *American Naturalist* 140:743-761

17. Trivers R L 1971 The evolution of reciprocal altruism. *Quarterly Review of Biology* 46 (4):35-57

18. Moran L 1997 *Random Genetic Drift*. Discussion Paper, http://www.talkorigins.org/faqs/genetic-drift.html

The Bivariate Marginal Distribution Algorithm

Martin Pelikan[1], **Heinz Muehlenbein**[2]

[1] *Dept. of Mathematics, Slovak Technical University, 81237 Bratislava, Slovakia, email: pelikan@svf.stuba.sk*
[2] *GMD Forschungszentrum Informationstechnik, D-53754 Sankt Augustin, Germany, email: muehlenbein@gmd.de*

Keywords: evolutionary algorithm, marginal distribution, dependency graph, decomposable problems

Abstract

The paper deals with the Bivariate Marginal Distribution Algorithm (BMDA). BMDA is an extension of the Univariate Marginal Distribution Algorithm (UMDA). It uses the pair gene dependencies in order to improve algorithms that use simple univariate marginal distributions. BMDA is a special case of the Factorization Distribution Algorithm, but without any problem specific knowledge in the initial stage. The dependencies are being discovered during the optimization process itself. In this paper BMDA is described in detail. BMDA is compared to different algorithms including the simple genetic algorithm with different crossover methods and UMDA. For some fitness functions the relation between problem size and the number of fitness evaluations until convergence is shown.

1. Introduction

Genetic algorithms work with populations of strings of fixed length. In this paper binary strings will be considered. From current population better strings are selected at the expense of worse ones. New strings are generated using the recombination/crossover operator and mutation. The recombination operator combines the information contained in two strings. Mutation performs a small perturbation to these strings in order to keep the population diverse and to introduce new information. The theory behind these operators is based on the schema theorem [1]. It is known that the operators often cause the disruption of schemata of large defining length. This may cause bad performance of genetic algorithms on problems where these schemata are needed to obtain the optimum. This has lead to new approaches for doing recombination. The first line of research uses different reordering methods and other methods that decrease the defining length of important schemata. In these methods, the values of bits on different positions are not the only thing that is optimized as it was with classical genetic algorithm. The order of bits and other features are optimized as well. The methods work quite well for decomposable problems although they require some prior knowledge about the problem and they are usually very memory and time consuming. This direction has lead to the GEMGA (Gene Expression Messy Genetic Algorithm) [2].

The second line of research is based on the estimation of probability distribution. The simplest way is to estimate the distribution using univariate marginal frequencies in the set of selected parents. This is what UMDA (Univariate Marginal Distribution Algorithm) [3] is based on. This algorithm works perfect for linear problems as is shown in [3]. But it also works very well for problems that don't contain significant dependencies. Performance of UMDA can be estimated with the use of the variance decomposition [3]. In general, the greater the rate of additive variance to the sum of higher order variances, the better UMDA performs.

For problems with dependencies among different genes this approach is not sufficient. The theory of UMDA has been extended to problems where the probability model is known. In FDA (Factorization Distribution Algorithm) it is assumed that the probability can be written as some product of marginal frequencies [4]. This is an ideal schema algorithm, because no important schemata can be disrupted. A more pragmatic way to extend UMDA is to use bivariate marginal distributions.

In the MIMIC algorithm [5] the distribution is assumed to be a simple chain-like product. All bits are ordered into a chain using a simple greedy algorithm first taking as input the univariate and bivariate marginal frequencies in the set of selected parents. For each new individual, the first bit is generated using its univariate marginal frequency. All other bits are generated using the conditional probability given the previous bit. Another approach, presented in [6], uses a tree structure for the probability model. The tree is constructed to maximize the sum of the so-called mutual information of genes that are connected. The value for the bit corresponding to the root of this tree is generated using its univariate frequency. The remaining bits are generated using conditional probability given the value of their parent in the dependency tree. Both univariate and bivariate frequencies are calculated incrementally. In our approach we combine these two methods. The mutual information dependency measure is replaced by Pearson's chi-square statistics, in order to be able to identify pairs that are independent with a certain probability. These pairs are mapped into a dependency graph. The algorithm will be precisely described in the next sections. Even though the algorithm and all terms are defined for chromosomes defined as binary strings of fixed length, these can be easily reformulated for any finite alphabet. The first position in a string is referred as the 0th position in this paper. This makes the equations simpler.

2. Marginal Frequencies and Pearson's statistics

Let us denote the length of chromosome as n. Let P be a set of binary strings of length n (the population). The size of P will be denoted as N. For each position $i \in \{0, \ldots, n-1\}$ and each possible value on this position $x_i \in \{0, 1\}$, we define the univariate marginal frequency $p_i(x_i)$ for set P as the frequency of strings that have x_i on ith position in set P. Similarly, for any two positions $i \neq j \in \{0, \ldots, n-1\}$ and any possible values on these positions $x_i, x_j \in \{0, 1\}$, we define the bivariate marginal frequency $p_{i,j}(x_i, x_j)$ for set P as the frequency of strings that have x_i and x_j on positions i and j, respectively. Sometimes the term of probability will be used instead of frequency. With the use of univariate and bivariate marginal frequencies, the conditional probability of appearance of the value x_i on ith position having given the value x_j on jth position can be calculated as

$$p_{i,j}(x_i|x_j) = \frac{p_{i,j}(x_i, x_j)}{p_j(x_j)} \tag{1}$$

Pearson's chi-square statistics [7] is defined by

$$X^2 = \sum \frac{(observed - expected)^2}{expected} \tag{2}$$

Here, for each pair of positions, the observed quantity is the number of possible pairs of values on these positions. If these two positions were independent, the number for each of these pairs of values could be easily calculated using the basic probability theory. This is the expected quantity. Then, in terms of univariate and bivariate frequencies and the total number of points taken into account, for positions $i \neq j$, we get [7]

$$X_{i,j}^2 = \sum_{x_i, x_j} \frac{\left(Np_{i,j}(x_i, x_j) - Np_i(x_i)p_j(x_j)\right)^2}{Np_i(x_i)p_j(x_j)} \tag{3}$$

If positions i and j are independent for 95%, then for Pearson's chi-square statistics following inequality holds [7]

$$X_{i,j}^2 < 3.84 \tag{4}$$

3. The Construction of a Dependency Graph

In this section the construction of a dependency graph will be described. The graph will be defined by three sets, V, E, and R, i.e. $G = (V, E, R)$. V is the set of vertices, $E \subset V \times V$ is the set of edges and R is a set containing one vertex from each of the connected components of G. In a dependency graph each node corresponds to a position in a string. There is one to one correspondence between the vertices and positions in a string. Thus, we can use the set of vertices $V = \{0, \ldots, n-1\}$, where vertex i corresponds to the

ith position. As it will be clear from the construction of a dependency graph, the graph does not have be connected. That means that it does not have to form a tree. The dependency graph is always acyclic. It can be seen as the set of trees that are not mutually connected. The generation of new strings does not depend on the number of connected components of the dependency graph. When talking about frequencies in this section, we always mean the set which is used for creation of a dependency graph. In the description of the algorithm it is always said which one is meant.

Let us denote A the set of vertices that have not been processed yet. At the start of the algorithm A is set to V. Then, successively, as edges are being added into E, A gets smaller. The algorithm ends up with A equal to an empty set what means that all vertices have been processed. Another set is denoted D. It is the set of all pairs from $V \times V$ that are not independent for 95% (see Equations 3 and 4), i.e.

$$D = \{(i,j)|i \neq j \in \{0,\ldots,n-1\} \wedge X^2_{i,j} \geq 3.84\} \tag{5}$$

Then, the algorithm for the construction of a dependency tree is defined as follows

Algorithm for the Construction of a Dependency Graph

1. set $V \leftarrow \{0,\ldots,n-1\}$
   set $A \leftarrow V$
   set $E \leftarrow \emptyset$
2. $v \leftarrow$ any vertex from A
   add v into R
3. remove v from A
4. if there are no more vertices in set A, finish
5. if in D there are no more dependencies of any v and v' where $v \in A$ and $v' \in V \setminus A$, go to 2
6. set v to the vertex from A that maximizes $X^2_{v,v'}$ over all $v \in A$ and $v' \in V \setminus A$
7. add edge (v, v') into the set of edges E
8. go to 3

The basic idea of the algorithm is very simple. It is similar to the well-known algorithm for obtaining the largest spanning tree. First, an arbitrary vertex is added to the graph. Then, the vertex with the greatest dependency with some of previously added vertices and the edge corresponding to this dependency are added to the graph. The last step is repeated until there is no dependency between already added vertices and the rest. If this is the case, an arbitrary vertex is added into graph and the process repeats. The whole process repeats until all vertices are added into the graph. The effect is that an acyclic graph with a maximal sum of chi-square statistics values over the connected vertices is constructed. Resulting graph does not have to be connected, as it was already mentioned above. As the first vertex from each component (that is successively created by adding vertices according to the dependencies) is added into the graph, it is added also into the set R (i.e., to the set of special vertices, one for each component of the resulting graph).

Since the set A is initialized to a finite set of vertices and in each cycle at least one vertex is removed from it, the algorithm does always finish and therefore it is well-defined. The time complexity of the described algorithm is $O(n^3)$.

4. Generation of New Individuals

To generate new individuals, a previously described dependency graph $G = (V, E, R)$ is used. For each individual the values for positions contained in R are generated by the univariate marginal frequencies. Then, if there exist a position v that is yet not generated and it is connected to some already generated position v' (according to the set of edges E), it is generated using the conditional probability (see Equation 1) for a position v having given the value on a position v'. The last step is repeated until values for all positions are generated.

In the following description of the algorithm for the generation of a new individual, one important set, among sets defining graph G, appears. It is denoted as K and it stands for the set of all positions that have been already generated. The individual is a string of length n and will be denoted by x. Its ith bit will be denoted as x_i.

Algorithm for the Generation of a New Individual

1. set $K \leftarrow V$
2. generate x_r for all $r \in R$ using univariate frequencies, i.e. set it to the value a with probability $p_r(a)$
   set $K \leftarrow K \setminus R$
3. if K is already empty, finish
4. choose k from K such that there exist k' from $V \setminus K$ connected to k in the graph G
5. generate x_k using conditional probability having given value for $x_{k'}$, i.e. set it to value a with probability $p_{k,k'}(a|x_{k'})$
6. remove k from the set K
7. go to 4

The set K is initialized as a finite set and in each cycle one vertex is removed from it. For each connected component, at least one vertex is generated first. The algorithm is therefore well-defined. The generation of one individual can be done in $O(n)$ steps. The generation of different individuals is independent. The algorithm for generation of new individuals is therefore well suited for parallelization.

5. Bivariate Marginal Distribution Algorithm

Having defined the algorithms for the construction of a dependency graph and the generation of new individuals, the bivariate marginal distribution algorithm (BMDA) can be described. In BMDA, the population is randomly generated first. From this population, the better individuals are selected. Univariate and bivariate marginal frequencies for these individuals are then calculated. Using these frequencies, a dependency graph is constructed as it is described in Section 3. Having given the dependency graph, new individuals are generated as described in Section 4. New individuals are then added into the old population from which the individuals were originally selected. They replace some of the old ones, usually the worst of them, so that the number of individuals in the population remains constant. From the new population, individuals are selected again. The process, starting off with the selection of better individuals and ending with adding the new individuals into the old population, repeats until the termination criteria are met. The termination criteria can cause the algorithm to stop if it has already found the optimum or the diversity of population is too low. The value of the optimum is usually unknown by the breeder. That is why the second condition is the more important one. When the diversity is too low almost all individuals in the population are the same. That means that there is not enough information in the population to create new individuals that would fit the problem better than already found ones.

Bivariate Marginal Distribution Algorithm

1. set $t \leftarrow 0$
   randomly generate initial population $P(0)$
2. select parents $S(t)$ from $P(t)$
   calculate univariate frequencies p_i and bivariate frequencies $p_{i,j}$ for the selected set $S(t)$
3. create dependency graph $G = (V, E, R)$ using the frequencies p_i and $p_{i,j}$
4. generate the set of new individuals $O(t)$ using dependency graph G and frequencies p_i and $p_{i,j}$
5. replace some of individuals from $P(t)$ with new individuals $O(t)$
   set $t \leftarrow t + 1$
6. if termination criteria are not met, go to 2

The termination criterion due to the lack of diversity is defined as follows: if all univariate frequencies are closer than $\epsilon > 0$ to 0 or 1, the algorithm is terminated. If this is the case, we say the algorithm ϵ-converged. In most of our experiments, we use this termination criterion.

The probability model used by BMDA is given by

$$p(x) = \prod_{r \in R} p_r(x_r) \prod_{i \in V \setminus R} p_{i,e(i)}(x_i | x_{e(i)}) \tag{6}$$

where $e(i)$ returns the vertex connected to the vertex i but added sooner than this vertex using the algorithm for the construction of a dependency graph from Section 3. To say it in another way, $e(i)$ is the vertex that is the next one on the way from ith vertex to the $r \in R$ that corresponds to the component where the vertex i is located. This factorization is a special case of the factorization considered in [4].

6. How Does It Work and Why?

The distribution of BMDA is based on the use of conditional probabilities for pairs of positions that seem to be dependent. This information is estimated from the current population of strings. Any binary function can be decomposed as follows [4]

$$f(x) = a + \sum_{i_1} a_{i_1} x_{i_1} + \sum_{i_1 < i_2} a_{i_1,i_2} x_{i_1} x_{i_2} + \ldots + a_{0,1,\ldots,n-1} x_0 x_1 \ldots x_{n-1} \quad (7)$$

A function is decomposable of order k if all coefficients of higher order are 0. Let us talk about decomposable functions of order at most two, first. For this class of functions, the best dependency graph can be constructed fairly easily. It can be done by connecting the vertices i and j just when the coefficient $a_{i,j}$ is not equal to zero. A graph constructed this way does not have to be acyclic. If this graph is acyclic or this can be achieved by deletion of only insignificant dependencies, then it serves as the best dependency graph that can be used for the generation of new individuals. If BMDA found this dependency graph and used it for generation of new individuals, it would perform very well. It would perform as well as UMDA does with linear functions. The performance of BMDA therefore relies on whether it is able to detect these dependencies having no problem-specific knowledge in an initial stage. In the empirical part of this paper it will be shown that most of the dependencies are usually found after only a few generations. If the mentioned graph is not acyclic and this property cannot be satisfied by deletion of significantly unimportant dependencies from it, the problem becomes impossible to solve within the use of this model. This problems could be possibly solved by the use of terms of conditional independence. Not only conditional probabilities for a position having given the value for another one but for a position having given values for a set of some of other positions as well would be taken into account. The problem with this approach is, as it was already mentioned above, that BMDA does not get any problem-specific knowledge about the problem that is solved. During optimization, it is learning the structure of the problem itself. From the beginning, the information about dependencies is very unclear and almost no pair of positions seems to be independent. The approximation gradually gets more and more accurate. There is no effective method for the prediction of the more complex model. Moreover, the fitness function does not have to be decomposable. A model that is simpler than the correct model may serve well in many cases. This can be demonstrated by the good performance of UMDA for many nonlinear problems. The right choice is somewhere between the accurate model and a simple model as univariate marginal distributions. The character of this class of algorithms does not allow very complex predictions of the model due to the amount of available information about the problem.

The problem arises when the optimized function is additively decomposable of order three or more. If this is the case, the given fitness distribution cannot be covered by the BMDA model. The best BMDA can do is to use the model that is as close as possible to the original distribution. The discovery of higher order dependencies is significantly more time and space complex and that is why it is hard to say if it was worthy to follow the way of enlarging the blocks taken into account. Moreover, there is again lack of information about the problem to predict the more complex models. If all the dependencies were known from the beginning, the best way would be to use them as in FDA [4]. Experiments show that BMDA performs well on problems of higher order dependencies as well. Higher order dependencies are usually substituted by chain-like dependency structures in our model.

7. Experiments

First experiments were done for a few different fitness functions. The choice of the fitness functions was done to make the behavior of the used algorithm more clear. Comparisons to some other methods were done. But the main purpose of this section is the explanation of the behavior of BMDA. The size of the paper does now allow us to take into account all other algorithms that could possibly compete with BMDA for solving these problems. For most of the problems the decomposition of the problem is shown too, to make things more clear.

In the experiments the so-called ordering parameter [8] is often used as a measure of convergence. The ordering parameter is defined as follows

$$\chi(p) = \frac{4}{n} \sum_{i=0}^{n-1} \left(p_i(1) - \frac{1}{2} \right)^2 \tag{8}$$

where p is the vector of univariate marginal frequencies $p_i(1)$. The closer the parameter is to one, the less diversity the population contains.

For some fitness functions a permutation of the variables will be allowed. The permutations will be denoted as π_k where $k \in \{1, \dots, n\}$. The permutation π_k is well-defined for any k such that n can be divided by k. It is defined as follows

$$\pi_k(i) = \left\lfloor \frac{n(i \bmod k) + i}{k} \right\rfloor \tag{9}$$

π_1 is the identity. For π_2 and $n = 12$ we get π_2 =(0, 2, 4, 6, 8, 10, 1, 3, 5, 7, 9, 11). Permutations will be used to change the order of the positions of a string to show the behavior of different algorithms with respect to the used permutation.

For all experiments a fixed selection method (Truncation selection [3] with $\tau = 50\%$) was used. The worse half of the old population was replaced by the new individuals.

7.1 Onemax fitness function

This fitness function is actually a simple linear function over the single bits with all coefficients equal to 1. That means it is just the sum of all bits in a string, i.e.

$$f_{onemax}(x) = \sum_{i=0}^{n-1} x_i \tag{10}$$

where x_i is the value on the ith position in string x. Onemax fitness function does not have a permutation as an input parameter because its value is the same for any permutation of bits in an input string. This function is decomposable of order one, as we can see from its basic form already. That means that BMDA should give good results for this function. As it was already shown in [3], UMDA works very well for this class of problems. For UMDA a very small population size is sufficient for its convergence into the optimum. BMDA uses a more complex distribution and to make this correct the dependencies that are present in the problem have to be discovered well. That is why the population size has to be enlarged. For each algorithm, the set of parameters is chosen to make it converge to the optimum in 100% of totally 30 independent runs. For all algorithms the ϵ-convergence termination criterion was used with $\epsilon = 0.05$.

As it is shown in Figures 1 and 2, BMDA converges slower than UMDA and the simple genetic algorithm with uniform crossover. This is caused by the use of statistical methods to discover dependencies that require to enlarge the population. UMDA takes the bits as independent from the beginning, so it uses the information BMDA has to discover! If BMDA had this information, it would perform exactly the same as UMDA. GA with uniform crossover and a great crossover rate performs similar to UMDA [3], i.e. it performs very well for linear problems. Genetic algorithm with onepoint crossover keeps the dependencies among neighboring bits so it has the same disadvantage with linear problems as BMDA. In Figure 1, an average number of fitness evaluations until convergence for 30 independent runs is shown. In Figure 2 the evolution of ordering parameter in a randomly chosen run is shown.

7.2 Quadratic Function

The quadratic fitness function used for comparisons in this section is defined as

$$f_{quadratic}(x, \pi) = \sum_{i=0}^{\frac{n}{2}-1} f_2(x_{\pi(2i)}, x_{\pi(2i+1)}) \tag{11}$$

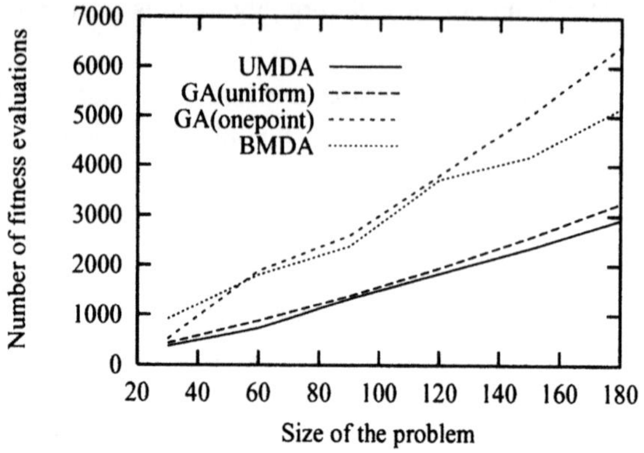

Figure 1. Number of fitness evaluations for f_{onemax}. The ranges of the used population size (for $n = 30$ to $n = 180$) were $50 - 170$ for UMDA, $32 - 100$ for GA (uniform), $32 - 160$ for GA (onepoint), and $120 - 260$ for BMDA. The truncation selection with $\tau = 50\%$ was used.

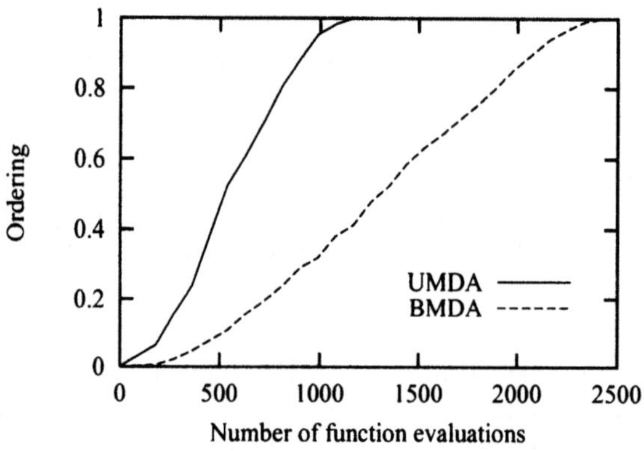

Figure 2. Ordering parameter for f_{onemax}, $n = 90$. The population sizes were 110 for UMDA and 180 for BMDA. The truncation selection with $\tau = 50\%$ was used.

where f_2 is defined as

$$f_2(u, v) = 0.9 - 0.9(u + v) + 1.9 \tag{12}$$

With both arguments equal to 0 we get $f_2(0, 0) = 0.9$. With different arguments we get $f_2(0, 1) = f_2(1, 0) = 0$. With both arguments equal to 1 we get $f_2(1, 1) = 1$. The optimum is clearly in the string with 1's on all positions. Criteria for convergence as well as the requirement for 100% convergence in 30 independent runs and the choice of optimal parameters for all algorithms are the same as in the last section. UMDA was not used for comparisons since it performed much worse than any of other methods. An average number of function evaluations until convergence over 30 runs is taken into account.

The function is not deceptive and that is why it should be not so hard to find the optimum for simple GA. It is an ideal function for the BMDA because it is decomposable of order 2 (see its definition) and the dependencies do not form cycles. First, the $f_{quadratic}$ with permutation π_1 (i.e., identity) will be discussed. Experiments have shown that although this function is not a big problem to solve by simple GA, BMDA performs much better. For GA with onepoint crossover the number of fitness evaluations seems to grow much faster than for BMDA with increasing size of a problem (see Figure 3). GA with uniform crossover performs significantly worse than GA with onepoint crossover (see Figure 4, notice that there is used a log-scaling for the number of fitness evaluations in this figure). The bad performance of GA is caused by the similar fitness values for both leading schemata that are opposite to each other and by the fact that their disruption significantly decreases the fitness. In both figures 4 an average number of fitness evaluations needed for convergence in 30 independent runs is shown.

Another permutation that was used for comparisons was π_2 that spreads pairs of positions $(2i)$ and $(2i+1)$ so that the distance between them is half of the size of a string. This reordering of bits has different effect on different algorithms. For BMDA (as well as it would be for UMDA), it actually does not affect anything because both algorithms are independent of positioning of bits. For simple GA with uniform crossover the situation is analogical. The problem arises for simple GA with onepoint crossover. In this case, this algorithm performs very poorly. Already for a problem of size $n = 30$, GA with onepoint crossover requires about 10 times more fitness evaluations than with π_1. This gap enlarges with the size of a problem. Results for BMDA and simple GA with uniform crossover are just slightly different from results with permutation π_1.

7.3 Deceptive function of order 3

Deceptive function is often used for comparisons of different optimization methods for its being deceptive. With deceptive problems, the average fitness of low order schemata present in optimum is lower than the average fitness of alternative ones. This property makes this class of functions hard to solve by the simple genetic algorithm as well as UMDA and many other evolutionary algorithms because these algorithms are based on the superior position of low order schemata that are matched by optimum. The fitness function is then defined as

$$f_{3deceptive}(x, \pi) = \sum_{i=0}^{\frac{n}{3}-1} f_3(x_{\pi(3i)} + x_{\pi(3i+1)} + x_{\pi(3i+2)}) \qquad (13)$$

where x is a bit string, π is any permutation of order n, and f_3 is defined as

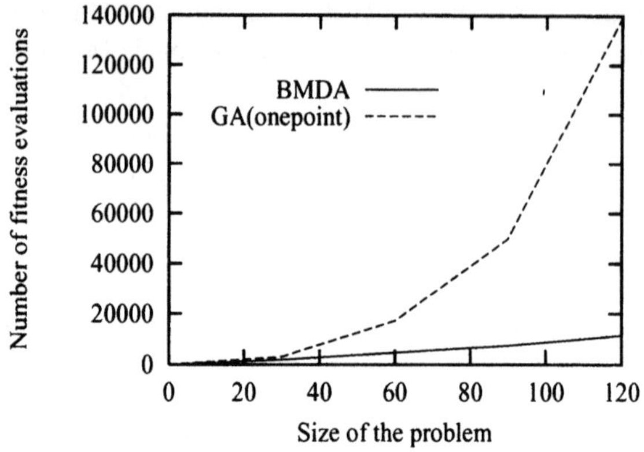

Figure 3. Number of fitness evaluations for $f_{quadratic}$. The ranges of the used population size (for $n = 30$ to $n = 120$) were 200 – 570 for BMDA and 260 – 2500 for GA (onepoint). The truncation selection with $\tau = 50\%$ was used.

$$f_3(u) = \begin{cases} 0.9 & \text{if } u = 0 \\ 0.8 & \text{if } u = 1 \\ 0 & \text{if } u = 2 \\ 1 & \text{otherwise} \end{cases} \tag{14}$$

Two different ordering permutations π_1 and π_3 (see Equation 9) are used. The π_3 operator mixes up the positions of the bits in a string so that the three neighboring bits $(3i)$, $(3i+1)$, and $(3i+2)$ are positioned so that the distance between each two of them is at least one third of the length of chromosome. For instance, for $n = 12$ we get $\pi_3 = (0, 4, 8, 1, 5, 9, 2, 6, 10, 3, 7, 11)$. The problem is actually the same for any permutation although for π_3 the dependencies are distributed more widely so that important schemata are of a greater defining length than originally.

Results for the problem of size $n = 30$ are shown in Table 1.

Table 1. Number of fitness evaluations for $f_{3deceptive}$. The population sizes for a permutation π_1 were 400 for GA (onepoint) and 1300 for BMDA. For permutation π_3 the population size for BMDA was 1300. GA (uniform) for both used permutations and GA (onepoint) for permutation π_3 did not achieve 100% convergence even for populations larger than 15000 (there is given a lower bound for the number of fitness evaluations in the table). The truncation selection with $\tau = 50\%$ was used.

	fitness eval. for π_1	fitness eval. for π_3
BMDA	17,550	17,420
GA (onepoint)	4,977	> 230,000
GA (uniform)	> 650,000	> 650,000

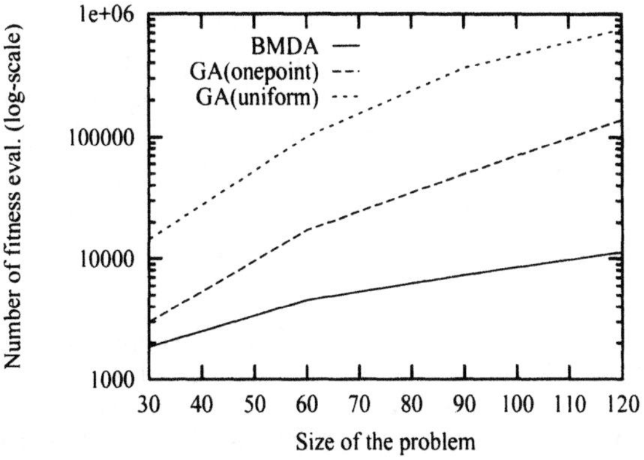

Figure 4. Number of fitness evaluations for $f_{quadratic}$ (log-scaling). The ranges of the used population size (for $n = 30$ to $n = 120$) were 200–570 for BMDA, 260–2500 for GA (onepoint), and 680 – 7000 for GA (uniform). The truncation selection with $\tau = 50\%$ was used.

7.4 NK fitness function

There are two important input parameters for the so-called NK fitness function [10], the length of an input string n, the number of neighbors to take into account k. When the function is initialized, for each position of a string there are chosen k other positions at random. In this fashion we get n groups of positions, each of length $k+1$. Each position is contained in at least one of these groups. Let us denote $\{p_{i,0}, \ldots, p_{i,k}\}$ to be the ith group. For each group there is randomly generated value for each of possible combinations of values on corresponding positions. That means that for ith group, there are generated 2^{k+1} values. Let us denote by f_i the function that returns the generated value for combination of values $\{x_{p_{i,0}}, \ldots, x_{p_{i,k}}\}$ on positions from ith group. Then for given matrix of positions $P = (p_{i,j})_{i,j=0}^{n-1,k}$ and vector of functions $F = (f_i)_{i=0}^{n-1}$ (both generated in the initial stage) the fitness function is defined as

$$f_{NK}(x) = \sum_{i=0}^{n-1} f_i(x_{p_{i,0}}, \ldots, x_{p_{i,k}}) \tag{15}$$

Since the neighbors are picked at random it makes no sense to use any permutation to reorder the positions. NK fitness function is clearly decomposable of order at most $k+1$. Random generation of both the neighbors and of the values can reduce the order, of course. However, the chance that this is the case is very small.

For $n = 50$ two different k were used, $k = 2$ and $k = 3$. For each of them, the matrix P and vector of functions F was generated first. Only one generated set of these parameters was used for each k. Results for various algorithms are presented in Table 2.

Since the optimum is not known for any of possible k, the best found number ever was taken as optimal and all algorithms were required to converge to this number. The problem space was explored by all algorithms with populations that should be large enough to find the real optimum. Even if this is not the case, i.e. the optimum was not found well, the best number ever found is taken into account, and even if the optimum were different from this number, this would not change anything on comparisons and their results.

Table 2. Number of fitness evaluations for f_{NK}. The population sizes for $n = 50$ and $k = 2$ were 210 for UMDA, 220 for GA (uniform), 340 for BMDA and 500 for GA (onepoint). The population sizes for $n = 50$ and $k = 3$ were 300 for GA (uniform), 700 for GA (onepoint), 2500 for BMDA and 8000 for UMDA. The truncation selection with $\tau = 50\%$ was used.

	$(n, k) = (50, 2)$	$(n, k) = (50, 3)$
BMDA	5,480	51,625
GA (onepoint)	11,010	17,440
GA (uniform)	4,288	7,826
UMDA	3,398	259.600

8. Conclusions

For linear and quadratic problems, BMDA works well what can be explained by the use of probabilistic distribution based on pair dependencies for the generation of new individuals. For linear problems UMDA and GA with uniform crossover perform better and GA with onepoint crossover performs similarly or worse than BMDA. The convergence of BMDA is slowed down by the need to discover the probabilistic model during optimization first. For quadratic problem, BMDA performs best among all compared algorithms. The GA with onepoint crossover performs best of all other algorithms but it fails when the defining length of important schemata is enlarged by different ordering of bits in a string. Experiments have shown that the gap between BMDA and other algorithms is for a quadratic function very large. The difference between this and other algorithms with linear function is insignificant in this context.

For BMDA, the problem arises with problems with dependencies of a higher order than two. For this class of problems, the used model of the search space is not sufficient. It is approximated somehow but this is often not sufficient. For deceptive fitness function of order 3, BMDA converged slower than GA with onepoint crossover. The GA with uniform crossover converged much worse than BMDA and UMDA was mislead in most of the cases even for huge population sizes. What is important though is that when the length of schemata is enlarged, GA with onepoint crossover performs very poor even for huge populations sizes and the number of fitness evaluations but the performance of BMDA remains almost the same. This, as well as the results for quadratic function, gives an evidence that GA with onepoint crossover does not effectively discover and use the schemata of a large defining length. It works very well for short building blocks. Uniform crossover is independent of the length of schemata but it disrupts dependencies very often. UMDA disrupts dependencies too. BMDA takes into account dependencies of order at most two and this causes that although it is independent of length of the schemata and it does not disrupt dependencies of order two, it might disrupt the dependencies of a higher order. The dependencies of a higher order are substituted by chain-like structures composed of the dependencies of order two but this is often not sufficient. The solution to this problem might be the use of Factorization Distribution Algorithm although this requires a problem specific knowledge in the initial stage. This is not required by any of UMDA, BMDA, or GA. If this were overcome, FDA would perform very well for all problems that are decomposable.

Acknowledgements

Authors would like to thank The Department of Mathematics of Slovak Technical University and The German National Center for Information Technology (GMD) for a technical support of the project. Martin Pelikan's stay at GMD was supported by Catholic Academic Program for Foreigners (KAAD, Bonn). Special thanks to Vladimir Kvasnicka and Jiri Pospichal for useful comments and help with the preparation as well as the completion of the paper.

References

1. Goldberg D E 1989 *Genetic Algorithms in Search, Optimization, and Machine Learning.* Addison-Wesley, Reading, MA

2. Kargupta H 1996 The Gene Expression Messy Genetic Algorithm. In: *Proceedings of the 1996 IEEE International Conference on Evolutionary Computation.* Nagoya, pp 631–636

3. Muehlenbein H 1998 The Equation for Response to Selection and its Use for Prediction. *Evolutionary Computation* 5: 303–346

4. Muehlenbein H, Rodriguez A O 1998 *Schemata, Distributions and Graphical Models in Evolutionary Optimization*, submitted for publication

5. De Bonet J S, Isbell Ch L, Viola P 1997 MIMIC: Finding Optima by Estimating Probability Densities. In: Mozer M, Jordan M, Petsche Th (Eds) 1997 *Advances in Neural Information Processing Systems 9.* MIT Press, Cambridge

6. Baluja S, Davies S 1997 Using Optimal Dependency-Trees for Combinatorial Optimization: Learning the Structure of the Search Space. Report Number CMU-CS-97-107, Carnegie Mellon University, Pittsburgh, PA

7. Marascuilo L A, McSweeney M 1977 *Nonparametric and Distribution-Free Methods for the Social Sciences.* Brooks/Cole Publishing Company, CA

8. Kvasnicka V, Pelikan M, Pospichal J 1996 Hill Climbing with Learning (An Abstraction of Genetic Algorithm). *Neural Network World* 5: 773-796

9. Baluja S 1994 Population-Based Incremental Learning: A Method for Integrating Genetic Search Based Function Optimization and Competitive Learning. Report Number CMU-CS-94-163, Carnegie Mellon University, Pittsburgh, PA

10. Kauffman S A 1993 *The Origins of Order: Self-Organization and Selection in Evolution.* Oxford University Press, Inc., NY

Part 9: New Frontier for Soft Computing

Papers:

Granular Computing using Neighborhood Systems

Y.Y. Yao

Department of Computer Science
University of Regina
Regina, Saskatchewan, Canada S4S 0A2
Email: yyao@cs.uregina.ca

Keywords: Fuzzy Sets, Granular Computing, Neighborhood Systems, Rough Sets

Abstract

A set-theoretic framework is proposed for granular computing. Each element of a universe is associated with a nonempty family of neighborhoods. A neighborhood of an element consists of those elements that are drawn towards that element by indistinguishability, similarity, proximity, or functionality. It is a granule containing the element. A neighborhood system is a family of granules, which is the available information or knowledge for granular computing. Operations on neighborhood systems, such as complement, intersection, and union, are defined by extending set-theoretic operations. They provide a basis of the proposed framework of granular computing. Using this framework, we examine the notions of rough sets and qualitative fuzzy sets.

1. Introduction

In many situations, it is impossible or unnecessary to distinguish individual objects or elements in a universe. For example, if a group of patients are described by using several symptoms, many patients would share the same symptoms, and hence are indistinguishable. This forces us to think a subset of the patients as one unit, instead of many individuals. In other words, one has to consider groups, classes, or clusters of elements. They are referred to as granules and can be either crisp or fuzzy. Elements in a granule may be drawn together by indistinguishability, similarity, proximity, or functionality [24]. Such a clustering of elements leads to information or knowledge granulation, which forms a basis of granular computing. There are at least two reasons for the study of granular computing [25, 26]. When a problem involves incomplete, uncertain, or vague information, it may be difficult to differentiate distinct elements and one is forced to consider granules [16].

Alternatively, although detailed information may be available, it may be sufficient to use granules in order to have an efficient and practical solution. Very precise solutions may in fact not be required for many practical problems.

The main objective of this paper is to propose a set-theoretic framework for granular computing using neighborhood systems [10, 21]. In this framework, an element of a universe is associated with a nonempty family of subsets of the universe. The family is called a neighborhood system of the element, and each subset is called a neighborhood of the element. Each neighborhood of an element may be regarded as a granule containing that element, and a neighborhood system of the element is a family of granules. Operations on neighborhood systems, such as complement, intersection, and union, will be defined by extending standard set-theoretic operations [21]. It is done in the manner used by Brink [3] for the development of power algebras. An ordering relation can also be defined from a neighborhood system. These notions establish a basis of proposed framework for granular computing.

To demonstrate the usefulness of the proposed framework, two problems will be discussed. The first problem deals with concept formulation and approximation. The solution is related to rough set approach [15]. The second problem is related to the interpretation of fuzzy sets [23]. A crisp set is represented by 1-neighborhood systems, i.e., each element has exactly one neighborhood. Set-theoretic operations are interpreted in terms of neighborhood system operations. Following the same argument, one may represent a fuzzy set by a nested neighborhood system. Each neighborhood corresponds to a particular α-cut of the fuzzy set. The notion of qualitative fuzzy sets is introduced. Such a notion may provide further insights to the understanding of fuzzy set membership functions and fuzzy set-theoretic operations. In summary, the proposed framework may provide a unified model for the study of two well known granular computing theories. It should also be pointed out that in the present study we only consider the set-theoretic facet of rough sets and fuzzy sets.

2. Overview of Neighborhood Systems

The concept of neighborhood systems is originated from studies of topological space and its generalization called Frechet (V)space [17]. Recently, Lin [9, 10] adopted this notion as a tool to describe relationships between objects.

2.1 Basic concepts

For an element x of a finite universe U, one associates with it a subset $n(x) \subseteq U$ called the *neighborhood* of x. Intuitively speaking, elements in a neighborhood of an element are somewhat indiscernible or at least not noticeably distinguishable to x. That is, elements in a neighborhood of x are drawns towards x by indistinguishability, similarity, or functionality [11]. In general, a neighborhood of x may or may not contain x. A neighborhood of x containing x is called a reflexive neighborhood. For the present study, we are only interested in reflexive neighborhoods of x, in order to accommodate the previous intuitive interpretation of neighborhoods. A neighborhood system $\mathrm{NS}(x)$ of x is a nonempty family of neighborhoods of x. A neighborhood system $\mathrm{NS}(x)$ of x groups the universe into classes. Distinct neighborhoods x consist of elements having different types of, or various degrees of, similarity to x. A neighborhood system may be formally interpreted as an operator from U to 2^{2^U} that maps each element to a family of subsets of the universe. Let $\mathrm{NS}(U)$ denote the collection of neighborhood systems for all elements in U. It determines a Frechet (V)space, written $(U, \mathrm{NS}(U))$. There is no additional requirements on neighborhood systems.

Example 2.1. Consider a universe $U = \{a, b, c, d\}$. We can have the following neighborhood systems for elements of U:

$$\begin{aligned} \mathrm{NS}(a) &= \{\{a\}, \{a, b\}\}, \\ \mathrm{NS}(b) &= \{\{a, b\}, \{a, b, c\}\}, \\ \mathrm{NS}(c) &= \{\{c\}\}, \\ \mathrm{NS}(d) &= \{\{d\}, \{b, d\}, \{a, b, d\}\}. \end{aligned}$$

All neighborhoods are reflexive. For elements a and d, a is in a neighborhood of d, but d is not in any neighborhood of a. This implies that neighborhood relationships may not be symmetric. Consider the neighborhood system of a. We may say that two types of similarity are represented: one suggests that a is only similar to itself, the other suggests that b is similar to a. The former may be considered to be a strong similarity and the latter a weak similarity.

The notion of neighborhood systems provides a convenient tool for representing relationship between objects in a universe. As shown by the following examples, it can be used to represent both quantitative and qualitative similarities.

Example 2.2. Consider a universe U and a distance function $D : U \times U \longrightarrow \mathcal{R}^+$, where $\mathcal{R}^+$ is the set of non-negative real numbers. For each number $d \in \mathcal{R}^+$, we may define the following neighborhood of $x \in U$:

$$n_d(x) = \{y \mid D(x, y) \leq d\}.$$

A neighborhood system of x is given by $\mathrm{NS}(x) = \{n_d(x) \mid d \in \mathcal{R}^+\}$. Conversely, we have $D(x, y) = \sup\{d \mid y \in n_d(x)\}$. That is, we may interpret quantitative similarity (expressed using a distance function) in terms of neighborhood systems. The use of a quantitative distance function implies that any two elements can be compared regarding their similarity to a particular element of U. In terms of neighborhood, we have $n_a(x) \subseteq n_b(x)$ for any two numbers $a, b \in \mathcal{R}^+$ with $a \geq b$. The neighborhood system $\mathrm{NS}(x)$ is a nested family of subsets of U, with each neighborhood representing a specific level of similarity to x.

Example 2.3. Suppose $R \subseteq U \times U$ is a binary relation on a universe U. Using R, for each element $x \in U$, we define four neighborhoods [20]:

$$\begin{aligned} R_s(x) &= \{y \mid xRy\}, \\ R_p(x) &= \{y \mid yRx\}, \\ R_{p\wedge s}(x) &= \{y \mid xRy \text{ and } yRx\}, \\ R_{p\vee s}(x) &= \{y \mid xRy \text{ or } yRx\}. \end{aligned}$$

They are referred to as the successor, predecessor, successor-and-predecessor, and successor-or-predecessor neighborhoods of x. The neighborhood system of x is given by $\mathrm{NS}(x) = \{R_s(x), R_p(x), R_{p\wedge s}(x), R_{p\vee s}(x)\}$. If R is a reflexive relation, all these neighborhoods are reflexive. If R is a symmetric relation, all neighborhoods become the same one. Obviously, unlike the distance function based neighborhood systems, two elements of U may not be comparable.

The representation of qualitative similarity using neighborhood systems may be more useful in some applications, where the meaning of a distance or a similarity function is not clear. It may also be case that the qualitative information, i.e., the order implied by the numeric values, is useful rather than the precise numeric values.

2.2 Operations on neighborhood systems

Let U be a set and $\circ$ a binary operation on U. One can define a binary operation $\circ^+$ on subsets of U as follows [3]:

$$X \circ^+ Y = \{x \circ y \mid x \in X, y \in Y\}, \tag{2.1}$$

for any $X, Y \subseteq U$. In general, one may lift any operation f on elements of U to an operation f^+ on subsets of U, called the power operation of f. Suppose $f : U^n \longrightarrow U$ $(n \geq 1)$ is an n-ary operation on U. The power operation $f^+ : (2^U)^n \longrightarrow 2^U$ is defined by [3]:

$$f^+(X_0, \ldots, X_{n-1}) = \{f(x_0, \ldots, x_{n-1}) \mid x_i \in X_i \text{ for } i = 0, \ldots, n-1\}, \tag{2.2}$$

for any $X_0, \ldots, X_{n-1} \subseteq U$. This provides a universal-algebraic construction approach. For any algebra $(U, f_1, \ldots, f_k)$ with the base set U and operations $f_1, \ldots, f_k$, its power algebra is given by $(2^U, f_1^+, \ldots, f_k^+)$. The power operation f^+ may carry some properties of f. For example, for a binary operation $f : U^2 \longrightarrow U$, if f is commutative and associative, f^+ is commutative and associative, respectively. If e is an identity for some operation f, the set $\{e\}$ is an identity for f^+. If an unary operation $f : U \longrightarrow U$ is an involution, i.e.,

$f(f(x)) = f(x)$, f^+ is also an involution. On the other hand, many properties of f are not carried over by f^+. For instance, if a binary operation f is idempotent, i.e., $f(x, x) = x$, f^+ may not be idempotent. If a binary operation g is distributive over f, g^+ may not be distributive over f^+.

A neighborhood system is a family of subsets of the universe U. By applying the idea of power algebras, we may lift set-theoretic operations on sets to neighborhood systems. For two neighborhood systems NS_1 and NS_2, the complement, intersection and union are defined by:

$$\begin{aligned} \neg\mathrm{NS}_1(x) &= \{\sim n_i(x) \mid n_i(x) \in \mathrm{NS}_1(x)\}, \\ \mathrm{NS}_1(x) \sqcap \mathrm{NS}_2(x) &= \{n_i(x) \cap n_j(x) \mid n_i(x) \in \mathrm{NS}_1(x), n_j(x) \in \mathrm{NS}_2(x)\}, \\ \mathrm{NS}_1(x) \sqcup \mathrm{NS}_2(x) &= \{n_i(x) \cup n_j(x) \mid n_i(x) \in \mathrm{NS}_1(x), n_j(x) \in \mathrm{NS}_2(x)\}. \end{aligned} \tag{2.3}$$

These operations can be interpreted as extensions of set-theoretic operations in a framework of set-based computations [22]. Different neighborhood systems may be provided by different experts or defined based on different features of objects. The above operations may be used for combining neighborhood systems.

It can be easily verified that the previous set-theoretic neighborhood system operations have the following properties:

$$\begin{aligned} \text{Commutativity}: \quad & \mathrm{NS}_1(x) \sqcap \mathrm{NS}_2(x) = \mathrm{NS}_2(x) \sqcap \mathrm{NS}_1(x), \\ & \mathrm{NS}_1(x) \sqcup \mathrm{NS}_2(x) = \mathrm{NS}_2(x) \sqcup \mathrm{NS}_1(x); \\ \text{Associativity}: \quad & (\mathrm{NS}_1(x) \sqcap \mathrm{NS}_2(x)) \sqcap \mathrm{NS}_3(x) = \\ & \mathrm{NS}_1(x) \sqcap (\mathrm{NS}_2(x) \sqcap \mathrm{NS}_3(x)), \\ & (\mathrm{NS}_1(x) \sqcup \mathrm{NS}_2(x)) \sqcup \mathrm{NS}_3(x) = \\ & \mathrm{NS}_1(x) \sqcup (\mathrm{NS}_2(x) \sqcup \mathrm{NS}_3(x)); \\ \text{De Morgan's law}: \quad & \neg(\mathrm{NS}_1(x) \sqcap \mathrm{NS}_2(x)) = \neg\mathrm{NS}_1(x) \sqcup \neg\mathrm{NS}_2(x), \\ & \neg(\mathrm{NS}_1(x) \sqcup \mathrm{NS}_2(x)) = \neg\mathrm{NS}_1(x) \sqcap \neg\mathrm{NS}_2(x); \\ \text{Double negation law}: \quad & \neg\neg\mathrm{NS}_1(x) = \mathrm{NS}_1(x); \\ \text{Unit element of } \sqcap: \quad & \mathrm{NS}_1(x) \sqcap \{U\} = \mathrm{NS}_1(x); \\ \text{Zero element of } \sqcup: \quad & \mathrm{NS}_1(x) \sqcup \{\emptyset\} = \mathrm{NS}_1(x). \end{aligned}$$

Operators $\sqcap$ and $\sqcup$ are not idempotent. Each of them is not distributive over the other. In general, $\mathrm{NS}(x) \sqcap \neg\mathrm{NS}(x)$ is not equal to $\{\emptyset\}$, and $\mathrm{NS}(x) \sqcup \neg\mathrm{NS}(x)$ is not equal to $\{U\}$. A detailed study of such a system can be found in a paper by Brink [2] on second-order Boolean algebras.

2.3 Orderings induced by neighborhood systems

In the last two subsections, the notion of neighborhood systems is formulated mainly from a purely mathematical point of view. A semantic interpretation is necessary, if we want to apply the theory to practical problems. In what follows, we interpret a neighborhood system by the more familiar notion of an ordering characterized by a "closer to" relation.

A neighborhood system models nearness between two elements in a space. It is originated from the abstraction of geometric notion, in which two points are "close to" or "approximate to" each other [11]. Each neighborhood of an element x of a universe U is a set of elements from U that are drawn towards x by their similarity to x. To make this interpretation more precise, we rely on the following intuitive results. An element must be close to itself. That is, all neighborhoods must be reflexive. For two neighborhoods with $n_i(x) \subset n_j(x)$, elements in $n_i(x)$ are closer to x than elements in $n_j(x) - n_i(x)$. This seems to be reasonable, especially for the nested neighborhood systems given in Example 2.2. By extending the argument to two arbitrary neighborhoods $n_i(x)$ and $n_j(x)$, one may say that elements in both neighborhoods are closer to x than those in only one neighborhood. That is, elements in $n_i(x) \cap n_j(x)$ are closer to x than elements in $n_i(x) \cup n_j(x) - n_i(x) \cap n_j(x)$. The same argument may be applied to a neighborhood system $\mathrm{NS}(x)$.

For a neighborhood system $\mathrm{NS}(x)$, the $\cap$-closure $\mathrm{NS}^*(x)$ is defined to be the minimum subset of 2^U, which contains $\mathrm{NS}(x)$ and the entire set U, and is closed under intersection $\cap$. The $\cap$-closure of a neighborhood system is a complete lattice under the ordering given by set inclusion. The meet is given by set intersection, but the join is different from the set union. It is commonly referred to as a closure system [4]. Based on the $\cap$-closure $\mathrm{NS}^*(x)$ of a neighborhood system $\mathrm{NS}(x)$, a binary relation $\prec$ on U is defined as follows:

$$a \prec b \iff \text{there exists } n_1(x), n_2(x) \in \mathrm{NS}^*(x) \text{ such that } n_1(x) \subset n_2(x), \quad a \in n_1(x), \text{ and } b \in n_2(x) - \bigcup_{n(x) \subset n_2(x)} n(x). \tag{2.4}$$

For two elements $a, b \in U$ with $a \prec b$, a is said to be closer to x than b. The relation $\prec$ shows the structure imposed by a neighborhood system. It can be verified that the relation $\prec$ is asymmetric and transitive. The binary relation defined by:

$$a \preceq b \iff a \prec b \text{ or } a = b, \tag{2.5}$$

is a partial order. In fact, a diagram for relation $\prec$ can be easily obtained from the Hasse diagram of the lattice $\mathrm{NS}^*(x)$: for a subset $n(x) \in \mathrm{NS}^*(x)$, deleting all elements in subsets in $\mathrm{NS}^*(x)$ proceeding $n(x)$.

Example 2.4. A nested neighborhood system of an element $x \in U$ is a family of subsets of U, $\mathrm{NS}(x) = \{n_1(x), \ldots, n_k(x)\}$, such that

$$n_1(x) \subset n_2(x) \subset \ldots \subset n_k(x).$$

The ∩-closure of $\mathrm{NS}(x)$ is itself if $n_k(x) = U$, otherwise it is $\mathrm{NS}(x) \cup \{U\}$. The following binary relation indicates the closeness of elements of U to x:

$$n_1(x) \prec n_2(x) - n_1(x) \prec \ldots \prec n_k(x) - n_{k-1}(x) \prec U - n_k(x).$$

In the above representation, we have implicitly extended the relation $\prec$ to subsets of the universe.

Example 2.5. Consider a universe $U = \{a, b, c, d, e, f\}$ and a neighborhood system of a defined by:

$$\mathrm{NS}(a) = \{\{a, e\}, \{a, b, c\}, \{a, b, d\}\}.$$

The ∩-closure of $\mathrm{NS}(a)$ is given by:

$$\mathrm{NS}^*(a) = \{\{a\}, \{a, b\}, \{a, e\}, \{a, b, c\}, \{a, b, d\}, U\}.$$

The lattice $\mathrm{NS}^*(a)$ and the induced relation $\prec$ are given by:

$$\{a\} \subseteq \begin{array}{c} \{a,b\} \subseteq \begin{array}{c} \{a,b,c\} \\ \{a,b,d\} \end{array} \\ \{a,e\} \end{array} \subseteq U \qquad \text{and} \qquad a \prec \begin{array}{c} b \prec \begin{array}{c} c \\ d \end{array} \\ e \end{array} \prec f.$$

From this example, we can see that e is not comparable with b, c, or d regarding their closeness to a.

3. Neighborhood Systems and Granular Computing

Based on the notion of neighborhood systems, we consider in this section problems of concept formation and representation. Our discussion centers on the understanding that each subset of a universe is a granule representing a certain concept. With specific interpretations of sets of granules, i.e., families of subsets of the universe, we may approximate a precise concept or represent a vague concept. The former is related to rough sets, and the latter is related to qualitative fuzzy sets.

3.1 Concepts and granules

In the study of formal concepts, every concept is understood as a unit of thoughts that consists of two parts, the extension and intension of the concept [14, 18]. The extension of a concept is the set of objects or entities which are instances of the concept. All objects in the extension have the same properties that characterize the concept. The intension (comprehension) of a concept consists of all properties or attributes that are valid for all those objects to which the concept applies. A concept is thus described jointly by its extension and intension, i.e., a set of objects and a set of properties. This formulation enables us to study formal concepts in a set-theoretic framework.

Consider a universe U. Each subset $A \subseteq U$ may be interpreted as the extension of certain concept. All elements of A are characterized by certain properties, namely, the intension of the concepts. In the following discussion, we only consider the extension of a concepts without explicitly referring to its intension. One may say that elements in A are similar or indistinguishable based solely on properties in the intension. That is, we regard the elements of A to be equivalent, as they are all instances of the same concept. Under this interpretation, we may characterize a concept trivially by a family of 1-neighborhood systems. With respect to the concept A, we define:

$$\mathrm{NS}_A(x) = \begin{cases} \{A\} & x \in A, \\ \{\emptyset\} & x \notin A. \end{cases} \tag{3.1}$$

That is, the set A is the neighborhood of every element of A, and the empty set $\emptyset$ is the neighborhood of all elements not in A. It is straightforward to verify that set-theoretic operations can be expressed as neighborhood system operations using such a representation. One may treat A as being a granule in the sense that elements in A are drawn together by their similarities, i.e., they are all instances of a certain concept. From the above discussion, we may use "neighborhoods", "concepts", "subsets", and "granules" interchangeably. The choice of a particular name is not as important as the intended meaning associated with each subset of the universe.

3.2 Rough set approximations

Let $E \subseteq U \times U$ be an equivalence relation on a finite and nonempty universe U. That is, the relation E is reflexive, symmetric, and transitive. The pair $apr = (U, E)$ is called a Pawlak approximation space. The equivalence relation E partitions the universe U into disjoint subsets called equivalence classes. Elements in the same equivalence class are said to be indistinguishable. Equivalence classes of E are called elementary sets. Let

$$[x]_E = \{y \mid xEy\}, \tag{3.2}$$

denote the equivalence class containing x. It may be interpreted as a granule containing x. If a precise concept $A \subseteq U$ is represented relative to granules $[x]_E$'s, it is no longer precise. A subset $A \subseteq U$ is approximated by a pair of sets as follows [20]:

$$\begin{aligned} \underline{apr}(A) &= \bigcup\{[x]_E \mid x \in U, [x]_E \subseteq A\}, \\ \overline{apr}(A) &= \bigcup\{[x]_E \mid x \in U, [x]_E \cap A \neq \emptyset\}. \end{aligned} \tag{3.3}$$

The lower approximation of A is the union of equivalence classes that are subsets of A, and the upper approximation is the union of equivalence classes that have a nonempty intersection with A. One may interpret $\underline{apr}, \overline{apr} : 2^U \longrightarrow 2^U$ as unary set-theoretic operators [19]. They are dual operators in the sense that $\underline{apr}(A) = \sim \overline{apr}(\sim A)$ and $\overline{apr}(A) = \sim \underline{apr}(\sim A)$. The system $(2^U, \sim, \underline{apr}, \overline{apr}, \cap, \cup)$ is called a Pawlak rough set algebra defined by the equivalence relation E. It may be viewed as an extension of the classical set algebra $(2^U, \sim, \cap, \cup)$.

Granules induced by an equivalence relation may be consider to be a special kind of neighborhoods. In general, one may use 1-neighborhood systems, in which each element has exactly one neighborhood. Such a neighborhood system can be conveniently constructed from a binary relation [20]. By extending Equation (3.3), we have the following two pairs of approximation operators [20]:

$$\begin{aligned} \underline{apr}'_n(A) &= \bigcup\{n(x) \mid x \in U, n(x) \subseteq A\} \\ &= \{x \in U \mid \exists y[x \in n(y), n(y) \subseteq A]\}, \\ \overline{apr}'_n(A) &= \sim \underline{apr}'_n(\sim A) \\ &= \{x \in U \mid \forall y[x \in n(y) \Longrightarrow n(y) \cap A \neq \emptyset]\}, \end{aligned} \tag{3.4}$$

and

$$\begin{aligned} \underline{apr}''_n(A) &= \sim \overline{apr}''_n(\sim A) \\ &= \{x \in U \mid \forall y[x \in n(y) \Longrightarrow n(y) \subseteq A]\}, \\ \overline{apr}''_n(A) &= \bigcup\{n(x) \mid x \in U, n(x) \cap A \neq \emptyset\} \\ &= \{x \in U \mid \exists y[x \in n(y), n(y) \cap A \neq \emptyset]\}. \end{aligned} \tag{3.5}$$

In the above definitions, we have replaced $[x]_E$ by $n(x)$ in either the lower or the upper approximation operator. The other one is defined by duality so that the two operators are dual to each other. By the first definition, the lower approximation is the union of granules which are subsets of A, while the upper approximation is the complement of the lower approximation of the complement of A. By the second definition, the upper approximation is the union of granules which have a nonempty intersection with A, while the lower approximation is the complement of the upper approximation of the complement of A. Each definition seems to be reasonable. If we only consider reflexive 1-neighborhood system, we have [20]:

$$\underline{apr}''_n(A) \subseteq \underline{apr}'_n(A) \subseteq A \subseteq \overline{apr}'_n(A) \subseteq \overline{apr}''_n(A). \tag{3.6}$$

One pair of approximations is tighter than the other pair of approximations. The set A lies between its lower and upper approximations. Additional properties of the approximation operators can be found in a recent paper the author [20].

In the above development of rough sets, we start from a precise concept defined by a subset of the universe U. Relative to granules, a precise concept becomes vague and must be approximated. This formulation clearly identifies one possible sources of uncertainty. Such a model may have many practical applications. In many tasks, such as classification, diagnosis, reasoning, and learning, we normally focus on subsets (i.e., granules) of the universe sharing some special features, rather than on individuals. The granules may be understood as being the basic concepts or elementary knowledge. All other concepts and knowledge must be represented using such granulated information and knowledge. Two different, but related, approximations have been proposed, each is intuitively appealing.

3.3 Qualitative fuzzy sets

A fuzzy set $\mathcal{A}$ of U is defined by a membership function: $\mu_{\mathcal{A}} : U \longrightarrow [0, 1]$. There are many definitions for fuzzy set complement, intersection, and union. With the min-max system proposed by Zadeh [23], fuzzy set operations are defined component-wise as:

$$\begin{aligned} \mu_{\sim\mathcal{A}}(x) &= 1 - \mu_{\mathcal{A}}(x), \\ \mu_{\mathcal{A}\cap\mathcal{B}}(x) &= \min[\mu_{\mathcal{A}}(x), \mu_{\mathcal{B}}(x)], \\ \mu_{\mathcal{A}\cup\mathcal{B}}(x) &= \max[\mu_{\mathcal{A}}(x), \mu_{\mathcal{B}}(x)]. \end{aligned} \tag{3.7}$$

In general, fuzzy set intersection and union may be defined in terms of t-norms and t-conorms [6]. By choosing different pairs of t-norms and t-conorms, one can derive distinct fuzzy set systems.

A crisp set can be regarded as a degenerated fuzzy set in which the membership function is restricted to the extreme points $\{0, 1\}$ of $[0, 1]$. In this case, the membership function is also referred to as a characteristic function. A fuzzy set can be related to a family of crisp sets through the notions of α-level sets. Given a number $\alpha \in [0, 1]$, an α-cut, or α-level set, of a fuzzy set is defined by:

$$\mathcal{A}_\alpha = \{x \in U \mid \mu_{\mathcal{A}}(x) \geq \alpha\}, \tag{3.8}$$

which is a subset of U. A strong α-cut is defined by:

$$\mathcal{A}_{\alpha^+} = \{x \in U \mid \mu_{\mathcal{A}}(x) > \alpha\}. \tag{3.9}$$

The 1-cut of a fuzzy set is called the core of the fuzzy set and the strong 0-cut is called the support. They are defined by:

$$\begin{aligned} core(\mathcal{A}) &= \{x \mid \mu_{\mathcal{A}}(x) \geq 1\}, \\ support(\mathcal{A}) &= \{x \mid \mu_{\mathcal{A}}(x) > 0\}. \end{aligned} \tag{3.10}$$

An element in the core may be interpreted as a typical instance of the fuzzy concept being modeled, an element not in the support is definitely not an instance, and an element in the support is somewhere in between. A fuzzy set is normal if the core is not empty. Using either α-cuts or strong α-cuts, a fuzzy set determines a family of nested subsets of U. Conversely, a fuzzy set $\mathcal{A}$ can be reconstructed from its α-level sets as follows:

$$\mu_{\mathcal{A}}(x) = \sup\{\alpha \mid x \in \mathcal{A}_\alpha\}. \tag{3.11}$$

This observation is commonly summarized by a representation theorem of fuzzy sets, which states that there is an one-to-one relationship between a fuzzy set and a family of crisp sets satisfying certain conditions [6, 12, 13]. An implication of the min-max system is that fuzzy set operations can be defined by set operations on α-level sets. They can be expressed by:

$$\begin{aligned} (\sim \mathcal{A})_\alpha &= \sim \mathcal{A}_{(1-\alpha)^+}, \\ (\mathcal{A} \cap \mathcal{B})_\alpha &= \mathcal{A}_\alpha \cap \mathcal{B}_\alpha, \\ (\mathcal{A} \cup \mathcal{B})_\alpha &= \mathcal{A}_\alpha \cup \mathcal{B}_\alpha. \end{aligned} \tag{3.12}$$

The α-level sets of fuzzy sets for intersection and union are obtained from the same α-level sets of the fuzzy sets involved. When an arbitrary pair of t-norm and t-conorm is used, it may be difficult to define such operations using set operations on α-level sets of fuzzy sets.

Unlike the rough set theory, the fuzzy set theory starts from vague concepts which cannot be represented by a single subset of the universe. The sources of such uncertainty are not considered in the model. Given a fuzzy set, one can easily find the set of its distinct α-cuts:

$$S(\mathcal{A}) = \{\mathcal{A}_\alpha \mid \alpha \in [0,1]\}. \tag{3.13}$$

Each α-cut may be considered as a granule representing the fuzzy concept under certain view. In this study, we are only interested the family of distinct α-cuts, instead of the actual membership values. Such a view of fuzzy sets is qualitative. This serves as a basis of the development of a theory of qualitative fuzzy sets. In order to distinguish them from the conventional fuzzy sets, we referred to the conventional fuzzy sets as quantitative fuzzy sets.

A qualitative fuzzy set is a subsystem of the power set of the universe. That is, a qualitative fuzzy set $\mathcal{A}$ is a subset of 2^U, namely, $\mathcal{A} \subseteq 2^U$. Intuitively, one may understand a qualitative fuzzy set as a neighborhood system of a typical instance (possibly imaginary) of the fuzzy concept being modeled. The core and support of a qualitative fuzzy set are defined by:

$$\begin{aligned} \mathit{core}(\mathcal{A}) &= \bigcap_{X \in \mathcal{A}} X, \\ \mathit{support}(\mathcal{A}) &= \bigcup_{X \in \mathcal{A}} X. \end{aligned} \tag{3.14}$$

A qualitative fuzzy set is normal if its core is not empty. By interpreting a qualitative fuzzy set as a neighborhood system of a typical instance (possibly imaginary) of the fuzzy concept, we can use neighborhood system operations to define fuzzy set operations. More specifically, for two fuzzy sets $\mathcal{A}$ and $\mathcal{B}$, from Equation (2.3) we have:

$$\begin{aligned} \neg\mathcal{A} &= \{\sim X \mid X \in \mathcal{A}\}, \\ \mathcal{A} \sqcap \mathcal{B} &= \{X \cap Y \mid X \in \mathcal{A}, Y \in \mathcal{B}\}, \\ \mathcal{A} \sqcup \mathcal{B} &= \{X \cup Y \mid X \in \mathcal{A}, Y \in \mathcal{B}\}. \end{aligned} \tag{3.15}$$

Properties of these operations are given in Section 2.2. With respect to the ordering relation introduced in Section 2.3, one may also define qualitative membership. Given a qualitative fuzzy set $\mathcal{A}$, we may define an ordering relation $\prec$ representing a qualitative belongingness [1]. For two elements $a, b \in U$ with $a \prec b$, we may say that a belongs to $\mathcal{A}$ "more than" b belongs to $\mathcal{A}$. In other words, a is closer to a typical instance of the fuzzy concept modeled by $\mathcal{A}$ than b.

The use of a family of crisp sets to represent a fuzzy concept has been proposed and studied by many authors. Klir [5] proposed a formulation of fuzzy sets based on modal logic. In this model, a vague concept is characterized by some, possibly different, crisp sets. Let $W = \{w_1, \ldots, w_n\}$ denote a set of n possible worlds or states. With respect to W, a vague concept is represented by n crisp sets:

$$\mathcal{A} = (A_{w_1}, \ldots, A_{w_n}). \tag{3.16}$$

In each world w_i, the vague concept is described precisely by a crisp set A_{w_i}. The vagueness is captured by distinct representations of the same concept in different worlds. The set of all families of n crisp sets is given by the n-fold Cartesian product of 2^U, namely, $\prod_n 2^U = 2^U \times \ldots \times 2^U$ (n repetitions). We define set-theoretic operations on $\prod_n 2^U$ component-wise as follows: for $\mathcal{A} = (A_{w_1}, \ldots, A_{w_n})$ and $\mathcal{B} = (B_{w_1}, \ldots, B_{w_n})$,

$$\begin{aligned} &\sim \mathcal{A} = (\sim A_{w_1}, \ldots, \sim A_{w_n}), \\ &\mathcal{A} \cap \mathcal{B} = (A_{w_1} \cap B_{w_1}, \ldots, A_{w_n} \cap B_{w_n}), \\ &\mathcal{A} \cup \mathcal{B} = (A_{w_1} \cup B_{w_1}, \ldots, A_{w_n} \cup B_{w_n}). \end{aligned} \tag{3.17}$$

The system $(\prod_n 2^U, \sim, \cap, \cup)$ may be interpreted as the n-fold product of classical set algebra $(2^U, \sim, \cap, \cup)$. A similar approach was used by Kruse, Schwecke, and Heinsohn [8], in which each possible world is referred to as a context in a framework consisting of layered contexts. Kononov [7] studied a similar system. Negoiţă and Ralescu [12] investigated a more general system based on the notion of L-flou subsets. Like the representation of fuzzy sets using α-cuts, in those studies vague concepts are represented by the same number of subsets (possibly the same) of universe. Operations are defined component-wise using standard set-theoretic operations. However, in our formulation of qualitative fuzzy sets, we consider distinct subsets of the universe. No constraints are imposed on the number of subsets in representing a qualitative fuzzy set. Although operations on qualitative fuzzy sets are defined based on standard set-theoretic operations, they are not defined component-wise.

For a given quantitative fuzzy set, one can obtain a qualitative fuzzy set through its α-cuts. Two different quantitative fuzzy sets may produce the same qualitative fuzzy set. In addition, the qualitative fuzzy set obtained from a quantitative fuzzy set consists of a family of nested subsets of U. In general, an arbitrary qualitative fuzzy set may not be obtainable from a quantitative fuzzy set. The relationships between quantitative and qualitative fuzzy sets may be extended to L-fuzzy sets, i.e., lattice-based fuzzy sets [12]. It will also be useful to study the structures of a qualitative fuzzy set system.

4. Conclusion

A reflexive neighborhood of an element in a universe is a granule consisting of elements drawn towards that element by similarity. A neighborhood system of the element is a nonempty family of such granules. By lifting set-theoretic operations to families of subsets, one may define set-theoretic operations on neighborhood systems. This established a basis of a set-theoretic framework for granular computing. Both rough sets and fuzzy sets may be understood within the proposed framework.

In the theory of rough sets, granules are the basic concepts or elementary knowledge. All other concepts and knowledge must be represented using such granulated information and knowledge. A precise concept is defined by a subset of the universe U. When a precise concept is viewed relative to granules, it becomes vague and must be approximated. This formulation clearly identifies one possible sources of uncertainty. Such a model may be applied to situations where generalization plays an important role. A generalization is normally made based on subsets (i.e., granules) of the universe sharing some special features. Two related but different rough set approximations may be used.

From a quantitative fuzzy set, one may obtain a qualitative fuzzy set through the α-cuts of the former. In general, a qualitative fuzzy set is defined as a family of nonempty subsets of the universe. It may be understood as a neighborhood system of a typical instance (possibly imaginary) of the fuzzy concept being modeled. Operations on qualitative fuzzy sets are defined based on operations on neighborhood systems. The ordering induced by a qualitative fuzzy set represents qualitative belongingness. In general, an arbitrary qualitative fuzzy set may not be obtainable from a quantitative fuzzy set. As future work, it may be interesting to examine the relationships between quantitative and qualitative L-fuzzy sets. It may also be useful to study the structures of a qualitative fuzzy set system.

References

1. Bollmann-Sdorra, P., Wong, S.K.M., and Yao, Y.Y., 1993, A Measurement-theoretic axiomatization of fuzzy sets, *Fuzzy Sets and Systems*, **60**, 295-308.
2. Brink, C., 1984, Second-order Boolean algebras, *Quaestiones Mathematicae*, **7**, 93-100.
3. Brink, C., 1993, Power structures, *Algebra Universalis*, **30**, 177-216.
4. Cohn, P.M., 1965, *Universal Algebra*, Harper and Row Publishers, New York.
5. Klir, G.J., 1994, Multivalued logics versus modal logics: alternative frameworks for uncertainty modelling, in: *Advances in Fuzzy Theory and Technology*, edited by P.P. Wang, Department of Electrical Engineering, Duke University, Durham, North Carolina, pp. 3-47.
6. Klir, G.J., and Yuan, B., 1995, *Fuzzy Sets and Fuzzy Logic, Theory and Applications*, Prentice Hall, New Jersey.
7. Kononov, B.P., 1989, A model of a multifaceted description of objects in terms of classes, *Automatic Documentation and Mathematical Linguistics*, **23**, 26-37.
8. Kruse, R., Schwecke, E., and Heinsohn, J., 1991, *Uncertainty and Vagueness in Knowledge Based Systems*, Springer-Verlag, Berlin.
9. Lin, T.Y., 1988, Neighborhood systems and approximation in relational databases and knowledge bases, *Proceedings of the 4th International Symposium on Methodologies of Intelligent Systems*, 1988.
10. Lin, T.Y., 1997, Neighborhood systems – application to qualitative fuzzy and rough sets, in: *Advances in Machine Intelligence and Soft-Computing*, edited by P.P. Wang, Department of Electrical Engineering, Duke University, Durham, North Carolina, USA, pp. 132-155.
11. Lin, T.Y., 1998, Granular computing on binary relations I: data mining and neighborhood systems, manuscript, Department of Mathematics and Computer Science, San Jose State University, San Jose, California, USA.
12. Negoiţă, C.V., and Ralescu, D.A., 1975, *Applications of Fuzzy Sets to Systems Analysis*, Birkhäuser Verlag, Basel.
13. Negoiţă, C.V., and Ralescu, D.A., 1975, Representation theorems for fuzzy concepts, *Kybernetes*, **4**, 169-174.
14. Orlowska, E., 1987, Reasoning about vague concepts, *Bulletin of the Polish Academy of Sciences, Mathematics*, **35**, 643-652.
15. Pawlak, Z., 1982, Rough sets, *International Journal of Computer and Information Science*, **11**, 341-356.
16. Pawlak, Z., 1998, Granularity of knowledge, indiscernibility and rough sets, *Proceedings of 1998 IEEE International Conference on Fuzzy Systems*, to appear.
17. Sierpenski, W., and Krieger, C., 1956, *General Topology*, University of Toronto, Toronto.
18. Wille, R., 1992, Concept lattices and conceptual knowledge systems, *Computers Mathematics with Applications*, **23**, 493-515.
19. Yao, Y.Y., 1996, Two views of the theory of rough sets in finite universes, *International Journal of Approximate Reasoning*, **15**, 291-317.
20. Yao, Y.Y., 1998, Relational interpretations of neighborhood operators and rough set approximation operators, *Information Sciences*, to appear.
21. Yao, Y.Y., and Chen, X.C., 1997, Neighborhood based information systems, *Proceedings of the 3rd Joint Conference on Information Sciences, Volume 3: Rough Set & Computer Science*, Research Triangle Park, North Carolina, USA, March 1-5, pp. 154-157.

22. Yao, Y.Y., and Noroozi, N., 1995, A unified model for set-based computations, *Soft Computing: 3rd International Workshop on Rough Sets and Soft Computing*, San Jose, California, USA, November 10-12, 1994, pp. 252-255.
23. Zadeh, L.A., 1965, Fuzzy sets, *Information and Control*, **8**, 338-353.
24. Zadeh, L.A., 1997, Some reflections on soft computing, granular computing and their roles in the conception, design and utilization of information/intelligent systems. Manuscript, Computer Science Division, University of California, Berkeley, USA.
25. Zadeh, L.A., 1997, Toward a restructure of the foundations of fuzzy logic (FL), Abstract of BISC Seminar, University of California, Berkeley, USA.
26. Zadeh, L.A., 1997, Towards a theory of fuzzy information granulation and its centrality in human reasoning and fuzzy logic, *Fuzzy Sets and Systems*, **19**, 111-127.

Toward Fuzziness in Natural Language Processing

Olgierd Unold

Institute of Engineering Cybernetics
Wroclaw University of Technology
Wyb. Wyspianskiego 27, 50-370 Wroclaw, Poland
phone: (48 71) 320 20 15, fax: (48 71) 21 26 77
e-mail:unold@ci.pwr.wroc.pl

Keywords: Natural Language Processing, Fuzzy Automata, Knowledge Representation

Abstract

This paper sketches a new approach to the use of fuzzy concepts in natural language processing (NLP). The main issue addressed is the fuzzification of the „automaton-driven" parser, which works in the stratificational knowledge representation system (SKRS).The proposed self-learning analyzer of a dialog system can learn the grammatical rules from the gained sentences imitating human behavior.

1. Introduction

The natural language processing (NLP) arised from research in artificial intelligence. There are two primary motivations for this type of research. First, the technological motivation is to build computer systems with the natural language access (so-called dialog systems), which have at present commercial applications, such as word processing and desktop publishing, spelling correction, dictionary access, translation aids, information management, retrieval, extraction, generation and so on [1]. Second motivation is to gain a better understanding of how the humans communicate by using natural language.

The goal of the NLP is to specify a theory of language comprehension and generation to such a level of detail that a person could write a computer program that can understand and generate natural language. Typical subareas of this field include the specification of parsing, the construction of knowledge representation formalisms and the modeling of reasoning processes.

The parsing phase states the syntactic dependecies among the different words in the sentence and gives one or more representations of the sentence structure trough syntactic trees. In general, the understanding of texts is achieved trough three phases of analysis: morphology, syntax and semantics. The morphology phase analyzes each word of the text, retrieving its morphologic features, its syntactic category, and lemma from which the word is derived. The syntax phase states the syntactic dependecies among the different words in the sentence and gives one or more representations of the sentence structure trough syntactic trees. The semantics phase represents the final step and the most complex part of a NLP system. It resolves syntactic ambiguities and recognizes semantic relationship among words, producing a representation of the sentence meaning trough the established formalism.

This paper presents the design of an adaptable sentence analyzer, which can be part of various dialog systems. This analyzer is based on the fuzzy automaton fPDAMS[2], which is the fuzzy version of the nondeterministic PushDown Automaton with associative Memory accesS (the PDAMS) [3-5]. Proper working of the PDMAS was checked in the stratificational knowledge representation system, the kind of the NLP system.

2. Stratificational Knowledge Representation System Overview

Stratificational knowledge representation system (SKRS) [4,5] is an attempt at formalizing the multilayer structure of natural language. In the stratificational model it is assumed that linguistic phenomena take place in various stratificationally arranged layers [6]. Thus, the SKRS is based on a multistratum semantic network whose nodes contain particular linguistic units (single words, the so-called structural-semantic components, sentences). The node-to-node links correspond to relations between particular linguistic units.

If a natural language text representing the acquired knowledge is to be mapped onto the stratificational semantic network, the particular linguistic units must be isolated in the analysed text, such as word groups and sentences. If we decide to limit our input texts to isolated clauses, we can limit the sentence decomposition task to two subtasks: one of word group decomposition into the particular words and one of sentence decomposition into word groups. In order to increase both knowledge integrity in the knowledge base and the efficiency of the search algorithm, an assumption has been made that it is not phrases, but the structural-semantic components (the components) that are the objective of the sentence decomposition, the components being word groups constituting semantic units which can no longer be decomposed, each of which has its own syntax characteristics [3]. Sentence decomposition into structural-semantic components in the SKRS is performed by a fuzzy automaton-driven parser, the fuzzy extension of the PDAMS. An automaton decomposition model in the SKRS can be used due to the reduction of the process of sentence decomposition into structural-semantic components to the process isolating appropriate substrings of symbols in the string of symbols representing lexical categories.

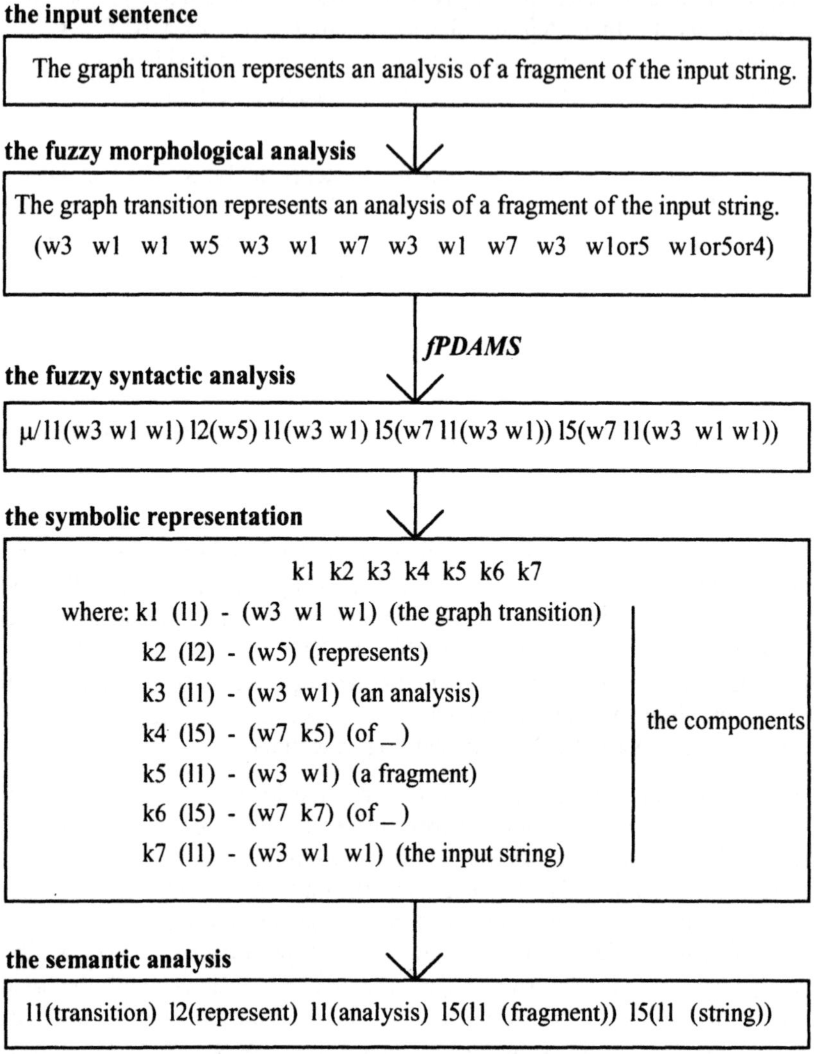

where: the l-symbols denote syntactic categories of substrings separated with the parentheses () (l1 - the nominal group, l2 - the verbal group, l3 - the adverbial, group, l5 - the prepositional group); the w1 stands for a noun, w3 for a determiner, w4 for an adjective, w5 for a verb, w7 for a preposition.

Figure 1. Example mapping between English sentence and knowledge components

After isolating in a sentence the knowledge particles corresponding to structural semantic components, the particles are transformed into appropriate symbolic expressions. Further, the expressions are put on the collective expressions that represents the already acquired knowledge in such a way that the same knowledge particle, which is included in its corresponding linguistic unit and which appears in many sentences, is represented by that same symbolic element. Thus only new knowledge elements are incorporated into the SKRS. The system construction enables the sentences in knowledge base (KB) to be reconstructed in the same form in which they have previously been input to the system. In addition, semantic kernels of the components are explicitly distinguished in their symbolic notation, which enables any utterance to be semantically analysed and, in effect, makes it possible to perfom a procedural reasoning on the knowledge base, basing on the matching operation [3]. The semantic sentence analysis is based on the systemic grammar mechanism [7] and on the process of associating the components with appropriate semantic categories [3].

3. Model of Language Processing in the SKRS

Knowledge expressed in natural language is acquired in the SKRS by moving through the stages shown in Figure 1. Lexical analysis, i.e. segmenting a sentence into words, is not a huge problem for a language like English, because it has explicit delimeters (blanks) between written words. Next, the lexical analyzer attaches part-of-speech information to each word identified during the process (the w_i symbol). Lets note that there can be the set of different lexical categories for the same word and the analyzer is forced to choose the only one (in the Figure 1 the symbol $w_{1or5or4}$ represents the set of part-of-speech information {noun, verb, adjective} connected with the word *string*). The output of the morphological-analysis component is a string of segmented words with the string of symbols representing lexical categories of words (so-called string (w)). Then, the fPDMAS splits the string (w) into structural-semantic components, i.e. produces the dependency structures, which represent various key linguistic relation. Decomposition is done with the certain characteristic functions value. In the last stage semantic kernels of the components are explicitly distinguished in their notation.

4. Fuzzy Pushdown Automaton with Associative Memory Access

In the construction and selection of operating principle of the fPDAMS there have been taken into account the following properties of decomposed English sentences:

i. two English sentences that can be decomposed into differing structural-semantic components can have the same initial substrings of their corresponding (w) strings whereas the lengths of those substrings cannot be determined in advance. For instance, although the four starting words of both following sentences have the same lexical categories (the substring $w_3 w_1 w_5 w_6$), the sentences can be decomposed into different structural-semantic components.

The train arrives early.

$(l_1(w_3\ w_1)\ l_2(w_5)\ l_6(w_6)w_0)$

The train will not arrive early.

$(l_1(w_3\ w_1)\ l_2(w_5\ w_6\ w_5)\ l_6(w_6)w_0)$

ii. two elemental structural-semantic components having different syntactical categories can have the same initial substrings of their corresponding (w) strings whereas the lengths of those substrings cannot be determined in advance. For instance, the following components have the same initial substring of lexical categories $(w_3 w_1)$ but they have different syntactical categories.

an hour ago $l_6(w_3 w_1 w_6)$

the student $l_1(w_3 w_1)$

iii. a single word can have more than one lexical interpretation. For instance, the word *string* has the following lexical description:

noun (w_1), adjective (w_4), verb (w_5).

The use of finite-state devices in natural language processing is mainly motivated by considerations of the speed and compactness of such representation. Their recent applications in NLP, which range from the morphological analysis and generation [8], parsing [9], speech processing [10] show the usefulness of automata in this area of artificial intelligence.

Fuzzification [11] of PDAMS is an attempt at handling „innate" uncertainty and imprecision of natural language and human knowledge expressed in natural language. Fuzzification of automaton relies in fact on replacing nonfuzzy transition relation by fuzzy relation [12]. The fuzzy relation is, in general, a fuzzy set in cartesian product of nonfuzzy or fuzzy universes. In case of fPDAMS we can observe both situations. The set of different lexical categories for the same word can be treated as fuzzy set (see the symbol $w_{1or5or4}$ in the Figure 1). Thus, process of transformation of an input sentence to a string of symbols representing lexical categories is burdened with all uncertainty. In fPDAMS each transition has its own „strength" - a grade of membership in a fuzzy set. The automaton chooses the next permissible move possessing the maximum value of membership. If we allow for a possibility of modification of memberhip function during parsing we will obtain the type of dynamic structure automaton or in other words the self-learning analyzer.

Figure 2 shows a graph of a subautomaton $fA(l_1)$ that isolates from the string (w) a substring having a syntactical category of the nominal group. There is another subautomaton $fA(l_3)$ incorporated into the $fA(l_1)$ subautomaton structure. The subautomaton $fA(l_3)$ is represented by the subgraph $fG(l_3)$ in the transition graph.

The label $w_p i_s$ means that a transition along the edge is possible when w_p is the input symbol that is being read.

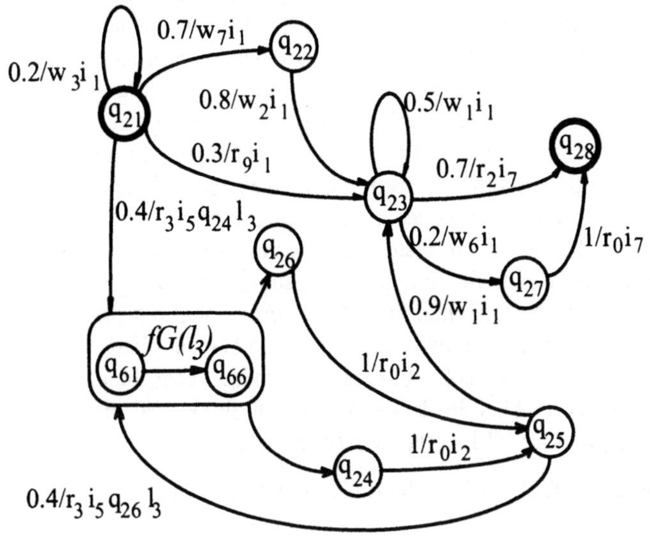

q_{21} - initial state of subautomaton

q_{28} - final state of subautomaton

Figure 2. The transition subgraph fG(l1) of the fuzzy subautomaton fA(l1)

The transition from the q_i to the q_j state at the input signal w_p can not only exist or not exist in the fPDAMS, but the transition has some degree of membership (usually denoted as $\mu_{ij} \in [0,1]$). There is the following role for the characteristic function μ_{ij} which describes the transition relation of the fPDAMS: for a given state q_i and input w_p of the automaton, the next state q_j of the automaton is determined so that to maximize the characteristic function value corresponding to the transition. The graph transition represents an analysis of a fragment of the input string (w), the analysis result being stored on one of the six stacks of the fPDAMS. The automaton operation is determined by a finite number of instructions. These instructions constitute the transition rules to allow the automaton to transit from one configuration to another. During the transition the instruction i_s is performed. The label $w_p i_s q_t l_u$ means that during the transition, the instruction i_s is performed with the parameters q_t l_u. The symbol r_k denotes a specific set of type w symbols; an edge marked with a label that contains the r_k symbol represents a bundle of edges each of which is marked with one of the symbols belonging the r_k set. The transition subgraph $fG(l_1)$ of the fuzzy subautomaton $fA(l_1)$ is described by the symbolic expression $fG^+(l_1)$.

$$fG^+(l_1) = {}^0(q_{21}\,{}^1(0.2/w_3 i_1 q_{21},\ 0.7/w_7 i_1 q_{22}\,{}^2(0.8/w_2 i_1 q_{23}\,{}^3(0.5/w_1 i_1 q_{23},\ 0.2/w_6 i_1 q_{27}\,{}^4(1/r_0 i_7 q_{28})^4,\ 0.7/r_2 i_7 q_{28})^3)^2,\ 0.3/r_9 i_1 q_{23},\ 0.4/r_3 i_5 q_{24} l_3 q_{61})^1,\ q_{24}\,{}^1(1/r_0 i_2 q_{25}\,{}^2(0.9/w_1 i_1 q_{23},\ 0.4/r_3 i_5 q_{61} l_3 q_{61})^2)^1,\ q_{26}\,{}^1(1/r_0 i_2 q_{25})^1)^0$$

The above mentioned English sentence properties to be modelled in the automaton are the reason why it has been decided to apply a transition relation instead of a transition function in the fPDAMS. Thus, in order to perform a deterministic simulation of a nondeterministic automaton we have to establish a deterministic transition-choice function for each set of the alternative automaton transitions. That deterministic transition-choice function for a nonfuzzy automaton (PDAMS) can correspond only to some arbitrarily accepted principle which would have little or nothing at all to do with the text and/or the text's grammar to be processed by the automaton. On the other hand, the lingual competence of a common language user (that includes us, among others) lets us think that there are language constructions used oftener than the other ones in the living (i.e. natural) language. In other words, when we linearly analyse the structure of an utterance we subconciously use some language templates and expect them to occur in the utterance. In most cases this leads to the expected result, for we can faster - and with lesser effort - understand the text just before we have actually completed reading, or listening to, it. In formal grammar terms it can be said that, indeed, not every possible parse tree of the analysed ambiguous sentence is equally likely to have been applied! However, the selection of a specific parse tree is by no means an unequivocal, binary one - it constitutes a fuzzy decision. This leads us around to think that the optimal automaton type to be applied for NLP is the fuzzy automaton.

The formal definition of the fPDMAS can be find in the Appendix.

5. Conclusions

The goal of our research is to design a computer system with the same capability of language and knowledge acquistion as human being. To this end, we have to first equip the NLP system with the "smart" analyzer, which can be taught while working. The basic advantage of applying a fuzzy automaton to process natural language lies in the existence of an established algorithm for the next state selection (more precisely, the next configuration selection) based on values of the characteristic functions, allowing the μ(.) value to change with time! With such an approach the fPDAMS conceptually corresponds to the deterministic automaton of a dynamic structure but, whilst the former is the operational equivalent of the latter, the fPDAMS is much less complex. The possibility to modify the characteristic function, which describes the fPDAMS transition relations is concerned with the possibility for the automaton to learn on the basis of the analysed sentences. In such a case an adaptive automaton must be connected in

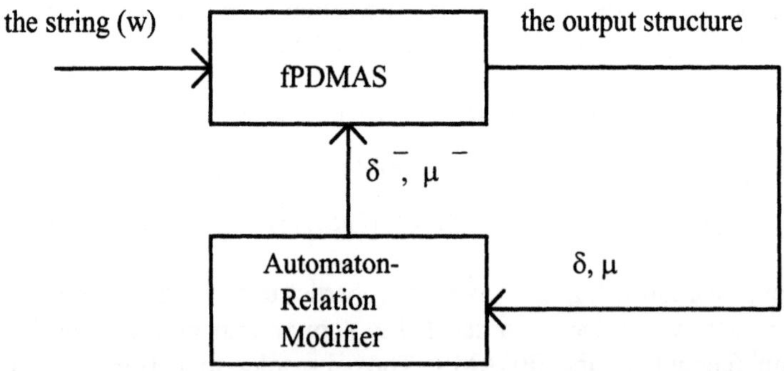

Figure 3. Structure of the control system of the self-learning analyzer

series, in a feedback loop, with an automaton-relation modifier. This solution exemplifies the Figure 3. There can be various algorithms for the modifier. The most trivial solution to this problem is to increase μ by an established value at all the automaton transitions which are situated in the graph on the path leading to the final state, each time the string (w) has been successfully decomposed. Such a modification of the fPDAMS transitions teaches the automaton, or "accustomes" it to, the syntax constructions occuring most frequently at its input, which behaviour simulates up to an extent the language behaviour in humans. The learning process of the proposed analyzer can also rely on inductive learning with genetic algorithms. In this case each graph path can be considered as a chromosome being subject to crossover and mutation. The analyzer increases μ by an established value at the paths which have the higher fitness value.

References

[1] Church W.E., Rau L.F. 1995 Commercial Applications of Natural Language Processing *Communications of the ACM* 38(11): 71-79.

[2] Unold O 1997 Application of Fuzzy Automaton in Natural Language Processing *Proc. of the KOWBAN'97* Wroclaw-Swieradow Zdroj: 523-528 (in Polish).

[3] Unold O 1993 *Knowledge Representation Expressed in the Sentences of Natural Language with Semantics and Reasoning,* PhD Thesis, Report No. 61/93, Wroclaw, ICT Wroclaw Technical University Press., Wroclaw. (in Polish).

[4] Unold O 1996 A Stratificational Knowledge Representation System, In: Vetulani Z, Abramowicz W(eds) *Language and Technology*, Academic Printing House PLJ, Warszawa: 177-181 (in Polish).

[5] Unold O 1997 Automatic Analysis of Natural Language Texts in Man-Machine Communication, In: Wojtkowski G at al (eds), *Systems Development Methods for the Next Century*, Plenum Publishing Corp., New York: 185-193.

[6] Lamb S 1966 *An Outline of Stratificational Grammar*, Georgetown University Press, Washington DC.

[7] Halliday M.A.K. 1967 Notes on transitivy and theme in English, *J. of Linguistics*, 3,4.

[8] Karttunen L., Kaplan R., Zaenen A. 1992 Two-level Morphology with Composition, *Proc. COLING-92*, France: 141-148.

[9] Tapanainen P., Voutilainen A. 1993 Ambiguity Resolution in a Reductionistic Parser, *Proc. 6th Conf. of the European Chapter of the ACL*, Utrecht: 394-403.

[10] Mohri M., Pereira F., Riley M. 1996 Weighted Automata in Text and Speech Processing, *ECAI-96 Workshop*, Budapest.

[11] Yager R.R., Filev D.P. 1994 *Essentials of Fuzzy Modeling and Control*, John Wiley&Sons, Inc.

[12] Kandel A., Lee S.C. 1979 *Fuzzy Switching and Automata: Theory and Applications*, Crane Russak, New York.

Appendix: Formal Definition of the fPDAMS

Formally we consider the fPDAMS as a 10-tuple

$$fPDAMS = (Q, \Gamma, F, \delta, z_0, q_0, q_f, L, T, p), \text{ where}$$

Q is a finite set of internal states ($q_i \in Q$),

$q_0 \in Q$ is a distinguished state so-called an initial state,

$q_f \in Q$ is a distinguished state so-called a terminal state,

$F \subset Q$ is a distinguished subset of the set Q so-called a set of terminal states of subautomata,

L is a finite set of l-markers ($L = \{l_1, l_2, l_3, l_4, l_5, l_6\}$),

T is a finite set of t-markers ($T = \{t_1, t_2\}$),

p is an element so-called a p-marker,

$\Gamma = W \cup Z_1 \cup Z_2 \cup Z_3 \cup Z_4 \cup Z_5 \cup Z_6$, where

W is an alphabet of an input tape ($W = \{w_0, w_1, w_2, w_3, w_4, w_5, w_6, w_7, w_8, w_9\}$),

Z_1 is an alphabet of a stack S_1 ($Z_1 = Q$),

Z_2 is an alphabet of a stack (Z_2 is a set of natural numbers),

Z_3 is an alphabet of a stack S_3 ($Z_3 = W \cup T \cup L$),

Z_4 is an alphabet of a stack S_4 ($Z_4 = Q$),

Z_5 is an alphabet of a stack S_5 ($Z_5 = W \cup T \cup L \cup p$),

Z_6 is an alphabet of a stack S_6 ($Z_6 = W \cup T \cup L \cup p$),

$z_0 = (z_{00}, z_{01}, z_{02}, z_{03}, z_{04}, z_{05}, z_{06}) \in W \times Z$, where $Z = Z_1 \times Z_2 \times Z_3 \times Z_4 \times Z_5 \times Z_6$, is a 7-tuple of initial symbols of the input tape and the stacks respectively.

Let $M_i \in (d \times \{m_l, m_h, m_r\})$ for $i = 1, ..., 6$ be intended to represent the shift of an i-th head, where m_h denotes inertness of the head, m_l -move the head d-th tape square to the left, m_r -move the head d-th tape square to the right. M_0 represents the shift of the input tape.

Let $M = M_0 \times M_1 \times M_2 \times M_3 \times M_4 \times M_5 \times M_6$ and $\Phi = Q \times W \times Z$.

Now the transition relation δ of fPDAMS is defined as follows:

$$\delta: \Phi \to 2^{\Phi \times M}$$

δ performs a mapping from the product space Φ to the set of the product space $\Phi \times M$. The product Φ is a 8-tuple consisting of a current state, an input symbol and six stack symbols. Each element of the set $\Phi \times M$ consists of a next state and seven pairs, each of which defines a new symbol of the input tape or stacks and both direction and length of head motion.

δ is the fuzzy relation with a membership grade

$$\mu: \Phi \times \Phi \times M \to [0,1]$$

Thus $\mu(q_i, w_{j1}, z_k, q_j, w_{j2}, z_p, M_r) \in [0, 1]$, where $q_i, q_j \in Q$, $w_{j1}, w_{j2} \in W$, $z_k, z_p \in Z$, $M_r \in M$.

In the fPDAMS each transition has its own „strength" - a grade of membership in a fuzzy set.

The grade of transition for an input sequence of length *m* is defined by an *m*-ary fuzzy relation. The membership grade is as follows.

For an input word $w_{j1}w_{j2}...w_{jm} \in W^*$

$$\mu(q_{i1}, w_{j1}, z_{k1}, q_{im}, w_{jm}, z_{km}, M_{rm}) = \max_{q_{i2}, q_{i3},...,q_{im-1} \in Q} \min \mu(q_{i1}, w_{j1}, z_{k1}, q_{i2}, \sigma_1, z_{k2}, M_{r1}), \mu(q_{i2}, \sigma_2, z_{k2}, q_{i3}, \sigma_3, z_{k3}, M_{r2}), ..., \mu(q_{im-1}, \sigma_j, z_{km-1}, q_{im}, w_{jm}, z_{km}, M_{rm}),$$

where $z_{k1}, z_{k2}, ..., z_{km} \in Z$, $M_{r1}, M_{r2}, ..., M_{rm} \in M$, $\sigma_1, \sigma_2, ..., \sigma_j \in W$.

Lets take notice of a fact, that the fuzzy transition relation δ does not have to change in one move neither internal state of the fPDAMS nor input element being read.

Formally an activity of fPDAMS is described by a configuration. We consider the configuration of fPDAMS as a 7-tuple

$K = (\alpha_0, \alpha_1, \alpha_2, \alpha_3, \alpha_4, \alpha_5, \alpha_6)$, where

α_0 is a triple (q_a, x, y), where q_a will stand for the current state of automaton, x represents the word written in input tape from the beginning to the current head position, y is the word written on the right of the head, while α_i (i = 1,..., 6) is a couple (q_a, v), where v denotes the word written in i-th stack (on the left of the head).

The binary relation $K_t \underset{\mu_{t,t+1}}{\Rightarrow} K_{t+1}$

stands for the transition from the configuration K_t to the configuration K_{t+1} in one step with a membership grade $\mu_{t,t+1}$, where the subscript *t* denotes the configuration K at the time *t*.

The notation $K_t \overset{*}{\underset{\mu_{t,t+j}}{\Rightarrow}} K_{t+j}$

stands for a configuration transition path

$$K_t \underset{\mu_{t,t+1}}{\Rightarrow} K_{t+1} \underset{\mu_{t+1,t+2}}{\Rightarrow} \dots \underset{\mu_{t+j-1,t+j}}{\Rightarrow} K_{t+j}$$

with a membership grade

$$\mu_{t,t+j} = \max_{t_{i+1}, t_{i+2},...,t_{i+j-1}} \min \mu_{t,t+1}, \mu_{t+1,t+2}, ..., \mu_{t+j-1,t+j}$$

The input word $w_{j1}w_{j2}...w_{jn}$ is accepted (splitted) by fPDAMS only if $K_t \overset{*}{\underset{\mu}{\Rightarrow}} K_f$ exists and $\mu > 0$, where $K_0 = ((q_0, z_{00}w_{j1}, w_{j2}w_{j3}...w_{jn}), (q_0, z_{01}),...,(q_0, z_{06}))$,

$K_f = (\alpha_0, \alpha_1, ..., \alpha_6)$ and q_f in α_i (i=0, ...,6).

The configuration K_0 is termed the initial configuration of fPDAMS, whereas the configuration K_f is called the terminal (final) configuration of fPDAMS. It should be noted that fPDAMS accepts the input word (divides it into subwords) as soon as it enters the terminal state.

The fPDAMS Instructions

The rules of the transitions from one configuration to the other one are clustered in the instructions below.

Let O, N, P, M, U denote the words over an alphabet of an input tape ($O, N, P, M, U \in W^*$),

A is a word over the stack alphabet S_1 ($A \in Q^*$), B is a word over the stack alphabet S_2 ($B \in Z_2^*$), C i J are words over the stack alphabet S_3 ($C,J \in (W \cup T \cup L)^*$), D is a word over the stack alphabet S_4 ($D \in Q^*$), E is a word over the stack alphabet S_5 ($E \in (W \cup T \cup L \cup p)^*$), G is a word over the stack alphabet S_6 ($G \in (W \cup T \cup L \cup p)^*$).

Let K^* represents the set of every possible configurations of fPDAMS

$$K^* = \{K \mid K_0 \overset{*}{\underset{\mu}{\Rightarrow}} K \text{ and } \mu > 0 \text{ and } K_0 = ((q_0, z_{00}w_{j1},w_{j2}O), (q_0, z_{01}),...,(q_0, z_{06})),$$

$K_0, K \in q \times z_{00}W^* \times W^* \times z_{00}Z_1^* \times z_{02}Z_2^* \times ... \times z_{06}Z_6^*$, where $q \in Q\}$.

Each instruction of fPDAMS belongs to one of seven types. Now, we will show the formal notation of the instructions:

1) type i_1: $(\exists K_1, K_2 \in K^*)$[if ($K_1 \underset{\mu_{1,2}}{\Rightarrow} K_2$ and $\mu_{1,2} > 0$) then ($K_1 = ((q_i, z_{00}Nw_{jk}, w_{jk+1}P), (q_i, z_{01}A), (q_i, z_{02}B), (q_i, z_{03}C), (q_i, z_{04}D), (q_i, z_{05}E), (q_i, z_{06}G))$,

   $K_2 = ((q_i, z_{00}Nw_{jk}w_{jk+1}, P), (q_i, z_{01}A), (q_i, z_{02}B), (q_i, z_{03}Cw_{jk}), (q_i, z_{04}D), (q_i, z_{05}E), (q_i, z_{06}G)))$]

{execution of i_1 performs a pushing down onto the stack S_3 the symbol read by the input tape head and moves the input tape head one tape square to the right}

2) type i_2: $(\exists K_1, K_2 \in K^*)$[if ($K_1 \underset{\mu_{1,2}}{\Rightarrow} K_2$ and $\mu_{1,2} > 0$) then ($K_1 = ((q_i, z_{00}Nw_{jk}, w_{jk+1}P), (q_i, z_{01}A), (q_i, z_{02}B), (q_i, z_{03}C), (q_i, z_{04}D), (q_i, z_{05}E), (q_i, z_{06}GRSX))$,

$K_2 = ((q_i, z_{00}Nw_{jk}, w_{jk+1}P), (q_i, z_{01}A), (q_i, z_{02}B), (q_i, z_{03}CSX), (q_i, z_{04}D), (q_i, z_{05}E), (q_i, z_{06}GR)), R, X \in (W \cup T \cup p)^*, S \in L^*)]$

{execution of i_2 pops a group of symbols from stack S_6 and pushes it onto stack S_3. This group comprises the symbols from the set W, the l-marker and the string of succeeding l-markers}

3) type i_3: $(\exists K_1, K_2 \in K^*)$[if ($K_1 \underset{\mu_{1,2}}{\Rightarrow} K_2$ and $\mu_{1,2} > 0$) then ($K_1 = ((q_i, z_{00}Nw_{jk}, w_{jk+1}P), (q_i, z_{01}A), (q_i, z_{02}B), (q_i, z_{03}C), (q_i, z_{04}D), (q_i, z_{05}E), (q_i, z_{06}G))$,

$K_2 = ((q_i, z_{00}Nw_{jk}, w_{jk+1}P), (q_i, z_{01}Aq_i), (q_i, z_{02}Bh), (q_i, z_{03}C), (q_i, z_{04}D), (q_i, z_{05}EpC), (q_i, z_{06}Gp))$, h is a natural number)]

{execution of i_3 stores the elements of current configuration in stack memory: the symbol of current state q_i is pushed down on top of stack S_1, the number of read input symbols h is pushed down on top of stack S_2, p-marker and next contents of stack S_3 are pushed on S_5, p-marker is pushed on S_6}

4) type i_4: $(\exists K_1, K_2 \in K^*)$[if ($K_1 \underset{\mu_{1,2}}{\Rightarrow} K_2$ and $\mu_{1,2} > 0$) then ($K_1 = ((q_i, z_{00}Nw_{jk}, w_{jk+1}P), (q_i, z_{01}Aq_j), (q_i, z_{02}Bh), (q_i, z_{03}C), (q_i, z_{04}D), (q_i, z_{05}EpJ), (q_i, z_{06}GpY))$,

$K_2 = ((q_j, z_{00}Mw_{jh}, w_{jh+1}U), (q_j, z_{01}A), (q_j, z_{02}B), (q_j, z_{03}J), (q_j, z_{04}D), (q_j, z_{05}E), (q_j, z_{06}G))$, h is a natural number, $Y \in (W \cup L \cup T)^*$)]

{execution of i_4 performs a return of the automaton to the configuration (denoted as K_a) popped from stack memory. From the top of stack S_1 the symbol q_j is popped, q_j becomes a current state of automaton. The contents of the stack S_3 on the ground of the contents of the stack S_5 is reconstructed. The number h of read-in input symbols in configuration K_a is popped from the top of stack S_2, the input tape head is shifted to the h-th tape square. At last the string of symbols stretching from the top of the stack to the first p-marker is popped from the stack S_6}

5) type i_5 (with arguments q_k i l_t): $(\exists K_1, K_2 \in K^*)$[if ($K_1 \underset{\mu_{1,2}}{\Rightarrow} K_2$ and $\mu_{1,2} > 0$) then

($K_1 = ((q_i, z_{00}Nw_{jk}, w_{jk+1}P), (q_i, z_{01}A), (q_i, z_{02}B), (q_i, z_{03}C), (q_i, z_{04}D), (q_i, z_{05}E), (q_i, z_{06}G))$,

$K_2 = ((q_i, z_{00}Nw_{jk}, w_{jk+1}P), (q_i, z_{01}A), (q_i, z_{02}B), (q_i, z_{03}Cl_tt_1), (q_i, z_{04}Dq_k), (q_i, z_{05}E), (q_i, z_{06}G)))$]

{execution of i_5 performs a pushing down onto the stack S_4 the symbol q_k and onto the stack S_3 l_t and t_1-markers}

6) type i_6: $(\exists K_1, K_2 \in K^*)$[if ($K_1 \underset{\mu_{1,2}}{\Rightarrow} K_2$ and $\mu_{1,2} > 0$) then (K_1 = ($(q_i, z_{00}Nw_{jk},$

$w_{jk+1}P), (q_i, z_{01}A), q_i, z_{02}B), (q_i, z_{03}C), (q_i, z_{04}Dq_k), (q_i, z_{05}E), (q_i, z_{06}G)),$

$K_2 = ((q_k, z_{00}Nw_{jk}, w_{jk+1}P), (q_k, z_{01}A), (q_k, z_{02}B), (q_k, z_{03}C), (q_k, z_{04}D),(q_k,$
$z_{05}E), (q_k, z_{06}G)))]$

{execution of i_6 changes the current state of automaton to a state popped from stack S_4. The instruction stands for the return of control from a subautomaton}

7) type i_7: $(\exists K_1, K_2 \in K^*)$[if ($K_1 \underset{\mu_{1,2}}{\Rightarrow} K_2$ and $\mu_{1,2} > 0$) then (K_1 = ($(q_i, z_{00}Nw_{jk},$

$w_{jk+1}P), (q_i, z_{01}A), (q_i, z_{02}B), (q_i, z_{03}C), (q_i, z_{04}D), (q_i, z_{05}E), (q_i, z_{06}G)),$

$K_2 = ((q_i, z_{00}Nw_{jk}, w_{jk+1}P), (q_i, z_{01}A), (q_i, z_{02}B), (q_i, z_{03}), (q_i, z_{04}D),(q_i, z_{05}E),$
$(q_i, z_{06}GCt_2)))]$

{execution of i_7 pushes down the contents of stack S_3 and next t_2-marker onto the stack S_6. Afterwards the stack S_3 is cleaned}

The fPDAMS Algorithm

In the beginning, fPDAMS is in the initial configuration K_0. There is the string $w_{j1}w_{j2}...w_{jn}$ written in the input tape. The current state of the automaton is denoted by the symbol q_a. The algorithm of working fPDAMS proceeds as follows:

(1) if $q_a = q_f$ then success.

{if the automaton entered the terminal state, then it stops. There is the splitted string (w) in the stack S_6}.

(2) if $q_a \in F$ then execute i_6 and goto step 1.

{if the automaton entered one of the terminal states of the subautomaton, then top of the stack S_4 indicates the next state}

(3) $K(q_a) = \{K \mid K^a \underset{\mu}{\Rightarrow} K$ and $\mu > 0$ and $K^a = ((q_a, z_{00}N, P), (q_a, z_{01}A), (q_a, z_{02}B),$

$(q_a, z_{03}C), (q_a, z_{04}D), (q_a, z_{05}E), (q_a, z_{06}G)),$

$K = ((q_x, z_{00}N', P'), (q_x, z_{01}A'), (q_x, z_{02}B'), (q_x, z_{03}C'), (q_x, z_{04}D'),(q_x, z_{05}E'), (q_x,$
$z_{06}G'))$ and $K^a, K \in K^*\}$

$\mu(q_a) = \{\mu \mid K^a \underset{\mu}{\Rightarrow} K$ and $\mu > 0$ and $K^a = ((q_a, z_{00}N, P), (q_a, z_{01}A), (q_a, z_{02}B),$

$(q_a, z_{03}C), (q_a, z_{04}D), (q_a, z_{05}E), (q_a, z_{06}G)),$

$K = ((q_x, z_{00}N', P'), (q_x, z_{01}A'), (q_x, z_{02}B'), (q_x, z_{03}C'), (q_x, z_{04}D'),(q_x, z_{05}E'), (q_x,$
$z_{06}G'))$ and $K^a, K \in K^*\}$

{the set $K(q_a)$ represents the set of every possible configurations of fPDAMS in the current state. The set $\mu(q_a)$ represents the set of the membeship functions of moves from the current state to the next ones}

(4) if $\|K(q_a)\| \geq 1$ then

(4a) choose the next configuration $K_{next} \in K(q_a)$, where $\mu_{next} = \max_{\mu} \mu(q_a)$, and

$K(q_a)=K(q_a) \setminus K_{next}$.

{if for the current state there is the set consisting of more than one next state, then choose one with the maximum membership grade and decrease the set of next states}

(4b) if $\|K(q_a)\| > 0$ then execute i_3.

{if the set of next states is still non-empty then store the elements of current configuration in stack memory}

(4c) perform $K^a \underset{\mu}{\Rightarrow} K_{next}$ and goto step 1.

(5) if stack S_1 is empty then fail.

{there is no configuration of the automaton with the non-empty set of next states. fPDAMS does not accept ,nor does it split, the input word}

(6) execute i_4 and goto step 4.

{return of automaton to the configuration popped from stack memory}.

A New Approach to Acquisition of Comprehensible Fuzzy Rules

Hiroshi Ohno[1], Takeshi Furuhashi[2]

[1] *Toyota Central Research & Development Labs., Inc, 41-1, Asa Yokomichi, Oaza Nagakute, Nagakute-cho, Aichi-gun, Aichi-ken, 480-1192, Japan, oono@milab.tytlabs.co.jp*
[2] *Nagoya University, Furo-cho, Chikusa-ku, 464-8603, Japan, furu@bioele.nuee.nagoya-u.ac.jp*

Keywords: Fuzzy modeling, Knowledge acquisition, Evolutionary programming, Fuzzy neural network

Abstract

This paper presents a new approach to acquisition of comprehensible fuzzy rules for fuzzy modeling from data. For the accuracy of the model, the identified model is probable to have many membership functions which are overlapped with each other. From the viewpoint of knowledge acquisition, it is desirable that the model has a smaller number of membership functions without the overlaps. Considering that the precision and the clarity of the fuzzy model are the trade-off, we propose an acquisition method of comprehensible fuzzy rules from the identified model which satisfies the desired accuracy. The precise model is to be obtained by, for example, fuzzy neural network(FNN) and using evolutionary algorithm. This paper proposes to use evolutionary programming(EP) to extract comprehensible rules from the fuzzy model. A numerical experiment is done to show the feasibility of the proposed method.

1. Introduction

In recent years, many researches on fuzzy modeling from data have been widely developed for numerous applications[1]. In these fuzzy modeling, parameters of membership functions and fuzzy rules are generated and tested by evolutionary algorithm. The resulting fuzzy model thus represents the nonlinear characteristics of unknown system. However, the model cannot always provide comprehensible fuzzy rules since many membership functions are generated and overlapped with each other to represent the nonlinear characteristics. These conventional approaches have put the emphasis on the precision of the acquired model. Comprehensibility of the fuzzy rules thus has been sacrificed for achieving the precision. Since comprehensible fuzzy rules discovered from unknown system are useful for the understanding of the system, a modeling method satisfying both the precision and the clarity of the resulting model is desirable.

In this paper we address the knowledge acquisition with clear linguisitic meanings for the fuzzy modeling and propose a new approach using EP[2], which consists of modeling phase and re-evaluation phase. The acquired fuzzy rules are re-evaluated for clarifying their linguistic meanings as the trade-off with the precision of the model. EP is a powerful technique for optimization, and can directly handle numerical data without the use of the binary code.

The purpose of another approaches described in the literature[3], [4], [5] aims to reduce the number of fuzzy sets or fuzzy rules in the model. These approaches do not always guarantee intrepretation and analysis of the rules after modeling.

In Section 2, our approach is presented. Section 3 illustrates computer experiments using a simple example from literature. In Section 4, a summary and conclusions complete the paper.

2. Proposed Algorithm

This section describes the proposed algorithm for the clarification of fuzzy rules identified by the FNN[6]. Another fuzzy modeling techniques can also be used[7]. The algorithm consists of modeling phase and re-evaluation phase.

2.1 Fuzzy modeling

In the modeling phase, we use the FNN with Gaussian-type membership functions depicted in Fig.2.1, and identify their parameters using EP from given input-output trainging data. We can also use an another modeling technique instead of EP, such as gradient descent method.

The fuzzy rules are represented as

$$\begin{aligned} R_i : If \;\; & x_1 \; is \; A_{i1} \;\; and \;\; x_2 \; is \; A_{i2}, \cdots, \\ & x_N \; is \; A_{iN} \\ & Then \;\; y = w_i \\ & y = \sum_{i=1}^{M} \nu_i \cdot w_i \\ \nu_i &= \frac{\prod_{j=1}^{N} \mu_{A_{ij}}(x_j)}{\sum_{j=1}^{M} \prod_{k=1}^{N} \mu_{A_{jk}}(x_k)} \end{aligned} \tag{2.1}$$

where $R_i (i = 1, 2, \cdots, M)$ denotes the ith fuzzy rule and $x_j (j = 1, 2, \cdots, N)$ is the input and y is the output of the fuzzy rule, w_i is the singleton, A_{ij} is the fuzzy variable with the Gaussian-type membership function given by

$$\mu_{A_{ij}}(x_j) = \exp(-(x_j - c_{ij})^2/2\sigma_{ij}^2), \tag{2.2}$$

where c_{ij} and σ_{ij} determine the central position and the width of the membership function.

Hence, c_{ij}, σ_{ij}, and w_i are unknown parameters to be identified by EP. The objective of EP is to reduce the squared error between the output of the fuzzy model and that of the training data.

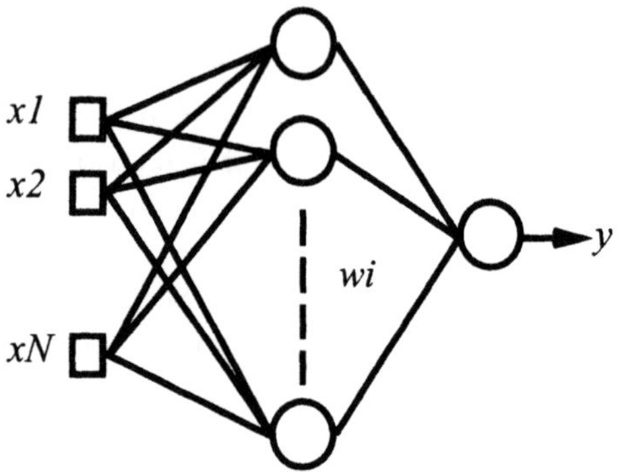

Figure 2.1. Architecture of fuzzy-neural network

2.2 Re-evaluation of fuzzy model

In the re-evaluation phase we simplify the membership functions for the clarity of the acquired knowledge in the following procedure: First, we select the types of membership functions in the antecedent such as trapezoidal or triangular with linguistic meanings, and decide the number of them. Second, we replace the original membership functions of the identified fuzzy model with a similar trapezoidal or triangular membership functions using EP for each antecedent variable under the constraints for the clarity. The objective of EP is to reduce the squared error between the value of each original membership function corresponding to the given input-output training data in the modeling phase and that of new membership function. The shape of the new membership functions is depicted in Fig.2.2. The unknown parameters of these membership functions are summarized in Table 2.1.

The constraints are imposed on the shape and the order of membership functions summarized in Table 2.2. These constraints are introduced for the clarity of the membership functions. If the constraints are not satisfied during the operation of EP, the parameters of the membership function are initialized and the EP operation is restarted.

Table 2.1. Parameters of membership functions

A	x, l
B	$x, l1, l2$
C	$x, l1, l2$
D	$x, l1, l2$

Finally, we evaluate the performance of the fuzzy model replaced by the new membership functions in terms of the input-output training data in the modeling phase.

The algorithm is summarized as follows:

Given the fuzzy model such as fuzzy-neural network constructed from data.

Step 1. Select the new membership functions for the antecedent variables.

Step 2. Create the training data which is the output values of the original membership functions corresponding to the input-otuput training data in the modeling phase.

Step 3. Adapt the new membership functions to the original membership functions of the fuzzy model using EP for each antecedent variable.

Step 4. Replace the membership functions with the new membership functions.

Step 5. Evaluate the new fuzzy model.

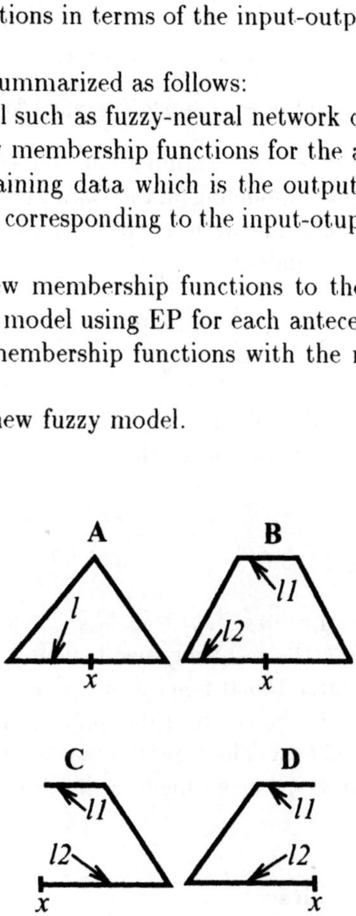

Figure 2.2. Shape of membership functions

Table 2.2. Conditions of the constraints

Shape	$0 < l$ for A	$0 < l1 < l2$ for B, C, D
Order	CBD	CABAD

3. Computer experiments

We used a simple example to demonstrate the proposed algorithm. Three-input and single-output function was chosen to be an unknown system for the modeling phase[6]. The function was represented as

$$y = (1 + x1^{0.5} + x2^{-1} + x3^{-1.5})^2. \tag{3.1}$$

Table 3.1 shows the input-output training data obtained from the function. Data 1~20 were used in the modeling phase. Data 21~40 were used for testing the generalization of model. The accuracy of fuzzy model was evaluated using the following performance index:

$$E = 1/L \sum_{i=1}^{L} \frac{|y_i - y_i^*|}{y_i} \times 100(\%), \tag{3.2}$$

where L denoted the number of data, y_i denoted the ith data corresponding to Table 3.1, and y_i^* was the output of the fuzzy model. For EP, population size was set to 100.

3.1 Modeling result

We set the initial random values of $c_{ij} \in [0, 5]$, $\sigma_{ij} \in [0.1, 1.0]$, and $w_i \in [0, 5]$. In Figs. 3.1, 3.2, and 3.3, the resulting membership functions corresponding to each input variable after 10000 generations with N = 3, M = 10, and ten membership functions are shown. In Table 3.2, the performance indexes of the resulting model are shown. The experimental result had little comprehensibility of the fuzzy rules, since the membership functions were overlapped.

3.2 Re-evaluation results

We examined two cases. Case 1: the number of the membership functions was A = 2, B = 1, C = 1, and D = 1, and Case 2: A = 0, B = 1, C = 1, and D = 1 for each input variable. The linguistic meanings in both cases were interpreted as C is "small," D is "big," B is "medium," and A is "small medium" or "medium big."

Table 3.3 shows the initial random values of the parameters of the membership functions assigned to A, B, C, and D. The maximum and minimum value of the input variable were 1 and 5, respectively.

Table 3.1. Training data

No.	x1	x2	x3	y	No.	x1	x2	x3	y
1	1	3	1	11.11	21	1	1	5	9.545
2	1	5	2	6.521	22	1	3	4	6.043
3	1	1	3	10.19	23	1	5	3	5.724
4	1	3	4	6.043	24	1	1	2	11.25
5	1	5	5	5.242	25	1	3	1	11.11
6	5	1	4	19.02	26	5	5	2	14.36
7	5	3	3	14.15	27	5	1	3	19.61
8	5	5	2	14.36	28	5	3	4	13.65
9	5	1	1	27.42	29	5	5	5	12.43
10	5	3	2	15.39	30	5	1	4	19.02
11	1	5	3	5.724	31	1	3	3	6.38
12	1	1	4	9.766	32	1	5	2	6.521
13	1	3	5	5.87	33	1	1	1	16
14	1	5	4	5.406	34	1	3	2	7.219
15	1	1	3	10.19	35	1	5	3	5.724
16	5	3	2	15.39	36	5	1	4	19.02
17	5	5	1	19.68	37	5	3	5	13.39
18	5	1	2	21.06	38	5	5	4	12.68
19	5	3	3	14.15	39	5	1	3	19.61
20	5	5	4	12.68	40	5	3	2	15.39

Table 3.2. Performance index(%)

learning(Data 1~20)	11.46
generalization(Data 21~40)	12.31

After 5000 generations, we obtained the membership functions as shown in Figs. 3.4, 3.5, and 3.6(Case 1) and 3.7, 3.8, and 3.9(Case 2). In the figures, A, B, C, and D are labeled on the membership functions. In Table 3.4, the performance indexes are shown. The result in Case 2 is better than that in Case 1. The performance indexes on both cases were degraded from that of the original fuzzy model shown in Table 3.2. This degradation may be regarded as the explanation cost for the comprehensibility.

In Figs. 3.4, 3.5, and 3.6 in Case 1, the membership functions assigned to B and D are not illustrated, because the another type membership functions, that is A or C, were automatically selected in Step 4 of the algorithm.

In both cases, the membership functions became more comprehensible than the original ones shown in Figs. 3.1, 3.2, and 3.3. If the objective of fuzzy modeling is on the performance of the model, the original model can be used. In the computer experiments, we needed to set carefully the initial values of parameters of EP for the purpose of avoiding local minimum of learning.

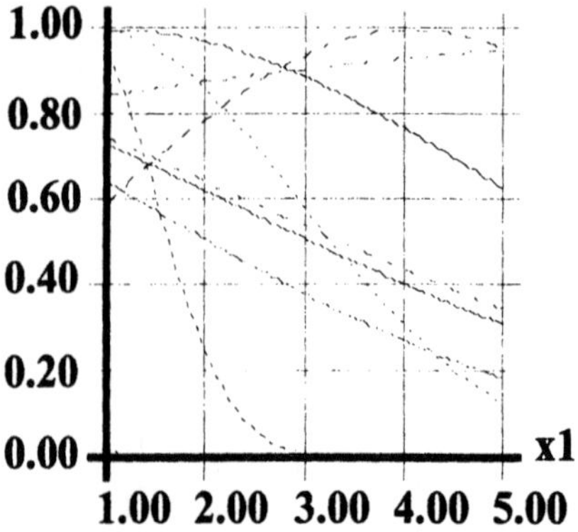

Figure 3.1. Membership functions of input variable(x1)

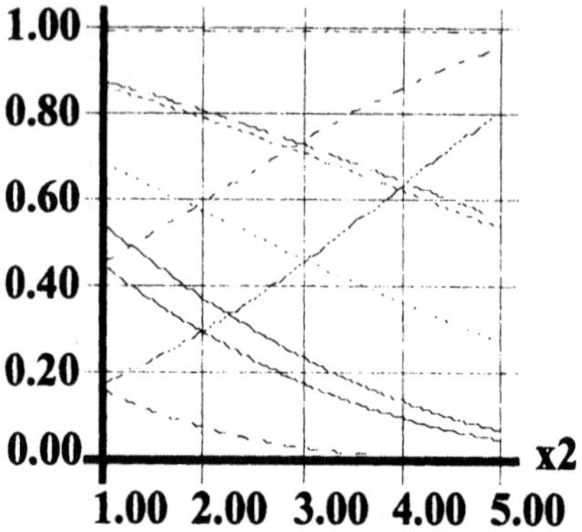

Figure 3.2. Membership functions of input variable(x2)

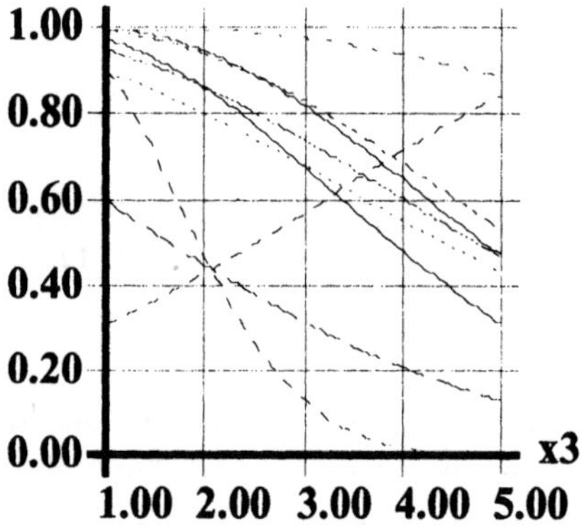

Figure 3.3. Membership functions of input variable(x3)

Table 3.3. Initial values of the membership functions

A	$x \in [3.8, 4.2]$, $[1.8, 2.2]$ $l \in [0.8, 1.2]$
B	$x \in [2.8, 3.2]$ $l1 \in [0.3, 0.7]$, $l2 \in [0.8, 1.2]$
C	$x \in [0.8, 1.2]$ $l1 \in [0.3, 0.7]$, $l2 \in [0.8, 1.2]$
D	$x \in [4.8, 5.2]$ $l1 \in [0.3, 0.7]$, $l2 \in [0.8, 1.2]$

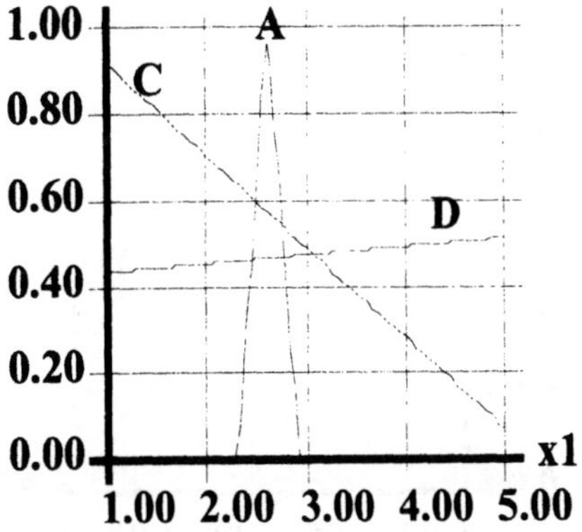

Figure 3.4. Membership functions for input variable(x1) in Case 1

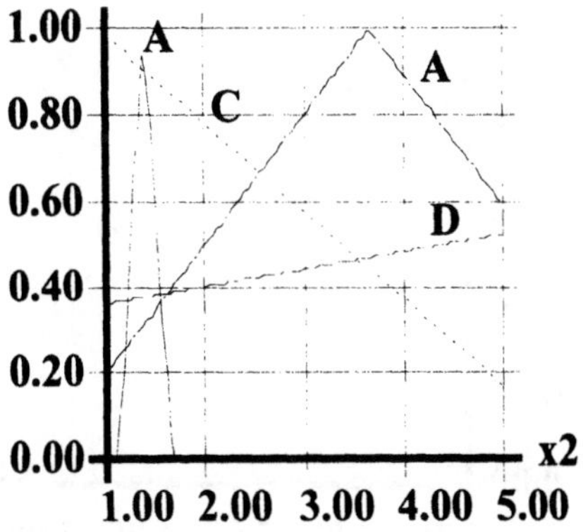

Figure 3.5. Membership functions for input variable(x2) in Case 1

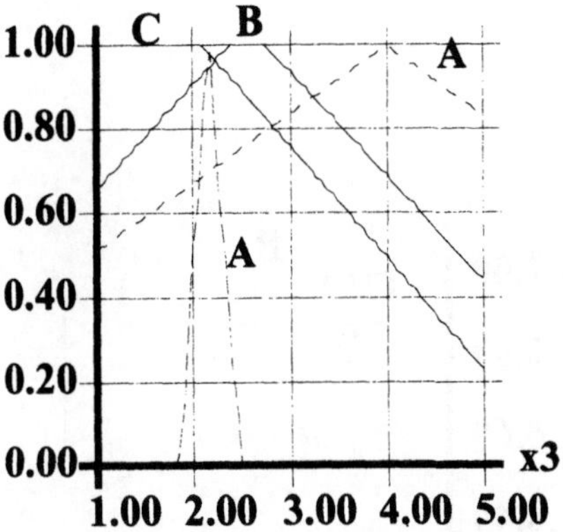

Figure 3.6. Membership functions for input variable(x3) in Case 1

Table 3.4. Performance index(%)

	Case 1	Case 2
learning	70.84	52.79
generalization	73.89	54.96

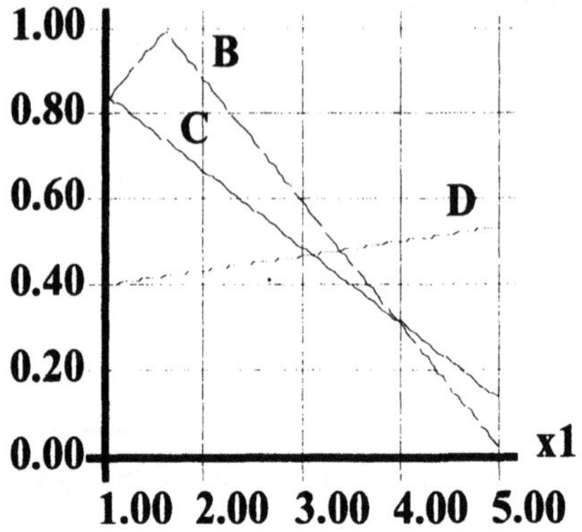

Figure 3.7. Membership functions for input variable(x1) in Case 2

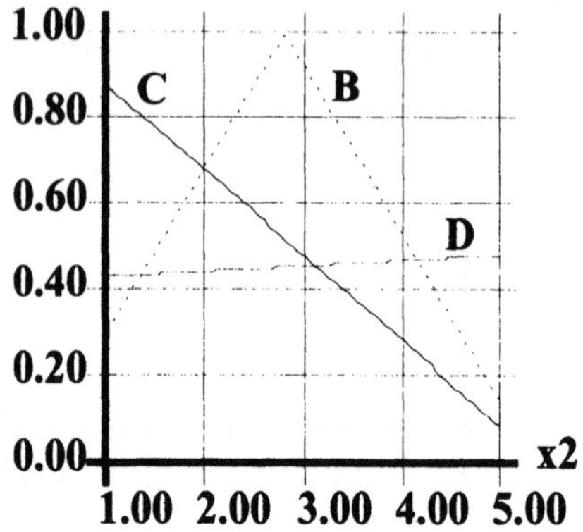

Figure 3.8. Membership functions for input variable(x2) in Case 2

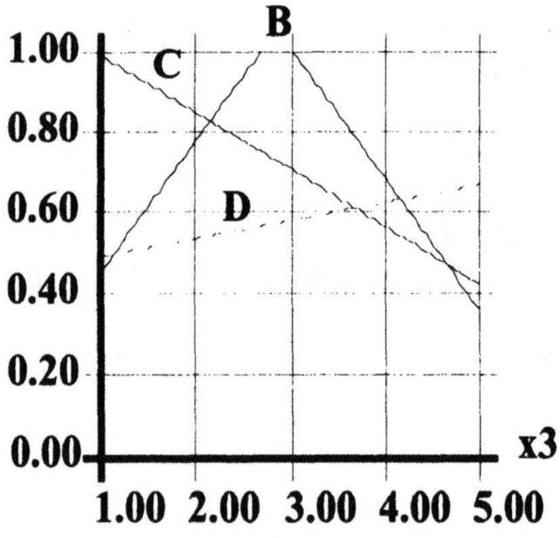

Figure 3.9. Membership functions for input variable(x3) in Case 2

4. Conclusions

In this paper, a new approach to the acquisition of comprehensible fuzzy rules from the FNN constructed from data was proposed and its feasibility was verified through computer experiments. This algorithm can also be used another fuzzy modeling techniques. The proposed algorithm using EP consists of two phases:modeling and re-evaluation. In the re-evaluation phase, we can control the explanation degree of knowledge acquisition. This is the feature that distinguishes the approach from conventional approaches. A numerical experiment was given to demonstrate the feasibility of the algorithm.

References

1. Bezdek J C 1993 Editorial:Fuzzy Models-What Are They, and Why ?. *IEEE Trans. Fuzzy Syst.*, vol.1, pp 1-6, Feb
2. Saravanan N, Fogel D B 1995 Evolving Neural Control Systems. *IEEE EXPERT*, 10(3), pp 23-27, June
3. Babuška R, Setnes M, Kaymak U, van Nauta Lemke H R 1996 Rule base simplification with similarity measures. In *Proc. Fifth IEEE Int. Conf. Fuzzy Syst.*, pp 1642-1647
4. Chao C T, Chen Y J, Teng C C 1996 Simplificaiton of fuzzy-neural systems using similarity analysis. *IEEE Trans. Syst., Man Cybern.*, vol.26, pp 344-354, April
5. Yen J, Wang L 1996 An SVD-based fuzzy model reduction strategy. In *Proc. Fifth IEEE Int. Conf. Fuzzy Syst.*, pp 835-841
6. Horikawa S, Furuhashi T, Uchikawa Y 1992 On fuzzy modeling using fuzzy neural networks with the back-propagation algorithm. *IEEE Trans. Neural Networks*, vol.3, pp 801-806, May
7. Sugeno M, Yasukawa T 1993 A fuzzy-logic-based approach to qualitative modeling. *IEEE Trans. Fuzzy Syst.*, vol.1, pp 7-31, Feb

Zero-Point Probability for Linear Source Separation

G. J. Scruby and S. J. Flockton

Signal Processing Group,
Physics Department,
Royal Holloway, University of London,
Egham, Surrey,
TW20 0EX, England
Email: G.Scruby@rhbnc.ac.uk, S.Flockton@rhbnc.ac.uk

Keywords: Blind, Signal, Separation

Abstract

"Blind" signal processing techniques have attracted a great deal of interest from the signal processing community in recent years. The term "blind" refers to the fact that very little information is known about the signals or processing in question. This paper deals with blind signal separation, the separation of unknown signals that have been mixed in an unknown way. In this paper the linear mixture model has been adopted; the signals in question are assumed to be linear mixtures of stationary sources. Techniques that address this problem using higher-order statistics rely on the accurate estimation of higher-order moments, either implicitly or explicitly. This can cause difficulties in real-time adaptive systems as anomalous outliers in the data can have disproportionate effects on the system. We introduce a novel on-line method for separating super-Gaussian signal mixtures that does not rely on higher-order moments, called "zero-point probability" (ZPP).

1. Introduction

Blind signal separation is an area of signal processing where much interest has been focused in recent years. The term "blind" is used because limited information is available about the manner of mixing or the type and number of sources in the mixture. Obviously, processing systems that can operate under such difficult conditions have a wealth of possible applications from real-time communications and the interpretation of on-line medical instrument readings to off-line data analysis in the social sciences and psychology. In its most general form this problem is intractable since some assumptions must be made in order to design a separation method. This has led to various directions of research within the blind separation field, each making different assumptions. In this paper we use the simplest model: linear, instantaneous mixtures of stationary sources. Although the simplest practical model, the linear model still has many uses and can often be applied successfully even when real-world conditions differ significantly from the constraints of the model. When the linear model is used the separation problem is often called Independent Component Analysis (ICA) because its solution can be likened to a higher-order generalisation of Principle Component Analysis (PCA). A common extension to this model is to assume that all the source signals have either a super-Gaussian or sub-Gaussian distribution and in this paper we assume all signals are super-Gaussian. Several techniques have been used to address this problem using higher-order statistics but all rely on the accurate estimation of higher-order moments, either implicitly or explicitly. This can cause difficulties in real-time adaptive systems as anomalous outliers in the data can have disproportionate effects on the system due to the high-order nature of the learning.

In this paper we describe a novel on-line method for separating super-Gaussian signal mixtures that does not rely on higher-order moments. The method uses a simple, constantly updating estimate of the central part of the probability distributions of the candidate mixed signals, the "zero-point probability" contrast, which can then be used to update the unmixing coefficients. The principle on which this contrast depends can be explained using the central limit theorem. This states that the limiting distribution of a standardised sum of independent, identically distributed summands as the number of terms increases, is a normal distribution. Therefore, if two signals that are "peakier" than a Gaussian signal are added together, the resulting signal will be more Gaussian than either of the two original signals. Thus by taking the zero-point probability as a contrast and attempting to maximise it, the least mixed version of the signals will be found. This method is simple to implement and its concentration on the central part of the probability distribution makes it insensitive to outliers in the data, in contrast with higher-order methods.

2. Statement of Problem

Because this paper deals with the separation of linear, instantaneously mixed signals with super-Gaussian probability distributions, no temporal information is available and higher-order statistics must be employed to separate the signals. This mixing model may be represented diagramatically as follows:

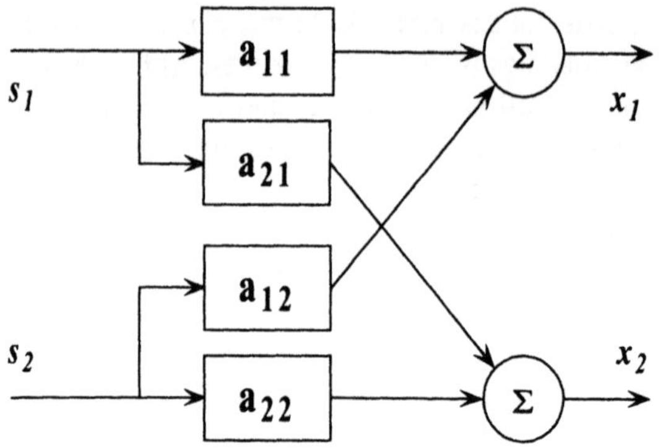

Figure 1. Linear, instantaneous mixture model

where s_1 and s_2 are the original independent signals. $\boldsymbol{a}_{11}$, $\boldsymbol{a}_{21}$, $\boldsymbol{a}_{12}$, $\boldsymbol{a}_{22}$ form the mixing matrix **A** and are the linear coefficients with which the signals are scaled and mixed. $\boldsymbol{x}_1$ and $\boldsymbol{x}_2$ are the only signals available to the observer: unknown linear mixtures of the sources. In this paper, as in much of the literature, the observed signals in question are considered to be zero mean, of unit standard deviation and are uncorrelated. Decorrelation is usually carried out in tandem with removal of the mean and normalisation of the standard deviation by means of pre-processing. This technique is known as whitening or sphering [1].

2.1 Nomenclature of Mixing Model

There are M scalar source signals written as an input vector.

$$\mathbf{s}(t) = \left[s_1(t), s_2(t).....s_M(t)\right]^{\mathrm{T}} \tag{1}$$

and the linear combination of the inputs is an L-dimensional output vector:

$$\mathbf{x}(t) = \left[x_1(t), x_2(t).....x_L(t)\right]^{\mathrm{T}} \tag{2}$$

where the subscript t represents a discrete time.

The output vector $\mathbf{x}_t$ is defined in terms of the input $\mathbf{s}_t$:

$$\mathbf{x}(t) = \mathbf{A}\mathbf{s}(t) \tag{3}$$

where

$$\mathbf{A} = \left[\mathbf{a}_1, \mathbf{a}_2.....\mathbf{a}_M\right] \tag{4}$$

where **A** is the $L\times M$ mixing matrix whose columns $\mathbf{a}_i$ are the basis vectors of ICA. Source separation requires an $M\times L$ separation matrix **K** which acts on $\mathbf{x}(t)$ to produce an estimate of the original source $\mathbf{s}(t)$.

$$\hat{\mathbf{s}}(t) = \mathbf{y}(t) = \mathbf{Kx}(t) \tag{5}$$

Equation (5) has no unique solution. If $\mathbf{y}(t)$ has independent components, the same property holds for the vector **DPy** where **P** is any permutation matrix and **D** is any non-singular, diagonal, scaling matrix. Therefore $\mathbf{s}(t)$ can only be determined within a scaling and a permutation [2]:

$$\mathbf{s}(t) \equiv \mathbf{DPs}(t) \tag{6}$$

Consequently it is necessary to find the separating matrix **K** such that:

$$\mathbf{KAs}(t) = \mathbf{DPs}(t) \tag{7}$$

In this paper we assume that the number of source signals M is equal to the number of received signals L. Consequently, the separating matrix **K** is a square matrix of dimension $n \times n$. One consequence of (7) is that among the n^2 unknowns of **K**, n elements cannot be determined. In order to obtain a well posed problem it is therefore necessary to impose n supplementary conditions. One way of doing this is to constrain the power of all the output components. When the problem has been simplified further by "whitening" or "sphering", the square separating matrix **K** reduces to an orthogonal transformation. "Sphering" can be easily achieved with conventional PCA [1].

3. Zero-Point Probability

In order to solve the linear problem we introduce the idea of "zero-point probability" (ZPP) as a contrast function. The contrast is formed by calculating the height of the probability density at zero for each channel and then forming the sum of these values. When it is maximised, separation has been achieved. In order for this to be a sufficient condition for separation the signals must be zero-mean and white, just as for many other separation methods. The principle on which this contrast depends can be explained using the central limit theorem. This states that the limiting distribution of a standardised sum of independent, identically distributed summands as the number of terms increases, is a normal distribution [3]. Therefore, if two signals that are "peakier" (have a higher kurtosis) than a Gaussian signal are added together, the resulting signal will be more Gaussian than either of the two original signals. Thus by taking the zero-point probability sum as a contrast and attempting to maximise it, the least mixed version of the signals will be found. The ZPP can be defined mathematically by use of the Dirac delta function.

$$\mathrm{ZPP} = \int_{-\infty}^{\infty} p(x)\delta(x)dx \tag{8}$$

given that $p(x)$ is continuous at $x = 0$.

3.1 Practical Considerations

In practice, because we are using a finite sample and because the Dirac delta function is only a theoretical entity, exact continuous integration is impossible; an approximation must be made. We choose a certain "window" about zero (of width γ) and use the proportion of data points that fall within this area as an approximation to the zero-point probability (Figure 2).

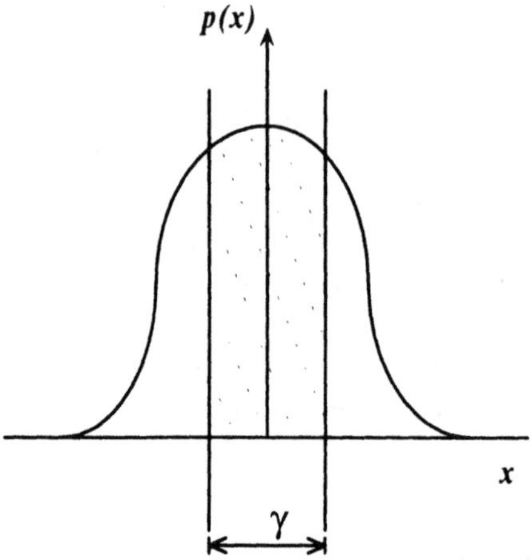

Figure 2. Window about zero for zero-point approximation

This practical ZPP value can be described mathematically by assuming the following approximation of the delta function during calculations.

$$\delta_{approx}(x) = \begin{cases} 0 & x < -\frac{\gamma}{2} \\ \frac{1}{\gamma} & -\frac{\gamma}{2} < x < \frac{\gamma}{2} \\ 0 & x > \frac{\gamma}{2} \end{cases} \tag{9}$$

Equation (9) defines a rectangular function with width γ, height γ^{-1} and area 1. It is an approximation of the delta function which provides a simple derivation of the integral property, such that when used in equation (8) it will integrate the shaded area shown in Figure 2. The value of γ will, of course, affect the performance of the separation algorithm. This can be seen below in a plot (Figure 3) that demonstrates the change in contrast with differing rotations and γ of a two channel mixture. The mixture consists of two 50000 point Gaussian cubed signals. The signals have been whitened and maximally mixed (effectively rotated by 45°). It can be seen that the contrast is greatest at 45° rotation. Thus a 45° rotation applied to the mixed signals would separate them again (as the surface is periodic with a period of 90° [4]).

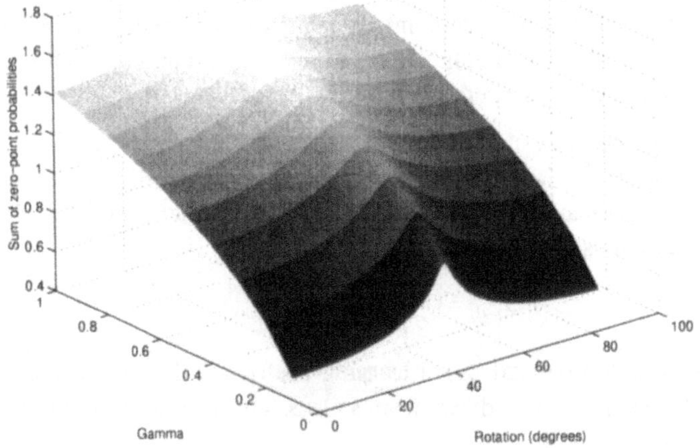

Figure 3. Performance surface of zero-point probability

It is clear that for large γ a very smooth performance surface will be obtained but with low resolution between minimum and maximum values whereas with small γ good definition will be obtained but with a loss of statistical reliability . The optimal choice of γ for a particular system is not a trivial problem. Optimisation requires *a priori* knowledge of the signal p.d.f.'s as well as the maximum contrast difference. This information is not usually available. The combination of γ and λ values will also determine the accuracy of the separation in a non-linear fashion dependent on the shape of the p.d.f.'s.

Figure 3 also shows that the contrast function is smoothly varying and uni-modal. The main reason for introducing the zero-point probability as a source separation contrast function, however, is its performance in on-line systems when adaptively estimating the contrast. A solution to the separation problem that uses 4th-order statistics has been found empirically to produce good results when operating on a batch of data [5]. This can be understood intuitively by considering that all information up to fourth order has been used optimally and that information above fourth order is both difficult to estimate accurately and has little effect on the solution. A problem arises, however, if an on-line separation system is required that can track changes in the mixing coefficients. Separation then becomes two distinct but equally important problems: the continuous estimation of the contrast function and the continuous update of the rotation. On-line estimation of the contrast causes difficulty due to the fourth-order nature of the function. Large data values have a disproportionate effect upon the contrast estimation because their fourth powers are used. This means that for signals with high kurtosis (very "peaky" signals) the contrast estimate will be very unstable and consequently so will the rotation estimate and thus the solution.

Since no fourth-order statistic is used with ZPP there are no adverse effects from outliers in the data. Indeed, zero-point probability ignores outliers altogether whether they are part of the underlying signal structure or simply spurious, as only the probability very near to the zero-point is considered.

A comparison between the learning of the zero-point probability and 4th-order moment is shown in Figure 4. It can be seen that disproportionate jumps in the 4th-order moment estimation are produced when large outliers appear in the data. The learning equation used in these examples was:

$$J_{t+1} = (1-\lambda)J_t + \lambda J_{current\ \ estimate} \tag{10}$$

where t is discrete time and λ is a learning coefficient. For the examples shown in Figure 4, the test signal was drawn from a Gaussian cubed distribution (with λ=0.001 in both cases).

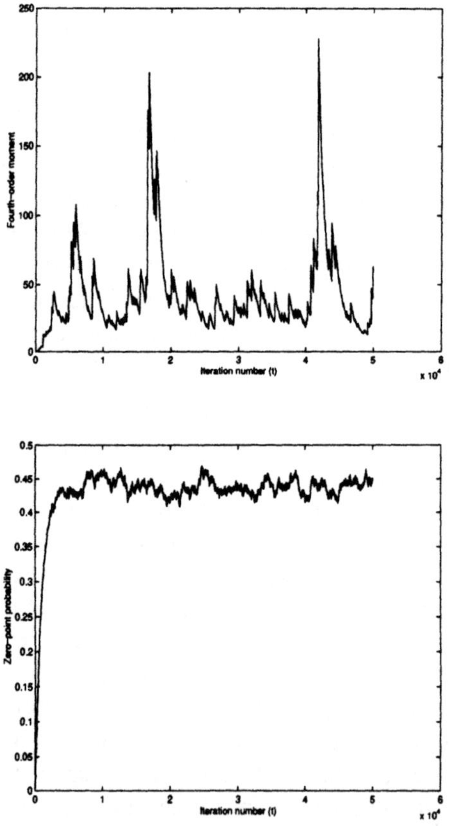

Figure 4. Comparison of estimators for 4th-order moment and zero-point probability

4. Simulation

An example of separation using ZPP can be seen in Figure 5. Two signals were drawn from a Gaussian distribution and then cubed to increase their kurtosis and the size of their outliers. The signals were then scaled to unit variance to ensure that when transformed orthogonally, they remained uncorrelated. The two independent, uncorrelated signals (***s***) were then mixed using a mixing matrix (**A**), as in equation 3, to produce mixed signals (***x***). The form of **A** used was an orthogonal transformation matrix: in this case a rotation of 10 degrees. The desired solution to this is a rotation of -10 degrees (or 80 degrees since the solution space repeats every 90 degrees [4]). The algorithm (the structure of which is outlined in the Appendix) consists of two parts: a continuously updating estimator of the ZPP sum (i.e. $J(x_1)+J(x_2)$ where J is found from equation 10) and a gradient ascent section that changes the rotation of the data every 500 iterations in order to maximise the ZPP sum.

A rotation step of 2 degrees was chosen empirically to ensure that in general the difference in the ZPP sum between rotation changes was greater than the error in the ZPP estimation. 2 degrees was found to be the smallest step size that gave consistently convergent results within the parameters of the learning algorithm in this simulation (λ=0.001).

A similar experiment using 4th-order moments was unstable for all step sizes. The value of the rotation after 60 rotation iterations was -12 degrees. After passing the results through a 20 point moving average filter to allow a more accurate and smoothly varying estimate of the answer, the rotation was found to be -9.8 degrees at the same point.

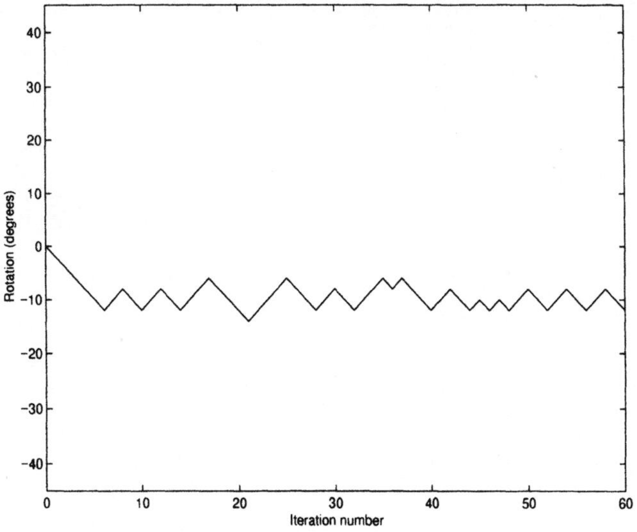

Figure 5. Separation of mixed signals using a zero-point probability contrast

5. Conclusion

We have introduced a new method for on-line source separation using a numerical contrast function, called the "zero-point probability", which has been derived from the central limit theorem. We have shown that during on-line estimation the zero-point probability is far more tolerant to outliers in the data than higher-order statistical methods and have demonstrated an algorithm for carrying out signal separation with this technique.

References

1. Karhunen J, Wang L, Vigario R 1995 Nonlinear PCA type approaches for source separation and independent component analysis. *Proceedings of the '95 IEEE International Conference on Neural Networks, ICNN '95,* Perth, Australia , 27th November 1st - December 1995, pp 995-1000

2. Oja E, Karhunen J 1995 Signal separation by nonlinear Hebbian learning. *Computational Intelligence: A Dynamic System Perspective,* IEEE Press, 83-97

3. Lindgren B W 1976 *Statistical Theory.* 3rd Edition, MacMillan Publishing Co. Inc.

4. Flockton S, Yang D, Scruby G 1996 Performance surfaces of blind source separation algorithms. *Proceedings of the International Conference on Neural Information Processing '96, ICONIP '96,* Hong Kong, pp. 1229-1234

5. Comon P 1994 Independent component analysis, a new concept?, *Signal Processing,* 36: 287-315

Appendix - Structure of Separation Algorithm

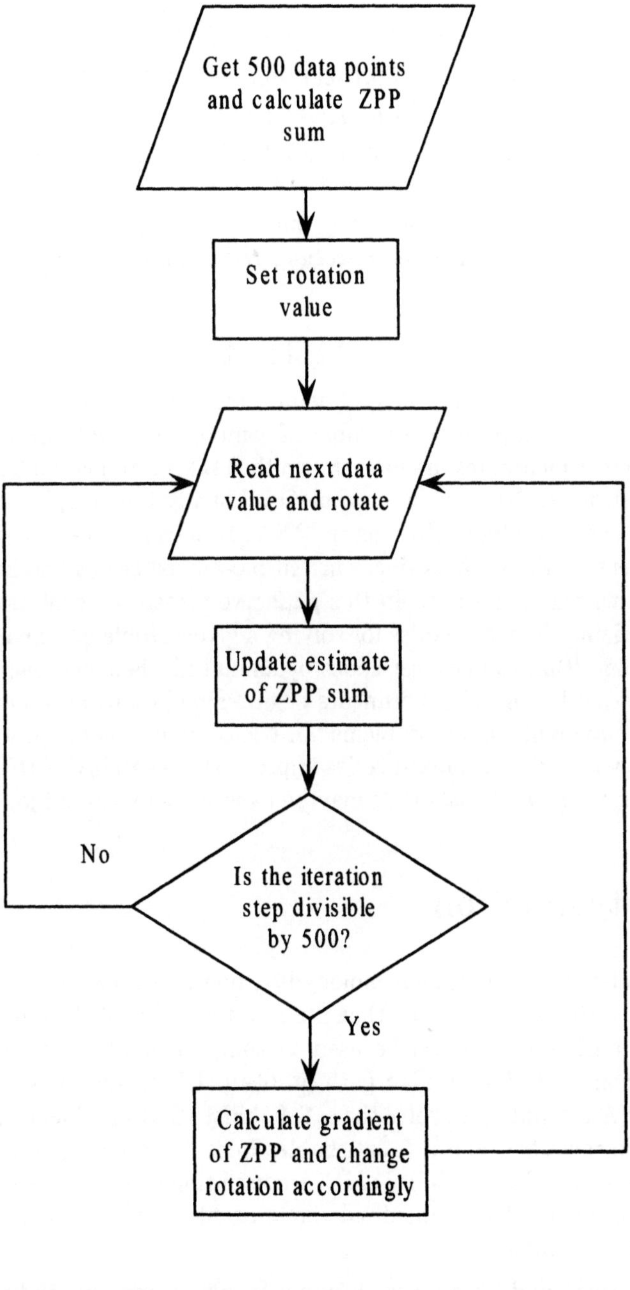

Code Optimization for DNA Computing of Maximal Cliques

Byoung-Tak Zhang, Soo-Yong Shin

Artificial Intelligence Lab (SCAI)
Dept. of Computer Engineering, Seoul National University
Seoul 151-742, Korea.
E-mail: {btzhang, syshin}@scai.snu.ac.kr

Keywords: *DNA computing, maximal clique problem, DNA code design, evolutionary optimization, NP-complete problems.*

Abstract

Since the Adleman's experimental demonstration of its feasibility, DNA computing has been applied to a number of combinatorial optimization problems. Several experimental results have shown that DNA strands can be used to compute solutions to NP-complete problems. Usually they employ random codes to represent candidate solutions in DNA. However, some codes have better error-tolerance than others since current bio-lab experiment techniques are involved with reaction errors. In this paper we present an evolutionary method for optimizing the DNA codes for solving a given problem. The method uses a genetic algorithm to find best codes by emulating chemical reaction processes before actual biological experiments begin. Simulations have been performed to solve maximum clique problems, an NP-complete search problem. The results show that the optimized codes improve the reliability of DNA computing and reduce time and costs that may be caused by repeated bio-experiments.

1 Introduction

Due to advances in molecular biology it is nowadays possible to create a soup of roughly 10^{18} DNA strands that fits in a test tube. Adleman [1] has shown that each DNA strand can be used to compute solutions to an instance of the NP-complete Hamiltonian path problem (HPP). Lipton [6] came up with using DNA computing to solve the satisfiability (SAT) problem: given a propositional formula decide if it is satisfiable. Recently, Ouyang et al. [7] provided further evidence for the ability of DNA computing to solve NP-complete search problems. They solved a maximal clique problem (MCP) by means of molecular biology techniques.

Both Adleman and Lipton used random encodings in their study. It was, how-

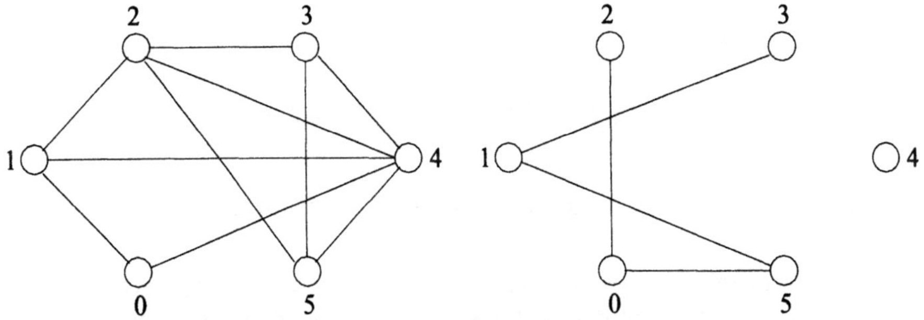

Figure 1: The maximum clique problem: (left) graph with 6 vertices, (right) its complementary graph.

ever, suggested that as the size of the problem grows, particular attention must be paid to errors and the formation of pseudopaths. In previous work [9] we studied an evolutionary method for further improvement of the efficiency and robustness of Adleman-style DNA computing by optimizing DNA encodings for representing problem instances. In this paper we extend this framework to solve a different combinatorial optimization problem, i.e. MCP, using a different DNA computing method suggested by Ouyang et al. [7].

The effectiveness of the proposed method is shown by simulations performed on two instances of the maximal clique problem. Our results indicate that using genetic algorithms to optimize the DNA codes improves robustness and fidelity of DNA computing. As a byproduct of our study in this paper we found that the new encoding scheme which uses restriction enzymes is in general better than the Adleman-style encoding in its fidelity.

The paper is organized as follows. In Section 2 we define the maximal clique problem and outline the algorithm to solve it by DNA computing. Section 3 describes the method of simulated DNA computing and its results. Section 4 presents the method for optimizing DNA codes their experimental results. Section 5 discusses the results.

2 DNA Solution of the Maximal Clique Problem

A clique is defined as a set of vertices in which every vertex is connected to every other vertex by an edge. The maximal clique problem is to find the largest clique, given a network containing N vertices and M edges. The graph of six vertices and 11 edges in Figure 1(left) defines such a problem. The vertices (5, 4, 3, 2) form the largest clique and thus the size of the largest clique in this network is four. Finding the largest clique has been proven to be an NP-complete problem [3].

Recently, Ouyang et al. [7] have solved this problem by means of molecular

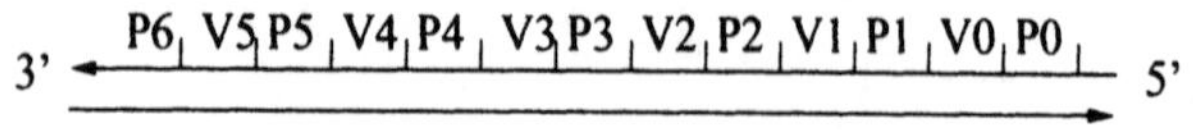

Figure 2: Encoding data in DNA.

biology techniques. Their experiment proceeds as follows.

1. For a graph with N vertices, each possible clique is represented by an N-digit binary number. A bit set to 1 represents a vertex in the graph, and a bit set to 0 represents a vertex out of the clique. For example, the clique (4,1,0) is represented by the binary number 010011. In this way, the complete set of possible cliques is transformed into an ensemble of all N-digit binary numbers. This is called the complete data pool.

2. Pairs of vertices in the graph are found that are not connected by an edge. The graph containing all edges missing in the original graph is called the complementary graph (Figure 1(right)). Any two vertices connected in the complementary graph are disconnected in the original graph (Figure 1(left)) and therefore cannot be members of the same clique; this means that the corresponding bits cannot both be set to 1.

3. All numbers containing connections in the complementary graph are then eliminated from the complete data pool. For the problem in Figure 1, the numbers $***1*1$, $1****1$, $1***1*$, and $**1*1*$ are removed ($*$ can be either 1 or 0). For the problem in Figure 4, the numbers $***1*1$, $1****1$, $**1*1*$, $**11**$, $*11***$ are removed. The remainder of the data pool corresponds to all cliques in the original graph.

4. The remaining data pool was sorted to find the data containing the largest number of 1's. Each of these ones represents a vertex in the corresponding clique; therefore, the clique with the largest number of 1's tells us the size of the maximal clique.

The possible solutions are encoded as double-stranded DNA molecules. Each bit in a binary number is represented by two DNA sections corresponding to the bit's value (V_i) and its position (P_i). For a DNA molecule representing a six-digit binary number, there are six value sections (V_0 to V_5) sandwiched sequentially between seven position sections (P_0 to P_6) (see Figure 2).

The last position section, P_6, is needed for polymerase chain reaction (PCR) amplification. Ouyang et al. report that P_i with a length of 20 base pairs worked well. The length of V_i was set to 0 bp if the value of $V_i = 1$, and to 10 bp if the value if $V_i = 0$. Therefore, the longest DNA has 200 bp corresponding to the number 000000, and the shortest DNA has 140 bp corresponding to the number of 111111. Each oligonucleotide consists of two position motifs and one value motif, $P_iV_iP_{i+1}$ for even i and complementary sequence $P_{i+1}V_iP_i$ for odd

Table 1: DNA Sequences for the experiments.

Index	Position Sequence	Value Sequence
0	CGTAGAATTCTGCGAACCTT	GACGCGGCAG
1	AAGAGTCACCTATCAGTAAG	TTCATAACCG
2	CTTCCTGAAGAGAGATTACT	CCGGTCACTT
3	AGTGCTAACAAGCCTGCGCA	CCTGACGATT
4	TGCAGTCTTTGAGATAAAGG	AAAAACCCAC
5	CCTACGTGTGAAATAGACTC	CCACCGATCT
6	GAGAGGGGCGGGATCCAGGG	

i. The data pool was digested with restriction enzymes. These enzymes break DNA at specific restriction sites, which were embedded within the sequences for $V_i = 1$. The broken strings were not amplified exponentially by PCR. To cut a connection in the complementary graph, the data pool was first divided into two test tubes, t_0 and t_1. In t_0 strings containing $V_0 = 1$ were cut with Afl II, in t_1 strings containing $V_2 = 1$ were cut with Spe I. Next, t_0 and t_1 were combined into test tube t, which did not contain $***1*1$.

The maximal clique in the graph of Figure 1 is 001111 or (2, 3, 4, 5), reading from left to right. The solution is

$$P_0 \rightarrow V_0 \rightarrow P_1 \rightarrow V_1 \rightarrow P_2 \rightarrow P_3 \rightarrow P_4 \rightarrow P_5 \rightarrow P_6 \qquad (1)$$

In DNA code form, the solution is represented as follows.

$$\begin{aligned} & CGTAGAATTCTGCGAACCTT \rightarrow GACGCGGCAG \\ & \rightarrow AAGAGTCACCTATCAGTAAG \rightarrow TTCATAACCG \\ & \rightarrow CTTCGTGAAGAGAGATTACT \\ & \rightarrow AGTGCTAACAAGCCTGCGCA \\ & \rightarrow TGCAGTCTTTGAGATAAAGG \\ & \rightarrow CCTACGTGTGAAATAGACTC \\ & \rightarrow GAGAGGGGCGGGATCCAGGG \end{aligned} \qquad (2)$$

3 Simulated DNA Computing

We have been developing a tool for simulating DNA computing and other molecular computation methods. The tool, named NACST (nucleic acid computing simulation toolkit), is used to design the biological experiments, such as code design for solving particular problems. Ouyang's method was implemented on NACST.

In simulations we made some modifications to the original experiment which seemed more reasonable from the computational perspective. In step 1, the

Table 2: Parameters for the experiments.

Parameters	Values
Task	6-MCP
Pool size	1,000,000
Reaction time	1,000,000
Code length for position P_i	20
Code length for value V_i	10
Hybridization error rate	0.003
Ligation error rate	0.003

method of parallel overlap assembly (POA) was modified to be compatible with the more common technique of ligation. The ligation process was emulated as realistic as biological processes. The second and third steps in the original experiment were merged into a single step. Thus in step 2 of simulation, restriction enzymes are used to cut and produce new graphs. Illegal graphs are then removed from the data pool. This process is the same as the original process. In step 3, the shortest sequence is chosen as the solution to the problem. The shortest sequence is the solution because in the presented encoding scheme the code length of V_i was set to 0 if the value of $V_i = 1$. We simulated this process by explicitly finding the sequence of 1's.

PCR plays an important role in DNA computing with biomolecules. They are used to purify the molecules to evolve useful ones. However, in simulations the process can be replaced by a more efficient search method. The use of restriction enzymes was also simulated. It was checked whether the position and value sections have the same codes as the enzymes.

Table 2 summarizes the parameters used in the simulations. The task was the 6-vertex maximal clique problem shown in Figure 1. We used the same encoding scheme as illustrated in the previous section. The code lengths for position P_i and value V_j were 20 and 10, respectively. The size of data pool was one million. The reaction time allowed was one million. Considering the huge storage capacity and parallelism of DNA molecules, this is a small amount yet sufficient to solve the problem. To simulate errors in chemical reactions we inserted point mutations during chemical reactions. Hybridization error rate was 0.3% and ligation error rate was 0.3%.

Figure 3 shows how many of the molecules survive each lab step simulated. The number of molecules survived all three steps are solutions. We studied the effect of reaction time on the reliability of computations. The results are summarized in Table 3. Given a reaction time of 10^6 the simulated DNA computing always found solutions.

Using reaction times of 5×10^5 and 2×10^5 the method also found more than one solution for each run on the average. However, given less reaction time

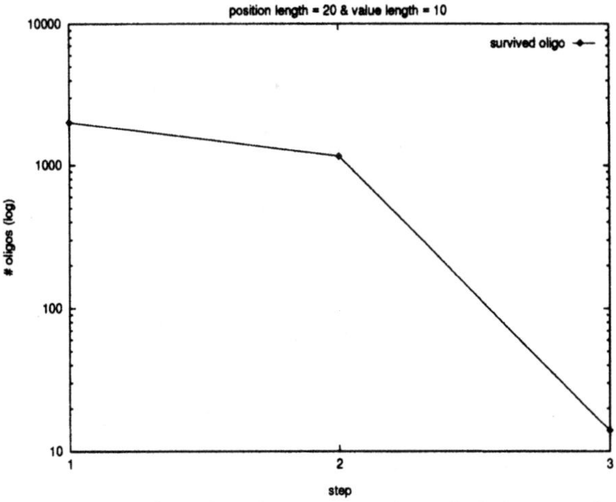

Figure 3: Number of molecules survived each lab step for simulated DNA computing.

Table 3: Comparison of the number of solutions for runs with varying reaction time.

Position Sequence Length	Value Sequence Length	Pool Size	Reaction Time	Num Runs	Succ Runs	Num Sol's
20	10	10^6	10^6	25	25	10.36
20	10	10^6	5×10^5	25	25	6.96
20	10	10^6	2×10^5	25	22	1.76
20	10	10^6	10^5	25	12	0.68

the number of solutions found reduced. In an extreme case of reaction time 10^5, the number of solutions are very small.

Simulations have also been made on a new problem instance. The graph of this new problem is depicted in Figure 4(left). Its complementary graph is shown in Figure 4(right). The solution found by simulated DNA computing is as follows.

$$\begin{aligned}
&CGTAGAATTCTGCGAACCTT \rightarrow GACGCGGCAG \\
&\rightarrow AAGAGTCACCTATCAGTAAG \\
&\rightarrow CTTCGTGAAGAGAGATTACT \\
&\rightarrow AGTGCTAACAAGCCTGCGCA \rightarrow CCTGACGATT \qquad (3) \\
&\rightarrow TGCAGTCTTTGAGATAAAGG \\
&\rightarrow CCTACGTGTGAAATAGACTC \\
&\rightarrow GAGAGGGGCGGGATCCAGGG
\end{aligned}$$

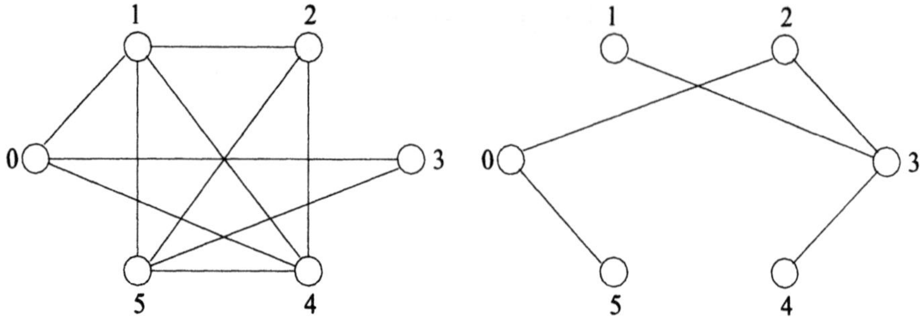

Figure 4: Another maximum clique problem: (a) graph, (b) complementary graph.

This corresponds to the representation

$$P_0 \to V_0 \to P_1 \to P_2 \to P_3 \to V_3 \to P_4 \to P_5 \to P_6 \quad (4)$$

which is equivalent to 011011 or $(1, 2, 4, 5)$; four vertices 1, 2, 4, and 5 construct the maximal clique for this graph.

It is interesting to note that given only 200,000 reaction steps a solution can be found. This is remarkable if we compare this result with our earlier results on Adleman-style DNA computing [9]. This is an example demonstrating that encoding schemes are very important for robustness and fidelity of DNA computing. In the following section we present yet another method that can further improve the performance of DNA computing.

4 Evolutionary Optimization of DNA Codes

Major errors in DNA computing described in Section 2 come from two sources. The first source of error is the production of single-stranded DNA (ssDNA) during PCR. This ssDNA cannot be cut by restriction enzymes. The second source of error is incomplete cutting by restriction enzymes, which also leads to incorrect answers.

If we carefully design the DNA codes to avoid the errors, the robustness of DNA computing can still be improved. In this section we describe a method that uses a genetic algorithm to find optimal codes for position sections P_i and value sections V_j in solving maximal clique problems.

Each encoding, i.e. a set of instantiated codes for P_i and V_i, builds an individual for the genetic algorithm. The fitness of each individual is determined by estimating the robustness of the codes when they are used in bio-lab steps. Both sources of errors are considered in determining the fitness of each code. Genetic operators are used to evolve fitter codes. Uniform crossover was used to exchange information between two codes. Point mutation was used to change the single nucleotides. The best encoding found by the genetic

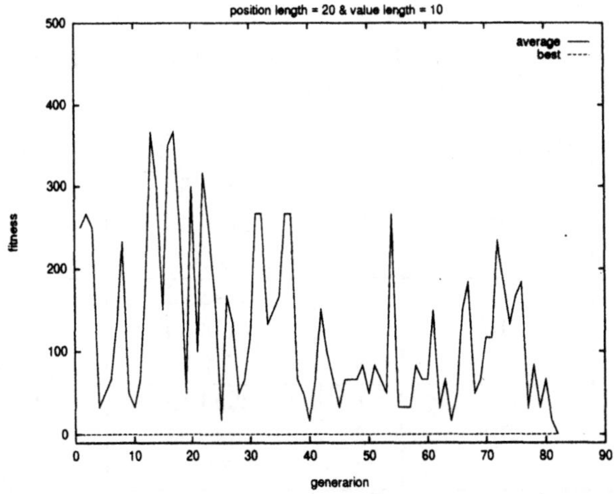

Figure 5: Fitness vs. generation for genetic code optimization.

algorithm is then used to run DNA computing with biomolecules.

Employing the genetic optimization phase, the entire molecular algorithm can be described as follows.

1. Generate a population of encodings for the problem.
2. While (generation $g \leq g_{max}$) do
   (a) Evaluate the fitness of each encoding.
   (b) Select fitter encodings.
   (c) Apply genetic operators to produce a new population.
3. Let the best code be the fittest encoding.
4. *Initialization*: Generate a pool of oligos using the best code.
5. While (cycle $c \leq c_{max}$) do
   (a) *Synthesis*: Produce candidate solutions by molecular operators.
   (b) *Separation*: Filter out infeasible solutions by lab-steps.

The evolutionary algorithm was run with population size of 10 for 100 generations. Crossover and mutation rates were 0.5 and 0.3, respectively. Other parameters were the same as the experiments in the previous section.

Figure 5 depicts the fitness evolution during the genetic code optimization process. It can be seen that encodings that can solve the problem are found in early generations. We also studied the effect of the code length on the fidelity of DNA computing. Experiments were performed with three varying values of position code length: 10, 20, 30.

Table 4: Number of molecules survived the lab steps: comparison with good (optimized) and bad (negatively optimized) encodings using position length $|P_i| = 10, 20, 30$ for reaction time = 1000000.

Position Length	Value Length	step	Good encoding	Bad encoding
10	10	1	2650.50	2649.20
		2	1748.40	1391.00
		3	3.60	0.25
20	10	1	2648.90	2648.90
		2	1433.50	759.00
		3	3.25	0.00
30	10	1	2653.85	2654.21
		2	1448.95	1258.37
		3	3.90	0.53

Table 5 shows a good (optimized) encoding. To see the difference, we also ran the genetic algorithm to find bad encodings by negatively optimizing the codes. Table 6 shows an example of bad (negatively optimized) codes found by the genetic search. One interesting feature of bad codes is that they have the same sequence as a restriction enzyme or sequences that can be the same if mutated. This leads to cutting wrong positions.

Table 4 compares the number of molecules that survived each simulation lab-step. This is a measure of reliability of DNA computing.

Both of the good and bad encodings well survived the first two steps, but DNA computing with good encodings resulted in a larger number of solutions than with bad encodings. The reason for no big difference between both encodings seems attributed to the fact that the presented encoding scheme gives so much constraints on the possibility of mismatches that there are no much room for finding better codes. This is contrasted with our own results on Adleman-style DNA encoding reported in [9], where the genetic optimization of codes have got more significant improvement in real experiments.

Significant errors are caused when a sequence has one or more subsequences that are the same as the codes for restriction enzymes. This leads to wrong cutting of the sequences. This type of error is affected by the length of position sequences. In case of position length 20, there are 10 to 20% of the cases where unnecessary restriction enzymes are placed within the position sequences. Even with mutations, the rate of wrong positions ranges between 20 and 30%. This suggests a possibility of error tolerance of the encoding scheme.

If we shorten the length of position sequences, say 10, then there are only four positions that may cause mismatches (remaining six are for two restriction enzymes for both ends of the sequence). This should improve the fidelity of DNA computing. The simulation results for $|P_i| = 10, 20, 30$ conform to this

Table 5: Good codes found by the code optimization process for $|P_i| = 20$.

Length P	Length V	DNA frag	Good Encoding
		$P_0V_0^0P_1$	CTCGTACAAATTGGGTTCTTgcggccttgc AAGCTGGTACCCGGACCAAG
		$P_0V_0^1P_1$	CTCGTACAAATTGGGTTCTTAAGCTGGTACCCGGACCAAG
		$\overline{P_2V_1^0P_1}$	AGTTGTACCTGCTCCCAAAGgccagatcgt CTTGGTCCGGGTACCAGCTT
		$\overline{P_2V_1^1P_1}$	AGTTGTACCTGCTCCCAAAGCTTGGTCCGGGTACCAGCTT
		$P_2V_2^0P_3$	CTTTGGGAGCAGGTACAACTtcgccccata AGTAGACCTACGCGAGAGCA
20	10	$P_2V_2^1P_3$	CTTTGGGAGCAGGTACAACTAGTAGACCTACGCGAGAGCA
		$\overline{P_4V_3^0P_3}$	CCTCTGCCGCGAGTAATGCAatgaatgttt TGCTCTCGCGTAGGTCTACT
		$\overline{P_4V_3^1P_3}$	CCTCTGCCGCGAGTAATGCATGCTCTCGCGTAGGTCTACT
		$P_4V_4^0P_5$	TGCATTACTCGCGGCAGAGGtcttgaggat CCTGGCAAGTCATTCGTCTC
		$P_4V_4^1P_5$	TGCATTACTCGCGGCAGAGGCCTGGCAAGTCATTCGTCTC
		$\overline{P_6V_5^0P_5}$	TACCAGGTCCGCGGACGCTCagagtaagcg GAGACGAATGACTTGCCAGG
		$\overline{P_6V_5^1P_5}$	TACCAGGTCCGCGGACGCTCGAGACGAATGACTTGCCAGG

Table 6: Bad codes found by the code optimization process for $|P_i| = 20$.

Length P	Length V	DNA frag	Bad Encoding
		$P_0V_0^0P_1$	CTCGAGTCAAGCACTGGCTTaacttaagtc AAGCACACGAAGCCCGCAAG
		$P_0V_0^1P_1$	CTCGAGTCAAGCACTGGCTTAAGCACACGAAGCCCGCAAG
		$\overline{P_2V_1^0P_1}$	AGTGAATTAATTCCCGAAAGagtagccaat CTTGCGGGCTTCGTGTGCTT
		$\overline{P_2V_1^1P_1}$	AGTGAATTAATTCCCGAAAGCTTGCGGGCTTCGTGTGCTT
		$P_2V_2^0P_3$	CTTTCGGGAATTAATTCACTtgacccctg AGTATACCTAATTAATGGCA
20	10	$P_2V_2^1P_3$	CTTTCGGGAATTAATTCACTAGTATACCTAATTAATGGCA
		$\overline{P_4V_3^0P_3}$	CCTCGTGGCTAACTGTTGCAttgtgtccac TGCCATTAATTAGGTATACT
		$\overline{P_4V_3^1P_3}$	CCTCGTGGCTAACTGTTGCATGCCATTAATTAGGTATACT
		$P_4V_4^0P_5$	TGCAACAGTTAGCCACGAGGcttcgtatgc CCTAATACTAGTACGGCCTC
		$P_4V_4^1P_5$	TGCAACAGTTAGCCACGAGGCCTAATACTAGTACGGCCTC
		$\overline{P_6V_5^0P_5}$	TCTACAACCCTCGCCCCCTCaagcaactaa GAGGCCGTACTAGTATTAGG
		$\overline{P_6V_5^1P_5}$	TCTACAACCCTCGCCCCCTCGAGGCCGTACTAGTATTAGG

analysis. The result shows the effect that restriction enzymes place wrong positions in the sequence. Thus care must be paid in interpreting the results in Table 4. In contrast to position sequences, restriction enzymes on value sequences have little effect on solutions. To understand why, recall that $V_i = 1$ if a value sequence exists and $V_i = 0$ otherwise. Though a value sequence has the same code as the restriction enzyme, if the solution does not contain the value sequence, then this has no effect on the solution at all. The only case affecting the result is the case that the value sequence has the same code as the restriction enzyme and this value sequence is required for the solution of the problem. For example, Table 6 includes restriction enzyme "CTTAAG" for the value sequence V_0 which is a part of the answer sequence. Hybridization errors have little effect since the restriction enzymes play the primary role in distinguishing sequences.

5 Conclusions

We presented an evolutionary method for the design of DNA codes in DNA computing. In the context of the maximal clique problem we demonstrated the effectiveness of the optimized codes. The general observation from our simulation results is that the encoding scheme used by Ouyang et al. are very robust against errors. This is contrasted with our earlier simulations of Adleman-style DNA computing. The main strength of the new encoding scheme lies in the use of restriction enzymes which rapidly cut out useless code fragments. Though the effect is less significant, using the genetic algorithm to optimize the DNA codes for the new encoding scheme still improves the performance of DNA computing.

Acknowledgments

This research was partially supported by grants from the Korea Science and Engineering Foundation (KOSEF 981-0920-107-2), the Research Foundation of Seoul National University, the Dongbu Foundation, and the IBM Korea.

References

[1] L.M. Adleman, Molecular computation of solutions to combinatorial problems, *Science*, 266:1021-1024, 1994.

[2] R. Deaton, R.C. Murphy, M. Garzon, D.R. Franceschetti, S.E. Stevens, Jr., Good encodings for DNA-based solutions to combinatorial problems, In *Proc. of Second DIMACS Workshop on DNA Based Computers*, American Mathematical Society, 1996.

[3] M.R. Garey and D. S. Johnson, *Computers and Intractability*, New York: Freeman, 1979.

[4] M. Garzon, P. Neathery, R. Deaton, R. C. Murphy, D.R. Franceschetti, and S.E. Stevens Jr., A new metric for DNA computing, *Proc. Second Annual Conf. on Genetic Programming*, Koza et al. (eds.), MIT Press, pp. 472-478, 1997.

[5] T. Haynes and A. Wu, *ICGA-97 Workshop on Exploring Non-coding Segments and Genetics-based Encodings*, East Lansing, Michigan, 1997.

[6] R.J. Lipton, DNA solution of hard computational problems, *Science*, 268:542-545, 1995.

[7] Q. Ouyang, P.D. Kaplan, S. Liu, and A. Libchaber, DNA solution of the maximal clique problem, *Science*, 278:446-449, 1997.

[8] B.T. Zhang and H. Mühlenbein, Balancing accuracy and parsimony in genetic programming, *Evolutionary Computation*, 3(1):17–38, 1995.

[9] B.T. Zhang and S.Y. Shin, Molecular algorithms for efficient and robust DNA computing, *Genetic Programming 1998*, Koza, J. R. *et al.* (eds.), Morgan Kaufmann, pp. 735-742, 1998

Part 10: Summary of Tutorials

Papers:

On Line Tutorial on Evolutionary Computation

http://www.lcc.uma.es/personal/cotta/semEC/ec_sem.html

Enrique Alba, Carlos Cotta
{eat, ccottap}@lcc.uma.es
Dpto. Lenguajes y Ciencias de la Computación
29071, Málaga, SPAIN

Abstract. This tutorial is intended to provide a comprehensive and broad overview of the main techniques, algorithms, application issues and software in relation with the growing area of evolutionary computation (EC). It includes in an on-line fashion a flexible access to different chapters containing explanations, examples and demos on the different algorithm families embedded in the EC-like search mechanisms. The tutorial is highly structured in chapters and sections, and also it offers a bibliography on the matter, as well as descriptions of concrete applications and theoretical details on evolutionary algorithms. The dynamic nature of such an info suggests the advantages of an on-line format. This allows a continuous improvement by the addition of results, graphs and new links to related material.

Outline of the Tutorial

The tutorial on evolutionary computing (EC) is a compendium of information about evolutionary algorithms, their applications and technical issues. It contains plain and graphical information, related bibliography, a demo section with an on-line executable algorithm, and multiple useful details on theoretical and empirical issues.

The on line tutorial on EC is structured in 8 chapters and 3 appendices. All the information inside the tutorial is organised through links to its contents and published material. The tutorial can be used both to be sequentially read or to be accessed from its home page for consulting a specific portion. It follows an outline of its contents.

The first chapter is focused on nature-inspired techniques, mainly addressing evolutionary algorithms: genetic algorithms, evolution strategies, and evolutionary programming. Also, classifier systems and genetic programming are dealt with in this and other chapters. The relationship with other non-EC techniques is outlined in this chapter. The field of EC is located in a taxonomy of modern search heuristics, and a structured revision on the applications is offered.

The second chapter explains the main characteristics and working details of genetic algorithms. The canonical genetic algorithm is revised and discussed with several working examples. A deep presentation on the main issues relating genotype and operator types is offered. Even implementation issues are discussed here.

The third chapter is devoted to the applications of evolutionary algorithms, but specially of genetic algorithms. A deeper presentation is used for the evolutionary training of neural networks and applications in combinatorial optimisation, due to our interest in these fields.

Chapter four contains a general overview of models and software concerning parallel genetic algorithms, since we believe this is the logical future of these techniques in which high computational resources are often needed. Both, coarse and fine grain parallel algorithms are discussed from a common point of view, as well as a comparison in the light of sequential models. Also, the impact of using object orientation is presented as a beneficial tool to enhance the resulting software systems and their possible combination/competition with other algorithms.

Chapter five describes some non-pure evolutionary techniques that are commonly found in EC applications or studies, as well as the benefits of using an adaptive parameterisation. In chapter six we discuss the genetic programming paradigm which uses genetic algorithms to evolve trees of symbolic information, usually representing computer programs performing a given task. A detailed case study is presented on the generation of fuzzy rules for a fuzzy control system.

In chapter seven we present some EC-related algorithms like tabu search in order to give the reader a good tour on the EC field as well as on interesting and related search mechanisms. Finally, chapter eight contains the description of several studies intended to hybridise evolutionary algorithms with other techniques to improve their performance. In particular, a hybrid genetic-branch and bound algorithm is addressed.

The first appendix contains a detailed bibliography on evolutionary computing. The second appendix contains a list of figures, since we believe in the advantages of a graphical knowledge transmission. The last appendix contains a Java applet that allows the user to change parameters and run a genetic algorithm in the net.

The contents of our tutorial is continuously being revised and extended. In the near future a version is going to appear in the "Flying Circus" CD demonstration used in EvoNet (european network of excellence for EC) to bring academic and industrial people an overview of the advantages of using evolutionary computation.

Discussion and Comments

The inclusion of the tutorial on EC in WSC3 has allowed an exchange of ideas on different matters. Among them we can mention the role of local search in genetic algorithms, memetic algorithms, their parallelisation and applications.

Some sections have been stripped off from the WSC3 version since they are in preparation. This is the case of evolution strategies and many results in the fields of genetic programming and the parallel resolution of complex problems.

Acknowledgements

The authors would like to thank the international committee that granted the "award to the best tutorial" to the *On Line Tutorial on EC* presented in WSC3.

Fuzzy Control Tutorial

Mariagrazia Dotoli[1] and Jan Jantzen[2]

(1) Politecnico di Bari, Dipartimento di Elettronica, Via Re David, 200,
I-70125 Bari, Italy, tel +39 80 5460 312, fax +39 80 5460 410
e-mail dotoli@poliba.it

(2) Technical University of Denmark, Dept. of Automation, bldg. 326,
DK-2800 Lyngby, Denmark, tel +45 4525 3561, fax +45 4588 1295,
e-mail jj@iau.dtu.dk

Keywords: Education, distance learning, automatic control, Matlab

Abstract. The tutorial concerns automatic control of an inverted pendulum, especially rule based control by means of fuzzy logic. A ball balancer, implemented in a software simulator in Matlab, is used as a practical case study. The objectives of the tutorial are to teach the basics of fuzzy control, and to show how to apply fuzzy logic in automatic control. The tutorial is distance learning, where students interact one-to-one with the teacher using e-mail.

1. Introduction

The case study concerns a cart-ball laboratory rig, consisting of a steel ball on a curved bridge on a moving cart. It is fairly easy to control the system such that the ball balances on the top of the bridge, but it is difficult to position the cart at the same time in a prespecified position.

Students design a fuzzy controller that balances the ball on top of the bridge. The software package PENDULUM, which runs under the Matlab environment, is used throughout the tutorial as a simulator.

The tutorial is an off-spring from a course on fuzzy control which runs on the Internet [www.iau.dtu.dk/~jj/learn/]. Development aspects and experiences from running the course are described in a paper from the WSC2 conference in 1997 [1]. Details about the implementation and organisation can also be found in the reference.

2. Method

When the student signs up for the tutorial he/she downloads the lecture notes [2, 3], the software simulator, and its user's guide [4]. The programs are Matlab m-files written for the Matlab Student Edition versions 4 or 5 for Windows; but they will also run under the professional edition and under DOS, UNIX, and MacIntosh.

If the student does not have access to Matlab, he/she must buy the student edition including the user's guide [5].

When ready, the student is asked to do the following for each lesson:

1. Download the lesson as a text file.
2. Include it inside an e-mail (not as an attachment), use copy and paste.
3. Work with it. Make notes about results and comments to pass to the instructor.
4. Write answers directly into the file. E-mail it to the instructor.

The instructor comments the answers and returns the e-mail note to the student, such that it is in the mail box the day after, at the latest. The students were required to submit two lessons per week.

The lessons to download had the following titles:

0. Tutorial introduction
1. The pendulum problem
2. Fuzzy controller design
3. Tutorial evaluation

The last lesson is a short evaluation report over the work. The report is partly for the benefit of the teacher, partly for the student.

3. Results

A total of five students signed up for the WSC3-sponsored course, of which one dropped out for lack of interest. Of the remaining four, one was a teacher, and the rest were PhD and MSc students. Not all managed to keep the schedule and finish the course on time. One student later signed up for the long course on the Internet.

The students were required to be familiar with fuzzy logic already.

There were 'negative' student comments regarding the readability of the textbook material (definition occurring after the concept was mentioned and used the first time, equations should be moved into the appendix). To the teacher, these comments are valuable feedback, and since the material is published on the Web, it is relatively easy to correct the textbook material.

Another student 'complaint' was that the students wanted more lessons and more material on topics such as fuzzy sets, fuzzy logic, tuning of controllers. This is part of the full course, but difficult to include in, effectively, two lessons on the core of the subject matter.

Positive student comments were: everything is downloadable; the teacher replied quickly, even on weekends; and the textbook material and the simulator were easy to understand. From the teacher's viewpoint, easily accessible material and fast teacher responses are regarded as the primary controllable factors for motivating the students.

4. Conclusions and Recommendations

An objective was to demonstrate that the design of a fuzzy controller can be founded on a so-called PID controller; and it seemed possible to do that in just two lessons. The students appreciated the design approach, which is not published elsewhere.

But two weeks is really too short for that type of tutorial; a suggestion is to extend the tutorial period to three or four weeks. At least the registration procedure should be over, when the tutorial is scheduled to start. The 'opening hours' are a bit short.

The technical content of the conference in general and the tutorials in particular is very broad. A suggestion is to narrow the technical scope and get a few top quality contributions.

It should be possible to attract good authors, because it is a fast and smooth way to publish, compared to conventional paper publishing.

References

[1] Jantzen, J. and Dotoli, M., 1998, "A Fuzzy Control Course on the Internet". In Chawdry, Roy and Pant (eds.), "Soft Computing in Engineering Design and Manufacturing", Springer Verlag London Ltd, pp. 122-130

[2] Jantzen, J., "Analysis Of A Pendulum Problem", Technical University of Denmark: Dept. of Automation, technical report no 98-E 863, 1998, 12 p.

[3] Jantzen, J., "Design Of Fuzzy Controllers", Technical University of Denmark: Dept. of Automation, technical report no 98-E-864, 1998, 27 p.

[4] Kvaran, H., and J. Jantzen, "Cart-Ball User's Guide", Technical University of Denmark: Dept. of Automation, technical report no 98-E-865, 1998, 5pp.

[5] The MathWorks Inc, "The Student Edition Of Matlab - v 4 for MS Windows" Englewood Cliffs, NJ: Prentice Hall, 1995, 833 pages + 5 diskettes (or the equivalent version for MacIntosh/PowerMac).

Part 11: Summary of Discussion

Summary of Discussion

The papers presented in the first nine parts of this book comprise of the papers presented in the 3rd On-line World Conference on Soft Computing in Engineering Design and Manufacturing (WSC3). This part summarises the discussion about the papers during the conference. The summary is divided into nine sections according to the first nine parts in the book, and most of the sections have been prepared by the session chair(s). The name of the person who has prepared the summary is given at the beginning of each section.

1. Keynote Papers

Prepared by: Prof. Erik Goodman, USA and Prof. Sung-Bae Cho, Korea

The paper, which drew the most active discussion, was by Sung-Bae Cho and Joo-Young Lee, about interactive judging of fitness by the user, applied to retrieval of images with particular user-judged emotional content. One participant sought more information on the wavelet coefficients representing particular emotions, with examples of images and their coefficients, and on the areas of application envisioned by the authors. The authors responded, indicating that space had precluded their providing images, and also that while the coefficients they found seemed useful, they would not claim that they were optimal characterisations of particular emotional expressions, at this point. They also provided several promising areas for application of their approach, such as the arts (in general), speech processing, flexible design, active noise control in dental drilling, virtual reality controllers, etc.

Another participant (from Russia, incidentally, which illustrates how effective a web-based conference may be in allowing participation without major travel expenditures) suggested the use of an entropy-based sorting of images for presentation to the user for fitness evaluation. The authors responded that they agreed, and that they and others had explored such approaches. The participant indicated that he would like to see how their approach works for him on a problem involving image restoration, where the user might be asked to judge which of a set of restorations looks "better."

Regarding the NIST design repository article, a participant inquired about the usefulness of the repository as a source of elements of innovation, doubting that it could produce real innovation. The authors responded that the design repository is not intended to produce solutions to problems, but rather to "represent, store, and retrieve design artefact knowledge." While it is interesting to speculate on the role of such archived knowledge in an "automatic" design system, it is not in the scope of their project.

As a promising direction, one of the chairs would like to emphasise the role of human in the soft computing. It has been actively exploited to combine the soft computing techniques such as fuzzy logic, neural networks and evolutionary computation to develop better systems that can match human ability, but more work is still required to be able to match human performance. In order to remedy this shortcoming, it would be a promising approach to incorporate human himself into the soft computing framework more actively, and a couple of preliminary studies have been presented at this session.

The breadth and interest of the keynote papers seems, to the session chairs, to indicate the viability of web-based conferences such as WSC3 as a good forum for communication of ideas and exchange of opinions. Its "barrier-free" nature also enhances its appeal as a global communications medium. This must be a challenging attempt to show the desirable model of future conferences.

2. Design Support Systems

Prepared by: Graham Jared, UK and Dr. Peter Bentley, UK

Peter Bentley began this discussion by thanking the authors for an interesting paper. He then raised two issues: firstly that he felt that GAs are capable of displaying creativity, but that creativity is very hard to define. Secondly he suggested that a creative designer is not necessarily a prerequisite to a Creative Evolutionary Design system. He proposed that the GA is capable of some astonishing feats of design, despite some very uncreative representations and fitness functions.

The author of the paper, William Crossley, agreed that creativity is difficult to define, and that any definition is largely subjective. He stated that he chose to use Ullman's description of a creative engineering designer, because this best matches his own point of view. He also expressed his agreement with the discussion on creativity in Dr. Bentley's paper. William felt that his work fitted into Dr. Bentley's categories of "Conceptual Evolutionary Design" and "Integral Evolutionary Design".

Prof. Crossley then went on to stress the need for creative designers with such computer systems. He suggested that in a GA-based design approach, the focus of the creative engineer shifts from conceptualising a design towards defining an appropriate design space. Using the stiffened composite panel from his paper as an example, he stated that it would be impossible, in its current definition, for the GA to determine that a "cruciform" stiffener provides the best stiffener shape - the GA cannot "think out of the box" defined for it. He also suggested that a designer must be aware of "holes" in fitness Peter Bentley replied by agreeing that that creativity can be required to arrive at suitable design representations and fitness functions. He then asked whether the additional computer literacy required by such systems could be considered an improvement for our designers. He also asked whether Prof. Crossely could foresee a time when computers will no longer need such creative guidance by us.

William Crossely responded by saying that he believes that for very complex engineering problems, where it is difficult for a human to mentally evaluate many different combinations, concepts, etc. that make up a possible design, using the GA to encourage creativity is an improvement. He then said that he believes that the need to be computer literate is (or will soon be) a prerequisite for virtually all engineers, and it is not too much of an issue. Graham Jared then "played devil's advocate" by suggesting that the previous discussions were tending towards a philosophy of "why bother thinking too much about the problem when you can let the computer search for a long time and find some useful answers anyway". He also asked whether a computer becomes more apparently creative when it is allowed some exercise with a good problem representation or even a "higher-level" or "more abstract" one?

Peter Bentley responded by saying that the point of such work is to use computers to suggest new concepts that, perhaps because of 'design fixation', human designers have not thought of yet. Another advantage, he said, is that a computer can churn out clever new designs non-stop, needing no coffee breaks or sleep - or pay. He finished his answer by saying that the best way to have a reliable Evolutionary Design system, is to get a creative person to design good representations, operators and fitness functions. Such people must think about the problem, he suggested.

William Crossley also responded to Graham's message, saying he hears "since the computer gave me this answer, so it must be right" from students nearly every semester. He felt that a key feature of the GA as a creative assistant for the designer is its ability to generate and evaluate combinations of defined design parameters beyond those combinations easily envisioned by a human designer. He continued by stating that he was not sure how the computer/algorithm could be directly measured as more or less creative for a good problem representation or a "higher level or more abstract" problem. He felt that this type of approach to design could be more creative, although this is difficult to measure.

Graham Jared replied by saying that he does agree / empathise with the two responses to his previous follow-up. He stated that early exploration of unconsidered and / or novel possibilities cannot be undervalued. However, he suggested that, in this context, the necessity to actually think about a problem be seen as bad news - by students and designers alike. He added that the message that if you do some thinking you will get a considerable payback in design innovation is equally poorly received or acted on.

Graham continued by proposing that perhaps those who are generally open to innovation anyway take up ideas such as Creative Evolutionary Design. He suggested that it may be contingent on researchers to provide more advice as to how to effectively apply effort to choosing representations, operators and fitness functions etc. Graham ended this discussion by saying that it is also clearly necessary, as P. Bentley points out in his paper, to learn to apply lessons and analogies across disciplines.

Peter Bentley opened discussion by enquiring whether the surface optimisation technique described in the paper was likely to be implemented in any commercially available CAD system. He also queried whether the authors had experimented with optimising the number of control points as well as their positioning.

The authors' response to the second point was that no experimentation had been made. However, a "representative" figure for the number of control points had been chosen on the basis of observation of industrial users. Furthermore, the intended application area was the final adjustment of high quality surfaces and hence the number of control points was effectively pre-determined before this optimisation stage.

On the first issue, the authors enlarged on the stated objectives of the work as reported in the paper and explained that one of the goals of the research was to explore the linkage with CAD. Specifically, it was hoped to eventually develop a specification for a generic API, which could be supported by many systems, and thus implementation within one particular CAD system ran counter to this goal.

3. Intelligent Control

Prepared by: Prof. Antonio A. R. Coelho, Brazil and
Dr. Frank Hoffmann, USA

The seven papers in the session on "Intelligent Control" are concerned with the architecture, learning methods and model based design of fuzzy control system. Soft Computing proposes that the integration of complimentary methodologies such as fuzzy logic, neural networks, evolutionary algorithms or probabilistic reasoning into hybrid systems, results in more powerful methods than using a single method alone. Many of the session papers follow this paradigm by enhancing a fuzzy system with an additional method, such as multi-agent systems, evolutionary algorithms, adaptive schemes or predictive control.

Janos Abonyi et al propose an adaptive fuzzy controller that utilises an inverse fuzzy model of the plant. Their method is suitable for online adaptation of fuzzy control systems and is able to compensate for changes and uncertainties in the plant dynamics.

Leandro dos Santos Coelho et al compare different fuzzy PID control structures in regard to performance and robustness to set-point changes and load disturbances.

Daijin Kim and Han-Pyul Lee propose a novel COG defuzzification method that reduces the approximation error and improves the performance of fuzzy control. They present an adaptation method that employs a combination of evolutionary and Lamarckian learning to optimise centre and width of fuzzy membership functions.

Peter Mikulecky et al present a multi-agent expert system for control and stabilisation of a glass furnace plant. The expert systems detect sensor failures, monitor the state of the plant and propose control actions to the human operators.

Jairo J. Espinosa and Joos Vandewalle propose a predictive control strategy that utilises a fuzzy model of the plant. The non-linear fuzzy model is derived from training examples of past plant states and inputs. The fuzzy knowledge representation offers the advantage that a human user is able to interpret the adapted process model.

Frank Hoffmann et al present an evolution strategy that automates the knowledge acquisition step in fuzzy control design. An incremental learning scheme gradually expands the structure and number of fuzzy rules, with the result that the optimisation task becomes more feasible for the evolutionary algorithm. Fuzzy techniques offer the advantage that existing domain knowledge can be easily incorporated into the learning process. The learning method was applied to design a helicopter autopilot.

Marjan Golob presents a novel approach to fuzzy real-time control that decomposes multi-variable rules into a set of one-dimensional fuzzy rules using an inference break-up method. The decomposition step reduces the overall number of rules and therefore facilitates the derivation of the rule base for the designer. The proposed decomposition method is applied to generate a fuzzy PID-controller for a magnetic suspension system. The experimental results demonstrate that the fuzzy PID-controller is robust towards changes of the nominal operating point and achieves the same tracking performance as an adaptive PID-controller.

The discussion in the section mainly focused on the issue of approximation accuracy versus rule base complexity. Increasing the number of membership functions to obtain a finer granulation of the input space can reduce the error of a fuzzy model. Nevertheless, the number of rules increases as well with the drawback that the linguistic integrity of the model deteriorates. A large number of rules also bear the risk of over-fitting the data, especially if the number of training examples is small. Cross-validation on a test data set is a common means to avoid over-fitting and to guarantee that the fuzzy model is a proper generalisation of the underlying process. Statistical information criteria provide a useful guideline to find a reasonable compromise between model accuracy and complexity [1].

Another discussion thread was concerned with the applicability of fuzzy relational models for model based predictive control. One of the discussants pointed out that a limited accuracy of the fuzzy model could play a negative role for predictions over a large time horizon. The interested reader is referred to [2] for more detailed discussion of this subject.

References:

[1] J. Yen and L. Wang, "Constructing Optimal Fuzzy Models Using Statistical Information Criteria", in Proc. ISAI/IFIS 96, pp. 394-402, Cancun, Mexico, 1996.

[2] B. Posthlethwaite, Buliding a Model-Basied Fuzzy Controller, Fuzzy Sets and Systems, 79, pages 3-13, 1997.

4. Identification and Modelling

Prepared by: Dr. Rajkumar Roy, UK

There are four papers in this session. Following are the questions and answer about the papers:

1. Experimental Evaluation of Intelligent Identification Algorithms Applied to a Wind Tunnel Process.
Leandro dos Santos Coelho and Antonio Augusto R. Coelho.

Q.-1 (Shaikh):
Can you please differentiate in terms of advantages/disadvantages between time series modelling using classical Box-Jenkins ARMA model and your presented modelling techniques.

A:
The paper evaluates hybrid genetic algorithm with simulated annealing (NARX and ARX modelling) and modified Elman neural networks (black box modelling) techniques.

An attractive property of the Box-Jenkins model is that it yields separate descriptions of the input-output relationship between input (u) and output (y) and the noise spectrum model described by C/D polynomials and {wk}. This is obviously an advantage as compared to ARMAX models, which contain a description only with the same A-polynomial for both the poles of the transfer function and for spectral peaks of the noise model.

The wind tunnel process (laboratory scale plant) presents features that can also be found in real process such as time delay and non-linearities. The conventional identification methods suffer from the following drawbacks: (i) some initial information on the system parameters is needed a priori for convergence, (ii) the estimated parameters may be biased if the noise is correlated, and (iii) they cannot easily be applied to non-linear processes.

Some advantages/disadvantages of intelligent modelling techniques are:

a) Hybrid genetic algorithm with simulated annealing, Genetic algorithms are probabilistic techniques that manipulate a search space, based on natural selection and genetic mechanisms. The basic characteristics that set genetic algorithms apart form the traditional optimisation methods are presented in section 2 in paper.

However, genetic algorithms and other evolutionary computation (GASA) methods, must be used with care. They cannot be guaranteed to produce an absolutely optimum solution. It is also difficult to predict how many generations it will take to produce a satisfactory result. In this paper, the genetic algorithm is fine-tuned by the incorporation of simulated annealing for a more accurate search and faster convergence. This technique presented, in the paper, any advantages of search to presence or absence of the NARX and ARX model terms and utilisation of acceleration of the evolution given by Cauchy distribution for simulated annealing algorithm.

Future works perspectives with GASA concerning to modified loss functions (fitness) that penalise both model misfit and high model order. Examples of this type are the Akaike information criterion, minimum description length, and Akaike final prediction error criteria. The basic idea is the utilisation of principle of parsimony (or Occam's razor) for linear/non-linear process identification.

b) Modified Elman Neural Networks (MELNN)

Neural Networks have shown to be a very promising model set for a large variety of different identification problems. The major features of neural networks are of parallel processing, associative and distributed memory, and learning, which are similar to those evident in biological neural networks.

Several studies in the literature have successfully demonstrated the power of neural networks when employed to model complex non-linear mechanisms. Some topologies of neural networks are universal approximators of any function with a given arbitrary precision within an appropriate interval.

MELNN introduce self-connections in the units of context to provide the Elman neural networks (ELNN) with units of inertia for avoid oscillations and instabilities in the process of training of ELNN. So, the MELLN is a power tool for non-linear process identification.

More details in terms of advantages/disadvantages see:

Kremer S C 1995 On the Computational Power of Elman-Style Recurrent Networks IEEE Trans on Neural Networks 6(4):1000-1004

Pham D T, Liu X 1992 Dynamic System Modelling Using Partially Recurrent Neural Networks J of System Engineering 2(2):90-97

Q.-2 (Shaikh):
Can you please elaborate on using Time Delay Neural Networks (Finite Impulse Response Neural Network) instead of Recurrent Neural Network. Which would be better and how.

In the first paper - see section 3 - 3rd paragraph (observation):

Feedforward neural network's do not have dynamic memory and the tapped-delay-line method is usually adopted for neural networks to represent dynamic systems. This method presents some disadvantages such as the learning speed and the high number of neurones in the input layer. These disadvantages are caused, if the order of the system to be identified is unknown. So, in this case, an overestimation is necessary and, consequently, a high number of neurones in the buffer input are considered to deal with high-order systems. This usually provides a feedforward neural networks with structure of high dimension and lower learning speed.

Q (Dote):
Is the GASA, generally speaking, more suitable for large-scale and complex system optimisation than the MELNN?

A:
Genetic algorithms (GAs) and other algorithms of evolutionary computation, such as evolutionary programming, evolution strategies, and genetic programming, have recently emerged as an increasingly popular family of methods for global optimisation. GAs have been applied to a wide range of optimisation problems, including well known NP-complete and NP-hard problems, scheduling and routing, configuration, layout design, multicriterion decision (multiobjective problems), multimodal problems, and other knowledge fields.

GAs can suffer from excessively slow convergence before providing an accurate solution because their fundamental requirement of using minimal a priori knowledge and not exploiting local information. So, the GA is fine-tuned by the incorporation of simulated annealing (SA) for a more accurate search and faster convergence with Cauchy distribution.

Since, according to the No Free Lunch theorems [1], there cannot exist any algorithm for solving all (e.g. optimisation) problems that is generally (on average) superior to any competitor, the question of whether evolutionary algorithms are inferior/superior to any alternative approach is senseless. What could be claimed solely is that evolutionary algorithms behave better that other methods with respect to solving a specific class of problems – with the consequence that they behave worse for other problem classes [2].

The artificial neural networks (ANN), as recurrent ANN (such as MELNN, Jordan ANN, and fully connected ANN), due to the learning characteristics, adaptability, inherent parallelism and fault tolerance features, have consolidated themselves as promising tools for pattern recognition, time signals (filtering, spectral analysis, classification, pattern recognition), time-series prediction, identification, intelligent and adaptive control.

Several researches have been realised with GA, SA and ANN techniques, but exist many open questions and directions for future research, such as convergence, auto-adaptation, Lamarckian evolution, Baldwin effect (evolutionary computation), stability, efficient numerical methods for training, pruning methods (ANN), comparative studies of computational intelligence (CI) methodologies (fuzzy systems, evolutionary computation, and artificial neural networks).

The development of CI by researchers of the more different areas of knowledge has yielded promising results not only in scientific and technologic environments, but also in economic, commercial, financial, political, social and medical environments.

References

[1] Wolpert, D. W. and W. G. Macready (1997). No Free Lunch Theorems for Optimization, IEEE Transactions on Evolutionary Computation, 1(1):67-82.

[2] Schwefel, H.-P. (1997). Advantages (and Disadvantages) of Evolutionary Computation over other Approaches), Handbook of Evolutionary Computation, T. Bäck, D. B. Fogel and Z. Michalewicz (eds), Institute of Physics Publishing and Oxford University Press, A1.3:1-A1.3.2.

2. Improvement of Membership Function Identification Method in Usability and Precision
Ayumi Yoshikawa

Q (Dote):
Is this method more efficient for membership function optimisation than a fuzzy-neural network approach?

A:
I think, there are two major approaches to MF identification methods. One is to express a MF for a fuzzy concept that she/he images, such as meaning of verbal tallness, very tall. In other words, the method gets out her/his unique MF "as is"

from her/him. It is close to a psychological approach. Other is to tune or adjust a MF for a fuzzy concept in order to work a system well as a whole, such as fuzzy labels in if-then rules for control systems.

The issue of MF optimisation is in the latter. From viewpoint of the optimisation, it is important that each MF works well with all other MFs and rules, rather than it expresses well his or her internal image of the fuzzy concept.

My method proposed in my paper IM-3, the BASE method, however, is in the former. Thus, the MF obtained from the BASE method is a MF embodied subjective meaning and difference adequately. Meanwhile, it might be the optimised one, and might not. It depends on other conditions where the MF is used.

5. Data Mining

Prepared by: Dr. Rajkumar Roy, UK

This session had four papers. Though there were many visitors to the session, but unfortunately there were no discussion.

6. Optimisation

Prepared by: Dr. Riccardo Poli, UK and Dipl.-Phys. Mario Koppen, Germany

The Optimisation session offered four papers for discussion:

1. M. Koeppen, S. Rudlof, "Multiobjective Optimisation by Nessy Algorithm"

This paper presented a new approach to the multiobjective optimisation problem by the modification of a former presented evolutionary algorithm referred to as Nessy algorithm. Due to the representation of this Nessy algorithm as a neural network, it is possible to incorporate the fulfilment of multiple optimisation goals by setting the number of output neurones to a value different from 1 (as it is the case for the standard Nessy algorithm). The case was considered using a test function from literature and comparing this algorithm with random search. Also, it was considered for the case of a superposed system for automated texture filter generation and showed fair performance. On request by R. Poli, the author gave further details of the algorithm and some references.

2. Sebastian Lozano et al., "Unconstrained Optimization Using Genetic Box Search"

This paper is concerned with a new algorithm, referred to as genetic box search (GBS). The algorithm replaces the standard local search strategy by a search restricted to a box, which is centred at the actual evaluation point within the search space. For each step, a genetic algorithm, the individuals of which are positionally bounded within the box, searches for the next box centre. The performance of this algorithm was proven, using two standard problems from literature, the 2-dimensional TRIG problem and the RAST problem. This paper was intensively discussed.

One discussion thread was about the performance on TRIG, and R. Poli compared GBS performance with a slightly modified GA. The GA showed a comparable performance. The authors gave further details about the test runs, and were encouraged to consider the 30-dimensional TRIG function, but this function was harder to solve.

A second discussion thread was concerned with the relations of GBS and Evolutionary Strategy (ES) and was initiated by M. Koeppen. The agreement of the discussants was that the sphere model of ES applies for the GBS as well, possibly disturbing the equal distribution of the magnitudes of box replacement distances. The question, whether or not certain evolutionary operators are subjected to the sphere model, remained open.

3. Fabio Abbattista et al., "The Scout Algorithm applied to the Maximum Clique Problem"

This paper introduces the scout algorithm, which is mainly based on the exploitation of the probability vector (PV) model. The components of a PV are values from [0,1]. At each algorithm step, a number of bit-strings are generated, using the PV components as partial probabilities. The bit-strings are evaluated and according to this outcome, the PV is modified. The discussion of the author F. Abbattista with M. Koeppen was about a possible extension of this algorithm in order to become more "evolutionary," and about general claims which could be made about using PVs (e.g. Schemata theorem, convergence). Because the Nessy algorithm uses PVs as well, both authors agreed to co-operate in this field in the future.

4. Jozef Kratica, "Improvement of Simple Genetic Algorithm for Solving the Incapacitated Warehouse Location Problem"

This paper gives a study on the application of a GA to the Warehouse Location Problem and introduces a minor modification of GA (caching individuals, removing duplicates) in order to gain better performance. On request by S.Etaner-Uyar, the author gave further details of his algorithm.

Comments:

All papers of this session presented special evolutionary algorithms. Two papers (no. 1 and 4 in this session) gave modified versions of already presented algorithms in order to get a better approach to the very nature of a special optimisation problem, the other two considered optimisation problems only for the purpose of benchmarking the newly proposed algorithms.

All approaches showed a good performance for the selected optimisation tasks. However, no general claims were made about evolutionary optimisation at all. The discussions were mainly concerned with this short-come. In order to present alternative ways for the consideration of algorithms, the discussion threats went on as follows:

- using other algorithms for the same problem, or
- relating the newly proposed algorithm to a well-known one, or
- judging about certain theoretical features of the algorithm.

From this, it should be advisable in the future to take more care for such claims and the more to refrain from the classical strategy of showing experimentally that the newly proposed algorithm on a carefully selected problem is better than a standard one.

As S. Lorenzo wrote at the end of the discussion: "The secret behind evolutionary computation success must be the power of the natural selection analogy. There is much debate among evolutionary biologist about whether natural selection by itself is enough to explain all the design and complexity that can be observed. Compared with real life threats and challenges individuals of every species face everyday, solving a structured optimisation problem must be easy enough. About the schema theorem, what it does is to tell us what are the characteristics of the solutions that proliferate due to the structure of the algorithm and the nature of the problem (i.e. the fitness landscape). When no schema theorem can be proved what happens is that we have no clear idea of what and why is proliferating but I have no doubt that the algorithm always performs a biased search concentrating efforts in certain regions and neglecting others."

7. Optimisation for Industrial Applications

Prepared by: Dr. Ivanoe De Falco, Italy and Dr. Fernando Guerrero, Spain

The session was composed of five papers, which described relevant Soft Computing Strategies for solving Industrial problems. In total seventeen interventions were presented in the session. At least each paper received one question. In the following we report a summary of the discussion related to those papers, consisting in questions by Conference attendees and in the answers provided by the authors.

The first paper, "Maximum entropy image restoration by evolutionary algorithm" by V. Promislov, proposes the Evolution Strategies algorithm for restoration of astrophysical images, and compares it against a gradient-based method.

Q. (De Falco): The problems you face in the paper are "astronomy-like" problems, rather than real astronomy images. Do you plan to run experiments on real images?
A. (Promislov): The quality of an image reconstruction in astronomy has very subjective meaning. Very important is the opinion of the astronomer himself what the image is. The paper presents only simulation result.
Q. (De Falco): About real images to be restored: how can people decide whether the best restoration is provided by ES or by gradient-based methods, if they do not know which the real image is?
A. (Promislov): The attempt to process images with a well known AIPS processing software based on Cornwell gradient algorithm was unsuccessful, while ES MEM takes less than 8 hours to converge for each image. So, my opinion is that for sparse, small data sets the ES algorithm is the only possible.

The second paper, "The finite element method and Soft Computing" by Larry Manevitz and Dan Givoli, reports on a programme to automate the finite element method.
Q. (De Falco): Is it possible to face the problems with Genetic Algorithms?
A. (Manevitz): Yes, of course any optimisation problem can be addressed by GAs. However we chose our approaches as to what seemed the best fit. So, for instance, the prediction problem is quite natural by the back-propagation style network. It might be quite interesting to try a GA on some of the problems I listed, including, say, the mesh placement. If this is done, one has a good way to test the effectiveness of GA versus NN on a non-trivial problem for example. To perform GA on the mesh problem, one has a natural measure of "goodness" that we defined and the encoding of the nodes is pretty natural. There are some further technical problems (e.g. making sure the mesh completely covers the domain and some others) but we also had to face this in the NN approach. If I get a good student in an advanced seminar, I might put him on the problem.
Q. (Guerrero): I would like to ask you if have results about the performance of your heuristic approaches (for example, comparing with optimal, pseudo-optimal, or other heuristic solutions).
A. (Manevitz): For the heuristic numbering approach, the best reference is the full paper by Manevitz, Givoli and Margi where there are some comparative studies. The exact reference is in the bibliography of the paper that is on-line here.
For the Kohonen style NN work, there is a full paper by Manevitz, Givoli and Yousef which is also referenced in the bibliography (there are also comparative studies there, mostly with the package PLTMG; a sample example appears in the current on-line paper).

The third paper, "A Tabu Search Approach for the Tool Assignment and Machine Loading problem in Flexible Manufacturing Systems" by F. Guerrero, S. Lozano, J. Racero, L. Onieva and J. Larraneta, presents a new Tabu Search approach to the solution of a mixed integer linear programming formulation of the tool assignment and machine loading problem.

Q. (De Falco): Have you tried to face the problem with Genetic Algorithms as well? If so, what can you tell us about TS vs. GA comparison?

A. (Guerrero): Of course, this could be made, but, by the moment, we don't have results.

Q. (De Falco): You vary Tabu List size randomly, according to "Robust Tabu Search" approach. There exist several other methods to do that, like "Reactive Tabu Search" or "Strict Tabu Search". Have you tried to use these TS methods too? If so, how about the efficiency of the several methods?

A. (Guerrero): No, we haven't tried those methods yet. We have tried static TS list size and Dynamic TS list size (which is presented in this paper), with the latter performing better on the problem.

The fourth paper, "Investigating evolutionary optimisation of constrained functions to capture shape descriptions from range data" by C. Robertson, R.B. Fisher, D. Corne, N. Werghi and A.P. Ashbrook, examines the application of the GENOCOP III evolutionary algorithm to the problem of fitting surfaces and lines to both 2D synthetic and real 3D range data.

Q. (De Falco): What can you tell us about the time needed for evolution in real cases?

A. (Robertson): In our test objects evolution can require tens of thousands of evaluations, taking the order of minutes. I am currently re-writing the code to do more off-line calculations, and this should help.

Q. (De Falco): What are the operators used?

A. (Robertson): They are the same as in Michalewicz's book. They are basically search-and-repair operations. Operators are selected and applied probabilistically from a selection. Crossover can be arranged to always create feasible offspring.

Q. (De Falco): Can you give us more details on the search parameters like crossover probability, mutation probability, and selection method?

A. (Robertson): This is a preliminary investigation, and optimisation is always possible. David Corne is considering looking at the crossover strategy to see if more can be done there. So yes, evolution can definitely be improved and I would be interested in suggestions. An example input file can be found at: http://www.dai.ed.ac.uk/deidb/staff/personal_pages/craigr/in.97

Q. (Promislov): The chromosomes (a,b,c). Is it really different objects? Because without special constraints the c chromosome covers (a,b) and b covers (a).

A: (Robertson): If I have a chromosome consisting of 3 plane equations genes G = (a,b,c), what stops each of the planes settling on the same data points? The geometric constraint that the normals of the planes have the relationships: a.b=0, b.c=0, c.a=0. This is easy to specify with Genocop. With these constraints in place, the best fit for the chromosome will be the correct one although there are obviously several local minima.

Q. (Guerrero): I would like to know if you have results comparing your Genetic approach with the NLP formulation (in solution quality and CPU requirements).

A. (Robertson): We are comparing this 'evolutionary approach' to the methods based on a Levenberg-Marquardt approach:
http://www.dai.ed.ac.uk/daidb/staff/personal_pages/naoufelw/Research/Constraints/geomconst.html

Hopefully there should be a journal paper detailing this in the near future. There are a couple of advantages to the evolutionary approach of which the most obvious is simplicity. Expressing constraints formally in order to get optim and optim2 to work can be a fairly laborious task. There is, of course, a great deal of cross-fertilisation in our group. In terms of CPU requirements, GAs are notoriously bad for serial CPUs and so the other methods win. I'm looking forward to the specialised GA hardware from Reading to become available:
http://www.cs.rdg.ac.uk/cs/people/imb/research.html

However, quality of solution is comparable and the GA approach is very robust. I've used some silly starting positions for the reference population and still achieved convergence whereas the other approaches will die rather horribly due to local minima. GA's can also use fairly relaxed starting criteria so long as they are feasible.

The fifth paper, "Optimal selection of pressure vessels" by C.M. Kalker-Kalkman and M. Shahtaji, proposes a general structure for the various variables a designer has to decide on for pressure vessels used for gases storage.
Q. (De Falco): Can you tell us something more about the GA used, like binary or real-valued, mutation and crossover probability, population size, selection mechanism?
A. (Kalker-Kalkman): I used a program written by myself that employs GAs and has a graphic interface. During runtime the values of the subsequent sub-optima are drawn on the screen and in an interactive way you can modify mutation rate, population size, search intervals for variables, scaling factors for the fitness function. In the version used the variables are represented by bit strings and a one-point-crossover is used. Mutation is done by randomly changing a bit from 0 to 1 or vice versa. I usually start with a mutation rate of 0.4% and increase this rate during runtime up to 10%. I have found that for some problems it helps speeding up the process of optimisation by modifying the search boundaries in the regions where feasible solutions are found. As selection, a roulette wheel is used.
Q. (De Falco): I have noticed that the population size increases; can you give us reasons why? Is this due to the GA model used?
A. (Kalker-Kalkman): Usually I start with a start population of 30 members and I increase this population up to 100 during runtime.
Q. (De Falco): Is the problem a "case study" or is it a real problem, presented to you by industries?
A. (Kalker-Kalkman): It is an imaginary case as the information you get from industry is highly confidential.

8. New Topics in EA Basics

Prepared by: Prof. Vladimir Kvasnicka, Slovakia

This section contains four papers. The first work of authors Kvasnicka and Pospichal is aimed on a study of Baldwin effect with inclusion of Dawkins memes. Dawkins memes are in the work understood as further stretch of Darwins evolution, when an information about the environment becomes so complex, that there is no chance for it to be incorporated into a genotype of an individual by Baldwin effect. In the presented approach a fitness of a chromosome is determined not only by a chromosome itself, but also by a meme, which can contain a useful information for further evolution in a population. The simulation calculation showed that Dawkins memes could serve as a substantial acceleration element of Darwinian evolution. The memes become important when evolution proceeds in an extremely complex environment, when the Baldwin mechanism becomes ineffective for an acceleration of evolution. The work was criticised in two points: (1) The method of calculation of fitness of chromosomes by Hamming distance of a current vector from an "ideal" vector generated in advance was considered as oversimplified, and (2) the inclusion of Dawkins memes as an integral part of calculation of fitness is considered very far from the "reality".

In the second paper of this section by Norenkov, Kosachevsky, and Pisarenko entitled "Approach to Structure Synthesis of Genetic Algorithms" some combinatorial problems known in operations research are studied. These include vehicle routing problem, application distribution problem (probably, in standard terminology, task allocation problem), and open-shop scheduling problem. The paper discusses various heuristics how to accelerate a convergence of genetic algorithm. Effectiveness of the proposed method is illustrated by model calculations. The work attracted very serious remarks concerning the presentation of the study, concretely the work was considered barely readable (if at all) due to usage of not clearly specified terms.

The third paper by Pospichal and Kvasnicka entitled "A study of altruism by genetic algorithm" discusses a simple model of altruism performed in a population of chromosomes. The chromosomes (individuals) are capable of stochastic altruistic interaction, where a chromosome - donor gives a part of its fitness to other chromosomes - recipients, which then have a higher chance to survive. Numerical simulations showed that altruistic behaviour is stable only if altruistic chromosomes are able to distinguish their "relatives" (property of chromosome inherited from one of its parents) and donate fitness preferably to them. The results are discussed in a framework of a simple probabilistic model. The interesting deduction show that in the presented model altruism works mainly because less fit individuals donate fitness to more fit individuals. However, this result is crucially influenced by a procedure recalculating fitness into probability of selection for reproduction.

In the fourth and the last paper by Pelikan and Muehlenbein entitled "The Bivariate Marginal Distribution Algorithm" an interesting problem is studied in a loose framework of genetic algorithms, where a generating procedure of a new population is based purely on a knowledge of distribution of 1- a 2-bit probabilities (probabilities of appearance 0 or 1 and probabilities determining correlation between couples of bit values) in chromosomes of the parent population. The proposed modification of a genetic algorithm is tested on deceptive binary functions with various degree of correlation between single bits. The described algorithm enables a generation of a set of binary vectors, so that the most important 1- and 2- bit probabilities would be nearly identical also for the new set. Critical remarks in discussion concerned absent suitable tests of the presented method on multimodal functions with real variables and comparison of such results with the standard genetic algorithm.

9. New Frontier for Soft Computing

Prepared by: Prof. Y. Y. Yao, Canada and Prof. Byoung-Tak Zhang, Korea

This session contains 5 papers on Soft Computing. The papers cover both theoretical work and applications of Soft Computing. There are questions for three papers, but one of the authors did not reply. Following are the questions and answers for each paper:

1. A new approach to acquisition of comprehensible fuzzy rules Hiroshi Ohno, and Takeshi Furuhashi

Q (Yao):
In the re-evaluation of fuzzy model, you used two possible orders of membership functions. The first order (CBD) implies that one has to use three membership functions. The second order (CABAD) requires five. In general, how to choose a particular order and the number of membership functions? What is the linguistic interpretation of the membership functions, if more membership functions are used? Can we use fuzzy numbers to interpret them?

In your paper, EP is used in finding membership functions. After looking at the Figures 3, 4, and 5, I am wondering if the following simple method may also work:

1. Determine the number of membership function required.

2. For each original membership function, use membership functions of shape A, B, C, and D as an approximation.

3. Repeatedly delete membership functions that are close enough to undeleted ones, until the required number of membership function is obtained. Of course, some kind of distance function may have to be defined.

A:

1. It may be very difficult for determining these issues automatically since the domain knowledge is needed. In our approach these issues are determined by the user which he is a domain expert.
2. This is a trade-off between the comprehensibility and the accuracy of the model. In our approach the linguistic interpretation level is predefined by the user in which the number of the membership functions is determined.
3. Yes. They are probably useful depending on the domain of the problem.

2. Zero-Point Probability for Linear Source Separation
G. J. Scruby and S. J. Flockton

Q (Yao):
Hello, I have some general questions on your work. I am only (very) slightly familiar with ICA and am trying to learn more. My questions are:

1) What is the main difference between ICA and factor analysis? Are they related?

2) Can ICA work in a non-linear signal environment (signals multiplied by each other)? If not, do you know of any technique to separate these types of signals out?

A:
I'll address your two questions individually:
1) Factor analysis and ICA are related in that they both produce a new set of non-orthogonal linear basis vectors in the data space, though the methods used to choose these bases are different. From my (limited!) knowledge of factor analysis I understand that the new bases are chosen either heuristically or by maximising the variance along a certain direction so that the structure of underlying latent variables is shown. Since variance is the principal measure used, the best that can be achieved is the production of uncorrelated variables (2nd order). When the bases are orthogonal and uncorrelated this produces the classic Principle Component Analysis (forgive me if I'm stating the obvious). A common misconception is that such variables are independent of each other; really they are only unrelated to 2nd order. For truly independent variables an Independent Component Analysis is required (which can be thought of as a higher order generalisation of PCA) to ensure the variables are independent at all orders above 2nd as well. To summarise: ICA is like factor analysis in that non-orthogonal bases can be chosen (unlike PCA) the mechanism of choice is strictly defined by information theory.

2) There's a million pound (dollar if you like) question! ICA as it stands uses a strictly linear model, and has taken about 10 years to get to where it is. For many real world applications this (surprisingly) works, though there are many situations where non-linear models are preferable. The study of "non-linear ICA" (the field

doesn't really have a name yet) is still in its infancy and the results produced so far use very restrictive models.
However, I envisage the same explosion in research over the next five years or so as has happened with standard ICA in the last five. If you want to know the current state of non-linear (and linear) ICA a good place to start is at the Salk Institute in the US. They have plenty of on-line papers and links to many other researchers. The URL is: http://cnl.salk.edu/~tewon/ica_cnl.html

3. Code Optimization for DNA Computing of Maximal Cliques
Byoung-Tak Zhang and Soo-Yong Shin

Q (yao):
Actually, although I'm not familiar with DNA computing, I'd like to ask basic questions on your work.

I'd like to know that what the size of search space of DNA codes is. The other is about a relationship between two criteria, that is, between the quality of solutions that will be found by molecular evolution and the robustness of DNA codes. Is it conflicting or not? Independent?

The merit of 'DNA' computing against 'electronic' computing is likely to handle optimisation problems of big size in reasonable computation time. It seems to me that your method, or DNA code optimisation by genetic algorithm might suffer from scale problem because most genetic algorithms are not good in terms of scalability.

Keyword Index

List of Reviewers

Prof. Zbigniew Michalewicz, University of North Carolina, USA
Prof. D T Pham, University of Wales Cardiff, UK
Prof. Norenkov Igor, Moscow State Technical University, RUSSIA
Prof. Rajkumar Pant, IIT Bombay, INDIA
Prof. Vladimir Kvasnicka, Slovak Technical University, SLOVAKIA
Prof. Prabhat Hajela, Rensselaer Polytechnic Institute, USA
Prof. Kay Hameyer, Katholieke Universiteit Leuven, BELGIUM
Dr. Tadashi Iokibe, Meidensha Corporation, JAPAN
Prof. Erik D. Goodman, Michigan State University, USA
Prof. John S. Gero, University of Sydney, AUSTRALIA
Prof. Peter Sackett, Cranfield University, UK
Prof. Witold Pedrycz, University of Manitoba, CANADA
Prof. Antonio A. R. Coelho, Univ. Federal de Santa Catarina, BRAZIL
Dr. Ip-Shing Fan, Cranfield University, UK
Prof. V.A. Zarubin, Samara State Aerospace University, RUSSIA
Dr. Frank Hoffmann, BISC, University of California, USA
Dr. Phil Culverhouse, University of Plymouth, UK
Dr. John Greene, University of Cape Town, South Africa
Prof. Asim Roy, Arizona State University, USA
Prof. Noel Sharkey, University of Sheffield, UK
Dr. Alex H B Duffy, University of Strathclyde, UK
Dr. Riccardo Poli, The University of Birmingham, UK
Dr. Antony Satyadas, Flexible Intelligence Group, USA
Prof. Okyay Kaynak, Bogazici University, TURKEY
Dr. Dr Xin Yao, Australian Defence Force Academy, AUSTRALIA
Mr. Hossein Soltan, Cranfield University, UK
Prof. Elie Sanchez, University of Marseille, FRANCE
Prof. SongDe MA, Chinese Academy of Sciences, CHINA
Mr. Steve Thornton, British Steel, UK
Ir.C.M.Kalker-Kalkman, Delft University, THE NETHERLANDS
Prof. Sung-Bae Cho, Yonsei University, KOREA
Prof. Vladislav Valkovsky, St.Petersburg Electrotechnical Univ., RUSSIA
Dr. George Bilchev, BT Laboratories, UK
Dr. Dorigo Marco, Universite' Libre de Bruxelles, BELGIUM
Dr. Michael A. Lee, INSEAD, FRANCE
Dr. Paolo Petta, Austrian Research Institute for AI, AUSTRIA
Dr. Richard Harris, Neural Computer Sciences, UK
Prof. Terry Fogarty, Napier University, UK
Prof. Nik Kasabov, University of Otago, NEW ZEALAND
Prof. Mark J. Embrechts, Rensselaer Polytechnic Institute, USA

Dr. Rajkumar Roy , Cranfield University, UK
Prof. Yasuhiko Dote, Muroran Institute of Technolgy, JAPAN
Prof. Takeshi Furuhashi, Nagoya University, JAPAN
Dr. Pravir Chawdhry, Univ. of Bath, UK
Dr. Ivanoe De Falco, IRSIP – CNR, ITALY
Dr. Wallace E. Kelly, III , Texas A&M University, USA

GPSR Compliance

The European Union's (EU) General Product Safety Regulation (GPSR) is a set of rules that requires consumer products to be safe and our obligations to ensure this.

If you have any concerns about our products, you can contact us on ProductSafety@springernature.com

In case Publisher is established outside the EU, the EU authorized representative is:

Springer Nature Customer Service Center GmbH
Europaplatz 3
69115 Heidelberg, Germany

Batch number: 10620157

Printed by Printforce, the Netherlands